edexcel

Edexcel GCSE

Mathematics A Linear Higher

Student Book

Series Director: Keith Pledger
Series Editor: Graham Cumming

Authors:
Chris Baston
Julie Bolter
Gareth Cole
Gill Dyer
Michael Flowers
Karen Hughes
Peter Jolly
Joan Knott
Jean Linsky
Graham Newman
Rob Pepper
Joe Petran
Keith Pledger
Rob Summerson
Kevin Tanner
Brian Western

ALWAYS LEARNING

PEARSON

Published by Pearson Education Limited, a company incorporated in England and Wales, having its registered office at Edinburgh Gate, Harlow, Essex, CM20 2JE. Registered company number: 872828

Edexcel is a registered trademark of Edexcel Limited

Text © Chris Baston, Julie Bolter, Gareth Cole, Gill Dyer, Michael Flowers, Karen Hughes, Peter Jolly, Joan Knott, Jean Linsky, Graham Newman, Rob Pepper, Joe Petran, Keith Pledger, Rob Summerson, Kevin Tanner, Brian Western and Pearson Education Limited 2010

The rights of Chris Baston, Julie Bolter, Gareth Cole, Gill Dyer, Michael Flowers, Karen Hughes, Peter Jolly, Joan Knott, Jean Linsky, Graham Newman, Rob Pepper, Joe Petran, Keith Pledger, Rob Summerson, Kevin Tanner and Brian Western to be identified as the authors of this Work have been asserted by them in accordance with the Copyright, Designs and Patent Act, 1988.

First published 2010

13 12 11
10 9 8 7 6 5

British Library Cataloguing in Publication Data
A catalogue record for this book is available from the British Library

ISBN 978 1 84690 083 9

Typeset by Tech-Set Ltd, Gateshead
Picture research by Rebecca Sodergren
Printed by Grafos S.A., Spain

Acknowledgements
The publisher would like to thank the following for their kind permission to reproduce their photographs:
(Key: b-bottom; c-centre; l-left; r-right; t-top)
Alamy Images: aerialarchives.com 237, dbimages 607, Roberto Herrett 397, Jochen Tack 211, The Photolibrary Wales 619; **Corbis:** Régis Bossu 528, LWA-Sharie Kennedy 473, Mike McGill 83, Paul Seheult 448, Terry Vine 194, Visuals Unlimited 28; **Getty Images:** Alberto Arzoz 593, Alan Copson 298–299, Creative Crop 293cr, Adrian Dennis 40, Marcus Lyon 579; **iStockphoto:** 374, Floyd Anderson 291, Alena Brozova 294, Natalia Lukiyanova 300, Antonio Jorge Nunes 295bl, 295br, Kirill Putchenko 292–293, Nicole S. Young 1; **NASA:** 105, 205; **Photolibrary.com:** Alfo Foto Agency 418, Fresh Food Images 324, Carson ganci 311l, Brian Lawrence 116, Steve Vidler 145, Kathy de Witt 289tr; **Reuters:** Luke MacGregor 222; **Rex Features:** Ray Roberts 60, Sipa 132, The Travel Library 504; **Robert Harding World Imagery:** Wally Herbert 357; **Science Photo Library Ltd:** 550, Mark Garlick 15, Kevin A Horgan 263, Us Department Of Energy 515; **Shutterstock:** 288bl, 289b, 294–295/2, 295r, 296, 296–297, 297, 297bl, 297br, Nicola Gavin 293cl, Marjan Veljanoski 295tl; **Wales on View:** 288br, 288–289.

All other images © Pearson Education.

We are grateful to the following for permission to reproduce copyright material:
Tables
Table in Exercise 16.H adapted from 'CO2 emissions 2003–2008', Crown Copyright material is reproduced with the permission of the Controller, Office of Public Sector Information (OPSI).; Table in Exercise 16.H adapted from 'Agricultural Survey of Cattle, England 2006–2008', Crown Copyright material is reproduced with the permission of the Controller, Office of Public Sector Information (OPSI).

Every effort has been made to trace the copyright holders and we apologise in advance for any unintentional omissions. We would be pleased to insert the appropriate acknowledgement in any subsequent edition of this publication.

Contents

About this book

All set to make the grade!

Section objectives show what you'll be learning.

Loads of practice to help you feel secure before you move on.

Crystal-clear worked examples – step-by-step guides to answering questions correctly, with helpful hints and reminders.

Non-calculator indicates questions where students must not use a calculator to find the answer. It does NOT indicate that the subject area covered by the question will only appear in the Non-Calculator paper of the exam.

Graded questions – so you know what you're achieving.

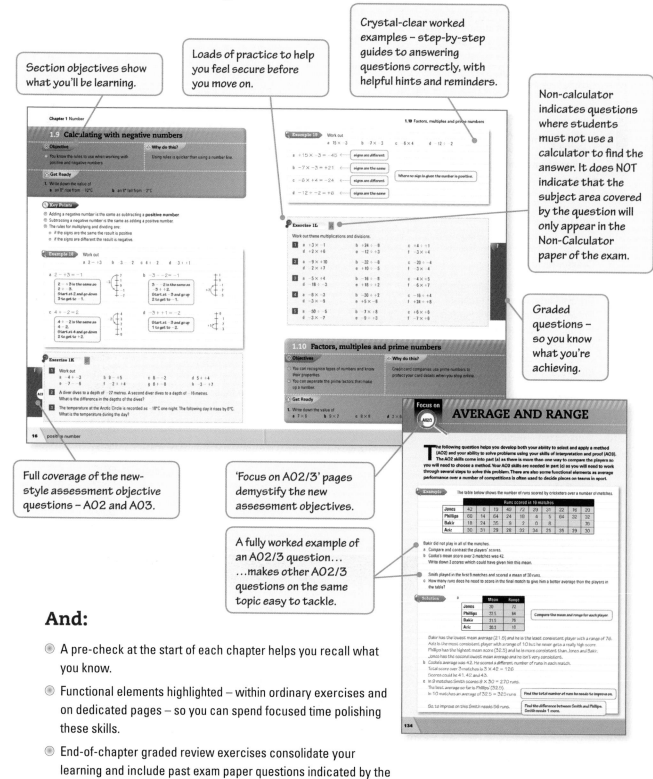

Full coverage of the new-style assessment objective questions – AO2 and AO3.

'Focus on AO2/3' pages demystify the new assessment objectives.

A fully worked example of an AO2/3 question...
...makes other AO2/3 questions on the same topic easy to tackle.

And:

- A pre-check at the start of each chapter helps you recall what you know.

- Functional elements highlighted – within ordinary exercises and on dedicated pages – so you can spend focused time polishing these skills.

- End-of-chapter graded review exercises consolidate your learning and include past exam paper questions indicated by the month and year.

About ActiveTeach

Use **ActiveTeach** to view and present the course on screen with exciting interactive content.

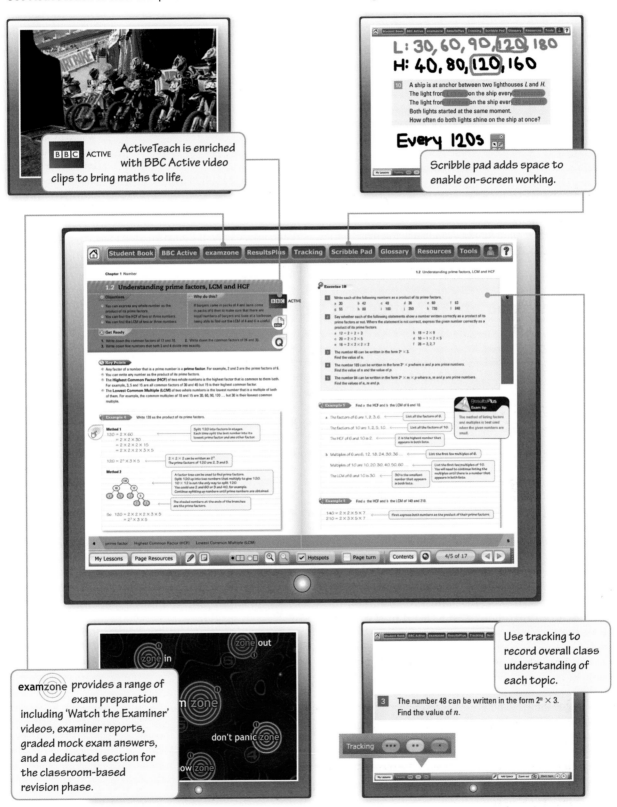

ActiveTeach is enriched with BBC Active video clips to bring maths to life.

Scribble pad adds space to enable on-screen working.

examzone provides a range of exam preparation including 'Watch the Examiner' videos, examiner reports, graded mock exam answers, and a dedicated section for the classroom-based revision phase.

Use tracking to record overall class understanding of each topic.

About Assessment Objectives

Assessment Objectives define the types of question that are set in the exam.

Assessment Objective	What it is	What this means	Range % of marks in the exam
AO1	**Recall** and use knowledge of the prescribed content.	Standard questions testing your knowledge of each topic.	45-55
AO2	**Select** and apply mathematical methods in a range of contexts.	Deciding what method you need to use to get to the correct solution to a contextualised problem.	25-35
AO3	**Interpret** and analyse problems and generate strategies to solve them.	Solving problems by deciding how and explaining why.	15-25

The proportion of marks available in the exam varies with each Assessment Objective. Don't miss out, make sure you know how to do AO2 and AO3 questions!

What does an AO2 question look like?

D **16** Katie wants to buy a car.
She decides to borrow £3500 from her father. She adds interest of 3.5% to the loan and this total is the amount she must repay her father. How much will Katie pay back to her father in total?

> This just needs you to
> (a) read and understand the question and
> (b) decide how to get the correct answer.

What does an AO3 question look like?

D **17** Rashida wishes to invest £2000 in a building society account for one year. The Internet offers two suggestions. Which of these two investments gives Rashida the greatest return?

> Here you need to read and analyse the question. Then use your mathematical knowledge to solve this problem.

CHESTMAN BUILDING SOCIETY	DUNSTAN BUILDING SOCIETY
£3.50 per month Plus **1% bonus** at the end of the year	**4%** per annum. Paid yearly by cheque

Focus on

We give you extra help with AO2 and AO3 on pages 274–287.

About functional elements

What does a question with functional maths look like?

Functional maths is about being able to apply maths in everyday, real-life situations.

GCSE Tier	Range % of marks in the exam
Foundation	30-40
Higher	20-30

The proportion of functional maths marks in the GCSE exam depends on which tier you are taking. Don't miss out, make sure you know how to do functional maths questions!

In the exercises...

D

AO3

20 The Wildlife Trust are doing a survey into the number of field mice on a farm of size **FS** 240 acres. They look at one field of size 6 acres. In this field they count 35 field mice.

a Estimate how many field mice there are on the whole farm.

b Why might this be an unreliable estimate?

> You need to read and understand the question. Follow your plan.
>
> Think what maths you need and plan the order in which you'll work.
>
> Check your calculations and make a comment if required.

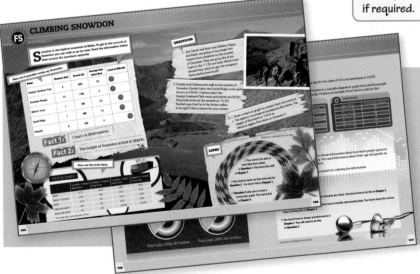

...and on our special functional maths pages: 288–299!

Quality of written communication

There will be marks in the exam for showing your working 'properly' and explaining clearly. In the exam paper, such questions will be marked with a star (*). You need to:

- use the correct mathematical notation and vocabulary, to show that you can communicate effectively
- organise the relevant information logically.

ResultsPlus

ResultsPlus features use exam performance data to highlight common pitfalls and misconceptions. ResultsPlus tips in the **student books** show students how to avoid errors in solutions to questions.

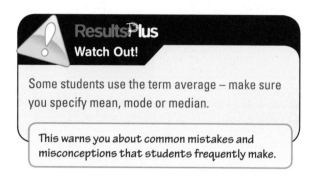

ResultsPlus
Watch Out!

Some students use the term average – make sure you specify mean, mode or median.

This warns you about common mistakes and misconceptions that students frequently make.

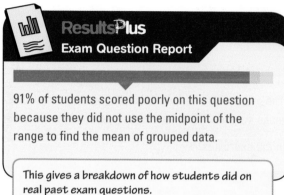

ResultsPlus
Exam Question Report

91% of students scored poorly on this question because they did not use the midpoint of the range to find the mean of grouped data.

This gives a breakdown of how students did on real past exam questions.

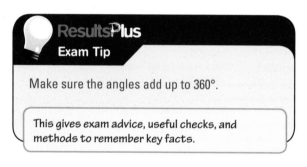

ResultsPlus
Exam Tip

Make sure the angles add up to 360°.

This gives exam advice, useful checks, and methods to remember key facts.

ResultsPlus in the **ActiveTeach** provides interactive practice for AO2 and AO3 questions…

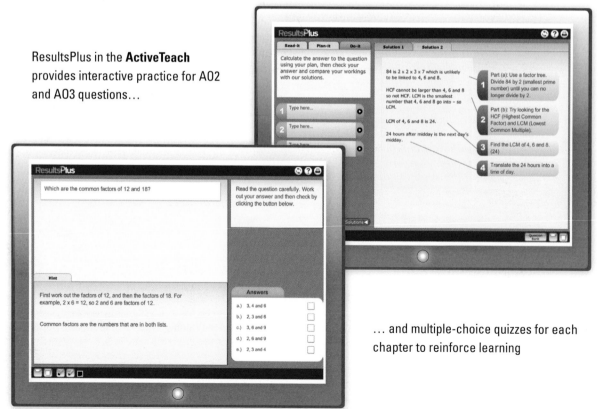

… and multiple-choice quizzes for each chapter to reinforce learning

1 NUMBER

We all use numbers as part of our everyday life. Some of the objects that you might use every day such as mobile phones and cash cards need to be protected by powerful codes. Most of the codes used involve very large prime numbers.

◎ Objectives

In this chapter you will:
- find the lowest common multiple and highest common factor of two numbers
- understand the meaning of square root and cube root
- know the correct order to carry out the different arithmetic operations
- use a calculator
- apply the laws of indices.

◇ Before you start

You need to:
- understand and use positive numbers and negative integers
- know how to use a number line
- know your multiplication tables
- know how to find factors and multiples of whole numbers
- be able to identify prime numbers
- understand index notation.

1.1 Understanding prime factors, LCM and HCF

◉ Objectives

○ You can express any whole number as the product of its prime factors.
○ You can find the HCF of two or three numbers.
○ You can find the LCM of two or three numbers.

◈ Why do this?

If burgers come in packs of 4 and buns come in packs of 6, being able to find out the LCM of 4 and 6 is useful to make sure that there are equal numbers of burgers and buns at a barbeque.

◈ Get Ready

1. State whether the following numbers are factors of 18, multiples of 18 or neither.
   a 4   b 6   c 9   d 36   e 12   f 3
2. State whether the following numbers are prime numbers or not.
   a 35   b 31   c 39   d 43   e 57

Key Points

◉ Any factor of a number that is a prime number is a **prime factor**. For example, 2 and 3 are the prime factors of 6.
◉ You can write any number as the product of its prime factors.
◉ The **Highest Common Factor (HCF)** of two whole numbers is the highest factor that is common to them both. For example, 3, 5 and 15 are all **common factors** of 30 and 45 but 15 is their highest common factor.
◉ The **Lowest Common Multiple (LCM)** of two whole numbers is the lowest number that is a multiple of both of them. For example, the **common multiples** of 10 and 15 are 30, 60, 90, 120, but 30 is their lowest common multiple.

Example 1   Write 120 as the product of its prime factors.

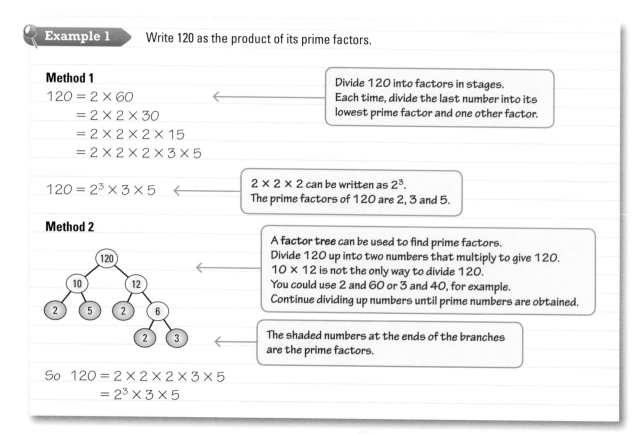

Method 1

$120 = 2 \times 60$
$ = 2 \times 2 \times 30$
$ = 2 \times 2 \times 2 \times 15$
$ = 2 \times 2 \times 2 \times 3 \times 5$

Divide 120 into factors in stages. Each time, divide the last number into its lowest prime factor and one other factor.

$120 = 2^3 \times 3 \times 5$

$2 \times 2 \times 2$ can be written as 2^3.
The prime factors of 120 are 2, 3 and 5.

Method 2

A factor tree can be used to find prime factors.
Divide 120 up into two numbers that multiply to give 120.
10×12 is not the only way to divide 120.
You could use 2 and 60 or 3 and 40, for example.
Continue dividing up numbers until prime numbers are obtained.

The shaded numbers at the ends of the branches are the prime factors.

So $120 = 2 \times 2 \times 2 \times 3 \times 5$
$ = 2^3 \times 3 \times 5$

prime factor   Highest Common Factor (HCF)   common factor   Lowest Common Multiple (LCM)

Example 2   Find **a** the HCF and **b** the LCM of 6 and 10.

a The factors of 6 are 1, 2, 3, 6. ←

List all the factors of 6.

The factors of 10 are 1, 2, 5, 10. ←

List all the factors of 10.

The HCF of 6 and 10 is 2. ←

2 is the highest number that appears in both lists.

b Multiples of 6 are 6, 12, 18, 24, **30**, 36 … ←

List the first few multiples of 6.

Multiples of 10 are 10, 20, **30**, 40, 50, 60 … ←

List the first few multiples of 10. You will need to continue listing the multiples until there is a number that appears in both lists.

The LCM of 6 and 10 is 30. ←

30 is the smallest number that appears in both lists.

Example 3   Find **a** the HCF and **b** the LCM of 140 and 210.

$140 = 2 \times 2 \times 5 \times 7$
$210 = 2 \times 3 \times 5 \times 7$

First express both numbers as the product of their prime factors.

Method 1
$140 = 2 \times 2 \times 5 \times 7$ ←
$210 = 2 \times 3 \times 5 \times 7$

Identify the common factors; the numbers that appear in both lists.

a HCF of 140 and 210 $= 2 \times 5 \times 7$ ←
$= 70$

Multiply the common factors together to get the HCF.

b LCM of 140 and 210 $= 70 \times 2 \times 3$ ←
$= 420$

Multiply the HCF by the numbers in both lists that were not highlighted to get the LCM.

Method 2

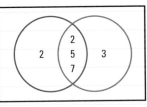

Put the prime factors into a Venn diagram. The prime factors of 140 are in the blue circle. The prime factors of 210 are in the red circle. The common factors of 140 and 210 are inside the part of the diagram where the two circles intersect.

a HCF of 140 and 210 $= 2 \times 5 \times 7$ ←
$= 70$

The HCF is the product of the numbers that are inside both circles.

b LCM of 140 and 210 $= 2 \times 2 \times 3 \times 5 \times 7$ ←
$= 420$

The LCM is the product of all the numbers that appear in the Venn diagram.

common multiple   factor tree

Exercise 1A

Questions in this chapter are targeted at the grades indicated.

C
A03

* **1** Can the sum of two prime numbers be a prime number?
Explain your answer.
[*Hint:* Try adding some pairs of prime numbers.]

2 The number 48 can be written in the form $2^n \times 3$.
Find the value of n.

3 The number 84 can be written in the form $2^n \times m \times p$ where n, m and p are prime numbers.
Find the values of n, m and p.

4 Find the HCF and LCM of the following pairs of numbers.
a 6 and 8   b 5 and 10   c 4 and 10   d 6 and 18

5 a Write 24 and 60 as products of their prime factors.
b Find the HCF of 24 and 60.   c Find the LCM of 24 and 60.

6 a Write 72 and 120 as products of their prime factors.
b Find the HCF of 72 and 120.   c Find the LCM of 72 and 120.

7 Find the HCF and LCM of the following pairs of numbers.
a 36 and 90   b 54 and 72   c 60 and 96   d 144 and 180

8 $x = 2 \times 3^2 \times 5$, $y = 2^3 \times 3 \times 7$
a Find the HCF of x and y.   b Find the LCM of x and y.

9 $m = 2^4 \times 3^2 \times 5 \times 7, n = 2^3 \times 5^3$
a Find the HCF of m and n.   b Find the LCM of m and n.

B

10 Bertrand's theorem states that 'Between any two numbers n and $2n$, there always lies at least one prime number, providing n is bigger than 1'. Show that Bertrand's theorem is true:
a for $n = 10$ b for $n = 20$ c for $n = 34$.

A03

11 A ship is at anchor between two lighthouses L and H.
The light from L shines on the ship every 30 seconds.
The light from H shines on the ship every 40 seconds.
Both lights started at the same moment.
How often do both lights shine on the ship at once?

A03

12 Burgers come in boxes of 8.
Buns come in packets of 6.
What is the smallest number of boxes of burgers and packets of buns that Mrs Moore must buy if she wants to ensure that there is a bun for every burger?

A
A03

* **13** Sally says that if you multiply two prime numbers then you will always get an odd number.
Is Sally correct? Give a reason for your answer.

1.2 Understanding squares and cubes

Objectives

○ You know how to find squares and cubes of whole numbers.
○ You understand the meaning of square root.
○ You understand the meaning of cube root.

Why do this?

If things are packed in squares you can quickly work out how many you have using square numbers, for example crates of strawberries or eggs.

Get Ready

1. Work out   **a** 6×6 **b** $2 \times 2 \times 2$ **c** -3×-3

Key Points

◉ A **square number** is the result of multiplying a whole number by itself.
The square numbers can be shown as a pattern of squares.

$1^2 = 1 \times 1 = 1$
1st square number

$2^2 = 2 \times 2 = 4$
2nd square number

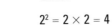

$3^2 = 3 \times 3 = 9$
3rd square number

$4^2 = 4 \times 4 = 16$
4th square number

◉ A **cube number** is the result of multiplying a whole number by itself then multiplying by that number again.
The cube numbers can be shown as a pattern of cubes.

$1^3 = 1 \times 1 \times 1 = 1$
1st cube number

$2^3 = 2 \times 2 \times 2 = 8$
2nd cube number

$3^3 = 3 \times 3 \times 3 = 27$
3rd cube number

◉ To find the **square** of any number, multiply the number by itself.
The square of $-4 = (-4)^2 = -4 \times -4 = 16$.
◉ $5 \times 5 = 25$, so we say that 5 is the **square root** of 25. It is a number that when multiplied by itself gives 25.
You can write the square root of 25 as $\sqrt{25}$. The square root of 25 can also be -5 because $-5 \times -5 = 25$.
◉ To find the **cube** of any number, multiply the number by itself then multiply by the number again.
The cube of $-2 = (-2)^3 = -2 \times -2 \times -2 = -8$.
◉ $-2 \times -2 \times -2 = -8$, so we say that -2 is the **cube root** of -8. It is a number that when multiplied by itself, then multiplied by itself again, gives -8. You can write the cube root of -8 as $\sqrt[3]{-8}$.

Example 4

Find   **a** the 6th square number
         b the 10th cube number.

a The 6th square number is $6^2 = 6 \times 6$
$= 36$
b The 10th cube number is $10^3 = 10 \times 10 \times 10$
$= 1000$

ResultsPlus
Exam Tip

You need to know the integer squares and corresponding square roots up to 15×15, and the cubes of 2, 3, 4, 5 and 10.

Exercise 1B

1 Write down:
  a the first 15 square numbers
  b the first 5 cube numbers.

2 From each list write down all the numbers which are:
  i square numbers   **ii** cube numbers.

  a 50, 20, 64, 30, 1, 80, 8, 49, 9
  b 10, 21, 57, 4, 60, 125, 7, 27, 48, 16, 90, 35
  c 137, 150, 75, 110, 50, 125, 64, 81, 144
  d 90, 180, 125, 100, 81, 75, 140, 169, 64

ResultsPlus
Exam Tip

You need to be able to recall
- integer squares from 2×2 up to 15×15 and the corresponding square roots
- the cubes of 2, 3, 4, 5 and 10.

Example 5   Find   **a** $(-3)^2$ **b** $\sqrt{100}$ **c** $(-4)^3 + \sqrt[3]{125}$

a $(-3)^2 = -3 \times -3 = 9$ ← Two signs the same so answer is positive.

b $\sqrt{100} = 10$

c $(-4)^3 = -4 \times -4 \times -4 = -64$
$\sqrt[3]{125} = 5$ ← $5^3 = 125$ so the cube root of 125 is 5.
$(-4)^3 + \sqrt[3]{125} = -64 + 5$
$\qquad = -59$

ResultsPlus
Exam Tip

Remember, when multiplying or dividing:
two signs the same give a +
two different signs give a −

Exercise 1C

1 Work out
  a 3^2 **b** 7^2 **c** 4^3 **d** 10^3 **e** 11^2

2 Write down
  a $\sqrt{36}$ **b** $\sqrt{16}$ **c** $\sqrt{81}$ **d** $\sqrt{1}$ **e** $\sqrt{64}$

3 Work out
  a $(-6)^2$ **b** $(-2)^3$ **c** $(-9)^2$ **d** $(-1)^3$ **e** $(-12)^2$

4 Write down
  a $\sqrt[3]{8}$ **b** $\sqrt[3]{-27}$ **c** $\sqrt[3]{-1}$ **d** $\sqrt[3]{64}$ **e** $\sqrt[3]{1000}$

5 Work out
  a $3^2 + 2^3$ **b** $\sqrt{4} \times 5^2$ **c** $5^2 \times \sqrt{100}$ **d** $\sqrt[3]{-8} + 4^2$
  e $\sqrt[3]{1000} \div \sqrt{100}$ **f** $4^3 \div 2^3$ **g** $(-1)^3 + 2^3 - (-3)^3$ **h** $4^2 + (-3)^3$
  i $\dfrac{6^2}{2^2}$ **j** $5^2 \times \dfrac{\sqrt{16}}{\sqrt[3]{8}}$ **k** $2^3 \times \dfrac{\sqrt{100}}{\sqrt{64}}$ **l** $\dfrac{4^2 - \sqrt[3]{-8}}{\sqrt{9}}$

1.3 Understanding the order of operations

⊙ Objective

- You know and can apply the order of operations.

◈ Get Ready

1. Work out   **a** 6×3 **b** 4^2 **c** $70 \div 7$

◈ Why do this?

When following a recipe, you need to add the ingredients in the right order. The same is true of calculations such as $3 \times 4 + 2 \times 5$. The operations must be carried out in the correct order or the answer will be wrong.

◈ Key Points

- **BIDMAS** gives the order in which each **operation** should be carried out.

- Remember that   **B**   **I**   **D**   **M**   **A**   **S**   stands for:

B rackets	If there are brackets, work out the **value** of the expression inside the brackets first.
I ndices	Indices include square roots, cube roots and **powers**.
D ivide	If there are no brackets, do dividing and multiplying before adding and subtracting, no
M ultiply	matter where they come in the expression.
A dd	
S ubtract	If an expression has only adding and subtracting then work it out from left to right.

◈ Example 6

$$10 \times 2^2 - 5 \times 3 = 10 \times 4 - 5 \times 3$$
$$= 40 - 15$$
$$= 25$$

Work out 2^2 first, then do all the multiplying before the subtraction.

◈ Example 7

$$(12 - 2 \times 5)^3 = (12 - 10)^3$$
$$= 2^3$$
$$= 8$$

The sum in the bracket is worked out first. Work out 2×5 and then do the subtraction.

⚙ Exercise 1D

1   Work out

   a $5 \times (2 + 3)$ **b** $5 \times 2 + 3$ **c** $20 \div 4 + 1$ **d** $20 \div (4 + 1)$

   e $(6 + 4) \div -2$ **f** $6 + 4 \div 2$ **g** $24 \div (6 - 2)$ **h** $24 \div 6 - 2$

   i $7 - (4 + 2)$ **j** $7 - 4 + 2$ **k** $5 \times 4 - 2 \times 3$ **l** $28 - 4 \times -6$

   m $14 + 3 \times 6$ **n** $6 + 3 \times 5 - 12 \div 2$ **o** $25 - 5 \times 4 + 3$ **p** $(15 - 5) \times (4 + 3)$

2   Work out

   a $(3 + 4)^2$ **b** $3^2 + 4^2$ **c** $3 \times (4 + 5)^2$ **d** $3 \times 4^2 + 3 \times 5^2$

   e $2 \times (4 + 2)^2$ **f** $3 \times \sqrt{25} + 2 \times 3^3$ **g** $\dfrac{(2 + 5)^2}{3^2 - 2}$ **h** $\dfrac{5^2 - 2^2}{-3}$

D

B

3 Work out

a $(2 + 3)^3 \div \sqrt{25}$

b $((15 - 5) \times 4) \div ((2 + 3) \times 2)$

c $2^3 + 6^2 \div \sqrt{9} - 4 \times 3$

d $(\sqrt[3]{-27} - 2)^2 + \sqrt{3^2 \times 2^2}$

1.4 Using a calculator

◎ Objectives

● You can use a calculator.

● You can find and use reciprocals.

❓ Why do this?

Many jobs require the accurate use of calculators, such as working in a bank or as an accountant.

◆ Get Ready

1. Work out an estimate for:　a 234×89　b $318.2 \div 2.98$　c $(7.2)^2$

◉ Key Points

◉ A scientific calculator can be used to work out arithmetic calculations or to find the value of arithmetic expressions.

◉ Scientific calculators have special keys to work out squares and square roots. Some have special keys for cubes and cube roots.

◉ To work out other powers, your calculator will have a $\boxed{y^x}$ or $\boxed{x^y}$ or $\boxed{\wedge}$ key.

◉ The inverse of x^2 is $\sqrt{x}$ or $x^{\frac{1}{2}}$, and the inverse of x^3 is $\sqrt[3]{x}$ or $x^{\frac{1}{3}}$.

◉ The **inverse operation** of x^y is $x^{\frac{1}{y}}$.

◉ You can use the calculator's memory to help with more complicated numbers.

Example 8　Work out　a $4.6^2 + \sqrt{37}$　b $\dfrac{1.2^3 + 12.5}{(3.7 - 2.1)^2}$

Give your answers correct to 3 significant figures.

a $4.6^2 + \sqrt{37}$　← Key in $4 . 6\, x^2 + \sqrt{\ } 3\, 7 =$

$4.6^2 + \sqrt{37} = 27.242\,762...$

$= 27.2$　← Round your answer to 3 significant figures.

b $\dfrac{1.2^3 + 12.5}{(3.7 - 2.1)^2}$　← Work out the sum on the top of the fraction.

$1.2^3 + 12.5 = 14.228$　← Key in $1 . 2\, x^3 + 1\, 2 . 5 =$

$(3.7 - 2.1)^2 = 2.56$　← Work out the sum on the bottom of the fraction. Key in $(3 . 7 - 2 . 1)\, x^2$

$14.228 \div 2.56 = 5.557\,8125$　← Divide your answers.

$= 5.56$　← Round the final answer to 3 significant figures.

ResultsPlus

Watch Out!

Do not round your numbers part way through a calculation; use all the figures shown on your calculator. Only round the final answer.

---

inverse operation

Exercise 1E

1 Work out:

  a $\sqrt{961}$ **b** $\sqrt{40.96}$ **c** $\sqrt[3]{4913}$ **d** $\sqrt[3]{3.375}$ **e** $\sqrt{1024}$

2 Work out:

  a $(3.7 + 5.9) \times 4.1$ **b** $3.1^2 + 4.8^2$ **c** $(-8.7 + 6.3)^2$ **d** $4.5^3 + 8^2$

3 Work out, giving your answers correct to one decimal place.

  a $3.2^3 \times 6.7$ **b** $\sqrt{24} + 6.7^3$ **c** $9.2^2 \div \sqrt{14}$ **d** $7.5^3 - \sqrt{120}$

4 Work out, giving your answers correct to three significant figures.

  a $\dfrac{5.63}{2.8 - 1.71}$ **b** $\dfrac{9.84 \times 2.6}{2.8 \times 1.71}$ **c** $\dfrac{6.78 + 9.2}{7.8 - 2.75}$ **d** $\dfrac{6.7^2}{5.6^2 - 2.1^2}$

5 Work out, giving your answers correct to three significant figures.

  a $\sqrt{11.62} - \dfrac{6.3}{9.8}$ **b** $\dfrac{5.63}{2.8} + \dfrac{1.7}{0.3}$ **c** $\dfrac{\sqrt{342}}{1.8 - 1.71}$ **d** $\left(\sqrt{\dfrac{56}{0.18}} + 657\right)^2$

6 Work out, giving your answers correct to three significant figures.

  a $\dfrac{\sqrt{45} + 6.3^2}{79.1 - 28.5}$ **b** $\sqrt{\dfrac{8.9 \times 2.3}{9.6 + 7.8}}$ **c** $\dfrac{4.2^3}{\sqrt{7.8^2 + 3.5^2}}$ **d** $\dfrac{(23.5 + 8.7)^2}{\sqrt{65^2 + 82}}$

Reciprocals

Key Points

⊙ The **reciprocal** of the number n is $\dfrac{1}{n}$. It can also be written as n^{-1}.

⊙ When a number is multiplied by its reciprocal the answer is always 1.

⊙ All numbers, except 0, have a reciprocal.

⊙ The reciprocal button on a calculator is usually $\boxed{1/x}$ or $\boxed{x^{-1}}$.

Example 9   Work out the reciprocal of   **a** 8   **b** 0.25   **c** $\dfrac{1}{4^3}$.

a $\dfrac{1}{8} = 0.125$

b $1 \div 0.25 = 4$

c 4^3

Exercise 1F

1 Find the reciprocal of each of the following numbers.

  a 4   **b** 0.625   **c** 6.4   **d** $\dfrac{2}{2^4}$

1.5 Understanding the index laws

◎ Objectives

- ● You can use index notation.
- ● You can use index laws.

❓ Why do this?

Using the index laws you can work out that you have $2^5 = 32$ great great great grandparents.

◆ Get Ready

1. Work out 2^5 **2.** Work out 5^3 **3.** Work out $27^4 \div 27^2$

Key Points

- ◎ A number written in the form a^n is an **index number**.
- ◎ The **laws of indices** are:

  $a^m \times a^n = a^{m+n}$ To multiply two powers of the same number add the indices.

  $a^m \div a^n = a^{m-n}$ To divide two powers of the same number subtract the indices.

  $(a^m)^n = a^{m \times n}$ To raise a power to a further power multiply the indices together.

  You will encounter negative and fractional indices in Chapter 25.

Example 10

Work out   **a** 3^4 **b** 2^6

a $3^4 = 3 \times 3 \times 3 \times 3$

     $= 81$

b $2^6 = 2 \times 2 \times 2 \times 2 \times 2 \times 2$

     $= 64$

Results Plus

Watch Out!

Remember that a^3 means that you multiply three as together. It does not mean $a \times 3$.

Example 11

Write each expression as a power of 5.   **a** $5^6 \times 5^4$ **b** $5^{12} \div 5^4$ **c** $(5^3)^2$

a $5^6 \times 5^4 = 5^{4+6}$ ← Use the index law $a^m \times a^n = a^{m+n}$

     $= 5^{10}$

b $5^{12} \div 5^4 = 5^{12-4}$ ← Use the index law $a^m \div a^n = a^{m-n}$

     $= 5^8$

c $(5^3)^2 = 5^{3 \times 2}$ ← Use the index law $(a^m)^n = a^{m \times n}$

     $= 5^6$

Example 12

Work out $\dfrac{4^7 \times 4}{4^5}$

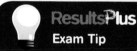

Results Plus

Exam Tip

'Work out' means 'evaluate' the expression, rather than leaving the answer as a power.

$\dfrac{4^7 \times 4}{4^5} = \dfrac{4^7 \times 4^1}{4^5}$

   $= \dfrac{4^8}{4^5}$ ← Simplify the top of the fraction, add 7 and 1.

   $= 4^3$

   $= 4 \times 4 \times 4 = 64$ ← As the question asks you to 'Work out', the final answer must be a number.

Results Plus

Watch Out!

Remember that a is the same as a^1.

Exercise 1G

1. Write as a power of a single number

   a $6^5 \times 6^7$ b $4^7 \div 4^2$ c $(7^2)^3$ d $5^9 \div 5^3$ e $3^8 \times 3^2$

2. Work out

   a $10^2 \times 10^3$ b $5^7 \div 5^4$ c $(2^3)^2$ d $3^4 \div 3^2$ e 4×4^2

3. Find the value of n

   a $3^n \div 3^2 = 3^3$ b $8^5 \div 8^n = 8^2$ c $2^5 \times 2^n = 2^{10}$ d $3^n \times 3^5 = 3^9$ e $2^6 \times 2^3 = 2^n$

4. Write as a power of a single number

   a $\dfrac{3^3 \times 3^5}{3^4}$ b $\dfrac{5^6 \times 5^7}{5^4}$ c $\dfrac{2^8 \times 2^5}{2^7}$ d $\dfrac{6^{15}}{6 \times 6^9}$ e $\dfrac{4^2 \times 4^7}{4^3 \times 4^4}$

5. Work out

   a $\dfrac{3^3 \times 3^5}{3^6}$ b $\dfrac{2^6 \times 2^2}{2^4}$ c $\dfrac{4^7}{4 \times 4^4}$ d $\dfrac{10^5 \times 10^6}{10^7}$ e $\dfrac{7^8 \times 7}{7^3 \times 7^4}$

6. Work out the value of n in the following

   a $40 = 5 \times 2^n$ b $32 = 2^n$ c $20 = 2^n \times 5$ d $48 = 3 \times 2^n$ e $54 = 2 \times 3^n$

Chapter review

Key Points

- Any factor of a number that is a prime number is a **prime factor**.
  You can write any number as the product of its prime factors.

- The **Highest Common Factor (HCF)** of two whole numbers is the highest factor that is common to them both.

- The **Lowest Common Multiple (LCM)** of two whole numbers is the lowest number that is a multiple of both of them.

- A **square number** is the result of multiplying a whole number by itself.

- A **cube number** is the result of multiplying a whole number by itself then multiplying by that number again.

- To find the **square** of any number, multiply the number by itself.

- The **square root** of 25 is a number that when multiplied by itself gives 25.
  You can write the square root of 25 as $\sqrt{25}$.
  The square root of 25 can also be -5 because $-5 \times -5 = 25$.

- To find the **cube** of any number, multiply the number by itself then multiply by the number again.

- The **cube root** of -8 is a number that when multiplied by itself, then multiplied by itself again, gives -8.
  You can write the cube root of -8 as $\sqrt[3]{-8}$.

- **BIDMAS** gives the order in which **operations** should be carried out.

◉ Remember that **BIDMAS** stands for:

Brackets   If there are brackets, work out the **value** of the expression inside the brackets first.

Indices   Indices include square roots, cube roots and **powers**.

Divide   If there are no brackets, do dividing and multiplying before adding and subtracting, no

Multiply   matter where they come in the expression.

Add   If an expression has only adding and subtracting then work it out from left to right.

Subtract

◉ A scientific calculator can be used to work out arithmetic calculations or to find the value of arithmetic expressions.

◉ Scientific calculators have special keys to work out squares and square roots. Some have a special key for cubes and cube roots.

◉ To work out other powers, your calculator will have a $\boxed{y^x}$ or $\boxed{x^y}$ or $\boxed{\wedge}$ key.

◉ The inverse of x^2 is $\sqrt{x}$ or $x^{\frac{1}{2}}$, and the inverse of x^3 is $\sqrt[3]{x}$ or $x^{\frac{1}{3}}$.

◉ The **inverse operation** of x^y is $x^{\frac{1}{y}}$.

◉ You can use the calculator's memory to help with more complicated numbers.

◉ The **reciprocal** of the number n is $\dfrac{1}{n}$. It can also be written as n^{-1}.

◉ When a number is multiplied by its reciprocal the answer is always 1.

◉ All numbers, except 0, have a reciprocal.

◉ The reciprocal button on a calculator is usually $\boxed{1/x}$ or $\boxed{x^{-1}}$.

◉ A number written in the form a^n is an **index number**.

◉ The **laws of indices** are:

$a^m \times a^n = a^{m+n}$ To multiply two powers of the same number add the indices.

$a^m \div a^n = a^{m-n}$ To divide two powers of the same number subtract the indices.

$(a^m)^n = a^{m \times n}$ To raise a power to a further power multiply the indices together.

 Review exercise   Except where indicated.

A03

1   Jim writes down the numbers from 1 to 100. Ben puts a red spot on all the even numbers and Helen puts a blue spot on all the multiples of 3.

    a What is the largest number that has both a red and a blue spot?

    b How many numbers have neither a blue nor a red spot?

    Sophie puts a green spot on all the multiples of 5.

    c How many numbers have exactly two coloured spots on them?

D

2   Find the missing numbers in each case.

    a $? \times 3 = -12$ **b** $(-20) \div (-5) = ?$ **c** $(-6) + ? = (-8)$

    d $(-5) \times ? = (-20)$ **e** $6 - ? = 8$

3   Find the missing numbers in each case.

    a $2 \times ? + (-3) = (-7)$ **b** $(-4) \times ? + 5 = (-3)$ **c** $? \div 2 + 4 = (-4)$

4 Neal works part time in a local supermarket, stacking shelves.

He has been asked to use the pattern below to advertise a new brand of beans.

This stack is 3 cans high.

a How many cans will he need to build a stack 10 cans high?

b If he has been given 200 cans, how many cans high would his stack be?

Next he is asked to stack cans of tomato soup in a similar shape, but this time it is two cans deep.

Use your answers to parts **a** and **b** to answer the following questions.

c How many cans will he need to build a stack 10 cans high?

d If he has been given 400 cans, how many cans high would his stack be?

5 A chocolate company wishes to produce a presentation box of 36 chocolates for Valentine's Day.

It decides that a rectangular shaped box is the most efficient, but needs to decide how to arrange the chocolates.

How many different possible arrangements are there:

a using one layer

b using two layers

c using three layers.

Which one do you think would look best?

6 The number 1 is a square number and a cube number. Find another number which is a square number and a cube number.

7 $4^2 \times 6^2 = 576$

Work out a $40^2 \times 60^2$ b $400^2 \times 6^2$ c $5760 \div 6^2$ d $4^2 \times 60^2$ e $4^3 \times 6^2$

8 Work out a $2 + 4 \div 4$ b $5^3 \div 5 + 5$ c $(2^2)^3 - (2^3)^2$

9 Simplify a $\dfrac{3^5 \times 3^3}{3^6}$ b $\dfrac{4^4 \times 4^7}{4^{10}}$ c $(2^4)^3$ d $\dfrac{5^{12}}{5^7 \times 5^3}$

10 a Express 252 as a product of its prime factors.

b Express 6×252 as a product of prime factors.

C A03

11 James thinks of two numbers.
He says 'The highest common factor (HCF) of my two numbers is 3.
The lowest common multiple (LCM) of my two numbers is 45'.
Write down the two numbers James could be thinking of.

ResultsPlus
Exam Question Report

75% of students answered this sort of question well because they chose the right method to answer the question.

June 2008

12 Write 84 as a product of its prime factors.
Hence or otherwise write 168^2 as a product of its prime factors.

A03

13 A car's service book states that the air filter must be replaced every 10 000 miles and the diesel fuel filter every 24 000 miles.
After how many miles will both need replacing at the same time?

14 Use your calculator to work out $\dfrac{\sqrt{19.2 + 2.6^2}}{2.7 \times 1.5}$
Write down all the figures on your calculator display.

B A03

15 $2^{30} \div 8^9 = 2^x$
Work out the value of x.

Nov 2007

A A03

16 Write whether each of the following statements is true or false. If the statement is false give an example to show it.
a   The sum of two prime numbers is always a prime number.
b   The sum of two square numbers is never a prime number.
c   The difference between consecutive prime numbers is never 2.
d   The product of two prime numbers is always a prime number.
e   No prime number is a square number.

17 a   Take a piece of scrap A4 paper.
If you fold it in half you create two equal pieces. Fold it in half again; you now have four equal pieces.
It is said that no matter how large and how thin you make the paper, it cannot be folded more than seven times. Try it.
If you fold it seven times, how many equal pieces does the paper now have?
b   In 2001, there were two rabbits left on an island.
A simple growth model predicts that in 2002 there will be four rabbits and in 2003, eight rabbits.
The population of rabbits continues to double every year.
How long is it before there are 1 million rabbits on the island?

Neptune was the first planet to be found by mathematical prediction. Scientists looked at the number patterns of the orbits of planets in the Solar System and correctly predicted Neptune's position to within a degree. Using the predicted position, Johann Galle identified Neptune almost immediately on 23 September 1846.

Objectives

In this chapter you will:
- distinguish the different roles played by letter symbols in algebra and use the correct notation in deriving algebraic expressions
- collect like terms
- use substitution to work out the value of an expression
- use the index laws applied to simple algebraic expressions and to algebraic expressions with fractional or negative powers
- generate terms of a sequence using term-to-term and position-to-term definitions
- derive and use the nth term of a sequence.

Before you start

You need to be able to:
- simplify an expression where each term is in the same unknown or unknowns
- use directed numbers in calculations
- use index laws with numbers.

2.1 Collecting like terms

◎ Objectives

- You can distinguish the different roles played by letter symbols in algebra and use the correct notation.
- You can manipulate algebraic expressions by collecting like terms.

❓ Why do this?

Waitresses use algebra to note people's orders and then collect like terms to make the order simple for the chef.

⬆ Get Ready

Simplify

1. $a + a + a + a$

2. $4c - c + 5c$

3. $3p^2 - 5p^2 + 4p^2$

🕐 Key Points

- ◉ $2x$, $3y$ and $2x + 3y$ are called **algebraic expressions**.
- ◉ Each part of an **expression** is called a **term** of the expression. $2x$ and $3y$ are terms of the expression $2x + 3y$.
- ◉ When adding or subtracting expressions, different letter symbols cannot be combined. For example $2x + 3y$ cannot be simplified further.
- ◉ The sign of a term in an expression is always written before the term.
  For example, in the expression $4 + 2x - 3y$ the '+' sign means add $2x$ and the '−' sign means subtract $3y$.
- ◉ The term x can be written as $1x$.
- ◉ In algebra, BIDMAS describes the order of operations when **collecting like terms** (see Section 1.3 for use of BIDMAS).

Example 1　　Simplify the expression $4p - 2q + 1 - 3p + 5q$.

$4p - 2q + 1 - 3p + 5q$
$= 4p - 3p - 2q + 5q + 1$　←　$-2 + 5 = +3$　so $-2q + 5q = +3q$
$= p + 3q + 1$

$4p - 3p = 1p$ which is written as just p.

Results Plus
Exam Tip

Rewrite each expression with the like terms next to each other.

Example 2　　Alfie is n years old. Bilal is 3 years older than Alfie. Carla is twice as old as Alfie.
Write down an expression, in terms of n, for the total of their ages in years.
Give your answer in its simplest form.

Alfie = n years
Bilal = $(n + 3)$ years　←　This can be written as $3 + n$.
Carla = $2n$ years　←　This can be written as $2 \times n$ or $n \times 2$ or $2n$.

Total = $n + (n + 3) + 2n$　←　This is a correct, un-simplified expression.
　　　= $n + n + 3 + 2n$　←　Remove the brackets.
　　　= $n + n + 2n + 3$
　　　= $4n + 3$ years　←　This is in its simplest form.

Exercise 2A

Questions in this chapter are targeted at the grades indicated.

1 Simplify

a $5x + 2x + 3y + y$ b $3w + 7w + 4z - 2z$ c $3p + q + p + 4q$

d $4a + 3b - a - 2b$ e $c + 2d + 5c - 4d$ f $3m - 7n - m + 4n$

g $5e - 3f - e - 4f$ h $2x + 8y - 3 + 2y + 5$ i $3p - q + 2 - 5p + 4q - 7$

j $9 + a - 2b - 5a + 4 - 3b$

2 Georgina, Samantha and Mason collect football stickers. Georgina has x stickers in her collection.
Samantha has 9 stickers less than Georgina. Mason has 3 times as many stickers as Georgina.
Write down an expression, in terms of x, for the total number of these stickers.
Give your answer in its simplest form.

3 The diagram shows a triangle.
Write down an expression, in terms of x and y, for the perimeter of
this triangle.
Give your answer in its simplest form.

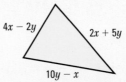

2.2 Using substitution

Objective

○ Given the value of each letter in an expression,
you can work out the value of the expression
by substitution.

Why do this?

In your science lessons you need to be able to
substitute into formulae when carrying out many
calculations.

Get Ready

Write expressions, in terms of x and y, for the perimeter of these rectangles:

1. length $2x + 4$, width $y + 2$ **2.** length $3y + 3$, width $x - 5$ **3.** length $4x + 5$, width $y - 2$.

Key Point

◉ If you are given the value for each letter in an expression then you can **substitute** the values into the
expression and **evaluate** the expression.

Example 3

Work out the value of each of these expressions when $a = 5$ and $b = -3$.

a $4a + 3b$ b $a - 2b - 8$ c $2a^2 + 4b$

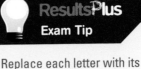

ResultsPlus

Exam Tip

Replace each letter with its
numerical value.

a $4a + 3b = 4 \times 5 + 3 \times (-3)$

$= 20 - 9$ ⟵ Positive × negative = negative.

$= 11$

b $a - 2b - 8 = 5 - 2 \times (-3) - 8$

$= 5 + 6 - 8$

$= 3$

⟵ Work out the multiplication first (BIDMAS).
Negative × negative = positive.

c $2a^2 + 4b = 2 \times (5)^2 + 4 \times (-3)$

$= 2 \times 25 - 12$

$= 50 - 12$ ⟵ It is only the value of a (=5) that is squared.

$= 38$

Exercise 2B

1 Work out the value of each of these expressions when $x = 4$ and $y = -1$.

    a $x + 3y$ **b** $x - y$ **c** $2x - 5y + 3$ **d** $4x + 1 + 2y$

2 Work out the value of each of these expressions when $p = -2$, $q = 3$ and $r = -5$.

    a $p + q + r$ **b** $2q + 3r + 5p$ **c** $2q - r + 3p$

    d $6 - q - 2r + p$ **e** $5p + 3q^2$ **f** $p^2 - 2q^2 + r^2$

2.3 Using the index laws

◎ Objective

- You understand and can use the index laws applied to simple algebraic expressions.

❓ Why do this?

To write large numbers, like the speed of sound, indices are often used to shorten the way the value is written.

⬆ Get Ready

1. Write as a power of a single number.

    a $4^3 \times 4^8$ **b** $\dfrac{7^8 \times 7^4}{7^5}$ **c** $(6^3)^2$

🔵 Key Point

- You can use the laws of indices to simplify algebraic expressions. See Section 1.5 for the index laws.

Example 4

    a Simplify $c^3 \times c^4$

    b Simplify $5y^3z^5 \times 2y^2z$

> **Results Plus**
> **Watch Out!**
>
> Group like terms together before attempting to use the laws of indices.

a $c^3 \times c^4 = c \times c \times c \times c \times c \times c \times c$

         $= c^7$ ⟵ Note: $3 + 4 = 7$.

b $5y^3z^5 \times 2y^2z = 5 \times y^3 \times z^5 \times 2 \times y^2 \times z$ ⟵ z is the same as z^1

            $= 5 \times 2 \times y^3 \times y^2 \times z^5 \times z^1$

              $= 10 \times y^{3+2} \times z^{5+1}$

               $= 10 \times y^5 \times z^6$ ⟵ Using $x^p \times x^q = x^{p+q}$

                $= 10y^5z^6$

Exercise 2C

1 Simplify
   a $m \times m \times m \times m \times m$ b $2p \times 3p$ c $q \times 4q \times 5q$

2 Simplify
   a $a^4 \times a^7$ b $n \times n^3$ c $x^5 \times x$ d $y^2 \times y^3 \times y^4$

3 Simplify
   a $2p^2 \times 6p^4$ b $4a \times 3a^4$ c $b^7 \times 5b^2$ d $3n^2 \times 6n$

4 Simplify
   a $5t^3u^2 \times 4t^5u^3$ b $2xy^3 \times 3x^5y^4$ c $a^2b^5 \times 7a^3b$
   d $4cd^5 \times 2cd^4$ e $2mn^2 \times 3m^3n^2 \times 4m^2n$

Example 5
a Simplify $d^5 \div d^2$

b Simplify $\dfrac{10x^2y^5}{2xy^3}$

a $d^5 \div d^2 = \dfrac{d^5}{d^2} = \dfrac{d \times d \times d \times d \times d}{d \times d}$

    $= d^3$ ← Note: $5 - 2 = 3$

b $\dfrac{10x^2y^5}{2xy^3}$ is the same as $10x^2y^5 \div 2xy^3$

$10x^2y^5 \div 2xy^3 = (10 \div 2) \times (x^2 \div x) \times (y^5 \div y^3)$

      $= 5 \times x^{2-1} \times y^{5-3}$

      $= 5 \times x \times y^2$ Using $x^p \div x^q = x^{p-q}$

      $= 5xy^2$

ResultsPlus
Exam Tip

Write fractions, such as $\dfrac{p^5}{p^3}$ as $p^5 \div p^3$.

Exercise 2D

1 Simplify
   a $a^7 \div a^4$ b $b^5 \div b$ c $\dfrac{c^8}{c^5}$ d $d^4 \div d^3$

2 Simplify
   a $6q^5 \div 3q^3$ b $12p^7 \div 4p^2$ c $8x^6 \div 2x^5$ d $\dfrac{20y^8}{2y}$

3 Simplify
   a $15a^5b^6 \div 3a^3b^2$ b $30p^3q^4 \div 6p^2q$ c $\dfrac{8c^4d^7}{2c^2d^3}$ d $\dfrac{6x^3 \times 2x^4}{4x^2}$
   e $\dfrac{5m^2n \times 4mn^2}{2mn^2}$

Example 6   Simplify $(2c^3d)^4$

Method 1

$(2c^3d)^4 = (2)^4 \times (c^3)^4 \times (d)^4$

$= 16 \times c^{3 \times 4} \times d^{1 \times 4}$

$= 16 \times c^{12} \times d^4$

$= 16c^{12}d^4$

> Using $(x^p)^q = x^{p \times q}$

ResultsPlus
Exam Tip

You must apply the power to number terms as well as the algebraic terms.

Method 2

$(2c^3d)^4$ can be written as $2c^3d \times 2c^3d \times 2c^3d \times 2c^3dd$

$= 2 \times 2 \times 2 \times 2 \times c^3 \times c^3 \times c^3 \times c^3 \times d \times d \times d \times d$

$= 16 \times c^{3+3+3+3} \times d^4$

$= 16 \times c^{12} \times d^4$

$= 16c^{12}d^4$

> Using $x^p \times x^q = x^{p+q}$

Exercise 2E

C

1 Simplify

  a $(a^7)^2$ b $(b^3)^5$ c $(c^3)^3$ d $(d^2)^8$

2 Simplify

  a $(2p^3)^2$ b $(3q^2)^4$ c $(5x^4)^2$ d $\left(\dfrac{m^4}{2}\right)^3$

B

3 Simplify

  a $(2x^3y^2)^4$ b $(7e^5f^3)^2$ c $(5p^5q)^3$ d $\left(\dfrac{2x^4y^2}{3xy^4}\right)^3$

2.4 Fractional and negative powers

◎ Objective

○ You can use the index laws applied to algebraic expressions with fractional or negative powers.

❓ Why do this?

To write very small numbers, like the radius of a molecule, negative powers of 10 are used.

◈ Get Ready

Simplify these expressions.

1. $(a^3)^6$ **2.** $(3y^5)^3$ **3.** $\left(\dfrac{4a^3b^2}{2a^2b^5}\right)^2$

Key Points

⊙ The laws of indices used so far can be used to develop two further laws.

$x^4 \div x^4 = x^{4-4} = x^0$

Also

$x^4 \div x^4 = 1$ since any term divided by itself is equal to 1.

Therefore $x^0 = 1$

In general

$x^0 = 1$

$x^3 \div x^4$

$= \dfrac{x \times x \times x}{x \times x \times x \times x} = \dfrac{1}{x}$

Also, using $x^p \div x^q = x^{p-q}$

$x^3 \div x^4 = x^{3-4} = x^{-1}$

Therefore $x^{-1} = \dfrac{1}{x}$

In general

$x^{-m} = \dfrac{1}{x^m}$

⊙ The laws of indices can be used further to solve problems with fractional indices.

The square root of x is written $\sqrt{x}$, and you know that:

$\sqrt{x} \times \sqrt{x} = x$

Using $x^p \times x^q = x^{p+q}$

$x^{\frac{1}{2}} \times x^{\frac{1}{2}} = x^{\frac{1}{2}+\frac{1}{2}} = x^1 = x$

and so, $x^{\frac{1}{2}} = \sqrt{x}$

Also, $x^{\frac{1}{3}} \times x^{\frac{1}{3}} \times x^{\frac{1}{3}} = x$, showing that $x^{\frac{1}{3}} = \sqrt[3]{x}$

In general

$x^{\frac{1}{n}} = \sqrt[n]{x}$

Example 7   Simplify $(3x^4y)^{-2}$

$(3x^4y)^{-2} = \dfrac{1}{(3x^4y)^2}$ ⟵ Using $x^{-m} = \dfrac{1}{x^m}$

$= \dfrac{1}{9x^8y^2}$ ⟵ Using $(x^p)^q = x^{p \times q}$

ResultsPlus

Exam Tip

Remember that a negative power just means 'one over' or 'the reciprocal of'.

Exercise 2F

1 Simplify

  a a^{-1}　　　　b $(b^2)^{-1}$　　　　c c^{-2}　　　　d $(d^3)^{-1}$

2 Simplify

  a $(e^3)^{-2}$　　　　b $(f^2)^{-4}$　　　　c $(x^{-1})^{-2}$　　　　d $(y^{-1})^{-1}$

3 Simplify

  a $(x^2y^7)^0$　　　　b $(2x^4y^5)^0$　　　　c $(5p^2q^4)^{-1}$　　　　d $(3c^3d)^{-3}$

  e $\left(\dfrac{2p^3q}{3r^2}\right)^{-2}$

B

A

Example 8 Simplify $(8x^6y^4)^{\frac{1}{3}}$

$(8x^6y^4)^{\frac{1}{3}} = 8^{\frac{1}{3}} \times (x^6)^{\frac{1}{3}} \times (y^4)^{\frac{1}{3}}$ ← Using $x^{\frac{1}{n}} = \sqrt[n]{x}$

$= \sqrt[3]{8} \times x^{6 \times \frac{1}{3}} \times y^{4 \times \frac{1}{3}}$

Using $(x^p)^q = x^{p \times q}$

$= 2 \times x^2 \times y^{\frac{4}{3}}$

$= 2x^2 y^{\frac{4}{3}}$

ResultsPlus
Exam Tip

Remember that the denominator of the index is the root.

Exercise 2G

A

1 Simplify

a $(9a^4)^{\frac{1}{2}}$ b $(16c^2)^{\frac{1}{4}}$ c $(27e^3f^{-9})^{\frac{1}{3}}$ d $(100x^3y^5)^{\frac{1}{2}}$

2 Simplify

a $(a^4)^{-\frac{1}{2}}$ b $(8c^3)^{-\frac{1}{3}}$ c $(32x^9y^5)^{-\frac{1}{5}}$ d $(x^2y^6)^{-\frac{1}{4}}$

2.5 Term-to-term and position-to-term definitions

◎ Objective

- You can generate terms of a sequence using term-to-term and position-to-term definitions of the sequence.

⦿ Why do this?

To recognise world trends in specific illnesses, patterns linking data are often used.

⬆ Get Ready

Continue these number patterns.

1. 2, 4, 6, 8, 10, …   **2.** 4, 9, 14, 19, 24, 29, …   **3.** 1, 3, 5, 7, 9, …

Key Points

- A **sequence** is a pattern of shapes or numbers which are connected by a **rule** (or definition of the sequence).
- The relationship between consecutive terms describes the rule which enables you to find subsequent **terms of the sequence**.

Here is a sequence of 4 square patterns made up of squares:

Pattern 1   Pattern 2   Pattern 3   Pattern 4

⊙ Each pattern above is a term of the sequence;

☐ is the 1st term in the sequence,

⊞ is the 2nd term in the sequence, etc.

⊙ The number of squares in each term form a sequence of numbers, 1, 4, 9, 16, …
⊙ The odd numbers form a sequence, 1, 3, 5, …
⊙ The even numbers form a sequence, 2, 4, 6, …
⊙ You can continue a sequence if you know how the terms are related: the **term-to-term rule**.
⊙ You can continue a sequence if you know how the position of a term is related to the definition of the sequence: the **position-to-term rule**.

Example 9   Find **a** the next term, and
b the 12th term of the sequence of numbers: 1, 4, 9, 16, …

1st term	2nd term	3rd term	4th term	5th term
1	4	9	16	

+3   +5   +7

> The difference between consecutive terms increases by 2.
> This is the term-to-term rule which enables you to find subsequent terms of the sequence.

a The difference between the 4th and the 5th term is $+9$ and so the 5th term is $16 + 9 = $ **25**.

b The 6th term $= 6^2 = 36$, the 7th term $= 7^2 = 49$, etc.

> The numbers 1 $(=1^2)$, 4 $(= 2^2)$, 9 $(= 3^2)$, 16 $(= 4^2)$ and 25 $(= 5^2)$ are the first five square numbers.

The 12th term $= 12^2 = $ **144**. ← In this way a term of the sequence can be found by the position of the term in the sequence.

Exercise 2H

Find **a** the term-to-term rule,
b the next two terms, and
c the 10th term for each of the following number sequences.

1   2   5   8   11

2   −4   2   8   14

3   19   12   5   −2

4   1   3   6   10

5   0   2   6   12

2.6 The nth term of an arithmetic sequence

Objectives

○ You can use linear expressions to describe the nth term of a sequence.
○ You can use the nth term of a sequence to generate terms of the sequence.

Why do this?

To be able to predict how many people might catch the flu, epidemiologists need to develop a general rule.

Get Ready

Find **a** the rule, **b** the next two terms, **c** the 10th term for each of the following number sequences.

1. 1, 4, 7, 10, …   **2.** −4, −1, 2, 5, 8, …   **3.** 124, 118, 112, 106, 100, …

Key Points

◉ An **arithmetic sequence** is a sequence of numbers where the rule is simply to add a fixed number.
For example, 2, 5, 8, 11, 14, … is an arithmetic sequence with the rule 'add 3'.
In this example the fixed number is 3.
◉ This is sometimes called the **difference** between consecutive terms.
◉ You can find the nth term using the result nth term $= n \times$ difference $+$ **zero term**.
◉ You can use the nth term to generate the terms of a sequence.
◉ You can use the terms of a sequence to find out whether or not a given number is part of a sequence, and explain why.

Example 10

Here are the first five terms of an arithmetic sequence: 2, 5, 8, 11, 14, …

a Write down, in terms of n, an expression for the nth term of the arithmetic sequence.
b Use your answer to part **a** to find the 20th term.

zero term	1st term	2nd term	3rd term	4th term	5th term
−1	2	5	8	11	14

+3   +3   +3   +3   +3

difference

a The zero term is the term before the first term.
Work out the zero term by using the difference of $+3$.
Zero term $= 2 - 3 = -1$

Inverse of $+3$.

The n**th term** $= n \times$ **difference** $+$ **zero term**
nth term $= n \times +3 + -1$
$\qquad = 3n - 1$

b For the 20th term, $n = 20$
When $n = 20$, $3n - 1 = 3 \times 20 - 1$
$\qquad = 60 - 1 = 59$
So the 20th term is 59.

ResultsPlus
Exam Tip

Always check your answer by substituting values of n into your nth term.
For example,
1st term, when $n = 1$, $3n - 1 = 3 \times 1 - 1 = 2$ ✓
2nd term, when $n = 2$, $3n - 1 = 3 \times 2 - 1 = 5$ ✓
3rd term, when $n = 3$, $3n - 1 = 3 \times 3 - 1 = 8$ ✓
etc.

arithmetic sequence   difference   zero term

C

Exercise 2I

1. Write down   i the difference between consecutive terms
              ii the zero term for each of the following arithmetic sequences.
   a 0, 2, 4, 6, 8, …
   b −7, −3, 1, 5, 9, …
   c 14, 9, 4, −1, −6, …

2. Here are the first five terms of an arithmetic sequence: 1, 7, 13, 20, 26, …
   a Write down, in terms of n, an expression for the nth term of this arithmetic sequence.
   b Use your answer to part **a** to work out the   i 12th term,   ii 50th term.

3. Here are the first four terms of an arithmetic sequence: 7, 11, 15, 19, …
   a Write down, in terms of n, an expression for the nth term of this arithmetic sequence.
   b Use your answer to part **a** to work out the   i 15th term,   ii 100th term.

4. Here are the first five terms of an arithmetic sequence: 32, 27, 22, 17, 12, …
   a Write down, in terms of n, an expression for the nth term of this arithmetic sequence.
   b Use your answer to part **a** to work out the   i 20th term,   ii 200th term.

5. Here are the first four terms of an arithmetic sequence: 18, 25, 32, 39, …
   Explain why the number 103 cannot be a term of this sequence.

   A03

* 6. Here are the first five terms of an arithmetic sequence:
   7   11   15   19   23
   Pat says that 453 is a term in this sequence. Pat is wrong.
   Explain why.

   A03

   Nov 2005

Chapter review

- $2x$, $3y$ and $2x + 3y$ are called **algebraic expressions**.
- Each part of an **expression** is called a **term** of the expression.
- When adding or subtracting expressions, different letter symbols cannot be combined.
- The sign of a term in an expression is always written before the term.
- The term x can be written as $1x$.
- In algebra, BIDMAS describes the order of operations when **collecting like terms**.
- If you are given the value for each letter in an expression then you can **substitute** the values into the expression and **evaluate** the expression.
- You can use the laws of indices to simplify algebraic expressions.
- The basic index laws can be used to develop further laws:
  $x^0 = 1$, for all values of x, $x^{-m} = \dfrac{1}{x^m}$ and $x^{\frac{1}{n}} = \sqrt[n]{x}$ where m and n are integers.
- A **sequence** is a pattern of shapes or numbers which are connected by a **rule** (or definition of the sequence).
- The relationship between consecutive terms describes the rule which enables you to find subsequent **terms of the sequence**.
- You can continue a sequence if you know how the terms are related: the **term-to-term rule**.
- You can continue a sequence if you know how the position of a term is related to the definition of the sequence: the **position-to-term rule**.
- An **arithmetic sequence** is a sequence of numbers where the rule is simply to add a fixed number. This is called the **difference** between consecutive terms.

⦿ You can find the nth term of an arithmetic sequence using the result nth term $= n \times$ difference $+$ **zero term**.

⦿ You can use the nth term of an arithmetic sequence to generate the terms of a sequence.

⦿ You can use the terms of a sequence to find out whether or not a given number is part of a sequence, and explain why.

⚙ Review exercise

1 Simplify   **a** $3x - 4y + 2x - y$ **b** $m - 7n + 5m - 3n$

2 Helen and Stuart collect stamps.
Helen has 240 British stamps and 114 Australian stamps.

 a Write down an algebraic expression that could be used to represent Helen's British and Australian stamps. Define the letters used.

 Stuart has 135 British stamps and 98 Australian stamps.

 b Using the same letters, write down an algebraic expression that could be used to represent the total of Helen's and Stuart's British and Australian stamps.

3 Work out the value of each of these expressions when $x = 2$, $y = -3$ and $z = -7$
 a $3x + y$ **b** $x - 2y$ **c** $x + 3y - 2z$ **d** $5xy$ **e** $x^2 + y^2 + z^2$

4 Simplify
 a $y \times y \times y$ **b** $x \times 3x$ **c** $z^3 \times z^5$ **d** $p \times p^6$ **e** $2a^2 \times 8a^5$

5 Simplify
 a $a^6 \div a^3$ **b** $b^9 \div b^4$ **c** $21p^4 \div 3p$ **d** $\dfrac{24x^5}{3x^2}$ **e** $16a^6b^3 \div 2a^5b^3$

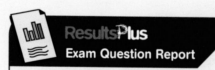

ResultsPlus
Exam Question Report

93% of students answered this sort of question well because they had learnt the rules for expressions involving indices.

6 Find   **a** the rule   **b** the next two terms   **c** the 12th term for this number sequence.
 102   99   96   93   90

7 Write down   **a** the difference between consecutive terms,
        b the zero term of this arithmetic sequence.
         -3 2   7   12   17

8 Here are the first four terms of an arithmetic sequence: 204, 192, 180, 168, …

 a Write down, in terms of n, an expression for the nth term of this arithmetic sequence.

 b Use your answer to part **a** to work out the   **i** 13th term   **ii** 99th term.

9 Here are the first four terms of an arithmetic sequence.
 5   8   11   14

 Is 140 a term in the sequence? You must give a reason for your answer.

10 Neal is asked to produce an advertising stand for a new variety of soup.
He stacks the cans according to the pattern shown.
The stack is 4 cans high and consists of 10 cans.

  a How many cans will there be in a stack 10 cans high?

  b Verify that the total number of cans (N) can be calculated by the formula

    $N = \dfrac{h(h+1)}{2}$ when h = number of cans high.

  c If he has 200 cans, how high can he make his stack?

11 Naismith, an early Scottish mountain climber, devised a formula that is still used today to calculate
how long it will take mountaineers to climb a mountain. The metric version states:
Allow one hour for every 5 km you walk forward and add on $\frac{1}{2}$ hour for every 300 m of ascent.

  a How long should it take to walk 20 km with 900 m of ascent?

A mountain walker's guide contains the following information for a particular walk.

> **Helvellyn Horseshoe**
> Glenridding to Helvellyn via the edges (circular walk)
> Length: 8.5 km
> Total ascent: 800 m
> Time: 4 hour round trip

  b Calculate how long this walk should take according to Naismith's formula. Give your answer to the nearest minute.

  c Suggest reasons why this time is different to the one in the guidebook.

12 Simplify

  a $(a^5)^4$ **b** $(3b^4)^2$ **c** $(3e^5f)^3$

13 The nth even number is $2n$.
Show algebraically that the sum of three consecutive even numbers is always a multiple of 6.

Nov 2008, adapted

14 The expression $\dfrac{6x^2 y}{4y^3}$ can never take a negative value. Explain why.

15 **a** Simplify $\left(\dfrac{9p^4}{4y^2}\right)^{\frac{1}{2}}$ **b** Simplify $(2q^3)^{-2}$ **c** Simplify $\left(\dfrac{12xy^3}{3x^5y}\right)^{\frac{1}{2}}$

16 A 4 by 4 by 4 cube is placed into a tin of yellow paint.

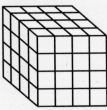

When it has dried, the 64 individual cubes are examined.
How many are covered in yellow paint on 0 sides, 1 side, 2 sides, 3 sides?
Extension: Repeat the question for an n by n by n cube, and show that your expressions add up to n^3.

3 FRACTIONS

Only one eighth of an iceberg shows above the surface of the water, which leaves most of it hidden. The largest northern hemisphere iceberg was encountered near Baffin Island in Canada in 1882. It was 13 km long, 6 km wide and had a height above water of about 20 m. It had a mass of over 9 billion tonnes – enough water for everyone in the world to drink a litre a day for over four years.

⦿ Objectives

In this chapter you will:
- add, subtract, multiply and divide fractions and mixed numbers
- find a fraction of a quantity
- solve problems involving fractions.

ⓘ Before you start

You need to be able to:
- find the highest common factor (HCF) of two numbers
- find the lowest common multiple (LCM) of two or three numbers
- simplify and order fractions
- convert between improper fractions and mixed numbers.

3.1 Adding and subtracting fractions and mixed numbers

⦿ Objectives

◉ You can add and subtract fractions.

◉ You can add and subtract mixed numbers.

? Why do this?

Measurements are not always given in whole numbers. You may need to find the total length of two distances given as fractions, for example, $2\frac{3}{4}$ km and $1\frac{1}{4}$ km.

✦ Get Ready

1. Write $\frac{32}{36}$ in its simplest form.

2. Change $2\frac{3}{8}$ to an improper fraction.

3. Change $\frac{47}{5}$ to a mixed number.

Key Points

◉ To add (or subtract) fractions, change them to **equivalent fractions** that have the same denominator. This new demoninator will be the LCM of the two denominators (see Section 1.1 for LCM). Then add (or subtract) the numerators but do not change the denominator.

◉ To add (or subtract) **mixed numbers**, add (or subtract) the whole numbers, then add (or subtract) the fractions separately.

Example 1 Work out $\frac{7}{8} - \frac{1}{4}$

$$\frac{1}{4} \xrightarrow{\times 2} = \xrightarrow{\times 2} \frac{2}{8}$$

> The LCM of 4 and 8 is 8.
> Convert $\frac{1}{4}$ to the equivalent fraction with a denominator of 8.

$$\frac{7}{8} - \frac{1}{4} = \frac{7}{8} - \frac{2}{8}$$

> Subtract the numerators only.

$$= \frac{5}{8}$$

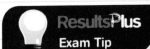
ResultsPlus
Exam Tip

If you could use several numbers as the new denominator, using the LCM of the two denominators means you won't need to simplify later.

Example 2 Work out $\frac{5}{6} + \frac{7}{10}$

Give your answer as a mixed number.

$$\frac{5}{6} \xrightarrow{\times 5} = \xrightarrow{\times 5} \frac{25}{30} \qquad \frac{7}{10} \xrightarrow{\times 3} = \xrightarrow{\times 3} \frac{21}{30}$$

> The LCM of 6 and 10 is 30.
> Change each fraction to its equivalent fraction with a denominator of 30.

$$\frac{5}{6} + \frac{7}{10} = \frac{25}{30} + \frac{21}{30}$$

$$= \frac{46}{30}$$

> Simplify the fraction.
> Divide each number in the fraction by 2.

$$= \frac{23}{15}$$

> Convert the mixed number to an improper fraction.
> $23 \div 15 = 1$ remainder 8.

$$= 1\frac{8}{15}$$

---

equivalent fractions   mixed numbers   **29**

Exercise 3A

D

Questions in this chapter are targeted at the grades indicated.

Give each answer as a fraction in its simplest form.

1 Work out

a $\frac{5}{11} + \frac{3}{11}$ b $\frac{1}{9} + \frac{4}{9}$ c $\frac{7}{15} + \frac{4}{15}$ d $\frac{3}{10} + \frac{1}{10}$

2 Work out

a $\frac{1}{5} + \frac{1}{2}$ b $\frac{1}{3} + \frac{1}{7}$ c $\frac{3}{7} + \frac{2}{5}$ d $\frac{5}{9} + \frac{1}{3}$

e $\frac{2}{5} + \frac{1}{4}$ f $\frac{1}{2} + \frac{2}{9}$ g $\frac{2}{9} + \frac{1}{6}$ h $\frac{7}{20} + \frac{2}{5}$

3 Work out

a $\frac{1}{2} - \frac{1}{4}$ b $\frac{1}{3} - \frac{1}{4}$ c $\frac{7}{8} - \frac{2}{5}$ d $\frac{8}{9} - \frac{2}{3}$

e $\frac{3}{4} - \frac{1}{2}$ f $\frac{2}{3} - \frac{1}{4}$ g $\frac{17}{20} - \frac{3}{4}$ h $\frac{5}{6} - \frac{7}{9}$

Give each answer as a fraction or a mixed number in its simplest form.

4 Work out

a $\frac{2}{5} + \frac{7}{8}$ b $\frac{3}{4} + \frac{4}{5}$ c $\frac{5}{6} - \frac{2}{3}$ d $\frac{7}{10} + \frac{1}{4}$

e $\frac{5}{6} + \frac{9}{10}$ f $\frac{3}{4} + \frac{7}{8}$ g $\frac{4}{5} - \frac{1}{2} + \frac{9}{10}$ h $\frac{7}{9} + \frac{2}{3} - \frac{1}{6}$

Example 3

Work out $5\frac{7}{10} + 4\frac{1}{2}$

$5\frac{7}{10} + 4\frac{1}{2} = 9\frac{7}{10} + \frac{1}{2}$ ←   Add the whole numbers.

$= 9\frac{7}{10} + \frac{5}{10}$ ←   Convert the fractions into equivalent fractions with a denominator of 10.

$= 9\frac{12}{10}$ ← $\frac{12}{10}$ is an improper fraction. Change this into a mixed number.

$= 10\frac{2}{10}$ ←   Simplify $\frac{2}{10}$.

$= 10\frac{1}{5}$

Exercise 3B

C

1 Work out   a $6\frac{3}{4} + 1\frac{1}{2}$ b $4\frac{4}{5} + \frac{1}{2}$ c $7\frac{5}{6} + 3\frac{2}{7}$ d $12\frac{3}{4} + 5\frac{2}{5}$

2 Becky cycled $2\frac{3}{4}$ miles to one village then a further $4\frac{1}{3}$ miles to her home. What is the total distance that Becky cycled?

3 A bag weighs $\frac{3}{7}$ lb. The contents weigh $1\frac{1}{5}$ lb. What is the total weight of the bag and its contents?

Example 4 Work out $7\frac{1}{4} - 2\frac{7}{10}$

Method 1

$7\frac{1}{4} - 2\frac{7}{10} = 5\frac{5}{20} - \frac{14}{20}$ ⟵ $\frac{5}{20} - \frac{14}{20}$ will give a negative result. Write $5\frac{5}{20}$ as $4 + 1\frac{5}{20} = 4\frac{25}{20}$.

$= 4\frac{25}{20} - \frac{14}{20}$

$= 4\frac{11}{20}$

Method 2

$7\frac{1}{4} - 2\frac{7}{10} = \frac{29}{4} - \frac{27}{10}$ ⟵ Convert the mixed numbers to improper fractions.

$= \frac{290}{40} - \frac{108}{40}$

$= \frac{182}{40}$

$= 4\frac{22}{40}$ ⟵ Convert the improper fraction to a mixed number. $182 \div 40 = 4$, remainder 22.

$= 4\frac{11}{20}$ ⟵ Simplify the fraction.

Exercise 3C

1 Work out   **a** $2\frac{1}{2} - 1\frac{1}{4}$ **b** $3\frac{7}{8} - 1\frac{1}{2}$ **c** $6 - 5\frac{1}{4}$ **d** $8 - 4\frac{2}{3}$

2 Work out   **a** $2\frac{1}{4} - 1\frac{1}{2}$ **b** $3\frac{1}{4} - 1\frac{2}{3}$ **c** $4\frac{2}{7} - 1\frac{3}{5}$ **d** $7\frac{1}{9} - 3\frac{2}{3}$

3 A box containing vegetables has a total weight of $5\frac{1}{4}$ kg. The empty box has a weight of $1\frac{7}{8}$ kg. What is the weight of the vegetables?

4 A tin contains $7\frac{1}{2}$ pints of oil. Julie pours out $4\frac{5}{8}$ pints from the tin. How much oil remains?

C

3.2 Multiplying fractions and mixed numbers

◎ Objectives

- ○ You can multiply fractions.
- ○ You can multiply mixed numbers.
- ○ You can find a fraction of a quantity.

⬦ Why do this?

Shops often advertise discounts as '$\frac{2}{3}$ off the normal price'. To work out the discount you will need to multiply by $\frac{2}{3}$.

⬆ Get Ready

1. Work out 4×8.   **2.** Work out 5×9.

3. Change $3\frac{4}{5}$ to an improper fraction.   **4.** Convert $\frac{14}{3}$ to a mixed number.

Key Points

- To multiply fractions:
  - Convert any mixed numbers to **improper fractions**.
  - Simplify if possible.
  - Multiply the numerators and multiply the denominators.

Example 5   Work out $\frac{2}{3} \times 7$.

$$\frac{2}{3} \times 7 = \frac{2}{3} \times \frac{7}{1}$$

⟵ Write 7 as an improper fraction.

$$= \frac{2 \times 7}{3 \times 1}$$

$$= \frac{14}{3}$$

$$= 4\frac{2}{3}$$

⟵ Write $\frac{14}{3}$ as a mixed number.

Example 6   Work out $\frac{5}{6}$ of 9 metres.

$$\frac{5}{6} \times 9 = \frac{5}{6} \times \frac{9}{1}$$

⟵ To find the fraction of a quantity, multiply the fraction by the quantity.

$$= \frac{5 \times 9^{3}}{{}_{2}6 \times 1}$$

⟵ Simplify by dividing the numerator and denominator by 3.

$$= \frac{15}{2}$$

$$= 7\frac{1}{2} \text{ metres}$$

Example 7   Work out $\frac{5}{14} \times \frac{7}{10}$

$$\frac{5}{14} \times \frac{7}{10} = \frac{5 \times 7}{14 \times 10}$$

$$= \frac{{}^{1}5 \times 7}{14 \times 10_{2}}$$

⟵ Simplify by dividing the numerator and denominator by 5.

$$= \frac{1 \times 7^{1}}{{}_{2}14 \times 2}$$

⟵ Simplify by dividing the numerator and denominator by 7.

$$= \frac{1}{4}$$

Example 8   Work out $2\frac{2}{3} \times 1\frac{4}{5}$

$$2\frac{2}{3} \times 1\frac{4}{5} = \frac{8}{3} \times \frac{9}{5}$$

⟵ Convert each mixed number into an improper fraction.

$$= \frac{8 \times 9^{3}}{{}_{1}3 \times 5}$$

⟵ Divide the numerator and denominator by 3.

$$= \frac{24}{5}$$

$$= 4\frac{4}{5}$$

⟵ Convert the improper fraction into a mixed number.

Exercise 3D

1. Work out
   a $\frac{3}{5} \times \frac{1}{2}$
   b $\frac{1}{4} \times \frac{3}{5}$
   c $\frac{10}{11} \times \frac{3}{5}$
   d $\frac{5}{6} \times \frac{4}{15}$
   e $\frac{2}{3} \times \frac{2}{7}$
   f $\frac{3}{4} \times \frac{3}{5}$
   g $\frac{9}{28} \times \frac{14}{15}$
   h $\frac{25}{36} \times \frac{27}{40}$

2. Work out
   a $2 \times \frac{1}{3}$
   b $3 \times \frac{1}{4}$
   c $\frac{9}{20} \times 8$
   d $\frac{3}{5} \times 25$

3. Work out
   a $\frac{3}{5}$ of 35 kg
   b $\frac{4}{9}$ of 15 m
   c $\frac{5}{8}$ of 12 litres
   d $\frac{3}{10}$ of 25 pints

4. Jomo delivers 56 newspapers on his round. On Fridays $\frac{3}{8}$ of the newspapers have a magazine supplement. How many supplements does he deliver?

5. Barry earns £130.60 in one week. He pays $\frac{1}{4}$ of this in tax. How much money does he pay in tax each week?

6. Work out
   a $1\frac{1}{4} \times \frac{1}{3}$
   b $1\frac{3}{5} \times \frac{1}{2}$
   c $3\frac{3}{4} \times 1\frac{1}{10}$
   d $1\frac{2}{3} \times 4\frac{1}{5}$
   e $1\frac{1}{3} \times 2\frac{1}{4}$
   f $3\frac{1}{2} \times 1\frac{1}{4}$
   g $6\frac{3}{7} \times 1\frac{5}{9}$
   h $8\frac{1}{3} \times 2\frac{7}{10}$

7. Kieran takes $2\frac{1}{4}$ minutes to complete one lap at the Go Kart Centre. How long will it take him to complete $6\frac{1}{2}$ laps?

8. A melon weighs $2\frac{1}{2}$ lb. Work out the weight of $8\frac{1}{4}$ melons.

C

3.3 Dividing fractions and mixed numbers

◎ Objectives

- ◎ You can divide fractions.
- ◎ You can divide mixed numbers.

◈ Why do this?

A carpenter may want to work out how many pieces of wood measuring $\frac{3}{4}$ m he can cut from a 5 m piece of wood.

◆ Get Ready

1. Work out $\frac{3}{7} \times \frac{2}{5}$.
2. Work out $\frac{2}{3} \times \frac{1}{4}$.
3. Write $3\frac{3}{7}$ as an improper fraction.

◗ Key Points

◉ To divide fractions:
   - ◉ Convert any mixed numbers to improper fractions.
   - ◉ Convert divide to multiply and invert the second fraction (**inverted** means turned upside down).
   - ◉ Multiply the top numbers and multiply the bottom numbers.

Example 9 Work out $\frac{4}{5} \div 3$

> Multiplying by $\frac{1}{3}$ is the same as dividing by 3. $\frac{1}{3}$ is called the reciprocal of 3.

$$\frac{4}{5} \div 3 = \frac{4}{5} \div \frac{3}{1}$$

> Write the whole number as an improper fraction.

$$= \frac{4}{5} \times \frac{1}{3}$$

> Change $\div$ to $\times$ and turn the second fraction upside down. Multiply the fractions.

$$= \frac{4}{15}$$

Example 10 Work out $\frac{5}{6} \div \frac{3}{4}$. Give your answer in its simplest form.

$$\frac{5}{6} \div \frac{3}{4} = \frac{5}{6} \times \frac{4}{3}$$

> Turn $\frac{3}{4}$ upside down to get $\frac{4}{3}$.

$$= \frac{5}{6_3} \times \frac{{}^2 4}{3}$$

> Divide the numerator and denominator by 2.

$$= \frac{10}{9}$$

$$= 1\frac{1}{9}$$

> Write the improper fraction as a mixed number.

Example 11 Work out $2\frac{4}{5} \div 2\frac{1}{10}$

$$2\frac{4}{5} \div 2\frac{1}{10} = \frac{14}{5} \div \frac{21}{10}$$

> Write the mixed numbers as improper fractions.

$$= \frac{14}{5} \times \frac{10}{21}$$

> Turn $\frac{21}{10}$ upside down to get $\frac{10}{21}$.

$$= \frac{{}^2 14}{5_1} \times \frac{{}^2 10}{21_3}$$

> Divide top and bottom by 7 and by 2.

$$= \frac{4}{3}$$

$$= 1\frac{1}{3}$$

> Write the improper fraction as a mixed number.

Exercise 3E

1 Work out

a $\frac{5}{6} \div 2$
b $\frac{3}{8} \div 2$
c $\frac{4}{5} \div \frac{3}{10}$
d $\frac{9}{16} \div \frac{3}{8}$

e $\frac{1}{4} \div \frac{1}{3}$
f $\frac{3}{5} \div \frac{1}{2}$
g $\frac{20}{21} \div \frac{8}{15}$
h $\frac{25}{32} \div \frac{15}{16}$

2 Work out

a $3\frac{1}{2} \div 7$
b $2\frac{4}{5} \div \frac{1}{10}$
c $3\frac{3}{4} \div 1\frac{4}{5}$
d $6\frac{2}{3} \div 2\frac{8}{9}$

e $1\frac{1}{2} \div \frac{3}{4}$
f $2\frac{4}{9} \div \frac{2}{3}$
g $7\frac{1}{2} \div 1\frac{1}{4}$
h $2\frac{1}{12} \div 1\frac{1}{9}$

3 A tin holds $10\frac{2}{3}$ litres of methylated spirit for a lamp. How many times will it fill a lamp holding $\frac{2}{3}$ litre?

4 A metal rod is $10\frac{4}{5}$ metres long. How many short rods $\frac{3}{10}$ metre long can be cut from the longer rod?

5 Tar and Stone can resurface $2\frac{1}{5}$ km of road in a day. How many days will it take them to resurface a road of length $24\frac{3}{5}$ km?

3.4 **Fraction problems**

⊙ Objective

○ You can solve problems involving fractions.

⊘ Why do this?

A vet may need to work out fractions of a dosage depending on the size of the animal in comparison to the standard.

⬙ Get Ready

1. Work out $\frac{3}{4} + \frac{1}{8}$

2. Work out $\frac{4}{9} \times \frac{1}{5}$

3. Work out $5\frac{7}{8} - 2\frac{1}{4}$

🌐 Key Point

◉ You can use your knowledge of fractions to solve problems from real life.

Example 12   In a cinema $\frac{2}{5}$ of the audience are women, $\frac{1}{8}$ of the audience are men. All the rest of the audience are children. What fraction of the audience are children?

$\frac{2}{5} + \frac{1}{8} = \frac{16}{40} + \frac{5}{40}$ ⟵   Add $\frac{2}{5}$ and $\frac{1}{8}$ to find the fraction of the audience who are women or men.

$= \frac{21}{40}$

$1 - \frac{21}{40} = \frac{40}{40} - \frac{21}{40}$ ⟵   Subtract $\frac{21}{40}$ from 1 to find the fraction of the audience who are children.

$= \frac{19}{40}$

$\frac{19}{40}$ of the audience are children.

Example 13   A school has 1800 pupils. 860 of these pupils are girls. $\frac{3}{4}$ of the girls like swimming. $\frac{2}{5}$ of the boys like swimming. Work out the total number of pupils in the school who like swimming.

$\frac{3}{4} \times 860 = 645$ ⟵   Work out the number of girls who like swimming.

$1800 - 860 = 940$ ⟵   Work out the number of boys in the school.

$\frac{2}{5} \times 940 = 376$ ⟵   Work out the number of boys who like swimming.

$645 + 376 = 1021$ ⟵   Work out the total number of pupils who like swimming.

1021 pupils like swimming.

Exercise 3F

D

1. Simon spends $\frac{1}{2}$ of his money on rent and $\frac{1}{3}$ of his money on transport.
   a What fraction of his money does he spend on rent and transport altogether?
   b What fraction of his money is left?

2. $\frac{8}{9}$ of an iceberg lies below the surface of the water. The total volume of an iceberg is 990 m³.
   What volume of this iceberg is below the surface?

3. DVDs are sold for £14 each. $\frac{2}{5}$ of the £14 goes to the DVD company.
   How much of the £14 goes to the DVD company?

A02

4. An MP3 player usually costs £130. In a sale all prices are reduced by $\frac{2}{5}$.
   Work out the sale price of the MP3 player.

A02 A03

5. A factory has 1710 workers. 650 of the workers are female.
   $\frac{2}{5}$ of the female workers are under the age of 30, $\frac{1}{4}$ of the male workers are under the age of 30.
   How many workers in total are aged under 30?

C

A03

6. There are 36 students in a class. Javed says that $\frac{3}{8}$ of these students are boys.
   Explain why Javed cannot be right.

7. Tammy watches two films. The first film is $1\frac{3}{4}$ hours long and the second one is $2\frac{1}{3}$ hours long.
   Work out the total length of the two films.

8. $\frac{2}{3}$ of a square is shaded. $\frac{3}{4}$ of the shaded part is shaded blue.
   What fraction of the whole square is shaded blue?

9. Alison, Becky and Carol take part in a charity relay race. The race is over a total distance of $2\frac{5}{8}$ km.
   Each girl runs an equal distance. Work out how far each girl runs.

A02 A03

10. In a book, $\frac{3}{8}$ of the pages have pictures on them.
    Given that 72 pages have a picture on, work out the number of pages in the book.

A02 A03

11. Alex spent $\frac{2}{7}$ of his pocket money on a computer game.
    He spent $\frac{1}{4}$ of his pocket money on sweets. He saved the rest.
    Given that Alex saved £4.50, work out how much pocket money he got.

Chapter review

- To add (or subtract) fractions, change them to **equivalent fractions** that have the same denominator. This new denominator will be the LCM of the two denominators. Then add (or subtract) the numerators but do not change the denominator.
- To add or subtract **mixed numbers**, add or subtract the whole numbers, then add or subtract the fractions separately.
- To multiply fractions, convert any mixed numbers to **improper fractions**, simplify if possible, then multiply the numerators and multiply the denominators.
- To divide fractions, convert any mixed numbers to improper fractions, convert divide to multiply and **invert** the second fraction, then multiply the numerators and multiply the denominators.
- You can use your knowledge of fractions to solve problems from real life.

Review exercise

1 Simplify these fractions

   a $\frac{12}{18}$ b $\frac{28}{35}$ c $\frac{32}{48}$ d $\frac{96}{144}$ e $\frac{84}{231}$

2 a Change $3\frac{2}{5}$ to an improper fraction.   b Change $\frac{31}{6}$ to a mixed number.

3 Many wage earners work a fixed number of hours, typically 35 hours a week.
   If they are required to work more than this, they are paid overtime. For example, overtime paid at
   'time and a half' would mean someone normally earning £8 per hour would receive £12 per hour for
   the extra hours.
   Copy and complete the table for the following workers.

Name	Hourly rate	Overtime at time and a half	Overtime at double time
Aaron	£8.50		
Chi	£12.00		
Mahmood	£14.40		

4 Work out

   a $\frac{2}{5} + \frac{1}{4}$

   b $\frac{5}{6} + \frac{3}{4}$

   c $2\frac{1}{4} + 3\frac{1}{3}$

   d $51\frac{7}{8} + 32\frac{2}{3}$

ResultsPlus
Exam Question Report

33% of students answered this sort of question
poorly because they simply added the numerators
and added the denominators.

5 a Using the table above find Aaron's weekly wage if he works 35 hours at £8.50 per hour and
   6 hours' overtime at time and a half.
   Chi normally works 36 hours a week and is paid for any overtime at time and a quarter.
   b One week she earned £522. How many extra hours did she work?

6 Work out

   a $\frac{3}{4} - \frac{2}{3}$

   b $5\frac{3}{5} - 2\frac{1}{6}$

   c $4\frac{3}{4} - 2\frac{5}{6}$

ResultsPlus
Exam Question Report

71% of students answered this sort of question
well because they dealt with the integers and
fractions separately.

7
   $1\frac{2}{3}$ m   Diagram **NOT** accurately drawn

   $2\frac{1}{2}$ m

   a Work out the area of this rectangle. *Hint:* Area of rectangle = length × width.
   b Work out the perimeter of this rectangle.
   c Work out the difference in lengths between the shortest and the longest side.

8

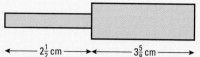

Diagram **NOT** accurately drawn

$2\frac{1}{2}$ cm $3\frac{5}{8}$ cm

The diagram represents a part of a machine.

In order to fit the machine, the part must be between $6\frac{1}{16}$ cm and $6\frac{3}{16}$ cm long.

Will the part fit the machine?

You must explain your answer.

June 2009

9 On a farm, $\frac{3}{8}$ of the land area is used to keep sheep.

Half of the rest of the land area on the farm is used to grow crops.

The land area of the farm used to grow grass is 600 000 m².

Work out the land area of the farm used to keep sheep.

June 2009

10 The distance from Granby to Hightown is $3\frac{2}{3}$ miles. The distance from Hightown to Isely is $2\frac{1}{2}$ miles.

Jim walks from Granby to Isely via Hightown. He stops for a rest when he has walked half the total distance. How far has he walked when he stops for his rest?

11

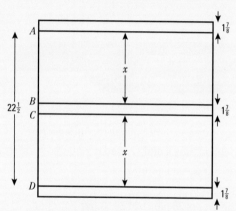

a Here is a design for a book case with two shelves. All the measurements are in inches.

The gap AB is the same as the gap CD. Work out the value of x.

b In another design the bookcase is the same except the middle shelf has been moved so that the gap AB is twice the gap CD.

Find the size of the gap AB.

12

a	$\frac{1}{15}$	$\frac{2}{5}$
c	b	$\frac{7}{15}$
d	$\frac{3}{5}$	$\frac{2}{15}$

This is a magic square. The sum of the three numbers in each row, each column and each diagonal is the same.

Work out the value of a, b, c and d.

13 A scientist wants to estimate how many fish there are in a large pond.

He catches 40 one day, tags them and puts them back into the pond.

Next week he again catches 40 fish, 25 of which are tagged.

Estimate how many fish there are in the pond.

14 There are 960 pupils in a school.

$\frac{5}{8}$ of the pupils are in lower school.

$\frac{7}{12}$ of the pupils in the lower school are girls.

Work out the number of girls in the lower school.

15 The diagram shows a square *ABCD*.

The points *E* and *F* are the midpoints of sides *AD* and *CD* respectively.

What fraction of the square are the triangles:

a *ABE*   b *DEF*   c *BCF*   d *BEF*?

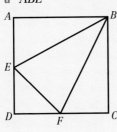

4 DECIMALS AND ESTIMATION

In athletics, sprinters often finish a race with only split seconds between them. Times are therefore given as decimals to one hundredth of a second in order to find the winner.

⊚ Objectives

In this chapter you will:
- convert between decimals and fractions
- carry out arithmetic using decimals
- write numbers correct to a given number of decimal places
- write numbers correct to a given number of significant figures
- work out estimates to calculations by rounding
- use one calculation to find the answer to another
- convert recurring decimals to fractions
- write down upper bounds and lower bounds
- calculate the bounds of expressions.

◈ Before you start

You need to be able to:
- understand and use fractions
- substitute numbers into an algebraic expression and work out its value.

4.1 Conversion between fractions and decimals

◎ Objectives

○ You can convert between decimals and fractions.
○ You can order integers, decimals and fractions.

❓ Why do this?

Parts of a whole can be given as fractions or decimals. You need to be able to convert between these to make a comparison. One and a half kilograms can be written as 1.5 kilograms.

⬆ Get Ready

1. Write this set of numbers in order of size, smallest first.
   8.092, 8.9, 8.02, 8.09, 8.2, 8.29, 8.92
2. Write these fractions in their simplest form.
   a $\frac{15}{45}$ **b** $\frac{42}{72}$ **c** $\frac{175}{200}$

🔑 Key Points

◉ Decimals are used as one way of writing parts of a whole number.
  The decimal point separates the whole number part from the part that is less than 1.
◉ A **terminating decimal** is a decimal which ends.
  0.34, 0.276 and 5.089 are terminating decimals.
◉ A **recurring decimal** is one in which one or more digits repeat.
  0.111 111..., 0.563 563 563..., 8.564 444... are all recurring decimals.
◉ All fractions can be changed into a decimal. To work out if a fraction will be represented by a terminating or a recurring decimal:
  ○ write the fraction in its simplest form
  ○ write the denominator of the fraction in terms of its prime factors (see Section 1.1)
  ○ if these prime factors are only 2s and/or 5s then the fraction will convert to a terminating decimal
  ○ if any prime number other than 2 or 5 is a factor then the fraction will convert to a recurring decimal.
◉ To convert a fraction to a decimal, either divide the numerator by the denominator, or, for a terminating decimal, create an equivalent fraction in lengths or hundredths. For example, $\frac{3}{5} = 3 \div 5 = 0.6$, or $\frac{3}{5} = \frac{6}{10} = 0.6$.
◉ Here are some common fraction-to-decimal conversions that you should remember.

Decimal	0.01	0.1	0.2	0.25	0.5	0.75
Fraction	$\frac{1}{100}$	$\frac{1}{10}$	$\frac{1}{5}$	$\frac{1}{4}$	$\frac{1}{2}$	$\frac{3}{4}$

Example 1 ▶ Work out whether these fractions will convert to terminating or recurring decimals.

a $\frac{9}{96}$ **b** $\frac{8}{30}$ **c** $3\frac{7}{20}$

a $\frac{9}{96} = \frac{3}{32}$ ← Simplify the fraction.

$32 = 2 \times 2 \times 2 \times 2 \times 2$ ← Write the denominator in terms of its prime factors.

The only prime factor is 2.

$\frac{9}{96}$ will convert to a terminating decimal.

b $\frac{8}{30} = \frac{4}{15}$ ← [Simplify the fraction.]

$15 = 3 \times 5$ ← [Write the denominator in terms of its prime factors.]
3 is a prime factor as well as 5.

$\frac{8}{30}$ will convert to a recurring decimal. ← [As 3 is a factor of 30.]

c $\frac{7}{20}$ ← [Just look at the fraction part.]

$20 = 2 \times 2 \times 5$ ← [Write the denominator in terms of its prime factors.]

$\frac{7}{20}$ will convert to a terminating decimal. ← [As only 2 and 5 are the factors.]

Example 2 Rearrange these numbers in order of size, smallest first.

$\frac{2}{5}, 0.34, \frac{3}{10}, 0.41$ ← [Convert the fractions to decimals.]

$\frac{2}{5} = \frac{4}{10} = 0.4$ ← [Multiply the numerator and denominator number by 2 to create an equivalent fraction as a tenth.]

$\frac{3}{10} = 0.3$ ← [As the denominator is already 10, the fraction can immediately be written as a decimal.]

$0.3, 0.34, 0.4, 0.41$ ← [Arrange all the decimals in order of size.]

$\frac{3}{10}, 0.34, \frac{2}{5}, 0.41$ ← [Write the original numbers in order of size.]

Exercise 4A [Questions in this chapter are targeted at the grades indicated.]

1 Write each of the set of numbers in order, starting with the smallest.
$\frac{9}{10}, 0.8, 0.85, \frac{86}{100}, 0.98$

2 By writing the denominator in terms of its prime factors, decide whether these fractions will convert to recurring or terminating decimals.

a $\frac{9}{40}$ b $\frac{17}{32}$ c $\frac{8}{45}$

d $\frac{13}{42}$ e $\frac{6}{125}$ f $\frac{37}{60}$

3 Linda says that $\frac{17}{48}$ can be converted to a terminating decimal.
Mitch says that the fraction converts to a recurring decimal.
Who is correct? You must give a reason for your answer.

4.2 Carrying out arithmetic using decimals

◉ Objectives

- ◉ You can add and subtract decimals.
- ◉ You can multiply and divide decimals by whole numbers.
- ◉ You can multiply numbers with up to two decimal places.
- ◉ You can divide numbers with up to two decimal places.

⑦ Why do this?

To work out how much your bill is in a café, you'll often need to add prices that have decimals in them.

⊕ Get Ready

1. Work out

    a $13.1 + 5.69$ **b** $8.6 - 3.42$ **c** $37 - 9.86 + 5.6$

◉ Key Points

- ◉ When adding or subtracting decimals, line up the decimal points first.
- ◉ To multiply by a decimal, do the multiplication with whole numbers and then decide on the position of the decimal point.
- ◉ To divide a number by a decimal, multiply both the number and the decimal by a power of 10 (10, 100, 1000 …) to make the decimal a whole number. It is much easier to divide by a whole number than by a decimal.

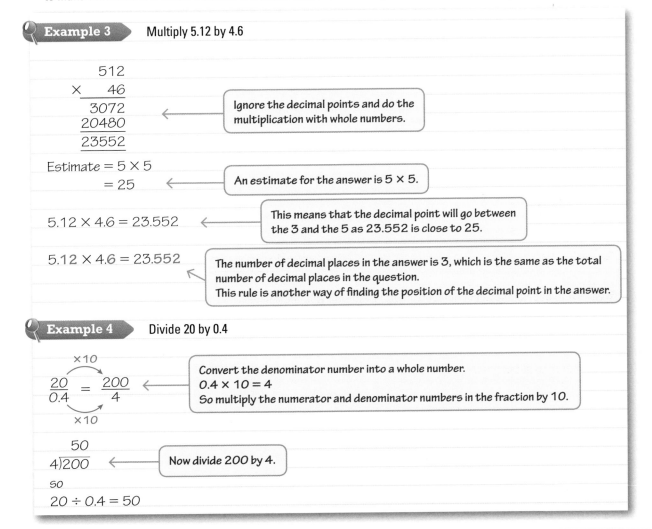

Example 3   Multiply 5.12 by 4.6

$$\begin{array}{r} 512 \\ \times\ 46 \\ \hline 3072 \\ 20480 \\ \hline 23552 \end{array}$$

Ignore the decimal points and do the multiplication with whole numbers.

Estimate = 5 × 5
           = 25

An estimate for the answer is 5 × 5.

$5.12 \times 4.6 = 23.552$

This means that the decimal point will go between the 3 and the 5 as 23.552 is close to 25.

$5.12 \times 4.6 = 23.552$

The number of decimal places in the answer is 3, which is the same as the total number of decimal places in the question.
This rule is another way of finding the position of the decimal point in the answer.

Example 4   Divide 20 by 0.4

$$\frac{20}{0.4} = \frac{200}{4}$$
(×10 / ×10)

Convert the denominator number into a whole number.
$0.4 \times 10 = 4$
So multiply the numerator and denominator numbers in the fraction by 10.

$$\begin{array}{r} 50 \\ 4\overline{)200} \end{array}$$

Now divide 200 by 4.

so
$20 \div 0.4 = 50$

Example 5 Divide 4.152 by 1.2

$$\frac{4.152}{1.2} = \frac{41.52}{12}$$

$$\frac{4.152}{1.2} \overset{\times 10}{\underset{\times 10}{=}} \frac{41.52}{12}$$

To make 1.2 into a whole number, multiply it by 10.

ResultsPlus
Exam Tip

It does not matter that the top number is still a decimal.

$$12\overline{)41.^5 5^7 2}\ \overset{3.\,4\,6}{}$$

Divide 41.52 by 12.

so $4.152 \div 1.2 = 3.46$

Exercise 4B

1. Work out
   a 0.3×0.4 b 0.006×0.2 c 0.8×0.05 d 0.09×0.07

2. Work out
   a 6.34×0.4 b 4.21×0.3 c 0.723×0.06 d 3.15×0.8
   e 3.1×4.2 f 0.36×1.4 g 0.064×0.73 h 0.095×3.4

3. Work out the cost of 0.6 kg of carrots at 25p per kilogram.

4. Work out the cost of 1.6 m of material at £4.29 per metre.

5. Work out
   a $12 \div 0.2$ b $4.2 \div 0.3$ c $19.2 \div 0.03$ d $26 \div 0.4$
   e $5 \div 0.2$ f $6.12 \div 0.003$ g $0.035 \div 0.7$ h $0.008\,28 \div 0.09$

6. Work out
   a $34.65 \div 0.15$ b $160.5 \div 0.25$ c $0.8673 \div 0.021$ d $9.706 \div 0.23$

7. Five people share £130.65 equally. Work out how much each person will get.

8. A bottle of lemonade holds 1.5 litres. A glass will hold 0.3 litres.
   How many glasses can be filled from the bottle of lemonade?

4.3 Rounding and decimal places

Objective

- You can write a number correct to a given number of decimal places.

Why do this?

When you find a fraction of an amount of money, you may need to round the answer to two decimal places.

Get Ready

1. Work out without a calculator
   a 9.62×0.06 b $9.62 \div 0.06$ c £9.87 ÷ 7

Key Points

◎ Decimals can be **rounded** to a given number of decimal places.

6.48 = 6.5 correct to 1 decimal place	Round up because 6.48 is closer to 6.5 than to 6.4
0.0748 = 0.07 correct to 2 decimal places	Round down because 0.0748 is closer to 0.07 than to 0.08
1.2475 = 1.248 correct to 3 decimal places	If the figure in the fourth decimal place is 5 or more then round up

Example 6   Write the following numbers correct to 2 decimal places.

   a   6.789   b   0.007   c   1.2999

a   6.79
b   0.01
c   1.30   ←   Note the difference between 1.3 (1 decimal place) and 1.30 (2 decimal places)

Exercise 4C

1   Write the following numbers correct to 1 decimal place (1 d.p.).
   a   6.38   b   5.66   c   16.949   d   0.067   e   0.99

2   Write the following numbers correct to 2 decimal places (2 d.p.).
   a   5.667   b   8.0582   c   0.125   d   3.044   e   0.076

3   Write the following numbers correct to 3 decimal places (3 d.p.).
   a   6.4458   b   0.0792   c   5.0792   d   6.0079   e   0.0199

4.4 **Significant figures**

Objective

◎ You can round a number correct to a given number of significant figures.

Why do this?

Sometimes you don't need to give the exact amount, you just need to give someone a round figure. For instance you might say that around 150 people were at a concert.

Get Ready

1. What is the value of the digit in the tenths column in these numbers?   a   6.38   b   4.07   c   3.99
2. Write three numbers with an '8' in the units column.

Key Points

◎ To write a number correct to 3 **significant figures** (3 s.f.), write down the first 3 figures, rounding up the last figure if the figure after it would be 5 or more. If necessary ignore any leading zeros.
◎ Leading zeros in decimals are not counted as significant.

Example 7   Round 436 to: **a** 2 significant figures
                                    b 1 significant figure.

a $436 = 440$ correct to 2 significant figures. ← | Round up because 436 is closer to 440 than to 430.

b $436 = 400$ correct to 1 significant figure. ← | Round down because 436 is closer to 400 than 500.

Example 8   Round 0.0258 to: **a** 2 s.f.   **b** 1 s.f.

a $0.0258 = 0.026$ correct to 2 significant figures. ← | The 8 means that the 5 will be rounded up to a 6.

b $0.0258 = 0.03$ correct to 1 significant figure. ← | The 5 means that the 2 will be rounded up to a 3.

Example 9   Write the following numbers correct to: **a** 3 significant figures   **b** 2 significant figures.
                  i 2788, **ii** 4.7084, **iii** 0.006 675

a **i** 2790 (3 s.f.)   **ii** 4.71 (3 s.f.)   **iii** 0.006 68 (3 s.f.)

b **i** 2800 (2 s.f.)   **ii** 4.7 (2 s.f.)   **iii** 0.0067 (2 s.f.)

Example 10   Write the following numbers correct to 2 significant figures.
                  a 7995   **b** 4.996   **c** 0.000 99

a 8000 (2 s.f.)

b 5.0 (2 s.f.) ← | Note the difference between 5 (1 significant figure) and 5.0 (2 significant figures).

c 0.0010 (2 s.f.)

ResultsPlus

Exam Tip

Although there may only be one non-zero figure in your answer, the numbers could still be correct to 2 significant figures.

Exercise 4D

1 Write the following numbers correct to 2 significant figures.
   a 3867   **b** 234.7   **c** 45.53   **d** 6.48   **e** 5.079   **f** -0.4318

2 Write the following numbers correct to 3 significant figures.
   a 2496   **b** 38.98   **c** 4.895   **d** 4.0899   **e** 0.010 96

3 Write the following numbers correct to 1 significant figure.
   a 3499   **b** 42.62   **c** 3.008   **d** 7.92   **e** 19.8   **f** 0.982

4.5 **Estimating calculations by rounding**

⊚ Objective

⊙ You can use rounding to 1 significant figure to work out an estimate for a calculation.

⊘ Why do this?

You can estimate the cost of buying 29 T-shirts at £5.99 using rounding.

⬆ Get Ready

1. Write correct to 1 significant figure

   a 4555   **b** 16.8   **c** −6.7

2. Work out **a** 20×30 **b** 50×300 **c** $600 \times 30 \times 30$

3. Work out **a** $\dfrac{600}{10}$ **b** $\dfrac{6000}{20}$ **c** $\dfrac{50}{2000}$

🔍 Key Point

⦿ A method of estimating the answer to a calculation is to write all numbers correct to 1 significant figure and then do the calculation.

Example 11   Work out an estimate for the area of a rectangle 38.6 cm by 12.2 cm.

Approximate area = $40 \times 10 = 400 \text{ cm}^2$ ⟵ $\begin{aligned} 38.6 &= 40 \text{ (1 sig fig)} \\ 12.2 &= 10 \text{ (1 sig fig)} \end{aligned}$

Example 12   **a** Calculate an estimate for the value of $\dfrac{38.9 \times 19.9}{20.3}$

b Explain why your answer is an overestimate of the true answer.

ResultsPlus
Watch Out!

Remember not to round to the nearest whole number as that is a mistake.

a $\dfrac{40 \times 20}{20} = \dfrac{800}{20} = 40$

b 38.9 and 19.9 have been rounded up, so the new numerator is larger.
20.3 has been rounded down, so the denominator is smaller and dividing by a smaller denominator results in a bigger answer.

⚙ Exercise 4E

1 Estimate the value of the following calculations.

   a 69×58 **b** 112×68 **c** 295×19 **d** 4897×38 **e** 788×109

2 Work out estimates for each of the following calculations.

   a $68 \div 1.9$ **b** $9.9 \div 4.9$ **c** $58.6 \div 6.1$ **d** $211.8 \div 39$ **e** $577 \div 97.8$

D

D **A03**

3 Work out estimates for each of the following. In each case state whether your answer is an overestimate or an underestimate of the true answer.

a 189×38 b $19.9 \div 5.1$ c $61.9 \div 5.92$ d $28.4 \times 1.89 \times 4.8$

C

4 Work out estimates for each of the following calculations.
State whether your answer is an underestimate or an overestimate.

a $\dfrac{48.9 \times 9.9}{11.3}$ b $\dfrac{203.8}{9.8 \times 4.9}$ c $\dfrac{999.8}{5.1 \times 5.3}$ d $\dfrac{9.55 \times 79.9}{11.8 \times 13.03}$

5 Work out an estimate for the value of 6.4×18.8^2.

4.6 Estimating calculations involving decimals

◎ Objective

● You can work out an estimate for a calculation which involves decimals by writing all numbers correct to 1 significant figure.

⊘ Why do this?

High-powered computers can perform calculations with numbers that have hundreds of decimal places. We need to round these numbers in order to work with them and estimate calculations.

⟡ Get Ready

1. Write correct to 1 significant figure

    a 0.33   b 0.0466   c 0.001 09

2. Work out

    a 0.2×30 b 0.5×300 c $0.6 \times 0.3 \times 0.3$

3. Work out

    a $\dfrac{600}{0.1}$ b $\dfrac{6000}{0.2}$ c $\dfrac{0.5}{0.02}$

◉ Key Point

● You can round figures in a calculation to a given number of significant figures to make an estimate of the answer. This can help you to check that your answer is reasonable.

Example 13   a Work out an estimate for the value of 0.399×208.8

          b Work out an estimate of the value of $\dfrac{4.89 \times 0.088}{0.0052}$

 **ResultsPlus**
Watch Out!

Remember not to make the mistake of rounding a decimal down to zero.

a $0.4 \times 200 = 80$

b $\dfrac{5 \times 0.09}{0.005} = \dfrac{0.45}{0.005} = \dfrac{450}{5} = 90$

> For division by a decimal, multiply the denominator and numerator by a number which produces a whole number in the denominator.

Exercise 4F

1 Work out estimates for the values of

a 6.4×0.38
b 0.49×0.33
c 12.1×0.128
d 0.089×0.021

2 Work out estimates for the values of

a $\dfrac{10.45}{0.49}$
b $\dfrac{20.8}{0.41}$
c $\dfrac{81.34}{0.81}$
d $\dfrac{0.43}{0.12}$
e $\dfrac{1.067}{5.49}$

3 Work out estimates for the values of the following.
State whether your answer is an overestimate or an underestimate.

a 5.4×0.32
b 0.48×0.38
c $\dfrac{1.22}{0.19}$
d $\dfrac{6.02}{0.028}$

4 Work out estimates for the values of the following.
State whether your answer is an overestimate or an underestimate.

a $\dfrac{9.8 \times 3.9}{0.14}$
b $\dfrac{6.8 \times 2.9}{0.11}$
c $\dfrac{12.1 \times 2.3}{0.83}$
d $\dfrac{206 \times 13.1}{0.48}$

5 $V = LWH$, $L = 0.046$, $W = 0.053$, $H = 122$. Work out an estimate for the value of V.

4.7 Manipulating decimals

⊙ Objective

● You can use one calculation to find the answer to another.

? Why do this?

If you know that \$1.50 is worth £1, then you could use this to calculate how many cents 10p is worth.

◈ Get Ready

1. Work out   a 20×3 b 200×3 c 2000×3
2. Work out   a $300 \div 10$ b $30 \div 10$ c $3 \div 10$

◉ Key Point

● Knowing the answer to one calculation can often be used to find the answer to a second calculation.

Example 14 Given that $\dfrac{3.46 \times 25.5}{3.4} = 25.95$, find the value of each of the following.

a $\dfrac{346 \times 25.5}{3.4}$
b $\dfrac{2.595 \times 0.34}{25.5}$

a **Method 1**

$$\dfrac{346 \times 25.5}{3.4} = \dfrac{3.46 \times 100 \times 25.5}{3.4}$$

$$= \dfrac{3.46 \times 25.5}{3.4} \times 100$$

$$= 25.95 \times 100$$

$$= 2595$$

> The final answer will be 100 times that of the given calculation as one of the numbers on the top of the fraction is 100 times the corresponding number in the original fraction.

Method 2

$$\frac{346 \times 25.5}{3.4} = \frac{300 \times 30}{3}$$

$$= \frac{9000}{3}$$

$$= 3000$$

$$\frac{346 \times 25.5}{3.4} = 2595$$

> Write down an approximation to the given sum. The answer is approximately 3000. The number closest to this gained by moving the decimal point in the answer to the given calculation is 2595 (rather than 259.5 or 25 950 etc.).

ResultsPlus
Exam Tip

Round each number to 1 significant figure so you can calculate an estimate quickly and easily.

b **Method 1**

$$\frac{2.595 \times 0.34}{25.5} = \frac{25.95 \div 10 \times 3.4 \div 10}{25.5}$$

$$= \frac{25.95 \times 3.4}{25.5} \div 100$$

$$= 3.46 \div 100$$

$$= 0.0346$$

> Two numbers on the top of the fraction have been divided by 10 so divide the answer to the rearranged calculation by 100.

Method 2

$$\frac{2.595 \times 0.34}{25.5} = \frac{3 \times 0.3}{30}$$

$$= \frac{0.9}{30}$$

$$= 0.03$$

$$\frac{2.595 \times 0.34}{25.5} = 0.0346$$

> From the original *rearranged* calculation, the number closest to 0.03 that can be obtained by moving the decimal point in 3.46 is 0.0346.

Exercise 4G

D

1 Given that $6.4 \times 2.8 = 17.92$ work out

a 64×28 b 640×2.8 c 0.64×28 d 0.64×0.028

2 Given that $18.3 \div 1.25 = 14.64$ work out

a $183 \div 1.25$ b $1.83 \div 1.25$ c $0.183 \div 1.25$ d $0.183 \div 12.5$

3 Given that $\frac{23.2 \times 5.1}{3.4} = 34.8$ work out

a $\frac{23.2 \times 51}{3.4}$ b $\frac{232 \times 51}{3.4}$ c $\frac{23.2 \times 51}{34}$ d $\frac{232 \times 51}{34}$

4 Given that $23 \times 56 = 1288$ work out

a 0.23×560 b $1288 \div 5.6$ c $12.88 \div 0.23$ d $1288 \div (23 \times 28)$

5 Given that $884 \div 34 = 26$ work out

a $8.84 \div 340$ b $884 \div 2.6$ c $8.84 \div 260$ d $884 \div (3.4 \times 2.6)$

6 Given that $\frac{1872}{1.2^2} = 1300$ work out

a $\frac{1872}{12^2}$ b $\frac{18.72}{1.2^2}$ c $\frac{187.2}{0.12^2}$ d $\frac{936}{120^2}$

4.8 Converting recurring decimals to fractions

◎ Objectives

○ You can convert between recurring decimals and fractions.
○ You understand and know how to use the recurring decimal proof.

◈ Get Ready

1. Determine whether these fractions will be represented by a terminating or recurring decimal.

   a $\frac{14}{64}$　　b $\frac{18}{35}$　　c $\frac{20}{44}$

◉ Key Points

◉ All recurring decimals can be converted to fractions.
◉ To convert a recurring decimal to a fraction:
  ◉ introduce a **variable**, usually x
  ◉ form an equation by putting x equal to the recurring decimal
  ◉ multiply both sides of the equation by 10 if 1 digit recurs, by 100 if 2 digits recur, by 1000 if 3 digits recur, and so on
  ◉ subtract the original equation from the new equation
  ◉ rearrange to find x as a fraction.

Example 15 Convert the recurring decimal $0.\overset{\bullet}{5}\overset{\bullet}{4}$ to a fraction. Give your fraction in its simplest form.

Let $x = 0.5454\ldots$ ← Put x equal to the recurring decimal.

$100x = 54.54\ldots$ ← Multiply both sides of the equation by 100 as 2 digits recur.
$-\quad x = \quad 0.5454\ldots$ ← Subtract the equations.
$\overline{99x = 54}$

$x = \frac{54}{99}$ ← Divide both sides by 99.

$= \frac{6}{11}$ ← Simplify the fraction.

$0.\overset{\bullet}{5}\overset{\bullet}{4} = \frac{6}{11}$

Example 16 Convert the recurring decimal $0.2\overset{\bullet}{3}7\overset{\bullet}{1}$ to a fraction.

Let $x = 0.237\,1371\ldots$ ← Put x equal to the recurring decimal. Care is needed here, the 2 does not recur.

$1000x = 237.1371\ldots$ ← Multiply both sides of the equation by 1000 as 3 digits recur.
$-\quad x = \quad 0.2371\ldots$
$\overline{999x = 236.9}$ ← Subtract the equations.

$x = \frac{236.9}{999}$ ← Divide both sides by 999.

$= \frac{2369}{9990}$

$0.2\overset{\bullet}{3}7\overset{\bullet}{1} = \frac{2369}{9990}$ ← Multiply both the numerator and denominator by 10 to convert the decimal in the numerator to an integer.

Exercise 4H

Convert each recurring decimal to a fraction. Give each fraction in its simplest form.
Do **not** use a calculator. You must use algebra.

A

| 1 | 0.777 77… | 2 | 0.343 434… | 3 | 0.915 915… | 4 | 0.1̇8̇ |

| 5 | 0.3̇17̇ | 6 | 0.05̇ | 7 | 0.32̇6̇ | 8 | 0.7̇01̇ |

| 9 | 0.23̇ | 10 | 6.8̇3̇ | 11 | 2.1̇06̇ | 12 | 7.35̇2̇ |

4.9 Upper and lower bounds of accuracy

◎ Objective

● You know the upper bound and the lower bound of a number given the accuracy to which it has been written.

⦾ Why do this?

If you knew how tall the Blackpool Tower was to 2 significant figures, then you could work out the maximum and minimum possible heights that it could be.

◈ Get Ready

1. Write the following numbers correct to 1 decimal place (1 d.p.).

a 6.05   **b** 6.99   **c** 6.49   **d** 6.51

2. Write the following numbers correct to 1 significant figure.

a 0.33   **b** 0.339   **c** 0.26   **d** 0.349

🔑 Key Points

◉ The **upper bound** of a number written to 1 decimal place is the highest value which rounds down to that number.

◉ The **lower bound** of a number written to 1 decimal place is the lowest value which rounds up to that number.

Example 17

a $x = 6.4$ (correct to 1 decimal place)
Write down the upper bound and the lower bound of x.

b $y = 248$ (correct to 3 significant figures)
Write down the upper bound and the lower bound of y.

ResultsPlus

Exam Tip

Remember that the upper bound is the same distance above x as the lower bound is below x.

a Upper bound of $x = 6.45$ ← For 1 decimal place the upper bound is 0.05 above the stated value.

    Lower bound of $x = 6.35$ ← For 1 decimal place the lower bound is 0.05 below the stated value.

b Upper bound of $y = 248.5$ ← 248.5 is the largest value which will round down to 248 correct to 3 significant figures.

    Lower bound of $y = 247.5$

Exercise 4I

1 Write down: **i** the upper bound and **ii** the lower bound of these numbers.
  a 84 (2 significant figures)   **b** 84.0 (3 significant figures)
  c 84.00 (4 significant figures)

2 Write down: **i** the upper bound and **ii** the lower bound of these numbers.
  a 0.9 (1 decimal place)   **b** 0.90 (2 decimal places)
  c 0.09 (2 decimal places)

3 The length of a line is 118 cm correct to the nearest cm. Write down:
  a the upper bound
  b the lower bound of the length of the line.
  Give your answers in cm.

4 The mass of a stone is 6.4 kg correct to the nearest one tenth of a kg. Write down:
  a the upper bound
  b the lower bound of the mass of the stone.
  Give your answers in grams.

5 The amount of fuel in a tank is 48.0 litres correct to the nearest tenth of a litre. Write down:
  a the upper bound
  b the lower bound of the amount of fuel in the tank.
  Give your answers in litres.

6 The length of a piece of wood is 1 metre correct to the nearest cm. Write down:
  a the upper bound
  b the lower bound of the length of the piece of wood.
  Give your answers in metres.

4.10 Calculating the bounds of an expression

Objective

You can work out the upper bound and the lower bound of an expression when the numbers in it are approximate.

Why do this?

If you had rough estimates for its length and width, you could work out the maximum and minimum possible areas of a basketball court.

Get Ready

1. Work out $6.45 + 3.65$.
2. $a = 3.4$ (1 decimal place), $b = 5.6$ (1 decimal place)
Work out the difference between the upper bound of a and the lower bound of b.
3. c, d, e and f are four positive numbers written in order of increasing size. Write down the pairs of numbers which will have the greatest:
  a sum   **b** difference   **c** product   **d** quotient.

Key Points

- Let $S = x + y$. Then upper bound of S = upper bound of x + upper bound of y.
- Let $D = x - y$. Then upper bound of D = upper bound of x − lower bound of y.
- Let $P = xy$. Then upper bound of P = upper bound of $x \times$ upper bound of y.
- Let $Q = \dfrac{x}{y}$. Then upper bound of Q = upper bound of $x \div$ lower bound of y.
- For the lower bounds of S and P, use the lower bounds of x and y.
- The lower bound of D = lower bound of x − upper bound of y.
- The lower bound of Q = lower bound of $x \div$ upper bound of y.

Example 18 $x = 3.4$ correct to 1 d.p. $y = 1.8$ correct to 1 d.p.

Find **i** the upper bound   **ii** the lower bound of

a $x + y$ **b** $x - y$ **c** xy **d** $\dfrac{x}{y}$

a i $3.45 + 1.85 = 5.30$
  ii $3.35 + 1.75 = 5.10$ ⟵ Add the upper bounds.

b i $3.45 - 1.75 = 1.7$
  ii $3.35 - 1.85 = 1.50$ ⟵ Lower bound − upper bound.

c i $3.45 \times 1.85 = 6.3825$
  ii $3.35 \times 1.75 = 5.8625$

d i $3.45 \div 1.75 = 1.971$ (3 d.p.)
  ii $3.35 \div 1.85 = 1.811$ (3 d.p.)

Example 19 Here is a formula from science.

$$H = \frac{V^2(\sin x)^2}{2g}$$

$V = 250$ correct to 3 significant figures
$x = 72$ correct to 2 significant figures
$g = 9.81$ correct to 2 decimal places

a Find the upper bound and the lower bound of H.
b Write the value of H correct to an appropriate number of significant figures.

a Upper bound $\dfrac{250.5^2 \times (\sin 72.5)^2}{2 \times 9.805} = 2910.5621$

Lower bound $= \dfrac{249.5^2 \times (\sin 71.5)^2}{2 \times 9.815} = 2851.897\,864\ldots$

b $H = 2900$ correct to 2 significant figures.

ResultsPlus
Exam Tip

The lower and upper bounds may agree when written correct to a certain number of significant figures. Make sure you choose an appropriate degree of accuracy.

Example 20 The length of a rectangle is measured as 8.3 cm to the nearest mm and its width as 3.6 cm to the nearest mm.

   a Write down the lower bound of the width.
   b Write down the lower bound for the perimeter.
   c Work out the lower bound of the area. Give your answer correct to 1 decimal place.

a 3.55 mm

b $2 \times 3.55 + 2 \times 8.25 = 23.6$ mm

c $3.55 \times 8.25 = 29.2875$
   $\qquad = 29.3$ mm^2 (1 d.p.)

Exercise 4J

1   $p = 480$ (3 s.f.), $q = 56$ (2 s.f.) Work out the lower bounds of
    a $p + q$ b pq c p^2

2   $c = 2.44$ (2 d.p.), $d = 4.45$ (2 d.p.) Work out the lower bounds of
    a $c + d$ b cd c $10c$

3   $r = 200$ (3 s.f.), $s = 250$ (3 s.f.), $t = 224$ (3 s.f.) Work out the upper bounds of
    a rst b $rs + st + tr$

4   $c = 42$ (2 s.f.), $d = 30$ (2 s.f.) Work out the lower bounds of
    a $c - d$ b $c \div d$

5   $e = 3.4$ (1 d.p.), $f = 2.5$ (1 d.p.) Work out the lower bounds of
    a $e - f$ b $\dfrac{e}{f}$

6   $p = q + rs$ $q = 18.7, r = -6.4, s = 7.7$ all correct to 1 d.p.
    a Find the lower bound of p.
    b Find the upper bound of p.
    c Write the value of p correct to an appropriate number of significant figures.

7   $E = mc^2$ $c = 3.0 \times 10^8$ (2 s.f.), $m = 2.4$ (2 s.f.)
    Calculate the lower bound of E. Calculate the upper bound of E.

8   $A = lw$
    a Suppose $l = 10, w = 5$, both exact. Work out the exact value of A.
    b Suppose $l = 10, w = 5$, both written correct to the nearest whole number.
    Work out the percentage difference between the upper bound of A and the exact value of A.

A02
A02
A02
A02
A02
A02
A03
A03
A03
A
A☆

Chapter review

- Decimals are used as one way of writing parts of a whole number.
- The decimal point separates the whole number part from the part that is less than 1.
- A **terminating decimal** is a decimal which ends. 0.34, 0.276 and 5.089 are terminating decimals.
- A **recurring decimal** is one in which one or more figures repeat. 0.111 111..., 0.563 563 563..., 8.564 444... are all recurring decimals.
- To work out if a fraction will be represented by a terminating or a recurring decimal:
  - write the fraction in its simplest form
  - write the denominator of the fraction in terms of its prime factors
  - if these prime factors are only 2's and/or 5's then the fraction will convert to a terminating decimal
  - if any prime number other than 2 or 5 is a factor then the fraction will convert to a recurring decimal.
- When adding or subtracting decimals, line up the decimal points first.
- To multiply by a decimal, do the multiplication with whole numbers and then decide on the position of the decimal point.
- To divide a number by a decimal, multiply both the number and the decimal by a power of 10 (10, 100, 1000 ...) to make the decimal a whole number. It is much easier to divide by a whole number than by a decimal.
- Decimals can be **rounded** to a given number of decimal places.
- To write a number correct to 3 **significant figures** (3 s.f.), write down the first 3 figures, rounding up the last figure if the figure after it would be 5 or more. If necessary ignore any leading zeros.
- Leading zeros in decimals are not counted as significant.
- A method of estimating the answer to a calculation is to write all numbers correct to 1 significant figure.
- You can round figures in a calculation to a given number of significant figures to make an estimate of the answer. This can help you to check that your answer is reasonable.
- Knowing the answer to one calculation can often be used to find the answer to a second calculation.
- All recurring decimals can be converted to fractions.
- To convert a recurring decimal to a fraction:
  - introduce a **variable**, usually x
  - form an equation by putting x equal to the recurring decimal
  - multiply both sides of the equation by 10 if 1 digit recurs, by 100 if 2 digits recur, by 1000 if 3 digits recur, and so on
  - subtract the original equation from the new equation
  - rearrange to find x as a fraction.
- The **upper bound** of a number written to 1 decimal place is the highest value which rounds down to that number.
- The **lower bound** of a number written to 1 decimal place is the lowest value which rounds up to that number.
- Let $S = x + y$. Then upper bound of S = the upper bound of x + upper bound of y.
- Let $D = x - y$. Then upper bound of D = upper bound of x − lower bound of y.
- Let $P = xy$. Then upper bound of P = upper bound of x × upper bound of y.
- Let $Q = \dfrac{x}{y}$. Then upper bound of Q = upper bound of x ÷ lower bound of y.
- For the lower bounds of S and P, use the lower bounds of x and y.
- The lower bound of D = lower bound of x − upper bound of y.
- The lower bound of Q = lower bound of x ÷ upper bound of y.

Decimal	Fraction
0.01	$\frac{1}{100}$
0.1	$\frac{1}{10}$
0.2	$\frac{1}{5}$
0.25	$\frac{1}{4}$
0.5	$\frac{1}{2}$
0.75	$\frac{3}{4}$

Review exercise

1 Here are the rates of pay in a company.

Grade	Basic pay for an hour's work	Overtime pay for an hour's work
Operative	£5.40	£8.10
Technician	£7.50	£11.25
Supervisor	£9.00	£13.50
Driver	£7.20	£10.80

Kaysha has a part-time job as an operative.
Last week Kaysha earned basic pay for 24 hours and overtime pay for 3 hours.
Work out Kaysha's total pay for last week.

June 2008, adapted

2 Write down which of the following, when written as decimals, are recurring.

$\frac{3}{4}$ $\frac{2}{3}$ $\frac{7}{8}$ $\frac{9}{24}$ $\frac{3}{5}$

3 Put these numbers in order. Start with the smallest number.

$\frac{3}{5}$ 0.47 $\frac{12}{25}$ $\frac{31}{50}$

4 Ethan has a '5p off per litre' voucher for use at a local petrol station.
He fills up his tank with 43 litres of petrol normally costing 104.9p per litre.
How much does he pay?

5 Write correct to 3 significant figures
  a 4778   **b** 106.74   **c** 3.228×10^{15} **d** 6996   **e** 56.97

6 Write correct to 2 significant figures
  a 45.87   **b** 30.72   **c** 0.0457   **d** 19.97   **e** 4.098

7 Write correct to 1 significant figure
  a 363   **b** 40.22   **c** 9.9×10^{17} **d** 0.005 48   **e** -3.056

8 Write the following numbers correct to 1 decimal place
  a 3.142   **b** 0.567   **c** 2.091   **d** 3.99

9 Use the information that

$$\frac{63 \times 99}{18^2} = \frac{77}{4}$$

Write down the value, as fractions or integers, of

a $\dfrac{6.3 \times 9.9}{18^2}$

b $\dfrac{6.3 \times 990}{18^2}$

c $\dfrac{63 \times 99}{0.18^2}$

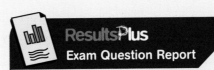

ResultsPlus
Exam Question Report

97% of students answered this sort of question well because they used the information provided.

10 Jim and Jenni are collecting money for charity. Jim collects a kilometre of 2p pieces. Jenni wants to collect 1p pieces. How many 1p pieces does she need to collect to get the same amount of money as Jim? (Diameter of 2p piece = 2.6 cm, diameter of 1p piece = 2 cm).

11 An electricity bill includes a standard charge and an amount depending on how much you use. Avery Energy charges 12.61p per unit for the first 400 units and 15.02p per unit for any units used above this amount.

AVERY ENERGY		Total units = 573	
400 units	@ 12.61p per unit	=	£50.44
173 units	@ 15.02p per unit	=	£25.98
Standing charge, 90 days	@ 14.01p per day	=	£12.61
	Total due	=	£89.03

a Check the amounts on the bill are correct.

Three electricity companies use different pricing structures.

AVERY ENERGY	
Standing charge	14.01p per day
First 400 units	@ 12.61p per unit
Additional units	@ 15.02p per unit

BRAWN POWER	
Standing charge	16.30p per day
All units charged	@ 13.60p per unit

CC ELECTRICS	
No standing charges	
First 100 units	@ 5.03p per unit
Additional units	@ 23.72p per unit

b Advise the following people which electricity company they should use. (Take 90 days as a quarter.)
i John is rarely in. He uses about 350 units a quarter.
ii Vijay works from home and uses 1261 units a quarter.

12 By writing each number correct to 1 significant figure, work out an estimate for the value of
a 49×59 **b** 79×51 **c** 4.1×5.9 **d** 499×691 **e** $6.1 \times 19.9 \times 2.8$

13 By writing each number correct to 1 significant figure, work out an estimate for the value of
a $199 \div 39$ **b** $19.9 \div 4.1$ **c** $411 \div 4.9$ **d** $4991 \div 21.8$ **e** $19.98 \div 20.8$

14 The table gives information about the length of time in minutes and seconds, the tracks on a CD last.

Track number	1	2	3	4	5	6
Time	7 m 56 s	3 m 5 s	4 m 8 s	5 m 58 s	3 m 57 s	11 m 48 s

Work out an estimate for the total length of time for all the tracks on the CD.
Give your answer in minutes.

15 Estimate in £s the cost of:
a an ice cream in Italy costing 3 euros
b a T-shirt in the USA costing $30
c a meal in Japan costing 1500 yen
d a drink in Turkey costing 10 lira
e a souvenir in Thailand costing 300 baht
f a camera in China costing 2000 yuan.

£1 buys you
1.1 euro
US$1.63
148 Japanese yen
2.45 Turkish lira
54.7 Thai baht
11.15 Chinese yuan

16 Rob's tariff for his mobile phone is shown in the box on the right.

  a Calculate his monthly bill if he made 100 minutes of calls and 60 texts.

  b In one particular month, the number of texts and calls were
    the same.

If his bill was £8, how many texts did he send?

No monthly fee
Calls
15p per minute anytime
Texts
10p per text to any network

17 Work out an estimate for the value of each of these. In each case state whether your answer is an overestimate or an underestimate.

  a $\dfrac{5.4 \times 3.2}{0.187}$ **b** $\dfrac{0.32}{0.001\,95}$ **c** $\dfrac{0.88 \times 0.37}{0.131}$ **d** $\dfrac{59 \times 36}{0.415}$ **e** $\dfrac{0.32 \times 320}{0.195 \times 0.012}$

18 The height of a room is 285 cm. The width of the room is 790 cm and the length is 880 cm.
A number of people are to work in the room. Building regulations state that for every person working in the room there must be 4.25 m^3.
Calculate an estimate for the number of people who could work in the room.

19 Jason hired a van.
The company charges £90 per day plus the cost of the fuel used.
The van can travel 6 miles for each litre of fuel used.
Fuel costs 98.9 p for 1 litre.
On Monday, Jason hired the van and drove from London to Cardiff.
On Tuesday, Jason drove from Cardiff to Edinburgh.
On Wednesday, Jason drove from Edinburgh back to London and returned the van.
Jason used this table for information about distances between cities.
Jason thought the total cost would be about £400.
Work out the total cost of hiring the van and the fuel used.

London			
155	Cardiff		
212	245	York	
413	400	193	Edinburgh

May 2009

20 Write the sum of the sequence $\frac{3}{10} + \frac{3}{100} + \frac{3}{1000} + \ldots$ as a fraction.
(Where the ... indicates the sequence goes on for ever).

21 $x = 4.9$ (1 decimal place), $y = 12.1$ (1 decimal place). Work out the upper bounds of
  a $x + y$ **b** $4x + y$ **c** xy **d** x^2

22 $p = 450$ (2 significant figures), $q = 240$ (2 significant figures) Work out the lower bounds of
  a $p - q$ **b** $2p - q$ **c** $p \div q$ **d** $2 - \dfrac{p}{q}$

23 The length of a metal rod is 200 cm correct to the nearest cm.
Explain why the rod may be able to fit into a slot with a stated length of 199.8 cm.

24 Katy drove for 238 km, correct to the nearest mile.
She used 27.3 litres of petrol, to the nearest tenth of a litre.
Work out the upper bound for the petrol consumption in km per litre for Katy's journey.
Give your answer correct to 2 decimal places.

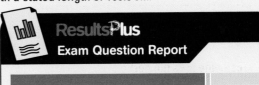

ResultsPlus
Exam Question Report

76% of students answered this question poorly because they selected the wrong bound for the numerator or the denominator.

June 2008, adapted

25 A cylinder has the label 'Contents 330 ml' printed on it. The diameter of the cylinder is 6.5 cm correct to the nearest mm and the height 10.0 cm to the nearest mm. Is the label correct?

A02
A03
C

A02

A02
A03

A

A03

A02
A03

A03

5 ANGLES AND POLYGONS

A curling bridge looks like a conventional bridge when it is extended. However, it curls up to form an octagon to allow boats through. This Rolling Bridge is in Paddington Basin in London, and curls up every Friday at midday.

⊙ Objectives

In this chapter you will:

- recognise and use corresponding angles and alternate angles
- use and prove angle properties of triangles and quadrilaterals
- recognise angles of elevation and depression
- give reasons for angle calculations
- recognise and know the names of special polygons
- know and use the interior and exterior angle properties of polygons
- find the bearing of one point from another.

◇ Before you start

You should know:

- how to measure and draw angles to the nearest degree
- how to find the sizes of missing angles on a straight line and at the intersection of straight lines
- how to recognise perpendicular and parallel lines, and vertically opposite angles
- that the angle sum of a triangle is 180°
- how to recognise scalene, isosceles, equilateral and right-angled triangles, and use their angle properties
- how to recognise acute, obtuse, reflex and right angles
- that a quadrilateral is a shape with four straight sides and four angles.

5.1 Angle properties of parallel lines

◎ Objectives

- ◎ You can mark parallel lines on a diagram.
- ◎ You can recognise corresponding and alternate angles.
- ◎ You can find the sizes of missing angles using corresponding and alternate angles.
- ◎ You can give reasons for angle calculations.

◈ Get Ready

1. Find the sizes of the marked angles.

a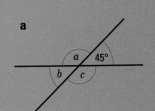

b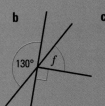

c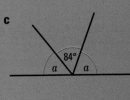

🔍 Key Points

- ◉ When two **parallel** lines are crossed by a straight line, as in the diagram, angles are formed.

  The green angles are in corresponding positions so they are called **corresponding angles**. Corresponding angles are equal.

  The blue angles are also corresponding angles.

  The orange angles are on opposite or alternate sides of the line so they are called **alternate angles**.

  The yellow angles are also alternate angles.

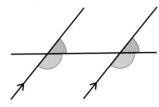 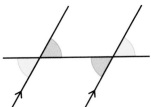

  The orange angles are on opposite or alternate sides of the line so they are called alternate angles. Alternate angles are equal.

  The yellow angles are also alternate angles.

🔍 **Example 1**   Find the size of angle a and angle b.

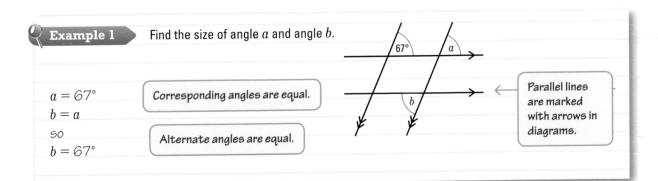

$a = 67°$

Corresponding angles are equal.

$b = a$

so

Alternate angles are equal.

$b = 67°$

Parallel lines are marked with arrows in diagrams.

**A02
A03**

Example 2 ▶ Explain why $a + b = 180°$.

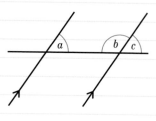

In the diagram
$\quad a = c$
$b + c = 180°$
so
$b + a = a + b = 180°$

Corresponding angles are equal.

Angles on a straight line add up to 180°.

✷ **Exercise 5A**

Questions in this chapter are targeted at the grades indicated.

In questions 1–6 find the size of each lettered angle. Give reasons for your answers.

D

1
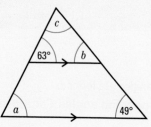

2
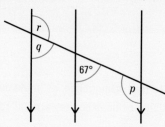

3
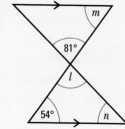

4
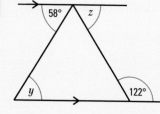

5
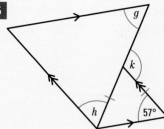

6
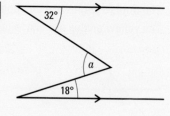

7 Here are two parallel lines crossed by a straight line.
  a List pairs of equal corresponding angles.
  b List pairs of equal alternate angles.
  c List pairs of angles which add up to 180°.
    Explain why the angles add up to 180°.

A03

8 ACE is a straight line.
  Explain why the lines AB and CD must be parallel.

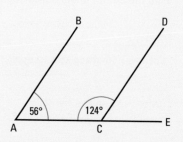

5.2 Proving the angle properties of triangles and quadrilaterals

◎ Objectives

◎ You can understand a proof that the angle sum of a triangle is 180°.

◎ You can understand a proof that an exterior angle of a triangle is equal to the sum of the interior angles at the other two vertices.

◎ You can explain why the angle sum of a quadrilateral is 360°.

◈ Why do this?

In mathematics it is important to be able to prove that results are always true. A demonstration only shows that the result is true for the chosen values.

◈ Get Ready

In questions 1 – 3, calculate the size of each lettered angle. Give reasons for your answers.

1.
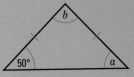

2.
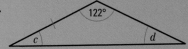

3.

◉ Key Points

◉ The angle marked e is called an **exterior angle**.

◉ The angle of the triangle at this vertex, i, is sometimes called an **interior angle**.

◉ $i + e = 180°$

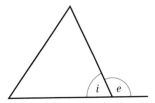

✑ Example 3

Prove that the angle sum of any triangle is 180°.

For any triangle, a straight line can be drawn through a vertex parallel to the opposite side, as shown in the diagram.

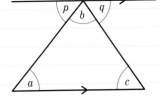

A03

$a = p$ ← Alternate angles are equal.

$c = q$ ← Alternate angles are equal.

$p + b + q = 180°$ ← Angles on a straight line add up to 180°.

So

$a + b + c = 180°$ ← The angle sum of a triangle is 180°.

✳ **Exercise 5B**

1 Here is a triangle with one side extended.

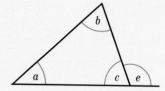

Complete the following proof that $e = a + b$ by giving a reason for each line of the proof.

$a + b + c = 180°$ ....

$e + c = 180°$ ....

so

$e = a + b$

An exterior angle of a triangle is equal to the sum of the interior angles at the other two vertices.

2 Here is a quadrilateral. A diagonal has been drawn to divide the quadrilateral into two triangles.

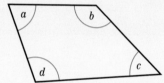 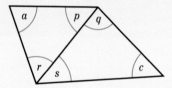

Copy and complete this proof that the angle sum of a quadrilateral is 360°.

$b = p + q$ ....

$d = r + s$ ....

$a + p + r = 180°$ ....

$c + q + s = 180°$ ....

Adding

$a + c + p + q + r + s = 360°$

$a + c + b + d = 360°$

so

$a + b + c + d = 360°$

The sum of the angles of a quadrilateral is 360°.

3

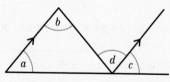

a Use properties of parallel lines to prove that

$a + b = c + d$

b Which angle property of triangles has this proved?

5.3 Using the angle properties of triangles and quadrilaterals

Objectives

○ You can use the property that an exterior angle of a triangle is equal to the sum of the interior angles at the other two vertices.
○ You can use the property that the angle sum of a quadrilateral is 360°.
○ You can use the angle properties of a parallelogram.
○ You can give reasons for angle calculations.

Why do this?

The angles in triangles are used in sports, for example in water-skiing. To ensure the longest jumps are made, the angle of the jump should be 14° to the water.

Get Ready

1. Calculate the size of the angle marked j.

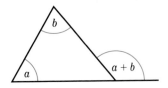

Key Points

◉ An exterior angle of a triangle is equal to the sum of the interior angles at the other two vertices.

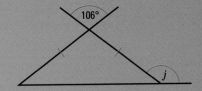

◉ The angle sum of a quadrilateral is 360°.
$a + b + c + d = 360°$

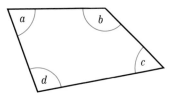

◉ Opposite angles of a parallelogram are equal. $a = c$
$b = d$

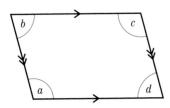

The two angles at the end of each side of a parallelogram add up to 180°.
$a + b = b + c = c + d = d + a = 180°$
Example 5 shows important angle properties of parallelograms.

Example 4

BCD is a straight line.

angle ACD = 123°

angle CBA = 58°

Work out the size of angle BAC.

$a + 58° = 123°$ ← Angle DCA is an exterior angle of the triangle.

$a = 123° - 58°$ ← Exterior angle is equal to the sum of the interior angles at the other two vertices.

$a = $ angle BAC = 65°

Example 5

ABCD is a parallelogram.

Find the size of each angle of the parallelogram.

$a = 60°$ ← Alternate angles on parallel lines BC and AD are equal.

$b + 60° = 180°$

$b = 180° - 60°$ ← Angles on a straight line add up to 180°.

$b = 120°$

$c = 60°$ ← Corresponding angles on parallel lines BA and CD are equal.

$a + b + c + d = 360°$ ← Angle sum of a quadrilateral is 360°.

$60° + 120° + 60° + d = 360°$

$240° + d = 360°$

$d = 120°$

Exercise 5C

D

1 BCDE is a quadrilateral.

ABE is an equilateral triangle.

Work out the size of angle ABC.

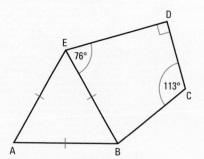

A02
A03

2 Work out the size of the angle e.

Give reasons for your working.

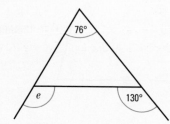

3 **a** Here is a kite. The diagonal shown dotted is an axis of symmetry.
Find the size of angle a and the size of angle b.
Give reasons for your working.

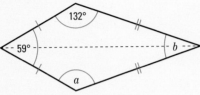

b Here is an isosceles trapezium.
The line shown dotted is an axis of symmetry.
The trapezium has an angle of 66° as shown.
Find the sizes of the three other angles of the trapezium.
Give reasons for your working.

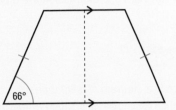

4 Here is a quadrilateral.
Work out the size of angle a.
Give reasons for your working.

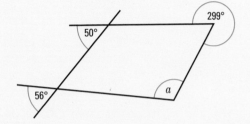

A03

A03

D

5.4 Angles of elevation and depression

◉ Objectives

- ◎ You can recognise an angle of elevation.
- ◎ You can recognise an angle of depression.
- ◎ You know that the angle of elevation of point A from point B is equal to the angle of depression of point B from point A.

◈ Why do this?

If you wanted to abseil down a building, you could work out the height of the building using the angle of elevation from a point on the ground to the top.

◈ Get Ready

1. 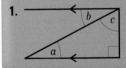   **a** Explain why $a = b$.
          b What is $b + c$?

🝔 Key Points

- ◉ The **angle of elevation** of point A from point B is the angle of turn above the horizontal to look directly from B to A.

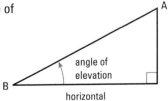

angle of elevation   **67**

● The **angle of depression** of point B from point A is the angle of turn below the horizontal to look directly from A to B.

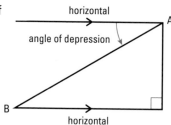

● Angles of elevation and depression are always measured from the horizontal.

Exercise 5D

D
A03

1 Here is a diagram of a lighthouse.
Explain why angle e = angle d.

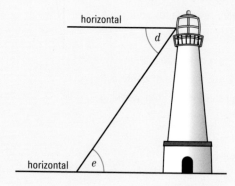

2 The angle of depression of a point A on horizontal ground from the top of a tree is 30°.
  a Show this information in a sketch.
  b On your sketch show and label the angle of elevation of the top of the tree from A.

5.5 Bearings

◉ Objectives

● You can use bearings for directions.
● You can find the bearing of one point from another point.
● You can work out the bearing of point B from point A when you know the bearing of point A from point B.

⟁ Why do this?

When giving directions, it is important to know the exact direction of A from B. This is useful for orienteering.

◈ Get Ready

Yuen lives in a town. The diagram shows the position of three places in the town in relation to Yuen's home.
Give the compass directions from Yuen's home of:
  a the cinema
  b the park
  c the school.

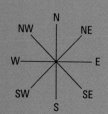

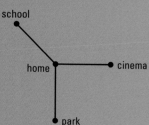

Key Points

- **Bearings** are angles measured clockwise from North.
- Bearings always have three figures.

Example 6   For each diagram, give the bearing of B from A.

a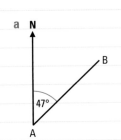

b

ResultsPlus

Watch Out!

Remember that a bearing has to have three figures and that is why some start with one or two zeros.

a At A, turn 47° clockwise from North to look towards B.
The bearing of B from A is 047°.

b The angle 351° is measured anticlockwise from North.
Clockwise angle = 360° − 351° = 9°
(as there are 360° in a complete turn).
The bearing of B from A is 009°.

Example 7   The bearing of B from A is 107°.
Work out the bearing of A from B.

Bearing of A from B   = 106° + 180°
                       = 286°

ResultsPlus

Exam Tip

Remember that bearings are always measured clockwise from the North.

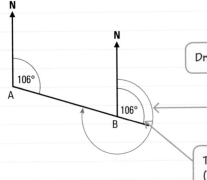

Draw a North line at B.

The angle marked in red is the angle that is the bearing of A from B.

The angle marked in blue is 106° (alternate angles).

⚙ **Exercise 5E**

1 In each of the following, give the bearing of B from A.

a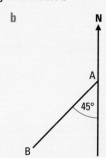

b

c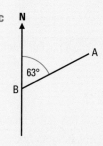

2 The diagram shows three towns A, B and C.

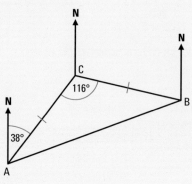

The bearing of C from A is 038°.

Angle ACB = 116°

CA = CB

Work out the bearing of

a B from A

b A from C

c B from C

3 The bearing of Norwich from Gloucester is 069°.
Work out the bearing of Gloucester from Norwich.

4 A plane flies from Skegness to Carlisle on a bearing of 132°.
Work out the bearing the plane needs to fly on for the return journey to Skegness.

5 The diagram shows the position of three towns P, Q and R.
Find the bearing of:

a R from Q

b P from Q.

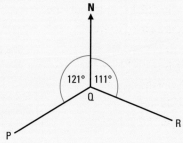

5.6 Using angle properties to solve problems

◈ Get Ready

1. Find the size of each lettered angle.
Give reasons for your answers.

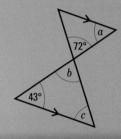
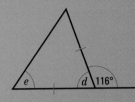
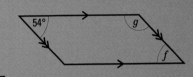

🔍 **Example 8** Work out the size of:

  i angle a

  ii angle b.

i $a + 75° + 120° + 103° = 360°$

        $a + 298° = 360°$

              $a = 62°$

Sum of the angles of a quadrilateral is 360°.

ii $b = a$

    $b = 62°$

Alternate angles are equal.

ResultsPlus
Watch Out!

Questions will rarely just ask you to work out the size of the unknown angle of a triangle or of a quadrilateral. In most cases you will need to use other angle properties.

⚙ **Exercise 5F**

1 AFB and CIGD are parallel lines.
EFGH is a straight line.
GH = GI
Angle GIH = 28°
Work out the size of:
  a angle DGF
  b angle EFA.

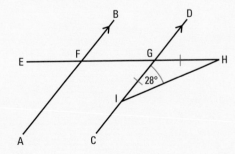

A02
A03
D

C

2   L, M and N are points, as shown, on the sides of triangle ABC.
    ML and AB are parallel.
    NL and AC are parallel.
    NM and BC are parallel.
    Angle BAC = 70°
    Angle ABC= 55°
    Work out the size of each angle of triangle LMN.

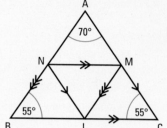

**A02
A03**

3   Work out the size of:
    a   angle p
    b   angle q.
    Give reasons for your working.

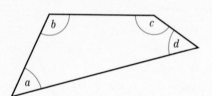

**A02
A03**

4   Here is a parallelogram.
    a   Explain why $a = c$.
    b   Hence prove that $a + b = c + d$.
        Give reasons for your working.
    c   What property of parallelograms have you proved?

**A02
A03**

5   Here is a quadrilateral.
    In this quadrilateral $a + c = 180°$.
    Prove that $b + d = 180°$.
    Give reasons for your working.

5.7 Polygons

◎ Objectives

- You know what a polygon is and what a regular polygon is.
- You know the names of special polygons.
- You know and can use the sum of the interior angles of a polygon.
- You know and can use the sum of the exterior angles of a polygon.
- You can answer problems on polygons involving angles.

⍰ Why do this?

Tiles are made in the shape of regular polygons. Knowing the properties of various polygons might help you make patterns with different shapes.

⬦ Get Ready

1. Write down the name of the shape that has three equal sides and three equal angles.
2. Use two words to complete the following sentence: A square is a quadrilateral with… sides and… angles.

Key Points

◉ A **polygon** is a closed two-dimensional shape with straight sides.

◉ A **regular** polygon is a polygon with all its sides the same length and all its angles equal in size.

◉ Here are the names of some special polygons.

Triangle	3-sided polygon	
Quadrilateral	4-sided polygon	
Pentagon	5-sided polygon	
Hexagon	6-sided polygon	
Heptagon	7-sided polygon	

ResultsPlus
Exam Tip

These are the names of the polygons that are needed for GCSE. Other polygons also have special names.

Octagon	8-sided polygon	
Decagon	10-sided polygon	

Angles of a polygon

Example 9   Here is a 7-sided polygon.

Work out the sum of the angles of this polygon.

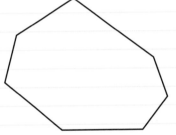

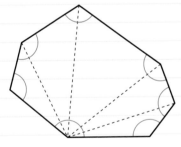

A 7-sided polygon is a heptagon.

All the diagonals from one vertex (corner) of the heptagon have been drawn.

There are 4 diagonals and the heptagon has been divided into 5 triangles.

The angle sum of each triangle is 180°.

Sum of the angles of a heptagon = 5 × 180° = 900°

As these angles are inside the polygon, they are also called the interior angles of the polygon.

Exercise 5G

D

1 **a** Copy and complete the following table.

Polygon	Number of sides (n)	Number of diagonals from one vertex	Number of triangles formed	Sum of interior angles
Triangle	3	0	1	180°
Quadrilateral	4	1	2	360°
Pentagon	5			
Hexagon	6			
Heptagon	7	4	5	900°
Octagon	8			
Nonagon	9			
Decagon	10			

 b For a polygon with n sides, write down:
   i the number of diagonals that can be drawn from one vertex
   ii the number of triangles that are formed
   iii the sum of the interior angles of the polygon.

A03 **2** A rhombus has sides that are the same length. Explain why a rhombus is, in general, not a regular polygon.

Sum of the interior angles of a polygon

Key Points

- A polygon can be divided into triangles when all diagonals are drawn from one vertex. For an n-sided polygon, the number of triangles will be $(n-2)$.
- A regular polygon will tesselate if the interior angle is an exact divisor of 360°. Regular triangles, squares and hexagons will therefore tessellate.
- Sum of the interior angles of a polygon with n sides $= (n - 2) \times 180°$
$$= (2n - 4) \text{ right angles}$$

Example 10

A polygon has 15 sides.
 a Work out the sum of the interior angles of the polygon.
 b Find the size of each interior angle of a regular polygon with 15 sides.

a $\qquad 15 - 2 = 13 \qquad \longleftarrow$ With $n = 15$, work out the number of triangles $= (n - 2)$

$\qquad 13 \times 180 = 2340 \qquad \longleftarrow$ Work out $(n - 2) \times 180$
Sum of interior angles $= 2340°$

b Each interior angle $= 2340° \div 15 \qquad \longleftarrow$ The regular polygon has 15 interior angles that are all the same size.

$\qquad = 156° \qquad \longleftarrow$ Divide the sum of the angles by 15.

Example 11 Here is a regular octagon with centre O.

    a Work out the size of:

       i angle x **ii** angle y.

    b Hence work out the size of each interior angle of a regular octagon.

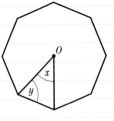

a   **i** $x = 360° \div 8 = 45°$ ⟵ | Joining each vertex of the polygon to the centre O will form in total 8 equal angles like angle x. These 8 angles make a complete turn of 360°.

    ii $y = \dfrac{180° - 45°}{2} = 67.5°$ ⟵ | The triangle shown is an isosceles triangle, with angle y as one of the two equal base angles.

b Each interior angle $= 2 \times 67.5° = 135°$ ⟵ | By symmetry, angle y is half an interior angle of the regular polygon.

Exercise 5H

1 John divides a regular polygon into 16 triangles by drawing all the diagonals from one vertex.

  a How many diagonals does John draw?

  b How many sides has the polygon?

  c What is the size of each of the interior angles of the polygon?

AO2 AO3 **D**

2 Work out the size of each interior angle of:

  a a regular hexagon

  b a regular decagon

  c a regular polygon with 30 sides.

AO2 AO3

3 Work out the size of each of the marked angles in these polygons. You must show your working.

AO3 **C**

  a

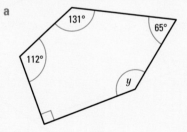

  b

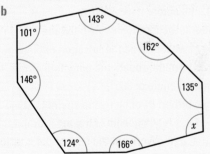

4 Explain why the size of the angle at the centre of a regular polygon cannot be 25°.

AO2 AO3

5 Here is an octagon.

  a Work out the size of each of the angles marked with a letter.

  b Work out the value of $a + b + c + d + e + f + g + h$

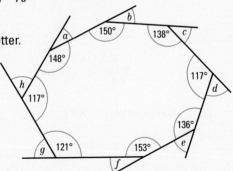

6 ABCD is a square. EFGHIJKL is an octagon.

AE = EF = FB = BG = GH = HC = CI = IJ = JD = DK = KL = LA

a Find the size of each interior angle of the octagon.

Give reasons for your answers.

b Tracy says that the octagon is a regular octagon.

i Why might Tracy think that the octagon is regular?

ii Explain why Tracy is wrong.

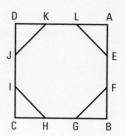

Exterior angles of a polygon

> **Key Points**

◉ When a side of a polygon is extended at a vertex, the angle between this extended line and the other side at the vertex is an exterior angle at this vertex.

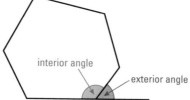

◉ The sum of angles on a straight line = 180°

So at a vertex, interior angle + exterior angle = 180°

◉ The sum of the exterior angles of any polygon is 360°.

ABCDEF is a hexagon.

Imagine a spider is at vertex A facing in the direction of the arrow.
The spider turns through angle a so that it is now facing in the
direction AB. The spider now walks to vertex B.

At B, the spider turns through angle b to face in the direction BC.
He continues to walk around the hexagon until he gets back to A.
The spider has turned through one complete circle, so it has
turned through an angle of 360°.

The total angle turned through by the spider is also
$a + b + c + d + e + f$, the sum of the exterior angles of the hexagon.
So $a + b + c + d + e + f = 360°$.
The same argument holds for any polygon.

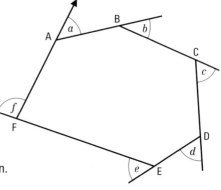

> **Example 12** A regular polygon has 20 sides.

a Work out the size of each exterior angle.

b Work out the size of each interior angle.

a Each exterior angle = 360° ÷ 20 = 18° ⟵ | The polygon is regular so the 20 exterior angles are equal in size. The sum of these 20 equal angles is 360°.

b Interior angle = 180° − 18° = 162° ⟵ | Exterior angle + interior angle = 180°

Example 13 The interior angle of a regular polygon is 160°.

Work out how many sides the polygon has.

Exterior angle = 180° − 160° = 20° ← Work out the size of an exterior angle.
Exterior angle + interior angle = 180°.

Number of sides = $\frac{360°}{20°}$ = 18 ← The polygon is regular so that all exterior angles are 20° with sum 360°.

Exercise 5I

1 One vertex of a polygon is the point P.
  a Work out the size of the interior angle at P when the exterior angle at P is: **i** 70° **ii** 37°.
  b Work out the size of the exterior angle at P when the interior angle at P is: **i** 130° **ii** 144°.

2 Work out the size of each exterior angle of:
  a a regular pentagon   **b** a regular octagon
  c a regular polygon with 12 sides   **d** a regular 25-sided polygon.

3 The size of each exterior angle of a regular polygon is 15°.
  a Work out the number of sides the polygon has.
  b What is the sum of the interior angles of the polygon?

4 The sizes of five of the exterior angles of a hexagon are 36°, 82°, 51°, 52° and 73°.
Work out the size of each of the interior angles of the hexagon.

5 A, B and C are three vertices of a regular polygon with 30 sides.

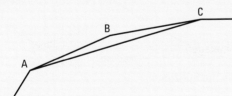

Work out the size of angle BCA.
Give reasons for your working.

* 6 The diagram shows three sides, AB, BC and CD, of a regular polygon with centre O.
The angle at the centre of the polygon is c.
The exterior angle of the polygon at the vertex C is e.

Explain why $c = e$.

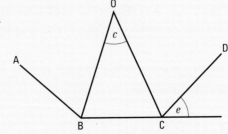

D

A02
A03

C

A03

A02
A03

A02
A03

Chapter review

◉ When two **parallel** lines are crossed by a straight line, as in the diagram, angles are formed. **Corresponding angles** are equal, and **alternate angles** are equal.

◉ The angle sum of a triangle is 180°.

◉ An **exterior angle** of a triangle is equal to the sum of the **interior angles** at the other two vertices.

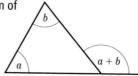

◉ The angle sum of a quadrilateral is 360°.
$a + b + c + d = 360°$

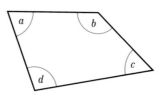

◉ Opposite angles of a parallelogram are equal.
$a = c$
$b = d$

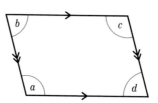

◉ The **angle of elevation** of point A from point B is the angle of turn above the horizontal to look directly from B to A.

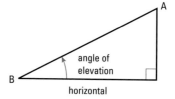

◉ The **angle of depression** of point B from point A is the angle of turn below the horizontal to look directly from A to B.

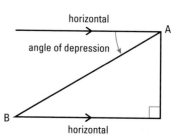

◉ Angles of elevation and depression are always measured from the horizontal.

◉ **Bearings** are angles measured clockwise from North.

◉ Bearings always have three figures.

◉ A **polygon** is a closed two-dimensional shape with straight sides.

◉ A **regular** polygon is a polygon with all its sides the same length and all its angles equal in size.

◉ Sum of the interior angles of a polygon with n sides $= (n - 2) \times 180°$
$= (2n - 4)$ right angles

◉ At a vertex of a polygon, interior angle + exterior angle = 180°

◉ The sum of the exterior angles of any polygon is 360°.

Review exercise

1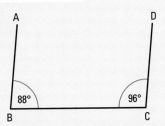

Diagram **NOT** accurately drawn

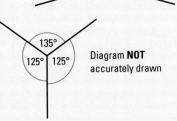

ResultsPlus

Exam Question Report

93% of students answered this question poorly because they did not justify their answers.

James says, "The lines AB and DC are parallel."

Ben says, "The lines AB and DC are **not** parallel."

Who is right, James or Ben?

Give a reason for your answer.

May 2009

2 AB is a straight line.

This diagram is wrong. Explain why.

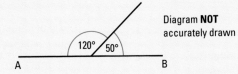

Diagram **NOT** accurately drawn

Nov 2008

3 a Write down the value of x.

b Give a reason for your answer.

Diagram **NOT** accurately drawn

This diagram is **wrong**.

c Explain why.

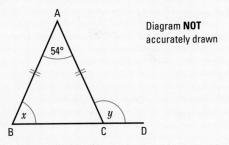

Diagram **NOT** accurately drawn

June 2008

4 ABC is an isosceles triangle.

BCD is a straight line.

AB = AC.

Angle A = 54°.

a i Work out the size of the angle marked x.

ii Give a reason for your answer.

b Work out the size of the angle marked y.

Diagram **NOT** accurately drawn

June 2007

A03

5 The diagram shows the position of two boats, P and Q.

The bearing of a boat R from boat P is 060°.

The bearing of boat R from boat Q is 310°.

Draw an accurate diagram to show the position of boat R.

Mark the position of boat R with a cross (✕). Label it R.

June 2009

D

D

6 AB is parallel to CD.
  a Write down the value of y.
  b Give a reason for your answer.

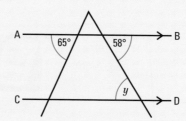

June 2008

* 7

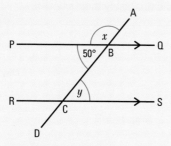

ABCD is a straight line.
PQ is parallel to RS.
Write down the size of angle x and y,
giving reasons for your answer.

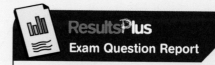

ResultsPlus
Exam Question Report

92% of students answered this question well
because they knew the difference between
alternate and corresponding angles.

March 2008

8 Work out the value of x.

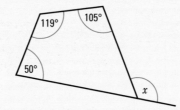

Nov 2007

9 In the diagram
AB = AC
Angle ABC = 52°
  a Work out the size of the angle marked x.
  b Give a reason for your answer.

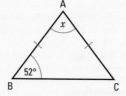

June 2009

10 ABCDEF is a regular hexagon and ABQP is a square.
Angle CBQ = x°.
Work out the value of x.

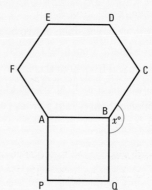

June 2007

11 In triangle ABC, angle BAC = 90°.
Work out the size of   **a** angle DAC   **b** angle DCA.

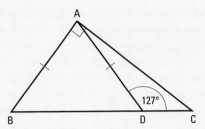

12 Work out the size of the angle p.
Give reasons for your working.

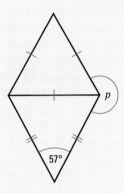

13

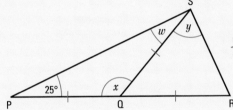

Diagram **NOT**
accurately drawn

PQR is a straight line.
PQ = QS = QR. Angle SPQ = 25°.
a   **i**   Write down the size of angle w.
    ii   Work out the size of angle x.
b   Work out the size of angle y.

ResultsPlus
Exam Question Report

84% of students answered this question poorly
because they did not use all of the information
given in the question.

Nov 2008

14   **a**   Find the bearing of B from A.
    b   On a copy of the diagram, draw a line
        on a bearing of 135° from A.

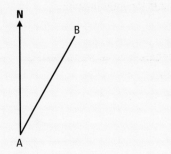

Nov 2006

15 The diagram shows part of a regular 10-sided polygon.
Work out the size of the angle marked x.

Nov 2008

C

16 In the diagram, ABC is a straight line and BD = CD.
a Work out the size of angle x.
b Work out the size of angle y.

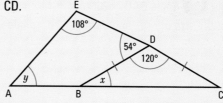

Nov 2006

17 In triangle ABC, AB = AC. Angle ABC = $(x + 20°)$.
Show that angle BAC = $(140° - 2x)$.
Give reasons for each stage of your working.

Diagram **NOT** accurately drawn

18 Prove that angle FED = $205° - 3x$
Give reasons for each stage of your working.

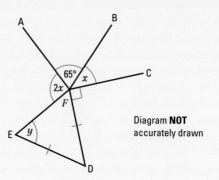

Diagram **NOT** accurately drawn

19 In a regular polygon each exterior angle is two thirds the size of each interior angle.
a Calculate the size of each interior angle.
b Calculate the number of sides of the polygon.

* **20** In the diagram, AC = BC .
Prove that angle BCD = $4(95° - x)$
Give reasons for each stage of your working.

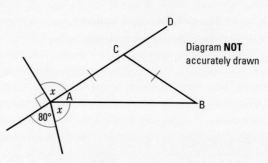

Diagram **NOT** accurately drawn

* **21** PQR is a triangle with PQ = PR.
Prove that PY = QY.

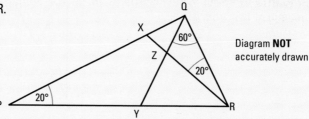

Diagram **NOT** accurately drawn

A
A02
A03

22 Just after 1 o'clock the hour and minute hands of a clock are pointing in the same direction, meaning that the angle between them is 0°. What time is this? (Answer to the nearest second).

6 COLLECTING AND RECORDING DATA

A local council wants to know whether the facilities for teenagers are adequate in the town.
How could it find out people's views?
How could these views be recorded and presented?

◉ Objectives

In this chapter you will:

- learn about the statistical problem-solving process and consider different types of data
- discover how to collect, record and interpret data
- look at various sampling methods
- learn how to identify possible sources of bias.

6.1 Introduction to statistics

◎ Objectives

- ● You can understand the stages of an investigation.
- ● You can formulate a question in terms of the data needed.
- ● You can classify data as qualitative (categorical) or quantitative (numerical).
- ● You can classify quantitative data as discrete or continuous.

❓ Why do this?

To find out how good teachers are at predicting the grades their students will get in an exam, you could carry out a statistical investigation.

◈ Get Ready

How can you find the following information?

a The average amount of lunch money for your classmates.

b What flights there are from Manchester to Washington D.C.

c How many people voted for the Green Party in the last election.

◷ Key Points

● **Statistics** is used to provide information. The statistical problem-solving process can be shown as a simple diagram:

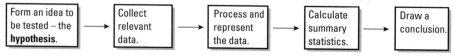

| Form an idea to be tested – the **hypothesis**. | → | Collect relevant data. | → | Process and represent the data. | → | Calculate summary statistics. | → | Draw a conclusion. |

● Data that you collect yourself is called **primary data**; data collected by other people is called **secondary data**.

● **Qualitative data** can be described in words. For example, the colours of shirts on sale in a shop.

● **Quantitative data** are numerical observations. There are two types:

  ◎ **Discrete data** can only take certain numerical values. For example the number of carriages on trains.

  ◎ **Continuous data** can take any numerical value. For example weights, times, lengths and temperatures are continuous.

⬗ Example 1

An estate agent collects the following information about houses for sale.

Type of house	Number of bedrooms	Garden area	Price
Detached	4	390 m²	£321 000
Semi-detached	3	170 m²	£184 000
Terraced	3	150 m²	£177 000
Flat	2	0	£196 000

Describe the data in each column as qualitative or quantitative. If quantitative, state whether it is discrete or continuous.

Type of house: qualitative
Number of bedrooms: quantitative and discrete
Garden area: quantitative and continuous
Price: quantitative and continuous

← See if the data item can be represented as a number. If it cannot it is qualitative data. If it can be given as a number it is quantitative. If quantitative ask yourself 'Can it only take certain values?' If it can it is discrete; otherwise it is continuous.

statistics   hypothesis   primary data   secondary data   qualitative data   quantitative data

Exercise 6A

Questions in this chapter are targeted at the grades indicated.

1 Write down whether each of the following is secondary or primary data.
   a Data collected by you from a government website.
   b Data collected by you from a newspaper.
   c Data collected by you questioning people in a shopping centre.

2 Write down whether the following are qualitative or quantitative data.
   a The numbers of students in classes.   b The colour of students' eyes.
   c The weight of dogs.   d The floor area of houses.

3 Write down whether the following are continuous or discrete data.
   a The number of trees in a wood.   b The time taken to run 100 m.
   c The length of flower stems.   d The number of animals in a zoo.

4 A medical researcher wants to find out how effective Drug A is at curing malaria.
   a Write down a hypothesis he could use.
   b What is the next thing that he would need to do?

A03 D

6.2 Sampling methods

◎ Objectives

◉ You can collect information about a population by using a sample.
◉ You can select a simple random sample.
◉ You know that in a simple random sample each member of the population has an equal chance of being selected.

◈ Why do this?

A city council wants to know how many people are likely to support the idea of building a swimming pool. They can't ask everybody in the city but they can ask a sample of people.

◈ Get Ready

1. If it takes 15 seconds for one student to answer a question, how long would it take to get answers from everyone in your class?

Key Points

◉ A small, but carefully chosen, number of people can be used to represent the **population** of a country. These chosen individuals are called a **sample** and the investigation itself is called a sample **survey**.

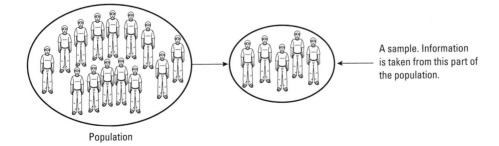

Population

A sample. Information is taken from this part of the population.

- The sample must be **representative** of all the people or items being investigated, with each member of the population having an equal chance of being selected. If it isn't it is **biased**. For example, 'adults only' would be biased. (See Section 6.7 for more information on bias.)
- To make a sample representative, each individual in the sample should be picked at random. This process is known as taking a simple **random sample.**
- To take a simple random sample:
  1. Each person or item in the population is given a number.
  2. If a sample of 10 is needed, then 10 numbers are selected. This can be either: from a random number table; by a random number generator on a calculator; by using a computer; or by putting the numbers in a hat. The people whose numbers are selected then form the sample.

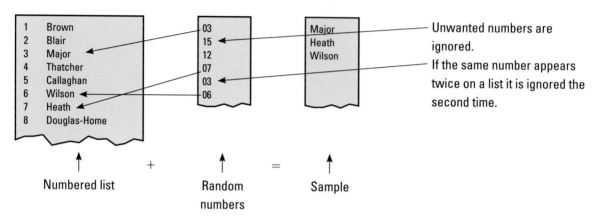

| Numbered list | + | Random numbers | = | Sample |

Example 2   John is collecting data from each of the 50 students in his year group about the number of brothers and sisters they have.

    a Give a reason why he might use a simple random sample.

This is an extract from a set of random numbers.

      335221170532482146321143059288234171412

    b Starting at 33 and working across, use the random numbers to give eight numbers less than 50.

    c Explain how John would use these numbers to take a sample of eight students.

a The number of students is large and it would take a long time to collect that data.

> This is one reason for taking a sample. Cost could be another reason.

b 33 21 17 05 32 48 46 11

> Start at 33 and take the digits in pairs. 05 counts as the number 5. Numbers like 21 which repeat are ignored when they appear a second time. If the population was 150 you would take the digits in threes.

c He would number the students and select the ones corresponding to these numbers.

C

B

Exercise 6B

1   Write down two ways in which you can generate random numbers.

2   Explain what is meant by a simple random sample.

3   A call centre has 60 workers. Eight are to be chosen for a new training scheme.
The manager decides to choose a simple random sample of eight.
He uses a calculator to generate random numbers. These are the first few numbers he generates.
21 32 67 54 89 78 90 34 26 45 78 54 35 64 22 …
Describe how he could use these numbers to get his sample of workers.

6.3 Stratified sampling

Objective

○ You can select and use a stratified sample.

Why do this?

A school has an equal number of boy and girl students. A simple random sample could contain more boys than girls. A stratified sample would contain an equal number of each.

Get Ready

1. Fifteen of a class of 25 students are girls.
    a What fraction are girls?   b What fraction are boys?
2. There are three classes in Year 11. There are 22 students in class A, 28 in class B and 30 in class C.
    a How many students are in Year 11?   b What fraction of the students in Year 11 are in class C?

Key Points

○ A population may contain groups in which the observation of interest is likely to differ. For example, if you are looking at the heights of students then the boys' heights are likely to be different to the girls' heights. These groups are called strata (singular stratum).

○ A **stratified sample** is one in which the population is split into strata, and a simple random sample is taken from each stratum. The number taken from each stratum should be in proportion to the total number in each stratum.

○ To find the number to be selected from a stratum:

1. Find what fraction of the population is in the stratum.

    $$\text{Fraction in stratum} = \frac{\text{number in stratum}}{\text{number in population}}$$

2. Multiply the fraction in the stratum by the total size of the sample.

    $$\text{The number sampled in a stratum} = \frac{\text{number in stratum}}{\text{number in population}} \times \text{total sample size}$$

Example 3

The table below shows the number in each year group of a school.
A sample of 60 students is to be taken.
How many students from each year group
should be in the sample?

Year	7	8	9	10	11
Number of students	150	150	100	100	100

ResultsPlus
Exam Tip

Always make sure your individual
samples total the required sample size.
Check: 15 + 15 + 10 + 10 + 10 = 60

The sample for Year 7 will be $\frac{150}{600} \times 60 = 15$

The sample from Year 8 will be the same size as for Year 7.

In each of Years 7 and 8 there are
150 students out of 600 students.

The sample for Year 9 will be $\frac{100}{600} \times 60 = 10$

Years 10 and 11 will also have a sample size of 10.

In each of Years 9, 10 and 11 there are
100 students out of 600 students.

Exercise 6C

B

1 A head teacher wants to find out what Year 7 students think about their first term at their new school.
He decides to ask a stratified sample of 50 students. The table shows the total number of boys and the
total number of girls in Year 7.

Boys	Girls
276	324

Work out the number of boys and the number of girls he should include in the sample.

A
A03

*2 A call centre allocates work according to the experience of its employees. Those with less than six months'
experience do the easier work; those with more than six months' experience do more difficult tasks.
There are 150 employees with less than six months' experience and 400 with more than six months'
experience. Describe exactly how you would find a stratified sample of 10% of the employees.

A02

*3 A factory owner wants to find out what his employees think about the parking facilities at his factory.
He decides to ask a stratified sample of 90 of his workers. The table shows how many people are in
each of the six strata he intends to use.

	Office workers	Factory floor workers	Managers
Females	50	250	10
Males	80	490	20

Calculate the number of workers he needs to ask in each strata and describe how he should pick the
individual members of each strata.

6.4 Collecting data by observation and experiment

◎ Objectives

○ You can design and use data collection sheets.
○ You can use tallying methods.
○ You can group data into class intervals.
○ You can collect data by observation, experiment or data logging.

⟡ Why do this?

If you are collecting data about the number of different types of vehicles passing by, it is easier to keep a record in the form of a data collection sheet rather than trying to remember each total.

⟐ Get Ready

1. What are the numbers given by each set of marks?   **a** |||| **b** ||||||
2. Is there a better way of grouping the marks in part **b**?

◉ Key Points

◎ There are a number of ways of collecting data.

  ○ **Collecting data by observation:** if you want to investigate whether a lot of traffic is caused by people taking students to school, you could observe how much traffic there is at school opening time and compare it with how much traffic there is at other times.

  ○ **Collecting data by experiment:** if you wish to find out how high a tennis ball bounces when dropped from different heights, you could drop a tennis ball from various heights and record how high it bounces. This would enable you to collect data on the bounce of tennis balls.

  ○ **Data logging:** when you go to a supermarket till each item is bar coded so that it can be identified at the checkout. The number of each item sold is automatically recorded and this enables the supermarket to know what items are popular and what they need to stock up on.

◎ When collecting data by observation a **data collection sheet** is used. The following diagram shows a data collection sheet for recording the different types of transport that might pass your door.

Vehicle	Tally	Frequency
Bicycle	ЖⱠ |||	8
Bus	|||	3
Car	ЖⱠ ||||	9
Lorry	ЖⱠ	5
Motorcycle	ЖⱠ		6
Van	ЖⱠ	5

Each time a bicycle passes a **tally** mark is put next to 'Bicycle'.

When the **survey** is complete the tally marks are added together to give the total number of each vehicle. This is known as the **frequency**.

The marks are grouped into fives with the fifth tally mark drawn through the other four.

Putting tally marks in fives makes totalling up easier.

◎ If data is numerical, and widely **spread**, you can group the data into **class intervals**. These class intervals do not have to be the same size.

◎ When dealing with continuous data you need to make sure the intervals do not overlap. For example, the class intervals for a variable such as weight (w) will be of the form:

$500 \text{ g} \leqslant w < 550 \text{ g}$ This means that w is greater than or equal to 500 g but less than 550 g.
or $500 \text{ g} < w \leqslant 550 \text{ g}$ This means that w is greater than 500 g but less than or equal to 550 g.
In this case, 500 g and 550 g are the lower and upper class **limits**, while the class size is $550 - 500 = 50 \text{ g}$.

Example 4　30 students are asked how many books they read in the Easter holiday. Each student's response is shown below. Draw and fill in a data collection sheet for this information.

1	5	9	9	6	13
6	8	4	5	5	7
9	6	11	3	8	9
14	7	7	2	12	9
3	2	1	4	0	5

Watch Out!

Make sure classes don't overlap.

Mark	Tally	Frequency
0–4	||||| ||||	9
5–9	||||| ||||| ||||| ||	17
10–14	||||	4
15–19		0

The marks have been grouped together into four equal class groups. The first class includes all the numbers between 0 and 4 inclusive.

In this example, different class intervals could have been chosen. For example:

Mark	Tally	Frequency
0–4	||||| ||||	9
5–7	||||| |||||	10
8–10	||||| ||	7
11–15	||||	4

Example 5　The tally chart below shows the age at marriage of a sample of men.

Age, a	Tally	Frequency
$16 < a \leqslant 20$	||	
$20 < a \leqslant 30$	||||| ||||| |||	
$30 < a \leqslant 40$	||||| ||||| ||||| ||||| |||||	
$40 < a \leqslant 50$	||||| ||	
$50 < a \leqslant 60$	|||	
$60 < a$			

  a Fill in the frequency column.

  b Write down the most popular age range in which men get married.

  c Work out how many men in total there were in the sample.

a

Age, a	Tally	Frequency
$16 < a \leqslant 20$	||	2
$20 < a \leqslant 30$	||||| ||||| |||	13
$30 < a \leqslant 40$	||||| ||||| ||||| ||||| |||||	25
$40 < a \leqslant 50$	||||| ||	7
$50 < a \leqslant 60$	|||	3
$60 < a$			1

Add together the tallies: $5 + 5 + 3 = 13$.

Look for the class with the highest frequency.

  b 30 to 40　　**c** 51　← Add together all of the frequencies: $2 + 13 + 25 + 7 + 3 + 1 = 51$.

tally　survey　frequency　spread　class intervals　limits

Exercise 6D

1. A road traffic controller keeps a record of the types of traffic using a busy junction during a two-minute rush-hour period. This data is listed below:

Car	Car	Bus	Car	Car	Car		HGV	Bike	Car	Car	Car	Bus
Bus	HGV	Car	Car	Car	Motorbike		Bike	Car	Car	Bus	Car	HGV
Bus	Car	Car	Car	Car	Bike		HGV	Car	Car	Car	Car	Bike

   a Draw a tally chart to show this data.
   b Write down the name of the least common type of traffic.
   c Write down the name of the most common type of traffic.

2. A shopkeeper asks 30 people entering her shop how many DVDs they have bought in the last three months. The responses are shown below:

3	5	8	9	2	7	4	10	12	3
6	2	4	9	12	13	1	7	7	11
14	3	6	5	8	1	2	7	4	3

   Draw and complete a data collection sheet showing this information. Use equal class intervals starting with the class 0−3.

3. A gardener weighs 24 tomatoes produced from plants in his greenhouse.
   The weights, in grams, are shown below:

60.5	65	64.5	59	67	61.5	67	69
58	59.3	57.2	67	68.5	63	64.2	69
57	57.8	62.4	65.5	67	58	70	75

Weight (w)	Tally	Frequency
$57 \leqslant w < 60$		
$60 \leqslant w < 63$		
$63 \leqslant w < 66$		
$66 \leqslant w < 69$		
$69 \leqslant w$		

   a Copy and complete the data collection sheet for this data.
   b Write down the most common class.
   c Write down the least common class.

6.5 Questionnaires

◎ Objectives

○ You can collect data by using a questionnaire.
○ You can criticise questions for a questionnaire.

◈ Why do this?

A restaurant may use a questionnaire to get feedback about its service, food and atmosphere, if it is looking to make improvements.

◈ Get Ready

Describe a good method for recording data on a data collection sheet.

⬤ **Key Points**

◉ A **questionnaire** is a list of questions designed to collect data. On questionnaires:
  ◉ keep questions short
  ◉ use words that are easily understood
  ◉ do not use **biased questions** that lead the respondent to a particular answer. For example, use 'Do you agree or disagree?', rather than 'You do agree, don't you?'
  ◉ write questions that address a single issue. For example, use 'Do you have a car?' rather than 'Do you have a petrol engine car?'.
◉ There are two types of question to use on questionnaires.
  ◉ An **open question** is one that has no suggested answers.
  ◉ A **closed question** is one that has a set of answers to choose from. It is easier to summarise the data from this type of question. Closed questions will often have an opinion scale to choose from. For example:

Statistics is an important subject.	These are **response boxes**.
☐	☐	☐	☐	☐	This allows for other answers.
Strongly agree	Agree	Disagree	Strongly disagree	Don't know

Sometimes a numerical scale is used. For example:

Tick one box to indicate your age group.	The categories do not overlap.
☐	☐	☐	☐	☐
Under 20	21 to 30	31 to 40	41 to 50	Over 50

◉ When designing questionnaires, it is important to ensure that possible answers are clear, do not overlap and cover all possibilities.

🔍 **Example 6** ▶ Write down what is wrong with each of these questions.

a Tick one box to indicate your age group.
| ☐ | ☐ | ☐ | ☐ |
| Under 20 | 20 to 30 | 31 to 40 | 40 to 50 |

b How often have you had a medical in the last 4 years? Tick one box.
| ☐ | ☐ | ☐ | ☐ | ☐ |
| Never | Seldom | Sometimes | Often | Very often |

c Do you agree that people who have regular medicals are less likely to have major illness that goes undetected?
| ☐ | ☐ |
| Yes | No |

a The categories overlap — 40-year-olds could go into two boxes.
   Other answers are not allowed for. Where does a 60-year-old tick?
b It is difficult to decide what these words mean.
c By asking 'Do you agree ...' you are inviting the answer 'Yes'. This is called a biased question.

questionnaire   biased question   open question   closed question   response boxes

Exercise 6E

1 A questionnaire includes the following question.
'Do you agree that we should build a new road?'

☐   ☐

Yes   No

Write down what is wrong with this question.

2 A local council wants to know whether or not the residents would like a new swimming pool in the town. It is decided to use a questionnaire. The following questions are suggested.

A: What do you think about the idea of a new pool being built?

B: Do you want a new pool?   Yes/No

C: Where should we build a new pool?

D: Is a pool a good idea? Yes/No

Which of the above are open questions and which are closed?

3 The management of a theme park have made some changes to the amusements. They want to use a questionnaire to find out what people think about the changes. The following questions are suggested. Write down what is wrong with each of them and design a new question for each that is more suitable.

a What do you think of the new amusements?

Very good ☐   Good ☐   Satisfactory ☐

b How much money would you normally expect to pay for each amusement?

£5−£7 ☐   £7−£8 ☐   More than £8 ☐

c How often do you visit the park each year?

Often ☐   Not very often ☐

*   **4** A supermarket manager wants to find out if people like the new layout. She decides to use a questionnaire. Write down a suitable question she could use.

6.6 **Two-way tables**

⊙ Objectives

○ You can design and use two-way tables.
○ You can use information to complete a two-way table.

Why do this?

You can use a two-way table to record results such as the drink preferences of boys and girls.

Get Ready

In a class of 30 students there are:

2 left-handed girls   13 right-handed girls
4 left-handed boys   11 right-handed boys.

a How many girls are there in the class?   b How many students are left-handed?

🔍 **Key Points**

◉ Sometimes we collect two pieces of information, for example gender and eye-colour. To record this we would use a **two-way table**. A two-way table shows the frequency with which data falls into two different categories.

	Blue	Brown	Green	Total
Boys	6	14	5	25
Girls	4	16	5	25
Total	10	30	10	50

This is the number of boys with brown eyes.
This is the number of girls with brown eyes.
This the total number with brown eyes.
This is the total number of boys and girls.

◉ Sometimes a table is incomplete and has to be filled in before you can answer a question.

A02

🔍 **Example 7** ▶ Students in Year 11 were asked to choose their favourite drink from a choice of three. Below are the boys' and girls' responses.

Girls

Tea	Coffee	Coffee	Tea	Soft
Tea	Coffee	Tea	Coffee	Tea
Soft	Soft	Tea	Tea	Soft
Coffee	Coffee	Soft	Soft	Coffee

Boys

Coffee	Coffee	Tea	Soft	Tea
Tea	Tea	Soft	Coffee	Coffee
Soft	Tea	Tea	Coffee	Coffee
Soft	Tea	Coffee	Coffee	Coffee

   a Show this information in a suitable table.
   b Write down the girls' top choice of drink.
   c Write down the boys' top choice of drink.
   d Write down the drink that was chosen by most of the students.

a

	Tea	Coffee	Soft drink	Total
Boys	7	9	4	20
Girls	7	7	6	20
Total	14	16	10	40

The most suitable table is a two-way table. Count up the number of boys that chose tea and enter it here. Do the same for the other drinks and the girls' drinks.

Total the rows and columns.

b Tea and coffee tied. ← Look for the highest number in the girls' row.

c Coffee ← Look for the highest number in the boys' row.

d Coffee ← Look for the drink which has the highest total.

Example 8 The following two-way table gives information about people's hair and eye colour.

		Eye colour			
		Brown	Green	Blue	Total
Hair colour	Brown/Black	4	4		16
	Fair	3		4	
	Ginger		1	1	4
	Total	9	8		30

a Complete the table.

b Which eye colour was most frequent?

c Which eye colour was least frequent?

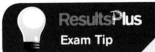
ResultsPlus
Exam Tip

Look for rows with only one number missing and fill these in first.

The numbers in each row must add up to the row total and the same goes for columns.

a

		Eye colour			
		Brown	Green	Blue	Total
Hair colour	Brown/Black	4	4	8	16
	Fair	3	3	4	10
	Ginger	2	1	1	4
	Total	9	8	13	30

The number of blue-eyed black-haired = $16 - 4 - 4 = 8$
The number of brown-eyed ginger-haired = $4 - 1 - 1 = 2$
The number of green-eyed fair-haired = $8 - 4 - 1 = 3$
The total number of fair-haired = $3 + 3 + 4 = 10$
The total number of blue-eyed = $30 - 8 - 9 = 13$

b Blue   c Green

Exercise 6F

1   A number of men and women were asked which type of crisps they liked best. A total of twelve people said Plain, of which seven were men. Six women liked Salt and Vinegar. Fourteen men and twelve women liked Cheese and Onion. There were 28 men in total.

   a Draw and complete a table of the data.

   b How many people liked Salt and Vinegar crisps best?

   c How many people were asked altogether?

2   In a supermarket survey 30 men and 30 women were asked whether they preferred orange juice or grapefruit juice. 22 men preferred orange juice. 12 women preferred grapefruit juice.

   a Draw up a two-way table to show this information.

   b How many people liked orange juice best?

3   A factory employs 12 supervisors, of which 2 are female; 14 office staff, of which 3 are male; and 120 shop floor workers, of which 38 are female.

   a Draw up a two-way table to show this information.

   b Write down the number of female employees.

   c Write down the total number of employees.

6.7 Sources of bias

Objectives

- You can identify possible sources of bias.
- You understand how different sample sizes may affect the reliability of any conclusions drawn.

Why do this?

If you want to accurately estimate the average height of students you need to collect reliable data. For example, if you include more boys than girls in your sample then you are likely to get a taller average.

Key Points

- When collecting data you should make sure that the data is representative of the population it is taken from. Data that does not do this is said to be biased. There are several types of bias.

Selection bias

- If you select only people who shop at a supermarket and ask them what they think about how that supermarket compares with a rival supermarket, you will get a biased opinion, since the people who use the other supermarket are not represented. This is called under-coverage bias.

- If you ask people to fill in a questionnaire and post it back to you, only a certain type of person will bother to respond. The respondents will not be representative of the general public. This is non-response bias.

- If you ask people to text a radio show about a controversial topic, you will get mainly people who have a strong opinion about that topic. This is voluntary response bias.

- Selection bias can be avoided by random selection and random allocation.

Measurement bias

- If you ask people if they are satisfied, dissatisfied or very dissatisfied, you are likely to get a biased opinion because there are two answers for dissatisfied and only one for satisfied.

- If you ask a question such as, 'You are satisfied, aren't you?' you are more likely to get 'Yes' as an answer. These are called **leading questions**.

- People like to present themselves in the best light, so if you ask them, 'Do you often behave unreasonably?' you are likely to get 'No' for an answer even if it should be 'Yes'.

Sampling error

- If you take two random samples you are unlikely to get exactly the same result (though they should be close to each other). This is called the sampling error.

- Increasing the sample size will reduce the sampling error. It is difficult to say how big a sample should be as this depends on how varied the population is. However, the larger the sample the more representative it will be of the population and the more accurate the information will be.

> **Example 9** Write down, with reasons, whether or not each of the following is biased.

a You want to find out what people think about a football team. You ask supporters as they enter the ground before a match.

b You wish to find out what proportion of the population has had flu in the last month. You interview people in the doctor's waiting room.

c You ask the first 10 people you meet, 'Do you agree that banning smoking in public places is a good thing?'

d You ask, 'Have you ever been convicted of drink driving?'

e You ask three people what they thought of the Eurovision Song Contest.

a Biased. Non-supporters are not represented.

b Biased. People who do not visit the doctor are not represented.

c Biased. Not everyone has an equal chance of being asked.
The question is leading the respondent to agree.

d Biased. This is a sensitive question. You are not likely to get a true answer.

e Biased. The sample is too small.

Exercise 6G

1 An examination board wants to get information on schools' views regarding how they respond to queries. They send a questionnaire to a sample of schools in the London area. Is this a biased sample? Give one reason for your answer.

2 Write down, with reasons, whether or not each of the following are biased.
A: A hospital wants to know how often people use A & E. They ask all the people attending A & E on one particular Wednesday.
B: An opinion poll company wants to find out how voters would vote if there were to be an election next week. They conduct a telephone poll of 20 voters in each of 10 towns.
C: A manufacturer of climbing ropes wants to see if his ropes are of the strength he advertises. He tests a sample. He tests every tenth rope made.
D: You ask 50 people using a recycling facility what they think about recycling.

* 3 One hundred people attend a rally on 'action for climate change'.
David says, 'That is a lot of people. They must be right.'
Jody says, 'I disagree.'
Discuss the views of David and Jody.

A03

C

6.8 Secondary data

Objective

○ You can extract data from lists and tables.

Why do this?

You want to find out how many accidents there were in your town last year. You can't count these yourself so you have to get the data from published sources.

Get Ready

If you needed a new mobile phone, where would you look to find one that best suited you at a price you could afford?

Key Points

○ Secondary data can be obtained relatively quickly and cheaply from a number of sources, including reference books, journals, newspapers and the internet. Remember, however, that the data may be inaccurate or out of date. Only use data from a reliable source and check the data against another source if possible.

○ It is also possible to obtain secondary data from a database. A **database** is an organised collection of information, usually stored on a computer.

○ The spreadsheet below shows part of a database kept on a computer. The entries at the top in red are fields. The entries below in black are the records. They can be easily changed to be arranged in numerical, alphabetical, gender or age order.

ID number	Surname	Forename	Gender	Age
01	Abbot	David	M	32
02	Adair	Jakie	F	27
03	Allison	Paul	M	45
04	Barber	Hassan	M	25
05	Baxter	Jenny	F	38

Example 10

Part of a database for second-hand Ford Mondeo cars is shown below.

Vehicle summary	Colour	Engine	Mileage	Price	Year
Ford Mondeo Edge	Black	2000cc petrol	11 549	£9 995	2006
Ford Mondeo Edge	Blue	2000cc petrol	14 100	£10 499	2008
Ford Mondeo Edge	Grey	2000cc petrol	10 400	£11 599	2008
Ford Mondeo Edge	Grey	2000cc petrol	12 654	£11 494	2007
Ford Mondeo Edge	Blue	2000cc petrol	7520	£11 999	2008
Ford Mondeo Zetec	Silver	2000cc petrol	10 078	£11 995	2008
Ford Mondeo Zetec	Silver	2000cc petrol	12 088	£14 995	2008
Ford Mondeo Titanium	Grey	2000cc petrol	11 555	£12 395	2008
Ford Mondeo Zetec	Black	2000cc petrol	5800	£12 895	2008
Ford Mondeo Zetec	Silver	2000cc petrol	12 123	£12 995	2008

a Which four fields could be used to order the data?

b What was the mileage of the car that cost over £13 000?

c What colour was the car that had driven the least number of miles?

d What was the maximum mileage driven by one of these cars?

a Mileage, price, year, colour. ← Mileage price and year could be put in numerical order. Colour could be put in alphabetical order.

b 12 088 ← Find the car that had a price greater than £13 000 and look in its mileage column.

c Black ← In the mileage column, find the car that had driven the least mileage then look across its row to the colour column.

d 14 100 ← Look for the largest number in the mileage column.

Exercise 6H

1 The following database gives some information about the CO_2 emissions, in thousand tonnes of carbon dioxide equivalent, in a certain country.

	Year					
	2003	**2004**	**2005**	**2006**	**2007**	**2008**
Buses	323	344	355	342	394	421
Cars	6280	6251	6163	6159	6063	6055
HGVs	2147	2154	2235	2295	2162	2221
Motorcycles	39	41	44	40	41	39
Railways	231	209	225	241	245	251

a What were the emissions for motorcycles in 2007?
b Which form of transport produced the most emissions?
c Write down the year when emissions for railways were lowest.
d Write down the method of transport for which the emissions have dropped each year.

2 The following database gives information about the weather in a certain town during the first six months of the year.

	Max temp °C	Min temp °C	Air frost days	Sunshine hours	Rainfall mm	Days of rainfall ≥ 1 mm
January	6.4	1.2	10.7	44.3	101.9	15.3
February	6.9	1.3	9.6	72.0	73.4	11.3
March	8.8	2.5	6.3	107.9	78.3	14.1
April	11.4	3.5	3.8	155.1	50.7	10.6
May	15.0	6.1	1.0	214.8	55.0	10.0
June	17.1	9.0	0	197.7	67.9	10.7

a How many days of air frost were there in March?
b Write down the month that had the least number of days of rainfall.
c Which was the sunniest month?
d Which month had the greatest difference between maximum and minimum temperatures?

C

3 The database below is part of an agricultural survey of cattle in England in 2006, 2007 and 2008. The numbers of cattle are given in thousands.

		Year	
	2006	**2007**	**2008**
Female cattle			
Aged 2 years or more	2550	2531	2475
Total breeding herd (cattle that have calved)	2043	2027	1994
Beef	767	768	758
Dairy	1276	1259	1236
Other female cattle (not calved)	507	504	481
Beef	224	220	216
Dairy	293	284	265
Aged between 1 and 2 years	825	799	778
Beef	512	502	497
Dairy	313	297	282
Male cattle			
Aged 2 years or more	217	217	217
Aged between 1 and 2 years	619	583	578

a Write down the number of female cattle aged between one and two years in 2007.

b What do you notice about the numbers of male cattle aged two years or more throughout the three years?

c Were there more female beef or more female dairy cattle in 2008?

A03

d What conclusions can you draw about the trend in the numbers of cattle over the three years?

Chapter review

◉ **Statistics** is used to provide information. The statistical problem-solving process can be shown as a simple diagram:

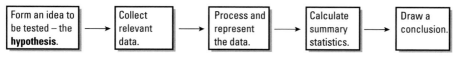

◉ **Primary data** is data you collect yourself.

◉ **Secondary data** is data that has been collected by others.

◉ **Qualitative data** can be described in words.

◉ **Quantitative data** are numerical observations.

◉ **Discrete data** can only take certain numerical values.

◉ **Continuous data** can take any numerical value.

◉ A **sample** is part of a **population** that is used to give information about the population as a whole in a sample **survey**. The sample must be **representative** of all the people or items being investigated, with each member of the population having an equal chance of being selected.

- A simple **random sample** is one where each person is given the same chance of being included.
- A **stratified sample** is one in which the population is split into groups called strata and a simple random sample is taken from each stratum. The number taken from each stratum is proportional to the size of the stratum.
- Data can be collected by **observation, experiment** or **data logging**.
- When collecting data by observation a **data collection sheet** is used.
- When dealing with continuous data you need to make sure the intervals do not overlap.
- If data is numerical, and widely **spread**, you can group the data into **class intervals**. These class intervals do not have to be the same size.
- A **questionnaire** is a list of questions designed to collect data.
- An **open question** is one that has no suggested answers.
- A **closed question** is one that has a set of answers to choose from.
- When designing questionnaires, it is important to ensure that possible answers are clear, do not overlap and cover all possibilities.
- A **two-way table** shows the frequency with which data falls into two different categories.
- **Biased** data is data that does not represent the population that it is taken from.
- A **database** is an organised collection of information.

Review exercise

1   James wants to find out how many text messages people send.
    He uses this question on a questionnaire.

    'How many text messages do you send?'

    1 to 10 ☐   11 to 20 ☐   21 to 30 ☐   more than 30 ☐

    a   Write down **two** things wrong with this question.

    James asks 10 students in his class to complete his questionnaire.

    b   Give **one** reason why this may not be a suitable sample.   *March 2009*

2   Poppy wants to find out how much time people use their computer for.
    She uses this questionnaire.

For how much time do you use your computer?	
0–1 hours ☐	3–4 hours ☐
1–2 hours ☐	4–5 hours ☐
2–3 hours ☐	5–6 hours ☐

    a   Write down **two** things that are wrong with this question.

    Poppy gives her questionnaire to all the students in her class. Her sample is biased.

    b   Give **one** reason why.   *Nov 2008*

D

C

A03

3 Naomi wants to find out how often adults go to the cinema.

She uses this question on a questionnaire.

> 'How many times do you go to the cinema?'
>
> ☐   ☐   ☐
>
> Not very often   Sometimes   A lot

a Write down **two** things wrong with this question.

b Design a better question for her questionnaire to find out how often adults go to the cinema.
   You should include some response boxes. *Nov 2008*

A03 * **4** Yolande wants to collect information about the number of e-mails the students in her class send.

Design a suitable question she could use on a questionnaire.

You must include some response boxes. *March 2008*

A03 **5** Melanie wants to find out how often people go to the cinema.

She gives a questionnaire to all the women leaving a cinema.

Her sample is biased.

Give **two** possible reasons why. *March 2008*

A03 * **6** Amberish is going to carry out a survey about zoo animals.

He decides to ask some people whether they prefer lions, tigers, elephants, monkeys or giraffes.

Design a data collection sheet that he can use to carry out his survey. *March 2006*

A02
A03 * **7** Angela asked 20 people in which country they spent their last holiday.

Here are their answers.

France	Spain	Italy	England
Spain	England	France	Spain
Italy	France	England	Spain
Spain	Italy	Spain	France
England	Spain	France	Italy

Design **and** complete a suitable data collection sheet that Angela could have used to show this information. *March 2004*

A03 **8** The manager of a Country Park asks the following two questions on a questionnaire.

'Do you go to the Country Park?' Sometimes ☐   Often ☐

'How old are you?' 0 to 10 years ☐   10 to 20 years ☐   Over 20 years ☐

a What is wrong with each of these questions?

b For both questions above, write a better version that the manager can use.

9 Write down, with reasons, whether or not each of the following is biased.

  a A call centre manager wants to know how easy it is to use the staff reference sheets when answering a call. He asks all the people working on the night shift.

  b A mobile phone company wants to find out what people think about their new pricing contract and randomly select 10% to ask.

  c A town council poses the question 'Do you agree that we are doing a good job in the area of recycling?'

10 The two-way table shows information about the number of students in a school.

	Year Group					Total
	7	**8**	**9**	**10**	**11**	
Boys	126	142	140	135	127	670
Girls	134	140	167	125	149	715
Total	260	282	307	260	276	1385

Robert carries out a survey of these students.

He uses a sample of 50 students stratified by gender and by year group.

Calculate the number of girls from Year 9 that are in his sample.   *June 2008*

11 The table shows the number of boys in each of four groups.

Group	A	B	C	D	Total
Number of boys	32	41	38	19	132

Jamie takes a sample of 40 boys stratified by group.

Calculate the number of boys from group B that should be in his sample.   *March 2008*

12 258 students each study one of three languages.

The table shows information about these students.

	Language studied		
	German	**French**	**Spanish**
Male	45	52	26
Female	25	48	62

A sample, stratified by the language studied and by gender, of 50 of the 258 students is taken.

  a Work out the number of male students studying Spanish in the sample.

  b Work out the number of female students in the sample.   *June 2009*

13

	Male	**Female**
First year	399	602
Second year	252	198

The table gives information about the numbers of students in the two years of a college course.

Anna wants to interview some of these students.

She takes a random sample of 70 students stratified by year and by gender.

Work out the number of students in the sample who are male and in the first year.   *Nov 2008*

ResultsPlus
Exam Question Report

82% of students answered this sort of question poorly.

A02

* 14 80 children went on a school trip.

They went to London or to York.

23 boys and 19 girls went to London.

14 boys went to York.

Draw and complete a suitable table of this information.

March 2009

In 1999, NASA spent $125 million on a space probe designed to orbit Mars. The mission ended in disaster after the probe steered too close to Mars and burned up whilst skimming the planet's thin atmosphere. Apparently, navigation commands to the probe's engines were provided in imperial rather then metric units even though NASA had been using metric units since at least 1990.

◎ Objectives

In this chapter you will:
◎ look at both the metric and imperial measurement systems and learn how to convert units between and within these systems
◎ consider compound measures and solve problems involving average speed and density.

◇ Before you start

You need to:
◎ know how to measure and draw lines to the nearest mm
◎ be able to make sensible estimates for a range of measures
◎ know that measurements given to the nearest whole unit may be inaccurate by up to one half in either direction
◎ know metric and imperial units of length, mass and volume and have an idea of their relative sizes.

7.1 Converting between units of measure

◎ Objectives

- You know the relationship between metric units and are able to convert between units in the metric system.
- You know the approximate metric equivalents of imperial units and are able to convert between metric units and imperial units.
- You are able to convert between units in the imperial system when you are given the relationship between the imperial units.

❓ Why do this?

It is important to be able to change from one unit to another when you are cooking as some measurements may be given in grams and some in kilograms.

⬆ Get Ready

1. The following sentences do not make sense because the wrong unit has been used. Rewrite each sentence using the correct unit.

   a The weight of a packet of biscuits is 150 kg.

   b The thickness of a book is 3 m.

   c The height of a giraffe is 5 km.

   d A teacup can hold 300 l of tea.

🕐 Key Points

- To **convert** from one metric unit to another metric unit it is necessary to know the following facts.

Length	Weight	Capacity/Volume
10 mm = 1 cm	1000 mg = 1 g	100 cl = 1 litre
100 cm = 1 m	1000 g = 1 kg	1000 ml = 1 litre
1000 mm = 1 m	1000 kg = 1 tonne	1000 cm³ = 1 litre
1000 m = 1 km		1000 l = 1 m³

You then only need to multiply or divide by 10, 100 or 1000 in order to convert between the different metric units.

- When you convert from a smaller unit to a larger unit, you need to divide.
- When you convert from a larger unit to a smaller unit, you need to multiply.

For example, for lengths:

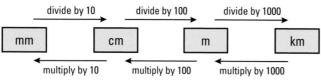

ResultsPlus

Watch Out!

The values in the table are not exact but they are the rough equivalents that need to be used in examinations.

- To convert between a metric unit and an imperial unit it is necessary to know the facts in this table.

Metric unit	Imperial unit
1 kg	2.2 pounds
1 litre (l)	$1\frac{3}{4}$ pints = 1.75 pints
4.5 l	1 gallon
8 km	5 miles
30 cm	1 foot
2.54 cm	1 inch

> **Example 1** a Convert 12 m into centimetres.
>
> b Convert 2670 g into kilograms.

a $12\,m = 12 \times 100\,cm = 1200\,cm$ ⟵ Centimetres are smaller than metres so there are more of them. As 100 cm = 1 m, multiply by 100.

b $2670\,g = 2760 \div 1000 = 2.76\,kg$ ⟵ Kilograms are larger than grams so there are fewer of them. As 1000 g = 1 kg, divide by 1000.

Exercise 7A

Questions in this chapter are targeted at the grades indicated.

1 Convert these lengths to centimetres.

a 6 m b 210 mm c 5.1 m d 0.84 m

e 59 mm f 483 mm g 3 km h 0.067 km

2 Convert these weights to kilograms.

a 3 tonnes b 8.2 tonnes c 6000 g d 900 g

e 430 g f 4700 g

3 Convert these volumes to litres.

a 2000 ml b 700 cl c 5900 ml d 45 000 ml

4 A bottle of lemonade contains 70 cl.

How many litres of lemonade are there in 10 of these bottles?

> **Example 2** a Convert 8 gallons into litres. b Convert 28 km into miles.

a $8\,gallons = 8 \times 4.5\,litres = 36\,litres$ ⟵ There are more litres than gallons so multiply.

b $8\,km = 5\,miles$
   $1\,km = 5 \div 8\,miles$
   $28\,km = 28 \times 5 \div 8\,miles$
   $= 140 \div 8 = 17.5\,miles$

⟵ There are fewer miles than km as a mile is longer than a km. Find 1 km. Multiply and then divide.

Exercise 7B

1 Convert 4 kg to pounds.

2 Convert 110 pounds to kilograms.

3 Convert 7 pints to litres.

4 Convert 36 litres to gallons.

5 Convert 10 litres to pints.

6 Convert 12 feet to centimetres.

7 Convert 96 km to miles.

8 Convert 60 miles to kilometres.

A02 A03 **9** The price of petrol is 130p per litre.
Work out the price of the petrol per gallon.

 Example 3 There are 12 inches in a foot.
a Convert 7 feet into inches.
b Convert 108 inches into feet.

a 7 feet = 7 × 12 = 84 inches ← A foot is longer than an inch so there are more inches than feet. So to change from feet to inches you multiply.

 ResultsPlus
Exam Tip

When converting between imperial units you will not be expected to know the relationship between the units as the conversions will be given.

b 108 inches = 108 ÷ 12 = 9 feet

To convert from inches to feet you divide.

 Exercise 7C

1 There are 16 ounces in a pound.
a Convert 5 pounds to ounces.
b Convert 96 ounces to pounds.

A02 **2** There are 12 inches in a foot and 3 feet in a yard.
Work out the number of inches in 5 yards.

A02 **3** There are 14 pounds in a stone. John's weight is 9 stones and 6 pounds.
a Work out John's weight in pounds.
b Work out John's weight in kilograms.

7.2 Compound measures

⊙ Objective

- ⊙ You can solve problems with compound measures, giving the correct units.

⊘ Why do this?

Compound measures are used when we want to see how a quantity changes in relation to another quantity, such as by how much the temperature of water increases each second when heated.

⊕ Get Ready

1. How many minutes are there in 1 hour?
2. What fraction of an hour is 15 minutes?
3. What fraction of an hour is 36 minutes? Give your answer as a decimal.
4. How many minutes is 0.3 hours?
5. Write 5.7 hours in hours and minutes.
6. Write 462 minutes in hours and minutes.

◉ Key Points

- ⊙ A **compound measure** is a measure which involves two units such as km per hour, litres per second or grams per cm³. Compound measures are often a measure of a **rate of change**.
  For example, a tank is filled with 3000 litres of water in 20 minutes. When the rate of filling is constant, this rate means that in 1 minute, $3000 \div 20 = 150$ litres of water would be added to the tank. If the rate is not constant, the average rate of filling the tank is 150 litres in 1 minute.
- ⊙ In a compound unit, the word 'per' means 'each' or 'for every'. So, the rate of filling the tank is 150 litres per minute or 150 litres/minute. The '/' is like a division sign showing that the rate is the amount of water divided by the time taken.

Example 4

Petrol is leaking from a tank at a rate of 5 litres/hour.

a Work out how much petrol leaks from the tank in i 30 minutes; ii 4 hours.

Initially there are 100 litres of petrol in the tank.

b Work out how long it takes for all this petrol to leak from the tank.

a i 30 minutes $= \frac{1}{2}$ hour ⟵ | 5 litres/hour means in 1 hour 5 litres of petrol leak from the tank.

Amount of petrol $= 5 \times \frac{1}{2} = 2.5$ litres. ⟵ | Amount = rate × time

ii Amount of petrol $= 5 \times 4 = 20$ litres.

b Time taken $= 100 \div 5 = 20$ hours. ⟵ | Time = amount ÷ rate

ResultsPlus
Exam Tip

The units of a compound measure will tell you what to do, so km/l will mean distance ÷ volume.

⚙ Exercise 7D

1 A car travels 260 km and uses 20 litres of petrol.
   a Work out the average rate of petrol usage. Give your answer in km/litre.
   b Estimate the amount of petrol that would be used when the car has travelled 78 km.

A03 C

C
A02

2 A line is turning at a rate of 12° per second.

 a Work out how many degrees the line turns through in 20 seconds.

 b How long does it take for the line to make one complete turn?

A02
A03

3 Water is flowing into a tank. In 5 minutes, 300 litres of water flows in.

 a Work out the average rate of flow of water into the tank. Give the units of your answer.

 b There were 90 litres of water in the tank immediately before the water started to flow in. When full, the tank holds 1200 litres of water. How long will it take for the tank to fill with water? Give your answer in minutes and seconds.

A03

4 On a long journey, a car travels 16 km per litre of petrol. Work out how many litres of petrol the car uses per kilometre.

7.3 Speed

◉ Objective

○ You can solve problems with average speed.

⟡ Why do this?

Speed is a compound measure that we use all the time, from speed limits for cars, to world record breaking running speeds.

◈ Get Ready

1. Work out the average rate of petrol usage if a car travels:

 a 252 km and uses 14 litres of petrol
 b 156 km and uses 24 litres of petrol
 c 84.48 km and uses 6.4 litres of petrol.

◉ Key Points

◉ Speed is a compound measure because it involves a unit of length and a unit of time, for example kilometres per hour, miles per hour or metres per second. We write kilometres per hour as km/h; the '/' is a sort of division sign showing that speed is distance divided by time.

◉ **Average speed** $= \dfrac{\text{total distance travelled}}{\text{total time taken}}$

 If the car travels at an average speed of 30 km/h, the car travels 30 km in 1 hour

$$30 \times 2 = 60 \text{ km in 2 hours}$$
$$30 \times 3 = 90 \text{ km in 3 hours}$$
$$\text{and so on.}$$

◉ Distance $=$ average speed $\times$ time

 The time the car takes to travel 90 km at 30 km/h is $\frac{90}{3} = 3$ hours.

 Therefore: time $= \dfrac{\text{distance}}{\text{average speed}}$

◉ The following diagram is a useful way to remember these results: D stands for distance, S stands for average speed and T stands for time.

$$D = S \times T$$
$$S = \frac{D}{T}$$
$$T = \frac{D}{S}$$

Example 5

The distance from Cardiff to Leeds is 335 km. Rhys drives from Cardiff to Leeds in 6 hours 15 minutes. Work out his average speed for this journey.

Time taken = 6.25 hours ← Average speed = $\frac{\text{total distance travelled}}{\text{total time taken}}$
Time must be in hours.

ResultsPlus
Watch Out!

It is important to be careful with time; it is best to use decimals and remember that there are 60 minutes in 1 hour.

Average speed = $\frac{335}{6.25}$ ← 15 minutes = $\frac{15}{60}$ = 0.25 hours
The distance is in km and the time is in hours so the speed is in km/h.

= 53.6 km/h.

Example 6

Michael decides to go for a cycle ride. He rides a distance of 80 km at an average speed of 24 km/h. Work out how long Michael's ride takes.

Time = $\frac{80}{24}$ = 3.3333... h ← Time = $\frac{\text{distance}}{\text{average speed}}$
Speed is in km/h and distance is in km so the time is in hours.

0.3333... × 60 = 20 ← 3.3333... h = 3 h + 0.3333... h
To change from hours to minutes, multiply by 60.

Time = 3 h 20 minutes

Exercise 7E

1. Paul takes part in a sponsored hike. He walks 18 km in $4\frac{1}{4}$ hours. What is his average speed? Give your answer correct to 3 significant figures.

2. Tim left his home at 11 am and went for a 20 km run. He arrived back at his home at 1 pm. Work out Tim's average speed.

3. A horse runs 12 km at an average speed of 10 km/h. How long, in hours and minutes, does this take?

4. Change a speed of 85 m/s into km/h.

5. In the 2008 Olympics, the men's 100 m race was won in a time of 9.69 s and the men's 200 m race was won in a time of 19.30 s. Which race was won with the faster average speed? You must give a reason for your answer.

D

C

7.4 Density

⊙ Objective

○ You can solve problems with density.

⟨?⟩ Why do this?

A submarine's ability to submerge and surface depends on its total density. It changes this by adjusting the amount of water in its ballast tanks.

⬦ Get Ready

1. Work out these calculations, giving your answers to one decimal place.

   a $50 \div 6$ **b** $4 \div 0.3$ **c** $400 \div 15.4$ **d** $347.1 \div 27$

Key Points

◉ Density is also a compound measure. To solve density-related problems, we can use the following equations:

$$\textbf{density} = \frac{\textbf{mass}}{\textbf{volume}}$$

$$\textbf{mass} = \textbf{density} \times \textbf{volume}$$

$$\textbf{volume} = \frac{\textbf{mass}}{\textbf{density}}$$

◉ The diagram below is a useful way to remember these equations: M stands for mass, D stands for density and V stands for volume.

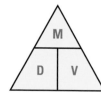

$$M = D \times V$$
$$D = \frac{M}{V}$$
$$V = \frac{M}{D}$$

◉ When the mass is measured in kilograms and the volume in cubic metres or m³, then density is measured in kg per m³ or kg/m³. Density can also be measured in g/cm³.

Example 7

A piece of silver has a mass of 42 g and a volume of 4 cm³. Work out the density of silver.

Density $= \dfrac{42}{4}$
$= 10.5 \, g/cm^3$ ⟵

Density $= \dfrac{mass}{volume}$
Divide the mass by the volume.
As the mass is in g and the volume is in cm³, the density is in g/cm³.

 Results**Plus**
Exam Tip

Usually in GCSE mathematics the term weight is used as it is easier to understand. However in problems involving density, the correct term, mass, is used.

Example 8   The density of steel is 7700 kg/m³.

  a A steel bar has a volume of 2.5 m³. Work out the mass of the bar.

  b A block of steel has a mass of 1540 kg. Work out the volume of the block.

a Mass = 7700 × 2.5   ←———   Mass = density × volume
Multiply the density by the volume.

= 19 250 kg   ←———   As the density is in kg/m³ and the volume is in m³, the mass is in kg.

b Volume = $\dfrac{1540}{7700}$ ←———   Volume = $\dfrac{mass}{density}$
Divide the mass by the density.

= 0.2 m³   ←———   As the mass is in kg and the density is in kg/m³ the volume is in m³.

Exercise 7F

1   A slab of concrete has a volume of 60 cm³ and a mass of 150 g. Work out the density of the concrete.

2   Gold has a density of 19.3 g/cm³. The gold in a ring has a mass of 15 g. Work out the volume of gold in the ring.

3   14.7 g of sulphur has a volume of 7.5 cm³. Work out the density of sulphur.

4   The density of aluminium is 2590 kg/m³. The density of lead is 11 400 kg/m³.
A block of aluminium has a volume of 0.5 m³. A block of lead has a volume of 0.1 m³.
Which of the two blocks has the greater mass and by how many kilograms?

C

A02
A03   B

Chapter review

◉ To **convert** from one metric unit to another metric unit it is necessary to know the following facts.

Length	Weight	Capacity/Volume
10 mm = 1 cm	1000 mg = 1 g	100 c*l* = 1 litre
100 cm = 1 m	1000 g = 1 kg	1000 m*l* = 1 litre
1000 mm = 1 m	1000 kg = 1 tonne	1000 cm³ = 1 litre
1000 m = 1 km		1000 *l* = 1 m³

You then only need to multiply or divide by 10, 100 or 1000 in order to convert between the different metric units.

◉ When you convert from a smaller unit to a larger unit, you need to divide.

◉ When you convert from a larger unit to a smaller unit, you need to multiply.

◉ To convert between a metric unit and an imperial unit it is necessary to know the facts in the following table.

Metric unit	Imperial unit
1 kg	2.2 pounds
1 litre (l)	$1\frac{3}{4}$ pints = 1.75 pints
4.5 l	1 gallon
8 km	5 miles
30 cm	1 foot
2.54 cm	1 inch

◉ A **compound measure** is a measure which involves two units such as km per hour, litres per second or grams per cm³. Compound measures are often a measure of a **rate of change**.

◉ In a compound unit, the word 'per' means 'each' or 'for every'. We write litres per minute as litres/minute; the '/' is like a division sign showing that the rate is the amount of liquid divided by the time taken.

◉ Speed is a compound measure because it involves a unit of length and a unit of time, for example kilometres per hour, miles per hour or metres per second.

◉ The following diagram is a useful way to remember the relationships between speed, distance and time: D stands for distance, S stands for average speed and T stands for time.

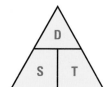

$$D = S \times T$$
$$S = \frac{D}{T}$$
$$T = \frac{D}{S}$$

◉ **Density** is also a compound measure. The following diagram is a useful way to remember the relationships between **mass**, density and **volume**: M stands for mass, D stands for density and V stands for volume.

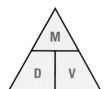

$$M = D \times V$$
$$D = \frac{M}{V}$$
$$V = \frac{M}{D}$$

◉ When the mass is measured in kilograms and the volume in cubic metres or m³, then density is measured in kg per m³ or kg/m³. Density can also be measured in g/cm³.

✳ Review exercise

1   **a**   Complete the table by writing a sensible **metric** unit for each measurement.

The length of the river Nile	6700 kilometres
The height of the world's tallest tree	110 ....
The weight of a chicken's egg	70 ....
The amount of petrol in a full petrol tank of a car	40 ....

   b   Convert 4 metres to centimetres.

   c   Convert 1500 grams to kilograms.

June 2008

2 Shalim says 1.5 km is less than 1400 m.

Is he right?

Explain your answer. *June 2007*

3 Stuart drives 180 km in 2 hours 15 minutes.

Work out Stuart's average speed. *Nov 2008*

4 The distance from London to New York is 3456 miles.

A plane takes 8 hours to fly from London to New York.

Work out the average speed of the plane. *June 2008*

5 John travelled 30 km in 1.5 hours.

Kamala travelled 42 km in 2 hours.

Who had the greater average speed?

You must show your working. *June 2009*

6 There are 40 litres of water in a barrel.

The water flows out of the barrel at a rate of 125 millilitres per second.

1 litre = 1000 millilitres

Work out the time it takes for the barrel to empty completely. *June 2009*

7 The density of juice is 4 grams per cm³.

The density of water is 1 gram per cm³.

315 cm³ of drink is made by mixing 15 cm³ of juice with 300 cm³ of water.

Work out the density of the drink.

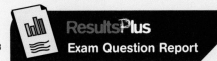

ResultsPlus
Exam Question Report

86% of students answered this question poorly because they could not remember the formula for calculating density.

June 2009

8 The volume of a gold bar is 100 cm³.

The density of gold is 19.3 grams per cm³.

Work out the mass of the gold bar. *Nov 2008*

8 CONGRUENCE, SYMMETRY AND SIMILARITY

The photo shows the Pyramide du Louvre in Paris. There are actually five pyramids, the large one, three smaller ones and an inverted pyramid which provides the entrance to the Louvre museum. The larger pyramid is made up of 603 diamond-shaped panes of glass with 70 triangular-shaped panes along the base of the pyramid.

◎ Objectives

In this chapter you will:
- prove two triangles are congruent
- learn about line symmetry and rotational symmetry
- learn about some special types of 2D shapes
- recognise similar shapes and use scale factors to find missing sides in similar triangles
- formally prove that triangles are similar.

◈ Before you start

You need to:
- know the angle properties of triangles and quadrilaterals
- know what a vertex and a diagonal of a shape are.

8.1 Congruent triangles

⊙ Objective

⊙ You will understand how to prove that two triangles are congruent.

❓ Why do this?

Designers, engineers and map makers often use scale drawings and plans. Using congruent and similar triangles enables them to find measurements for inaccessible lengths and angles.

◈ Get Ready

1. If two triangles have the same angles, are the triangles the same?
2. Given the lengths of all three sides, is it possible to draw two different triangles?
3. Given lengths of two sides and the size of the included angle, is it possible to draw two different triangles?

◷ Key Points

⊙ Two triangles are **congruent** if they have exactly the same shape and size.
⊙ For two triangles to be congruent one of the following **conditions of congruence** must be true.
  ⊚ The three sides of each triangle are equal (SSS).
  ⊚ Two sides and the **included angle** are equal (SAS).
  ⊚ Two angles and a corresponding side are equal (AAS).
  ⊚ Each triangle contains a right angle, and the hypotenuses and another side are equal (RHS).

 Example 1

ABCD is a quadrilateral.
AD = BC.
AD is parallel to BC.
Prove that triangle ADC is congruent to triangle ABC.

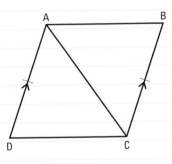

 **Results**Plus
Watch Out!

The only properties that can be used to prove congruence are those given in the question.

AD = BC (given)
Angle DAC = angle ACB (alternate angles) ← Each statement for a congruence proof must be justified.
AC is common to both triangles.
Hence triangle ADC is congruent to triangle ABC
(two sides and the included angle).

Exercise 8A

Questions in this chapter are targeted at the grades indicated.

A
A03

* **1** Prove that triangles PQS and QRS are congruent.

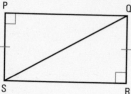

A03

* **2** Prove that triangles XYZ and XVW are congruent.
Hence prove that X is the midpoint of YW.

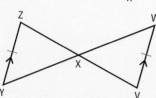

* **3** PQR is an isosceles triangle. S and T are points on QR.
PQ = PR, QS = TR.
Prove that triangle PST is isosceles.

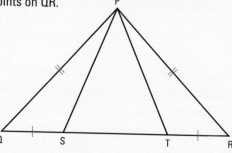

* **4** LMN is an isosceles triangle with LM = LN. Use congruent triangles to prove that the line from L which cuts the base MN of the triangle at right angles also bisects the base.

* **5** ABC is a triangle. D is the midpoint of AB. The line through D drawn parallel to the side BC meets the side AC at E. A line through D drawn parallel to the side AC meets the side BC at F.
Prove that triangles ADE and DBF are congruent.

8.2 Symmetry in 2D shapes

Objectives

- You can recognise line symmetry in 2D shapes.
- You can draw lines of symmetry on 2D shapes.
- You can recognise rotational symmetry in 2D shapes.
- You can find the order of rotational symmetry of a 2D shape.
- You can draw shapes with given line symmetry and/or rotational symmetry.

Why do this?

There are examples of 2D symmetry in the man-made and natural world, such as wheels, flowers and butterflies.

Get Ready

1. Trace this star.
   Fold your tracing along the dotted line. What do you notice?
   Place your tracing over the star and turn the tracing paper clockwise.
   Keep turning the tracing paper until you get back to the starting position.
   What do you notice?

line symmetry   line of symmetry

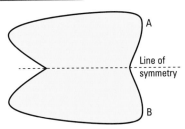

Key Points

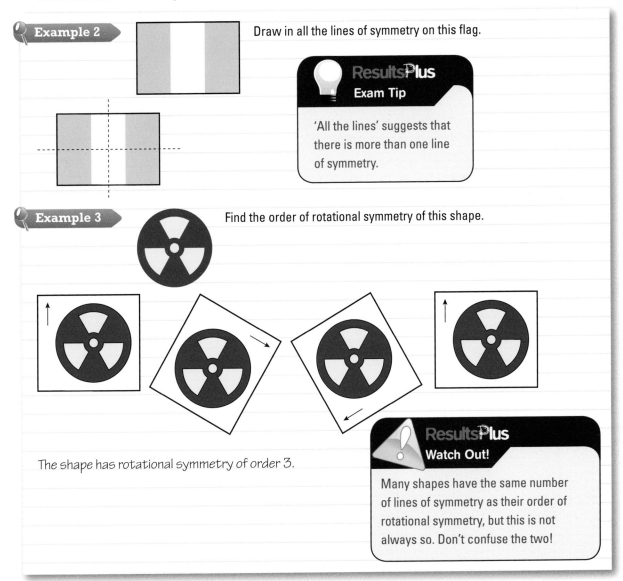

- A shape has **line symmetry** if it can be folded so that one part of the shape fits exactly on top of the other part.

- Every point of the shape on one side of the **line of symmetry** has a corresponding point on the **mirror image** the other side of the line. Notice that the point A and its corresponding point B are the same distance from the line of symmetry.

- If a mirror were placed on the line of symmetry, the shape would look the same. This is why line symmetry is sometimes called **reflection symmetry** and the line of symmetry is sometimes called the **mirror line**.

- A shape has **rotation symmetry** if a tracing of the shape fits exactly on top of the shape in more than one position when it is rotated.

- A tracing of a shape with rotation symmetry will fit exactly on top of the shape when turned through less than a complete turn.

- The number of times that the tracing fits exactly on top of the shape is called the **order of rotational symmetry**.

- Some two-dimensional shapes do not have any symmetry.

Example 2   Draw in all the lines of symmetry on this flag.

ResultsPlus
Exam Tip

'All the lines' suggests that there is more than one line of symmetry.

Example 3   Find the order of rotational symmetry of this shape.

The shape has rotational symmetry of order 3.

ResultsPlus
Watch Out!

Many shapes have the same number of lines of symmetry as their order of rotational symmetry, but this is not always so. Don't confuse the two!

Example 4

Copy and complete the drawing of the shape so that it has line symmetry.

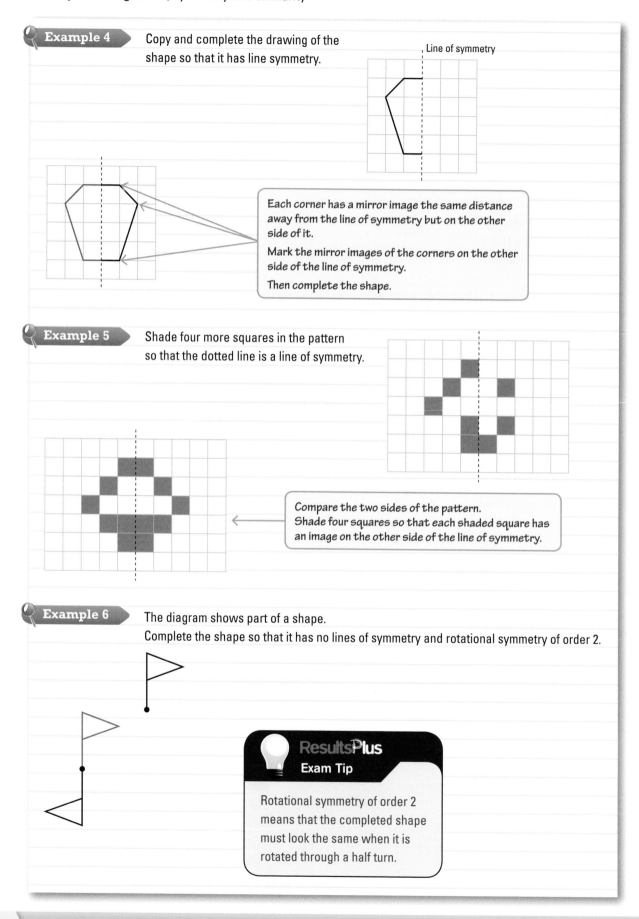

Line of symmetry

Each corner has a mirror image the same distance away from the line of symmetry but on the other side of it.

Mark the mirror images of the corners on the other side of the line of symmetry.

Then complete the shape.

Example 5

Shade four more squares in the pattern so that the dotted line is a line of symmetry.

Compare the two sides of the pattern.
Shade four squares so that each shaded square has an image on the other side of the line of symmetry.

Example 6

The diagram shows part of a shape.
Complete the shape so that it has no lines of symmetry and rotational symmetry of order 2.

ResultsPlus
Exam Tip

Rotational symmetry of order 2 means that the completed shape must look the same when it is rotated through a half turn.

Exercise 8B

1. For each shape, write down if it has line symmetry or not. If it has symmetry, copy the diagram and draw in all the lines.

   a   b 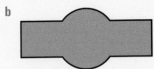   c

   d   e   f

2. Using tracing paper if necessary, state which of the following shapes have rotational symmetry and which do not have rotational symmetry.
   For the shapes that have rotational symmetry, write down the order of the rotational symmetry.

   a   b   c

   d   e   f

3. a Copy and complete this shape so that it has line symmetry.
   b Write down the name of the complete shape.

   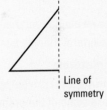

   Line of symmetry

4. Each diagram shows an incomplete pattern.
   For part **a**, copy the diagram and shade six more squares so that both dotted lines are lines of symmetry of the complete pattern. For part **b**, shade three more squares so that the complete pattern has rotational symmetry of order 4.

   a 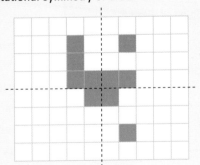   b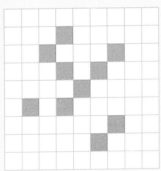

5 **a** Draw a shape that has two lines of symmetry and rotational symmetry of order 2.

  b Draw a shape with one line of symmetry and no rotational symmetry.

  c Draw a shape that has no lines of symmetry and rotational symmetry of order 4.

8.3 Symmetry of special shapes

◎ Objectives

- You know the symmetries of special triangles.
- You can recognise and name special quadrilaterals.
- You know the properties of special quadrilaterals.
- You know the symmetries of special quadrilaterals.
- You know the symmetries of regular polygons.

? Why do this?

Many architectural designs are symmetrical in some way. The Taj Mahal, the Pyramids and the Greek Parthenon have impressive and beautiful uses of symmetry.

⟡ Get Ready

1. a What is **i** an isosceles triangle **ii** an equilateral triangle?

  b Is an equilateral triangle an isosceles triangle?

2. What is a quadrilateral?

Key Points

◉ **Triangles**

A **triangle** is a polygon with three sides.

Here is an isosceles triangle. It has two sides the same length.

An isosceles triangle has one line of symmetry.

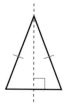

Here is an equilateral triangle. All its sides are the same length.

An equilateral triangle has three lines of symmetry and rotational symmetry of order 3.

◉ **Quadrilaterals**

A **quadrilateral** is a polygon with four sides. Some quadrilaterals have special names.

Here are some of the properties of special quadrilaterals.

Square

All sides equal in length.

All angles are 90°.

4 lines of symmetry and rotational symmetry of order 4.

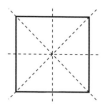

Rectangle

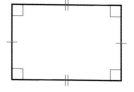

Opposite sides equal in length.

All angles are 90°.

Rhombus

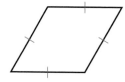

All sides equal in length.

Opposite sides parallel.

Opposite angles equal.

2 lines of symmetry and rotational symmetry of order 4.

Parallelogram

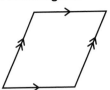

Opposite sides equal in length and parallel.
Opposite angles equal.
No lines of symmetry and rotational symmetry of order 2.

Trapezium

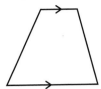

One pair of parallel sides.
No lines of symmetry and no rotational symmetry.

Isosceles trapezium

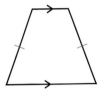

One pair of parallel sides.
Non-parallel sides equal in length.
One line of symmetry and no rotational symmetry.

Kite

Two pairs of **adjacent** sides equal in length.
(Adjacent means 'next to'.)
One line of symmetry and no rotational symmetry.

Exercise 8C

1   a   On squared paper, draw a right-angled triangle that has one line of symmetry.
Draw the line of symmetry on your triangle.

b   Write down what is special about this right-angled triangle.

A02 A03

2   Janine says, 'I am thinking of a quadrilateral. It has opposite sides that are parallel.'

a   Is there enough information to know what the quadrilateral is? Give reasons for your answer.

Janine now says, 'It has rotational symmetry of order 2.'

b   Is there now enough information to know what the quadrilateral is? Give reasons for your answer.

Janine now says, 'It has two lines of symmetry.'

c   Is there now enough information to know what the quadrilateral is? Give reasons for your answer.

Janine now says, 'It has sides that are not all the same length.'

d   What quadrilateral is Janine thinking of?

A03

3   Draw a non-regular polygon which has a line of symmetry.

A02 A03

4   Draw a non-regular polygon which has rotational symmetry.
State the order of rotational symmetry of your polygon.

A02 A03

8.4 Recognising similar shapes

⊙ Objectives

○ You can recognise similar shapes.
○ You can find missing sides using facts you know about similar shapes.

⌕ Why do this?

Similar shapes allow us to calculate missing dimensions from plans which may be difficult to measure on the real objects.

⬥ Get Ready

1. Which of these triangles are congruent?

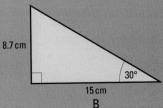

5.8 cm   8.7 cm   60°   10 cm

30°   10 cm   30°   15 cm

A   B   C

Key Points

◎ Shapes are **similar** if one shape is an enlargement of the other.

⊙ The corresponding angles are equal.

⊙ The corresponding sides are all in the same ratio.

A03

Example 7 Show that the parallelogram ABCD is not similar to parallelogram EFGH.

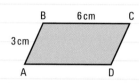 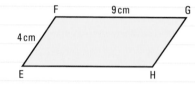

$$\frac{AB}{EF} = \frac{3}{4} = 0.75$$ ← Work out the ratios of the corresponding sides.

$$\frac{BC}{FG} = \frac{6}{9} = 0.66667$$

The lengths of the corresponding sides are not in the same proportion so the parallelograms are not similar.

Exercise 8D

D

1 State which of the pairs of shapes are similar.

a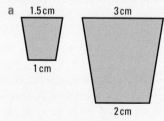

b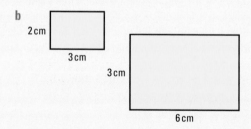

B

2 Show that pentagon ABCDE is similar to pentagon FGHIJ.

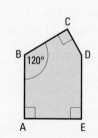 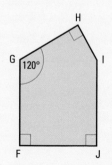

Example 8 These two rectangles are similar. Find the length L of the larger rectangle.

4 cm

2 cm

L cm

3 cm

The widths of these rectangles are in the ratio $2:3$.

→ Consider the ratio of the widths of the rectangles.

The lengths must be in the same ratio.

→ The rectangles are similar so the lengths must be in the same ratio.

$$\frac{\text{small}}{\text{large}} = \frac{2}{3} = \frac{4}{L}$$

$2L = 12$

$L = 6$ cm

ResultsPlus

Exam Tip

Make sure you keep corresponding sides together by stating which rectangle they come from.

Exercise 8E

1 A large packet of breakfast cereal has height 35 cm and width 21 cm.
A small packet of cereal is similar to the large packet but has a height of 25 cm.
Find the width of the small packet.

A02
A03

B

2 The diagram shows a design for a metal part.
The sizes of the plan are marked on diagram A.
Diagram B is marked with the actual sizes.
Calculate the value of:

a x

b y.

A02
A03

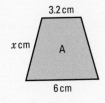

3.2 cm

x cm   A

6 cm

y m

0.72 m   B

0.84 m

3 These cylinders are similar. The height of the smaller cylinder is 5 cm.
Find the height of the larger cylinder.

A

6 cm

2 cm

8.5 Similar triangles

◎ Objectives

○ You can use scale factors to find missing sides in similar triangles.

○ You can formally prove that triangles are similar.

⊘ Why do this?

Using the fact that triangles are similar can help us to measure lengths and distances which we cannot measure practically.

⬙ Get Ready

1. Copy these diagrams and mark the pairs of corresponding angles.

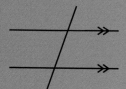 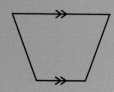

🔍 Key Points

◎ Two triangles are similar if any of the following is true.

  ◎ The corresponding angles are equal.

  ◎ The corresponding sides are in the same ratio.

  ◎ They have one angle equal and the adjacent sides are in the same ratio.

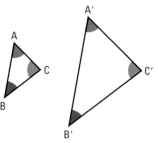

AB and A'B', AC and A'C', and BC and B'C' are corresponding sides.

$$\frac{A'B'}{AB} = \frac{A'C'}{AC} = \frac{B'C'}{BC}$$

A03

🔍 **Example 9**   Show that triangle ABC is not similar to triangle DEF.

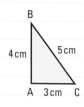

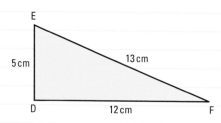

$$\frac{AB}{DE} = \frac{4}{5} = 0.8$$

$$\frac{AC}{DF} = \frac{3}{12} = 0.25$$

> 💡 **ResultsPlus**
> **Exam Tip**
>
> Make sure you look carefully at the parallel lines in a diagram, as they can give you a lot of information about the angles.

The lengths of the corresponding sides are not in the same proportion so the triangles are not similar.

Example 10 ABC is a triangle.

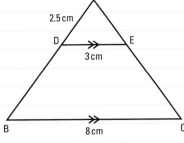

DE is parallel to BC.

a Show that triangle ABC is similar to triangle ADE.

b Find the length of BD.

a ∠ADE = ∠ABC (corresponding angles).

∠AED = ∠ACB (corresponding angles).

∠DAE = ∠BAC (common to both triangles).

> Give reasons from what you know about parallel lines.

All angles are equal so triangle ABC is similar to triangle ADE.

b $\dfrac{BC}{DE} = \dfrac{8}{3} = \dfrac{AB}{2.5}$

> The corresponding sides are in the same ratio.

$3 \times AB = 8 \times 2.5$

$3AB = 20$

$AB = 6.67\,cm$

$BD = 6.67 - 2.5$

> BD is only part of the side of the triangle.

$= 4.17\,cm$

Exercise 8F

1 For each pair of similar triangles:

  i name the three pairs of corresponding sides

  ii state which pairs of angles are equal.

a 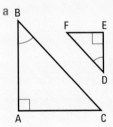 b c 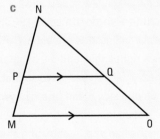 d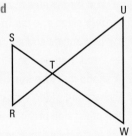

2 Triangle ABC is similar to triangle DEF.

∠ABC = ∠DEF

Calculate the length of:

a EF

b FD.

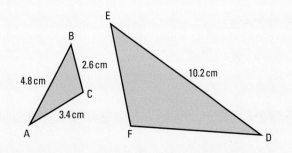

C

A03 A

3 The diagram shows triangle ABC which has a line DE drawn across it.

∠ACE = ∠DEB

a Prove that triangle ABC is similar to triangle DBE.

b Calculate the length of AB.

c Calculate the length of AD.

d Calculate the length of EC.

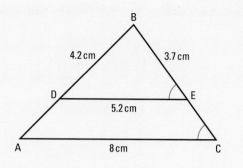

4 The diagram shows triangle ABC which has a line DE drawn across it. ∠CAD = ∠BDE

a Prove that triangle ACB is similar to triangle DEB.

b Calculate the length of DE.

c Calculate the length of BC.

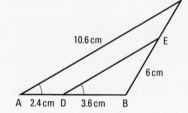

5 In the diagram AB is parallel to CD.

a Prove that triangle ABM is similar to triangle CDM.

b AC has length 20 cm.

   Calculate the lengths of:

   i AM ii MC.

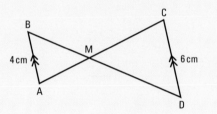

Chapter review

◉ Two triangles are **congruent** if they have exactly the same shape and size.

◉ For two triangles to be congruent one of the following **conditions of congruence** must be true.

  ◉ The three sides of each triangle are equal (SSS).

  ◉ Two sides and the **included angle** are equal (SAS).

  ◉ Two angles and a corresponding side are equal (AAS).

  ◉ Each triangle contains a right angle, and the hypotenuses and another side are equal (RHS).

◉ A shape has **line symmetry** if it can be folded so that one part of the shape fits exactly on top of the other part.

◉ Every point of the shape on one side of the **line of symmetry** has a corresponding point on the **mirror image** the other side of the line. Notice that corresponding points are the same distance from the line of symmetry.

◉ If a mirror were placed on the line of symmetry of a shape, the shape would look the same in the mirror. This is why line symmetry is sometimes called **reflection symmetry** and the line of symmetry is sometimes called the **mirror line**.

◉ A shape has **rotation symmetry** if a tracing of the shape fits exactly on top of the shape in more than one position when it is rotated.

◉ A tracing of a shape with rotation symmetry will fit exactly on top of the shape when turned through less than a complete turn.

◉ The number of times that the tracing fits exactly on top of the shape is called the **order of rotational symmetry**.

- Some shapes do not have any symmetry.
- Shapes are **similar** if one shape is an enlargement of the other.
  - The corresponding angles are equal.
  - The corresponding sides are all in the same ratio.
- Two triangles are similar if any of the following is true.
  - The corresponding angles are equal.
  - The corresponding sides are in the same ratio.
  - They have one angle equal and the adjacent sides are in the same ratio.

Review exercise

1 **a** On the diagram below, shade **one** square so that the shape has exactly **one** line of symmetry.

b On the diagram below, shade **one** square so that the shape has rotational symmetry of order 2.

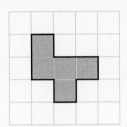

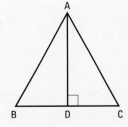

Nov 2008

2 Which of these triangles are similar?

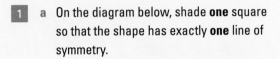

3 Which of these rectangles are similar?

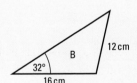

4 ABC is an equilateral triangle.
D lies on BC. AD is perpendicular to BC.
a Prove that triangle ADC is congruent to triangle ADB.
b Hence, prove that $BD = \frac{1}{2}BC$.

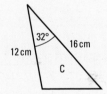

June 2009

D

A

129

A

5 In the diagram, AB = BC = CD = DA.
Prove that triangle ADB is congruent to triangle CDB.

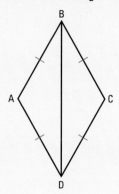

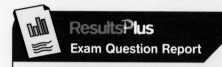

92% of students answered this sort of question poorly because they did not justify their answers or prove the conditions for congruency.

Nov 2008

6 AB is parallel to DE.
ACE and BCD are straight lines.
AB = 6 cm
AC = 8 cm
CD = 13.5 cm
DE = 9 cm

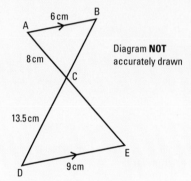

Diagram **NOT** accurately drawn

a Work out the length of CE.
b Work out the length of BC.

Nov 2005

7 Parallelogram P is similar to parallelogram Q.

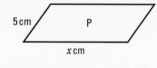

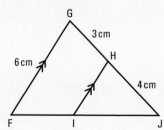

Calculate the value of x.

8 In triangle FGJ, a line IH is drawn parallel to FG.

a Prove that triangle HIJ is similar to triangle GFJ.
b Calculate the length of HI.

9 BE is parallel to CD.
AB = 9 cm, BC = 3 cm, CD = 7 cm, AE = 6 cm.

a Calculate the length of ED.

b Calculate the length of BE.

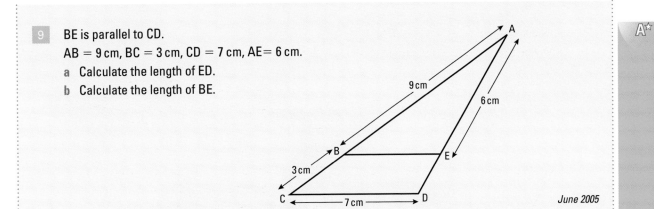

June 2005

9 EXPANDING BRACKETS AND FACTORISING

Algebra is regularly used by Formula One teams to maximise the performance of their cars when racing. For example, new rules introduced into Formula One in 2009 have given teams a booster button which gives the car extra power and can be pushed for a maximum of 6.7 seconds during a race. In order to maximise the benefit of the 'boost', F1 teams use algebra to work out the best moment for the driver to use it.

◎ Objectives

In this chapter you will:
- ◎ expand brackets
- ◎ factorise algebraic expressions
- ◎ simplify complicated algebraic expressions.

◈ Before you start

You should be able to:
- ◎ simplify algebraic expressions by collecting like terms
- ◎ use the index law $x^m \times x^n = x^{m+n}$
- ◎ add, subtract, multiply and divide directed numbers.

9.1 Expanding brackets

◎ Objective

◎ You can expand expressions which have a single pair of brackets.

⬆ Get Ready

1. Simplify

  a $5 \times 2x$ **b** $3x \times (-4x)$ **c** $(-x) \times (-2x)$

  d $4x - 5 - 3x + 1$ **e** $x^2 + 2x + 3x + 6$ **f** $x^2 - x - 2x + 2$

◗ Key Points

◎ When there is a number outside a bracket there is a hidden multiplication sign. So $20(n + 3) = 20 \times (n + 3)$.

◎ In algebra, **expand** usually means multiply out.

◎ To expand a bracket you multiply each term inside the bracket by the term outside the bracket.

Example 1   Expand $20(n + 3)$. ⬅ *Remember to multiply both terms inside the bracket by 20.*

$$20(n + 3) = 20 \times n + 20 \times 3$$
$$= 20n + 60$$ ⬅ *Write your answer in its simplest form.*

Example 2   Expand $3(2x + 1)$. ⬅ *Multiply both $2x$ and 1 by 3.*

$$3(2x + 1) = 3 \times 2x + 3 \times 1$$
$$= 6x + 3$$

Example 3   Expand $p(p + q - 5)$.

$$p(p + q - 5) = p \times p + p \times q - p \times 5$$ ⬅ *$p \times 5$ is usually written as $5p$.*
$$= p^2 + pq - 5p$$

Example 4   Expand $-2x(3x + 1)$. ⬅ *Multiply both terms by $-2x$.*

$$-2x(3x + 1) = -2x \times 3x + -2x \times 1$$ ⬅ *For each term, negative × positive = negative.*
$$= -6x^2 - 2x$$

⚙ Exercise 9A

Questions in this chapter are targeted at the grades indicated.

1   Expand

  a $2(x + 3)$ **b** $3(p - 2)$ **c** $4(m + n)$ **d** $3(5 - q)$

  e $2(2x + y - 3)$ **f** $5(2c + 1)$ **g** $4(x^2 - 2)$ **h** $3(n^2 - 2n + 1)$

D

2 Expand

a	$y(y + 2)$	b	$g(g - 3)$	c	$2x(x + 5)$	d	$n(4 - n)$
e	$a(b + c)$	f	$s(3s - 4)$	g	$3t(2t + 1)$	h	$4x^2(x - 3)$

3 Expand

a	$-2(m + 3)$	b	$-3(2x + 2)$	c	$-m(m + 5)$	d	$-4y(2y + 3)$
e	$-5(p - 2)$	f	$-3q(1 - q)$	g	$-2s(s - 3)$	h	$-3n(4m + n - 5)$

Example 5 Expand and simplify $3(2a + 1) + 2(3a + 5)$.

$3(2a + 1) + 2(3a + 5) = 6a + 3 + 6a + 10$ ← Expand each bracket separately.

$= 12a + 13$ ← Collect like terms.

Example 6 Expand and simplify $3x(y - 2) - 2y(x - 3)$.

$3x(y - 2) - 2y(x - 3) = 3xy - 6x - 2xy + 6y$ ← For the last term, negative × negative = positive.
$= xy - 6x + 6y$

Example 7 Expand and simplify $6p + 3p(2p - 7) + 4$.

ResultsPlus
Watch Out!

$6p + 3p(2p - 7) + 4 = 6p + 6p^2 - 21p + 4$
$= 6p^2 - 15p + 4$

You must multiply out the brackets before you collect like terms.
Check your signs.

Exercise 9B

1 Expand and simplify

a	$3(t - 1) + 5t$	b	$6p + 3(p + 2)$	c	$6(w + 1) + 5w$
d	$3(d + 2) + 4(d - 2)$	e	$3a + b + 2(a + b)$	f	$2(5x - y) + 5(y - x + 1)$

2 Expand and simplify

a	$3(y + 10) - 2(y + 5)$	b	$6(2a + 1) - 3(a + 4)$	c	$x - 5(x + 3)$
d	$q(q + 3) - 3(q + 1)$	e	$2n(n - 2) - n(2n + 1)$	f	$3m(2 + 5m) - 4m(1 + m)$

3 Expand and simplify

a	$5(t - 4) - 4(t - 1)$	b	$3(x + 3) - 2(x - 5)$	c	$2g(g + 1) - g(g + 1)$
d	$6c(2c - 3) - c(4 - c)$	e	$4s(s + 3) - 2(1 - s)$	f	$p(p + q) - q(p - q)$

4 Expand and simplify

a	$7s - 4(s + 1)$	b	$12m + 3(m + 2)$	c	$8f^2 - 3f(f + 1)$
d	$5n + n(n - 1)$	e	$2x - x(x - y)$	f	$7p - 2p(1 - p)$

9.2 Factorising by taking out common factors

Objectives

- You can take out common factors.
- You can factorise expressions by taking out common factors.

Get Ready

1. What is the HCF (Highest Common Factor) of the following pairs?

   a 6 and 8   **b** 10 and 25   **c** $8x$ and 12   **d** $9y$ and $15y$

Key Points

- **Factorising** is the opposite of expanding brackets, as you will need to put brackets in.
- To factorise an expression, find a common factor of the terms, take this factor outside the brackets, then decide what is needed inside the brackets.
- You can check your answer by expanding the brackets.
- Common factors are not always single terms such as 2, $5x$, $3a^2b$.
- Sometimes a common factor can have more than one term, for example $x + 2$ or $2a - b$.

Example 8   Factorise $12b + 8$.

$$12b + 8 = 4(\quad)$$
$$= 4(3b + 2)$$

The common factor of $12b$ and 8 is 4.
Note that you would not usually write the 4( ) but it is there to remind you to find the common factor first.

Check this multiplies out to give $12b + 8$.

Example 9   Factorise $2 - 6y$.

$$2 - 6y = 2(\quad)$$
$$= 2(1 - 3y)$$

Pick out the common factor first.
1 is needed as the first term in the bracket.

Example 10   Factorise $x^2 + 3x$.

$$x^2 + 3x = x(\quad)$$
$$= x(x + 3)$$

The common factor of x^2 and $3x$ is x.
Remember to check by multiplying out.

Example 11   Factorise $15p - 10q - 20pq$.

$$15p - 10q - 20pq = 5(\quad)$$
$$= 5(3p - 2q - 4pq)$$

Find the common factor of all three terms.

Example 12 Factorise completely $6a^2b + 9ab^2$.

$6a^2b + 9ab^2 = 3ab(\quad)$

$ = 3ab(2a + 3b)$

The common factor of $6 \times a \times a \times b$ and $9 \times a \times b \times b$ is $3 \times a \times b$.

Exercise 9C

1 Factorise

a $3x + 6$

b $2y - 2$

c $5p + 10q$

d $14t - 7$

e $8s + 2t$

f $9a + 18b$

g $15u + 5v + 10w$

h $xt - yt$

i $ac - c$

j $6x^2 + 9x + 3$

k $2p^2 - 2p$

l $q^2 - q$

m $4x^2 + 3x$

n $2h - 5h^2$

o $p^3 + 2p$

p $s^2 + s^3$

2 Factorise completely

a $5xy + 5xt$

b $3ad - 6ac$

c $6pq + 4hp$

d $8xy - 4y$

e $4pq + 2ps + 8pt$

f $mn - kmn$

g $2x^2 + 4x$

h $12s^2 - 24s$

i $6f^2 + 2f^3$

j $y^4 + y^2$

k $3cd^2 - 5c^2d$

l $a^3b + ab^3$

m $8pqr + 10prs$

n $14a^2b - 7ab^2 + 21ab$

o $15x^2y - 35x^2y^2$

p $(3y)^2 + 3y$

Example 13 Factorise $5(x + 2)^2 - 3(x + 2)$.

$5(x + 2)^2 - 3(x + 2) = (x + 2)[\quad]$

$ = (x + 2)[5(x + 2) - 3]$

$ = (x + 2)[5x + 10 - 3]$

$ = (x + 2)(5x + 7)$

$(x + 2)$ is a common factor.

Simplify the expression inside the square bracket.

Example 14 Factorise completely $12(s + 2t) - 4(s + 2t)^2$.

$12(s + 2t) - 4(s + 2t)^2 = 4(s + 2t)[\quad]$

$ = 4(s + 2t)[3 - (s + 2t)]$

$ = 4(s + 2t)(3 - s - 2t)$

The common factor is $4(s + 2t)$.

This cannot be simplified further.

Exercise 9D

1 Factorise

a $(x + 3)^2 + 2(x + 3)$

b $x(x - y) + y(x - y)$

c $p(p + 4) - 3p$

d $(2t + s)(2t - s) + (2t - s)$

e $(a - 5)^2 - 2(a - 5)$

f $(2d + 1)^2 + (2d + 1)$

2 Factorise completely

a $2(y + 2)^2 + 4(y + 2)$

b $15(x - 1)^2 - 10(x - 1)$

c $8(p + 5)^2 + 10(p + 5)$

d $9(q + 1) + 6(q + 1)^2$

e $7(a - b)(a + b) - 14(a + b)$

f $4x^2(x + 1) - 6x(x + 1)$

9.3 Expanding the product of two brackets

Objective

◉ You can multiply out the product of two brackets.

Get Ready

1. Work out the area of this rectangle.

6 cm

4 cm

2. Write down an expression, in terms of x, for the area of this rectangle.

$x + 2$

x

Key Points

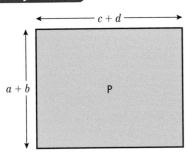

Area of rectangle P $= (a + b) \times (c + d)$

$= (a + b)(c + d)$

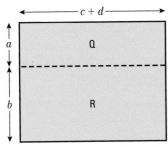

Area of rectangle Q $= a(c + d)$

Area of rectangle R $= b(c + d)$

Area of rectangle P $=$ Area of rectangle Q $+$ Area of rectangle R

$(a + b)(c + d) \quad = a(c + d) + b(c + d)$

$= ac + ad + bc + bd$

◉ To multiply out the product of two brackets:

  ◉ multiply each term in the first bracket by the second bracket

  ◉ expand the brackets

  ◉ simplify the resulting expression.

◉ An alternative method is the **grid method** (see Example 15).

Example 15   Expand and simplify $(x + 2)(x + 3)$.

Method 1

$(x + 2)(x + 3) = x(x + 3) + 2(x + 3)$

$= x^2 + 3x + 2x + 6$

$= x^2 + 5x + 6$

Take each term in the first bracket, in turn, and multiply it by the second bracket.
Expand the brackets.
Collect the like terms.

Method 2 — the grid method

	x	$+3$
x	x^2	$+3x$
$+2$	$+2x$	$+6$

← Each term in the first bracket is multiplied by each term in the second bracket.

$(x + 2)(x + 3) = x^2 + 3x + 2x + 6$ ← Add the four terms highlighted.

$\qquad\qquad\quad = x^2 + 5x + 6$ ← Collect like terms.

Example 16 Expand and simplify $(m + 2)^2$.

$(m + 2)^2 = (m + 2)(m + 2)$ ← Write out $(m + 2)^2$ in full.
$\qquad\quad = m(m + 2) + 2(m + 2)$
$\qquad\quad = m^2 + 2m + 2m + 4$
$\qquad\quad = m^2 + 4m + 4$

ResultsPlus
Watch Out!

Note that $(a + b)^2$ is not equal to $a^2 + b^2$.

Example 17 Expand and simplify $(2t - 1)(3t - 2)$.

Method 1
$(2t - 1)(3t - 2) = 2t(3t - 2) - 1(3t - 2)$ ← Check your signs are correct.
$\qquad\qquad\qquad = 6t^2 - 4t - 3t + 2$
$\qquad\qquad\qquad = 6t^2 - 7t + 2$

Method 2

	$3t$	-2
$2t$	$6t^2$	$-4t$
-1	$-3t$	$+2$

$(2t - 1)(3t - 2) = 6t^2 - 4t - 3t + 2$
$\qquad\qquad\qquad = 6t^2 - 7t + 2$

Exercise 9E

B

1 Expand and simplify

a $(x + 3)(x + 4)$ b $(x + 1)(x + 2)$ c $(x + 2)(x - 5)$

d $(y - 2)(y + 3)$ e $(y + 1)(y - 2)$ f $(x - 2)(x - 3)$

g $(a - 4)(a - 5)$ h $(x + 2)^2$ i $(p + 4)^2$

j $(k - 7)^2$ k $(a + b)^2$ l $(a - b)^2$

2 Expand and simplify

a $(x + 1)(2x + 1)$ b $(x - 1)(3x + 1)$ c $(2x + 3)(x + 4)$

d $(y - 3)(3y + 1)$ e $(2p + 1)(p + 3)$ f $(2t + 1)(3t + 2)$

g $(3s + 2)(2s + 5)$ h $(2x - 3)(2x + 5)$ i $(3y + 2)(4y - 1)$

j $(2a - 1)(3a - 2)$ k $(3x + 2)^2$ l $(2k - 1)^2$

3 Expand and simplify

a $(x + y)(x + 2y)$ b $(x - y)(x + 2y)$ c $(x + y)(x - 2y)$

d $(x - y)(x - 2y)$ e $(2p + 3q)(3p - q)$ f $(3s - 2t)(2s - t)$

g $(2a + 3b)^2$ h $(2a - 3b)^2$

9.4 Factorising quadratic expressions

◎ Objectives

- ◉ You can factorise quadratic expressions of the form $x^2 + bx + c$.
- ◉ You can recognise and factorise the difference of two squares.
- ◉ You can factorise quadratic expressions of the form $ax^2 + bx + c$.

◈ Get Ready

1. Write down all possible pairs of numbers whose product is

   a -6 b 15.

2. Find a pair of numbers whose product is 10 and whose sum is 7.

3. Find a pair of numbers whose product is 15 and whose sum is -8.

◈ Key Points

- ◉ Factorising is the reverse process to expanding brackets so, for example, factorising $x^2 + 5x + 6$ gives $(x + 2)(x + 3)$.

- ◉ To factorise the quadratic expression $x^2 + bx + c$
  - ◉ find two numbers whose product is $+c$ and whose sum is $+b$
  - ◉ use these two numbers, p and q, to write down the factorised form $(x + p)(x + q)$.

- ◉ To factorise the quadratic expression $ax^2 + bx + c$
  - ◉ work out the value of ac
  - ◉ find a pair of numbers whose product is $+ac$ and sum is $+b$
  - ◉ rewrite the x term in the expression using these two numbers
  - ◉ factorise the first two terms and the last two terms
  - ◉ pick out the common factor and write as the product of two brackets.

- ◉ Any expression which may be written in the form $a^2 - b^2$, known as the difference of two squares, can be factorised using the result $a^2 - b^2 = (a + b)(a - b)$.

Example 18 Factorise $x^2 + 7x + 12$.

The pairs of numbers whose product is 12 are:

$$+1 \times +12 \qquad -1 \times -12$$
$$+2 \times +6 \qquad -2 \times -6$$
$$+3 \times +4 \qquad -3 \times -4$$

ResultsPlus
Exam Tip

You may find it helpful to start by writing down all the pairs of numbers whose product is $+12$.

$$+3 \times +4 = +12 \leftarrow$$
$$+3 + +4 = +7$$
$$x^2 + 7x + 12 = (x + 3)(x + 4) \leftarrow$$

Find two numbers whose product is $+12$ and whose sum is $+7$.

Put into factorised form using the numbers $+3$ and $+4$.

Example 19 Factorise $x^2 - 10x + 25$.

The pairs of numbers whose product is $+25$ are:

$$+1 \times +25 \qquad -1 \times -25$$
$$+5 \times +5 \qquad -5 \times -5$$

ResultsPlus
Exam Tip

You can check your answer by expanding the brackets.

$$-5 \times -5 = +25$$
$$-5 + -5 = -10$$
$$x^2 - 10x + 25 = (x - 5)(x - 5) \leftarrow$$

This may also be written as $(x - 5)^2$.

Exercise 9F

1 Write down a pair of numbers:

a whose product is $+15$ and whose sum is $+8$

b whose product is $+24$ and whose sum is -10

c whose product is $+18$ and whose sum is -9

d whose product is -8 and whose sum is $+2$

e whose product is -8 and whose sum is -2

f whose product is -9 and whose sum is 0.

2 Factorise

a $x^2 + 8x + 15$ b $x^2 + 8x + 7$ c $x^2 + 9x + 20$

d $x^2 + 6x + 9$ e $x^2 - 6x + 5$ f $x^2 - 2x + 1$

g $x^2 + 3x - 18$ h $x^2 - 3x - 18$ i $x^2 + 3x - 28$

j $x^2 - x - 12$ k $x^2 + 2x - 24$ l $x^2 - 4$

m $x^2 - 81$

Example 20 Factorise $x^2 - n^2$.

$$x^2 - n^2 = x^2 + 0x - n^2$$

The pair of numbers whose product is $-n^2$ and whose sum is 0 is $+n \times -n$:

$$x^2 - n^2 = (x + n)(x - n)$$

Example 21 Factorise $x^2 - 100$.

$$x^2 - 100 = x^2 - 10^2$$
$$= (x + 10)(x - 10)$$

Substitute $a = x$ and $b = 10$ into $a^2 - b^2 = (a + b)(a - b)$.

ResultsPlus
Exam Tip

It will help you in the examination if you learn $a^2 - b^2 = (a + b)(a - b)$.

Example 22
a Factorise $p^2 - q^2$.
b Hence, without using a calculator, find the value of $101^2 - 99^2$.

a $p^2 - q^2 = (p + q)(p - q)$
b $101^2 - 99^2$
$\quad = (101 + 99)(101 - 99)$
$\quad = 200 \times 2$
$\quad = 400$

Use the result $a^2 - b^2 = (a + b)(a - b)$.
Substitute $p = 101$ and $q = 99$ in the answer to part (a).

Work out each bracket.

Example 23 Factorise $(x + y)^2 - 4(x - y)^2$.

$$= (x + y)^2 - [2(x - y)]^2$$

Write $(x + y)^2 - 4(x - y)^2$ in the form $a^2 - b^2$.

$$= [(x + y) + 2(x - y)][(x + y) - 2(x - y)]$$

Substitute $a = (x + y)$ and $b = 2(x - y)$ into $a^2 - b^2 = (a + b)(a - b)$.

$$= [x + y + 2x - 2y][x + y - 2x + 2y]$$

Expand and simplify the expression in each square bracket.

$$(x + y)^2 - 4(x - y)^2 = (3x - y)(-x + 3y)$$

Note that alternatively this answer may be written as $(3x - y)(3y - x)$.

Exercise 9G

1 Factorise
a $x^2 - 36$
b $x^2 - 49$
c $y^2 - 144$
d $25 - y^2$
e $w^2 - 2500$
f $10\,000 - a^2$
g $(x + 1)^2 - 4$
h $81 - (9 - y)^2$
i $(a + b)^2 - (a - b)^2$

2 Without using a calculator, find the value of:
a $64^2 - 36^2$
b $7.5^2 - 2.5^2$
c $0.875^2 - 0.125^2$
d $1005^2 - 995^2$

A

A02

141

A

3 Factorise these expressions, simplifying your answers where possible.

    a $4x^2 - 49$ **b** $9y^2 - 1$ **c** $121t^2 - 400$

    d $1 - (q + 2)^2$ **e** $(2t + 1)^2 - (2t - 1)^2$ **f** $(p + q + 1)^2 - (p + q - 1)^2$

    g $100(p + \frac{1}{2})^2 - 4(q + \frac{1}{2})^2$ **h** $25(s + t)^2 - 25(s - t)^2$

4 Factorise completely

    a $3x^2 - 12$ **b** $5y^2 - 125$ **c** $10w^2 - 1000$

    d $4p^2 - 64q^2$ **e** $12a^2 - 27b^2$ **f** $2(x + 1)^2 - 2(x - 1)^2$

Example 24   Factorise $3x^2 - 7x + 4$.

$a = +3, b = -7, c = +4$ ⟵ | Find two numbers whose product is $+12$ and whose sum is -7.
$ac = 12, b = -7$

$-3 \times -4 = +12$ | Replace $-7x$ with $-3x - 4x$.
$-3 + -4 = -7$

$3x^2 - 7x + 4 = 3x^2 - 3x - 4x + 4$ ⟵ | Factorise by grouping.

$\qquad = 3x(x - 1) - 4(x - 1)$ ⟵ | Pick out the common factor and write as the product of two brackets.

$\qquad = (x - 1)(3x - 4)$
$3x^2 - 7x + 4 = (x - 1)(3x - 4)$

Exercise 9H

A

1 Factorise

    a $5x^2 + 16x + 3$ **b** $2x^2 + 11x + 5$ **c** $3x^2 + 4x + 1$ **d** $8x^2 + 6x + 1$

    e $6x^2 + 13x + 6$ **f** $6x^2 - 7x + 1$ **g** $5x^2 - 7x + 2$ **h** $12x^2 - 11x + 2$

    i $8x^2 + 2x - 3$ **j** $2x^2 - 7x - 15$ **k** $7x^2 - 19x - 6$ **l** $3x^2 - 10x - 8$

    m $4y^2 + 12y + 5$ **n** $6y^2 - 13y + 2$ **o** $6y^2 - 25y + 25$

2 Factorise completely

    a $6x^2 + 14x + 8$ **b** $6y^2 - 15y + 6$ **c** $5x^2 + 5x - 10$

A★

3 Factorise

    a $x^2 + xy - 2y^2$ **b** $2x^2 + 7xy + 5y^2$ **c** $6x^2 + 5xy - 6y^2$

Chapter review

- When there is a number outside a bracket there is a hidden multiplication sign.
- In algebra, **expand** usually means multiply out.
- To expand a bracket you multiply each term inside the bracket by the term outside the bracket.

- **Factorising** is the opposite of expanding brackets, as you will need to put brackets in.
- To factorise an expression, find the common factor of the terms, take this factor outside the brackets, decide what is needed inside the brackets.
- You can check your answer by expanding the brackets.
- Common factors are not always single terms.
- To multiply out the product of two brackets:
  - multiply each term in the first bracket by the second bracket
  - expand the brackets
  - simplify the resulting expression.
- An alternative method is the **grid method**.
- Factorising is the reverse process to expanding brackets.
- To factorise the quadratic expression $x^2 + bx + c$
  - find two numbers whose product is $+c$ and whose sum is $+b$
  - use these two numbers, p and q, to write down the factorised form $(x + p)(x + q)$.
- Any expression which may be written in the form $a^2 - b^2$, known as the difference of two squares, can be factorised using the result $a^2 - b^2 = (a + b)(a - b)$.

Review exercise

1   Expand and simplify $2(x - 4) + 3(x + 2)$

87% of students answered this sort of question well because they remembered all of the necessary multiplications.

June 2009

2   **a**   Factorise $5m + 10$ **b**   Factorise $y^2 - 3y$

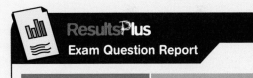

50% of students answered this question poorly because they did not put factors in the right place.

Nov 2008

3   Factorise **a** $ax + by + bx + ay$ **b**   Factorise $ac - bd + ad - bc$

4   Expand and simplify $(x + 4)(x - 3)$ *June 2009*

5   Expand **a** $(a + 2)^2$ **b** $(c - 3)^2$ **c** $(d + 1)^2$ **d** $(x + y)^2$

6   Expand and simplify
   a $(x + 5)(x + 10)$ **b** $(y + 9)^2$ **c** $(x - 4)(x + 2)$ **d** $(x + 2)(x - 3)$ **e** $(t - 1)(t - 6)$
   f $(x + 4)(2x + 3)$ **g** $(3p - 1)(2p + 1)$ **h** $(2c + d)(2c - d)$ **i** $(4y - 1)^2$

C

B

B

7 Factorise

a $t^2 + 11t + 30$ b $x^2 + 14x + 49$ c $p^2 + 2p - 15$

d $y^2 - 12y + 36$ e $x^2 - 5x + 4$ f $s^2 - 64$

8 a Factorise $x^2 + 8x + 7$ b Express 187 as the product of 2 prime numbers.

A02 A03

9 Jamie is planting flowers in a local park.

$$
\begin{array}{ccccc}
R & R & R & R & R \\
R & Y & Y & Y & R \\
R & R & R & R & R \\
\end{array}
$$

When he plants three yellow flowers, he surrounds them with twelve red flowers, as shown in the diagram.

a How many red flowers (R) does he plant with ten yellow flowers (Y)?

b Write your answer in both factorised and unfactorised forms.

A

10 Factorise

a $x^2 - 400$ b $9t^2 - 4$ c $100 - y^2$ d $25 - 4p^2$

A02 A03

11 Use some of your answers to question 7 to work out the values of the following expressions without using a calculator.

a $21^2 - 20^2$ b $10^2 - 9.9^2$ c $5^2 - 3^2$

12 Work out $1002^2 - 998^2$ using algebra.

A03

13 For any three consecutive numbers show that the difference between the product of the first and second and the product of the second and third is equal to double the second number.

14 Factorise fully $3(x + 2)^2 - 3x(x + 2)$

A03

15 a In the group stage of the Champions League, four teams play each other both at home and away. Prove that this requires 12 matches in total.

b Similarly, in the Premier League all the teams play each other twice. There are 20 teams. How many games are there altogether?

c How many games are there in a league with a teams? Write your answer in a factorised and unfactorised form.

A

16 Factorise

a $2x^2 + 5x + 2$ b $2w^2 + 5w - 3$ c $3a^2 + 14a + 8$

d $30z^2 - 23z + 2$ e $8y^2 + 23y - 3$ f $6p^2 - pq - q^2$

A03

17

1	2	3	4	5	6
7	8	9	10	11	12
13	14	15	16	17	18

$3 \times 8 - 2 \times 9 = 6$ and $12 \times 17 - 11 \times 18 = 6$

Show that for any 2 by 2 square of numbers from the grid, the difference of the products of numbers from opposite corners is always 6.

10 AREA AND VOLUME 1

The Tower of Pisa is a circular bell tower. Construction began in the 1170s, and the tower started leaning almost immediately because of a poor foundation and loose soil. It is 56.7 metres tall, with a diameter at the base of 15.5 metres, and there are 297 steps to the top. The tower continues to sink about 1 mm each year.

◎ Objectives

In this chapter you will:
- solve problems involving perimeters and areas
- know and use the formulae for the circumference and area of a circle
- draw the nets, elevations and plans for a variety of 3D shapes
- work out the volume of cubiods, prisms and cylinders.

◈ Before you start

You need to know:
- how to measure or calculate the perimeters of rectangles and triangles
- how to use the formula for the area of a rectangle
- what a circle, semicircle and quarter circle are, and be able to name the parts of a circle and related terms
- how to draw circles and arcs to a given radius.

10.1 Area of triangles, parallelograms and trapeziums

◉ Objectives

- You know and can use the formula for the area of a triangle.
- You know and can use the formula for the area of a parallelogram.
- You know and can use the formula for the area of a trapezium.

? Why do this?

Zoologists at game reserves need to know the areas of different sections of their reserve, so that they know how many animals it can accommodate.

◈ Get Ready

1. The diagram shows a rectangle.
   The length of the rectangle is 9 cm.
   The perimeter of the rectangle is 28 cm.
   Work out the width and the area of
   the rectangle.

9 cm

9 cm

🕐 Key Point

- The **area** of a 2D shape is a measure of the amount of space inside the shape.

Area of a triangle

🕐 Key Points

- The diagram below shows triangle ABC. A rectangle has been drawn around the triangle.
  The inside of the rectangle has been split into four triangles.

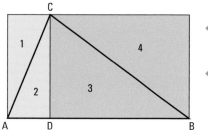

Triangles 1 and 2 are congruent so area triangle 1 = area triangle 2.

Also area triangle 3 = area triangle 4.

- The length of the rectangle is the
  base of the triangle and the width
  of the rectangle is the **perpendicular
  height** of the triangle.

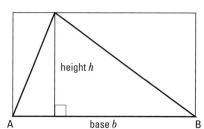

This means that the area of triangle ABC is half the area of the rectangle.

Area of the rectangle = base × height

So to find the area of a triangle, work out a half of its base × its height.

- Area of a triangle $= \frac{1}{2} \times$ base $\times$ height

  $A = \frac{1}{2}bh$

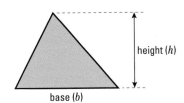

height (h)

base (b)

Example 1 Work out the area of the triangle.

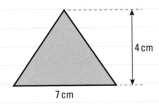

4 cm

7 cm

> Area of a triangle $= \frac{1}{2} \times$ base $\times$ height
>
> $7 \times 4 = 28$ $\frac{1}{2} \times 28 = 14$

> Do not forget to put the units of the answer.

ResultsPlus
Exam Tip

The height of a triangle is its vertical or perpendicular height.

Area $= \frac{1}{2} \times 7 \times 4 \, \text{cm}^2$

$= 14 \, \text{cm}^2$

Area of a parallelogram

Key Points

Here are two congruent triangles.

The triangles can be put together to form a parallelogram. The two triangles have equal areas so the area of the parallelogram is twice the area of one of the triangles.

Area of one triangle $= \frac{1}{2} \times$ base $\times$ height

Area of parallelogram $= 2 \times \frac{1}{2} \times$ base $\times$ height $=$ base $\times$ height

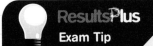

- Area of a parallelogram $=$ base $\times$ height

  $A = bh$

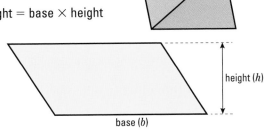

height (h)

base (b)

Example 2 Work out the area of the parallelogram.

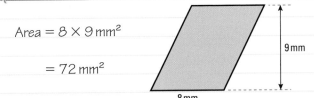

Area $= 8 \times 9 \, \text{mm}^2$

$= 72 \, \text{mm}^2$

9 mm

8 mm

> Area of a parallelogram $=$ base $\times$ height.

> As the lengths are in millimetres, the units of the area are mm^2.

D

Exercise 10A

Questions in this chapter are targeted at the grades indicated.

1 Work out the areas of these triangles and parallelograms.

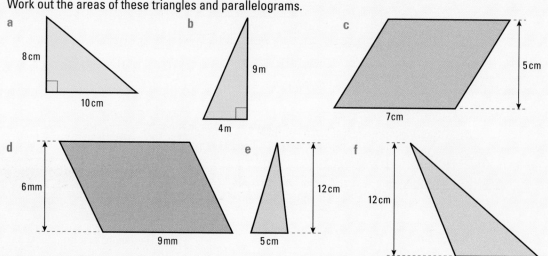

a 8 cm, 10 cm

b 9 m, 4 m

c 5 cm, 7 cm

d 6 mm, 9 mm

e 12 cm, 5 cm

f 12 cm, 9 cm

2 Copy and complete this table.

Shape	Base	Height	Area
Triangle	6 cm	5 cm	
Triangle	5 cm	10 cm	
Triangle		8 cm	24 cm²
Parallelogram	8 cm	4 cm	
Parallelogram	7 cm		56 cm²

A03

3 a A rectangle has a length of 7 cm and an area of 35 cm². Work out the width of the rectangle.

b A square has an area of 144 cm². Work out the length of side of the square.

Area of a trapezium

Key Points

Here is a trapezium. The trapezium is split into two triangles by a diagonal.
Area of trapezium = area of yellow triangle + area of pink triangle.

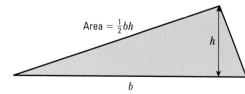

Area = $\frac{1}{2}bh$

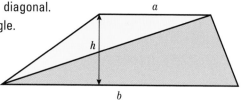

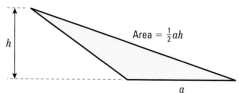

Area = $\frac{1}{2}ah$

Area of trapezium =
$\frac{1}{2}ah + \frac{1}{2}bh = \frac{1}{2}(a + b)h$

◉ Area of a trapezium $= \frac{1}{2} \times$ sum of parallel sides $\times$ distance between them.

$A = \frac{1}{2}(a + b)h$

Example 3 Work out the area of the trapezium.

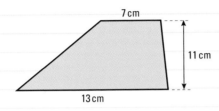

> **ResultsPlus**
> **Exam Tip**
>
> Remember that unless the question tells you to take measurements from a diagram you should not do so as diagrams are not accurately drawn.

Area $= \frac{1}{2} \times (7 + 13) \times 11$

⟵ Area of a trapezium $= \frac{1}{2} \times$ sum of parallel sides $\times$ distance between them.

$= \frac{1}{2} \times 20 \times 11 = 10 \times 11$

Work out the brackets first.

$= 110 \, cm^2$

Exercise 10B

1 Work out the area of each of these trapeziums.

a

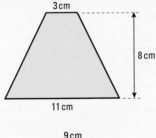

3 cm
8 cm
11 cm

b

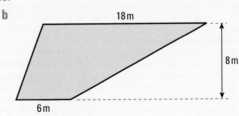

18 m
8 m
6 m

c

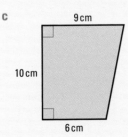

9 cm
10 cm
6 cm

d

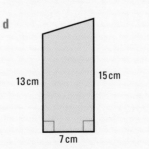

13 cm
15 cm
7 cm

C

10.2 Problems involving perimeter and area

Objectives

- You can find the area and perimeter of a more complicated shape made from simpler shapes.
- You can solve problems involving perimeters and areas.

Why do this?

A lot of houses seen from the side are a pentagon shape, so a painter would need to work out the area of a pentagon to get the right amount of paint.

Get Ready

1. Write down the formula for the area of:

    a a rectangle   **b** a square   **c** a triangle   **d** a parallelogram   **e** a trapezium.

Key Point

- The perimeter or area of a compound shape can be found by splitting the shape into its simpler parts.

A02 A03

Example 4 Work out the area of this pentagon.

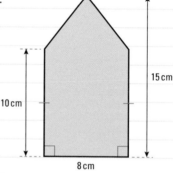

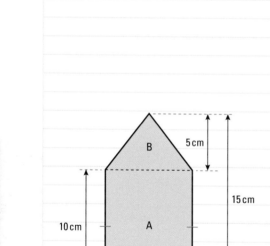

Split the pentagon into a rectangle A and a triangle B.

The height of the triangle is $15 - 10 = 5$ cm.
The base of the triangle is 8 cm.

The rectangle has length 8 cm and width 10 cm.

Area of rectangle A $= 8 \times 10 \quad = 80 \, cm^2$
Area of triangle B $\quad = \frac{1}{2} \times 8 \times 5 = 20 \, cm^2$
Area of pentagon $\quad = 80 + 20 \quad = 100 \, cm^2$

Area of a rectangle $=$ length $\times$ width
Area of a triangle $\quad = \frac{1}{2} \times$ base $\times$ height
Area of pentagon $\quad =$ area of A $+$ area of B

Example 5   A rectangular wall is 450 cm long and 300 cm high. The wall is to be tiled.
The tiles are squares of side 50 cm. How many tiles are needed?

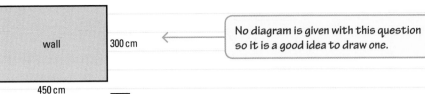

A02
A03

No diagram is given with this question
so it is a good idea to draw one.

wall   300 cm

450 cm

50 cm | tile
50 cm

Method 1

Number of tiles needed for the length $= \dfrac{450}{50} = 9$

One way to answer questions like
this is to work out how many tiles
are needed for the length and how
many are needed for the height.

Number of tiles needed for the height $= \dfrac{300}{50} = 6$

So there are 6 rows of
tiles, each with 9 tiles.

Number of tiles needed $= 9 \times 6$

$= 54$

Number of tiles = number of tiles in
each row × number of rows.

Method 2

Area of wall $= 450 \times 300 \text{ cm}^2 = 135\,000 \text{ cm}^2$

Area of a tile $= 50 \times 50 = 2500 \text{ cm}^2$

The other way to answer this question is to
divide the area of the wall by the area of a tile.

Number of tiles $= \dfrac{135\,000}{2500} = 54$

But remember that you should not use a calculator
and the arithmetic is easier in the first method.

Exercise 10C

1   The diagram shows the floor plan of a room.
   a   Work out the perimeter of the floor.
      Give the units of your answer.
   b   Work out the area of the floor.
      Give the units of your answer.

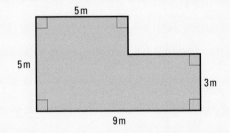

5 m

5 m

3 m

9 m

D

A02

2   Karl wants to make a rectangular lawn in his garden. He wants the lawn to be 30 m by 10 m.
   Karl buys rectangular strips of turf 5 m long and 1 m wide.
   Work out how many strips of turf Karl needs to buy.

A02
A03

3   A wall is a 300 cm by 250 cm rectangle. The wall is to be tiled.
   The tiles are squares of side 50 cm. Work out how many tiles are needed.

A02
A03

4   A rectangle is 9 cm by 4 cm. A square has the same area as the rectangle.
   Work out the length of side of the square.

A02
A03

D

5 Keith is going to wallpaper his living room and his bedroom.
   Here are the floor plans of these rooms.

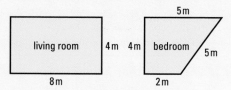

   a Work out the area of the floor in:
      i Keith's living room
      ii Keith's bedroom.
   b Work out the perimeter of the floor in Keith's living room.

To work out the number of rolls of wallpaper he needs, Keith uses this chart.
Keith is going to use standard rolls of wallpaper.

Standard rolls of wallpaper are approx 10 m long							
How many rolls for the walls							
	Distance around the room including doors & windows						
Wall height	10 m – 33 ft	12 m – 39 ft	14 m – 46 ft	16 m – 52 ft	18 m – 59 ft	20 m – 66 ft	22 m – 72 ft	24 m – 79 ft
2 – 2.3 m 7' – 7'6"	5	5	6	7	8	9	10	11
2.3 – 2.4 m 7'6" – 8'	5	6	7	8	9	10	10	11
2.4 – 2.6 m 8' – 8'6"	5	6	7	9	10	11	12	13
2.6 – 2.7 m 8'6" – 9'	5	6	7	9	10	11	12	13
2.7 – 2.9 m 9' – 9'6"	6	7	8	9	10	12	12	14

The height of the walls in Keith's living room is 2.5 m.
   c Find how many rolls of wallpaper Keith needs for his living room.
The height of the walls in Keith's bedroom is 2.6 m.
   d Find the number of rolls of wallpaper Keith needs for his bedroom.

C

6 Here is a quadrilateral.

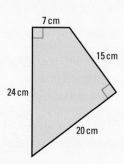

   a Work out the perimeter of the quadrilateral.
   b Work out the area of the quadrilateral.

7 Work out the area of the yellow shaded region in this diagram.

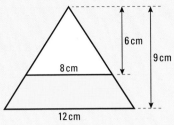

6 cm

9 cm

8 cm

12 cm

A02
A03

C

8 A kite has diagonals of length 10 cm and 20 cm.
Work out the area of the kite.

A02
A03

10.3 **Circumference and area of a circle**

◎ Objectives

○ You can work out the circumference of a circle.

○ You can work out the area of a circle.

○ You can solve problems involving circles, including semicircles and quarter circles.

⑦ Why do this?

To fit a new tyre on the wheel of your bike, you may need to know the circumference of the wheel to find the correct size.

◈ Get Ready

1. Draw a circle of radius 5 cm. For this circle, draw and label clearly:

a a radius **b** a diameter **c** a chord **d** a sector **e** an arc **f** a segment **g** a tangent.

🔍 Key Points

◉ For all circles $\dfrac{\text{circumference of circle}}{\text{diameter of circle}} = \pi$ (pi).

This value cannot be found exactly.

To 3 decimal places, $\pi = 3.142$.

circumference of circle $= \pi \times$ diameter of circle

$C = 2\pi r$

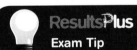

$C = \pi \times d$
$d = C \div \pi$

ResultsPlus
Exam Tip

Calculator exam papers have the following instruction about π, 'If your calculator does not have a π button, take the value of π to be 3.142 unless the question instructs otherwise.'

ResultsPlus
Watch Out!

It is important not to confuse the diameter with the radius.

Example 6

Work out the circumference of
a circle with:

a diameter 8.7 cm

b radius 3.1 m.

Give your answers correct to
3 significant figures.

> **ResultsPlus**
> **Exam Tip**
>
> Remember that the circumference
> is approximately 3 times the
> diameter or 6 times the radius.

a $C = \pi \times 8.7$ ← Use $C = \pi d$ with $d = 8.7$ cm.

    $= 27.3318...$

  Circumference $= 27.3$ cm ← Use the π button or 3.142.
Write down at least 4 figures of the calculator display.
Give the answer correct to 3 significant figures. The
units are the same as the diameter (cm).

b $C = 2 \times \pi \times 3.1$ ← The diameter can be worked out from $d = 2r$
so $d = 2 \times 3.1 = 6.2$ and then use $C = \pi d$.
Or use $C = 2\pi r$ with $r = 3.1$ m.

    $= 19.47787...$

  Circumference $= 19.5$ m The units are the same as the radius (m).

Example 7

The circumference of a circle is 84.3 cm. Work out the radius of the circle.
Give your answer correct to 3 significant figures.

$84.3 = 2 \times \pi \times r = 2\pi \times r$ ← Use $C = 2\pi r$ with $C = 84.3$ cm as the radius is given in the question.
Divide both sides by 2π and write down at least 4 figures of the
calculator display.

$r = 84.3 \div (2\pi)$

   $= 13.4167...$

> **ResultsPlus**
> **Watch Out!**

Radius $= 13.4$ cm ← Give the answer correct to
3 significant figures.
The units are the same as
the circumference (cm).

Be careful when dividing by 2π
on a calculator. It is best to use
brackets.

Exercise 10D

In this exercise, if your calculator does not have a π button, take the value of π to be 3.142. Give answers
correct to 3 significant figures unless a question says differently.

D

1 Work out the circumference of a circle with diameter:

  a 7 cm   b 12.9 mm   c 5.6 cm   d 40 cm   e 21.9 m

2 The radius of a basketball net hoop is 23 cm.

  a Work out the circumference of a basketball net hoop.

  A netball hoop has a radius of 19 cm.

  b Work out how much longer is the circumference of a basketball net hoop than the circumference of
     a netball hoop.

3 The circumference of a CD is 37.7 cm. Work out the radius of the CD.

4 The diameter of the front wheel of Michael's bicycle is 668 mm.
   a Work out the circumference of the wheel.
   Give your answer in cm correct to the nearest cm.

   Michael rides his bicycle.
   b Work out the distance cycled when the wheel makes 1000 complete turns.
   Give your answer in km correct to 2 significant figures.

   The distance Michael rides his bicycle is 6 km.
   c Work out the number of complete turns made by this wheel.

5 The length of the minute hand of a watch is 1.2 cm.
   a Work out the distance moved by the point end of the hand in 1 hour.
   b Work out the distance moved by the point end of the hand in: **i** 6 hours **ii** 20 minutes.

6 A circular table has a radius of 65 cm.
   a Work out the circumference of the table.
   The circumference of a circular tablecloth is 5 m.
   The tablecloth is put symmetrically on the table so that the distance
   from the table to the edge of the tablecloth is the same all around the table.
   b Work out the distance from the table to the edge of the tablecloth.

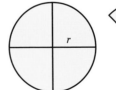

7 The diagram shows a shape made from a
semicircle, a rectangle and an equilateral
triangle.
The rectangle has length 18 cm and width 10 cm.
Work out the perimeter of the shape.

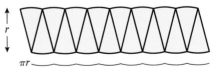

Area of a circle

Key Points

● To find the area of a circle means to find the area enclosed by the circle.
Here is a circle that has been divided into four equal
wedges or sectors. The sectors are then arranged as
shown to form a parallelogram-like shape.

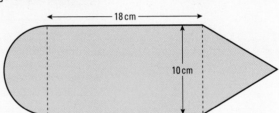

The length shown as πr is half the
circumference, $2\pi r$, of the circle.
The area of the circle is the same as the
area of the shape.
Here is what happens when the circle is
divided into more sectors.

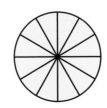

The shape looks more like a parallelogram and as the number of sectors increases the parallelogram becomes more like a rectangle.

The width of this rectangle is equal to half of the circumference of the original circle and the height of the rectangle is equal to the radius of the circle.

Area of circle = area of rectangle = $\pi r \times r = \pi r^2$

Taking A as the area of a circle and r as the radius of the circle, $A = \pi r^2$
That is Area = $\pi \times$ radius $\times$ radius

Example 8   Work out the area of a circle with: **a**   a radius of 9 cm   **b**   a diameter of 12.8 m.
Give your answers correct to 3 significant figures.

a $A = \pi \times 9^2$ $\leftarrow$

      $= 254.4690\ldots$

> Use $A = \pi r^2$ with $r = 9$ cm.
> Write down at least 4 figures of the calculator display.

Area = 254 cm^2 $\leftarrow$

> Give the answer correct to 3 significant figures.
> As the units of the radius are cm, the units of the area are cm^2.

b   Radius = $12.8 \div 2$ m $\leftarrow$

      $= 6.4$ m

> Divide the diameter by 2 to get the radius.
> Write down at least 4 figures of the calculator display.
> Give the answer correct to 3 significant figures.

$A = \pi \times 6.4^2$

    $= 128.6796\ldots$

Area = 129 m^2 $\leftarrow$

> As the units of the radius are m, the units of the area are m^2.

ResultsPlus
Exam Tip

When the diameter of a circle is given, to work out the area of the circle first find the radius by dividing the diameter by 2.

Example 9   Work out the radius of a circle with area 46 cm^2.

$46 = \pi \times r^2$ $\leftarrow$

> Use $A = \pi r^2$ with $A = 46$ cm^2.

$r^2 = 46 \div \pi = 14.64225\ldots$ $\leftarrow$

> Work out the value of r^2 by dividing both sides by π.

$r = \sqrt{14.64225\ldots} = 3.8265\ldots$ $\leftarrow$

> Take the square root to find the value of r.

Radius = 3.83 cm

Exercise 10E

In this exercise, if your calculator does not have a π button, take the value of π to be 3.142.
Give answers correct to 3 significant figures unless the question says differently.

1 Work out the area of a circle with radius:
    a 8 cm   **b** 12.7 cm   **c** 28.5 mm   **d** 9.72 cm   **e** 12.6 m

2 Work out the area of a circle with diameter:
    a 24 cm   **b** 8.3 cm   **c** 0.95 m   **d** 58.4 mm   **e** 18.26 cm

3 The diagram shows a pond surrounded by a path.
    a Work out the area of the blue region of the pond.
    b Work out the area of the path.
    c The path is made of shingle that costs £1.95 per
       square metre of path. Work out the cost of the
       shingle to make the path.

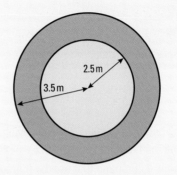

4 The diagram represents the plan of a sports field. The field is a rectangle with semicircular ends.
The rectangle has length 100 m and width 70 m. The semicircles have diameter 70 m.
    a Work out the area of the field.

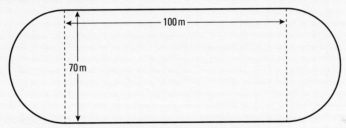

The field is to be covered in fertiliser that costs 23p per square metre.
    b Use your answer to part **a** to work out the cost of the fertiliser for the field.

5 A circle of diameter 8 cm is cut from a piece of yellow card.
The card is in the shape of a square of side 11 cm.
The card shown yellow in the diagram is thrown away.
Work out the area of the card thrown away.

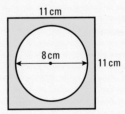

6 A, B and C are three circles. Circle A has radius 5 cm and circle B has radius 12 cm. Circle C is such that
area of circle C = area of circle A + area of circle B. Work out the radius of circle C.

7 The diagram shows a star made by removing four identical
quarter circles from the corners of a square of side 30 cm.
Work out the area of the star.

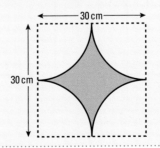

10.4 Drawing 3D shapes

◎ Objective

○ You can recognise and draw the net of a 3D shape.

⑦ Why do this?

A manufacturer of chocolate boxes would have to consider the nets of different sizes of boxes in order to see how best to package their product.

⬦ Get Ready

1. Sketch these shapes.
   a a triangular prism
   b a square-based pyramid
   c a cylinder
   d a triangular-based pyramid

🔍 Key Points

◉ Isometric paper will help you to make scale drawings of **three-dimensional** objects.
◉ Isometric paper must be the right way up i.e. vertical lines down the page and no horizontal lines.
◉ A **net** of a 3D shape is a 2D shape that can be folded to make the 3D shape.
◉ A 3D shape can have more than one net.

This cube has sides of length 2.

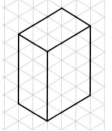

This **cuboid** has height 4, length 3 and width 2.

This prism has a triangular face.

Shapes can be joined together

🔍 Example 10

Draw two different nets for this cuboid.

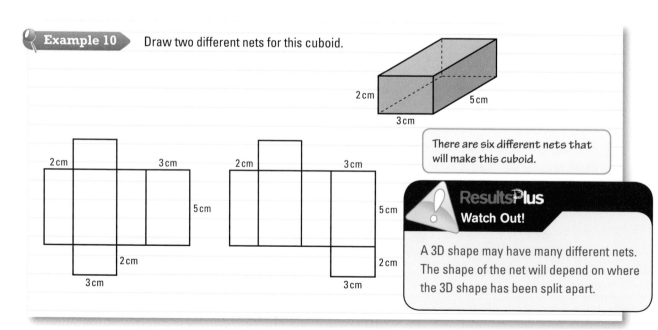

There are six different nets that will make this cuboid.

ResultsPlus
Watch Out!

A 3D shape may have many different nets. The shape of the net will depend on where the 3D shape has been split apart.

Exercise 10F

1. Use isometric paper to draw a cuboid with height 2 cm, width 4 cm and length 3 cm.

2. Sketch six different nets that will make a cube.

3. Here are the nets of some 3D shapes. Identify the shapes.

a  b c d
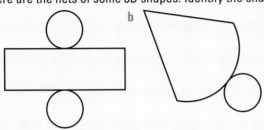
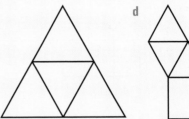

4. Draw an accurate net for each of these.

a 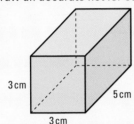 b
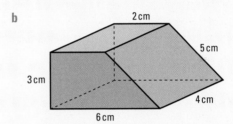

10.5 Elevations and plans

◉ Objective

● You can draw elevations and plans of 3D shapes.

⟐ Why do this?

Architectural proposals will usually contain plans and elevations of the proposed building, to give people an idea of what the building will look like from each side.

⟐ Get Ready

1. What would the shapes in question 4, above, look like if drawn from above, the side and the front.

⬤ Key Points

● The **front elevation** is the view from the front.
● The **side elevation** is the view from the side.
● The **plan** is the view from above.

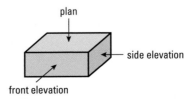

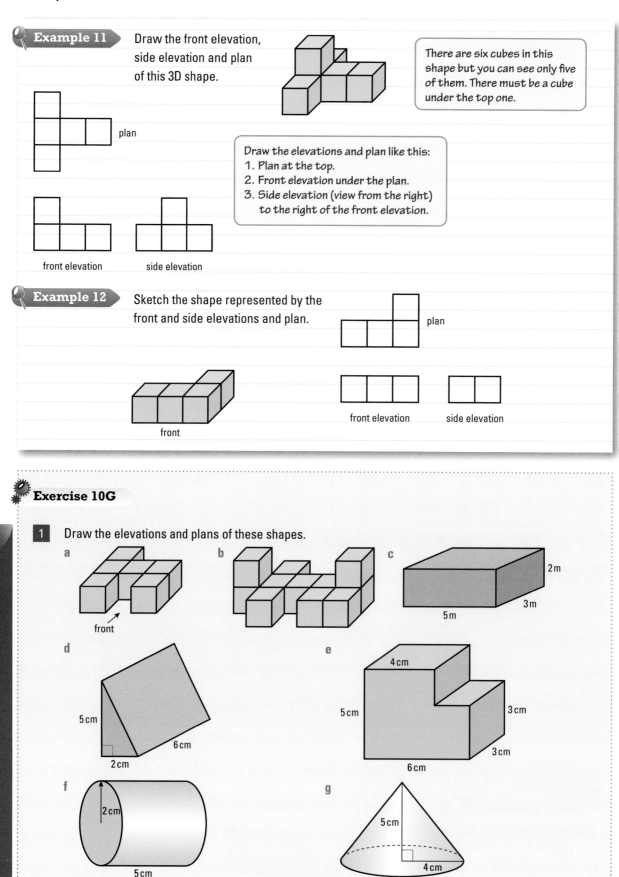

Example 11 Draw the front elevation, side elevation and plan of this 3D shape.

There are six cubes in this shape but you can see only five of them. There must be a cube under the top one.

plan

Draw the elevations and plan like this:
1. Plan at the top.
2. Front elevation under the plan.
3. Side elevation (view from the right) to the right of the front elevation.

front elevation   side elevation

Example 12 Sketch the shape represented by the front and side elevations and plan.

plan

front

front elevation   side elevation

Exercise 10G

1 Draw the elevations and plans of these shapes.

a
front

b

c
2m
3m
5m

d
5 cm
6 cm
2 cm

e
4 cm
5 cm
3 cm
3 cm
6 cm

f
2 cm
5 cm

g
5 cm
4 cm

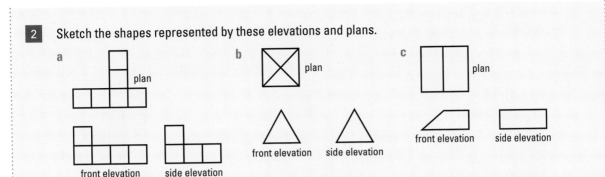

2 Sketch the shapes represented by these elevations and plans.

a   plan

front elevation   side elevation

b   plan

front elevation   side elevation

c   plan

front elevation   side elevation

D

10.6 Volume of a cuboid

Objective

You can work out the volume of a cuboid and shapes made from cuboids.

Why do this?

If you were filling a swimming pool you might first have to consider its volume in order to work out how much water you would need.

Get Ready

1. Work out the volumes of these cuboids. Give the units with your answers.

a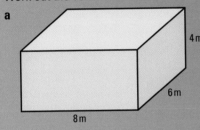
4 m
6 m
8 m

b
6 cm
8 cm
12 cm

Example 13 This shape is made from two cuboids. Work out the total volume of the shape.

9 m
4 m
3 m
2 m
3 m
2 m

Work out the volume of each cuboid. Use volume of cuboid = $l \times w \times h$.

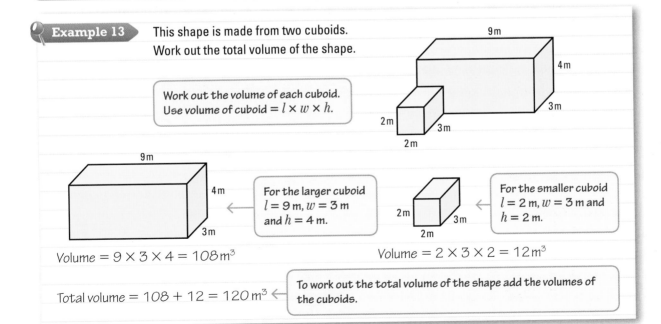

9 m
4 m
3 m

For the larger cuboid
$l = 9$ m, $w = 3$ m and $h = 4$ m.

2 m
3 m
2 m

For the smaller cuboid
$l = 2$ m, $w = 3$ m and $h = 2$ m.

Volume = $9 \times 3 \times 4 = 108 \text{ m}^3$

Volume = $2 \times 3 \times 2 = 12 \text{ m}^3$

Total volume = $108 + 12 = 120 \text{ m}^3$

To work out the total volume of the shape add the volumes of the cuboids.

Exercise 10H

D

1 These shapes are made from cuboids. Work out the volumes of the shapes.

a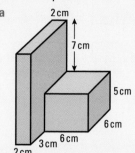

b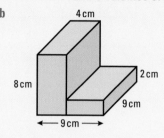

c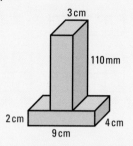

A03

2 Here is a net of a cuboid. Work out the volume of the cuboid.

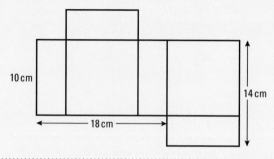

10.7 Volume of a prism

Objective

○ You can work out the volume of a prism.

Why do this?

Sandwiches are often sold in packs that are triangular prisms, so you can work out how much sandwich you are getting.

Get Ready

1. Work out the volume of these shapes.

a

b

c Find the volume of half shape **b**.

Key Point

◉ Volume of **prism** = area of cross-section × length

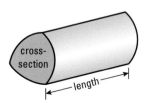

Example 14

The area of the cross-section of this prism is 25 cm². The length of the prism is 10 cm.
Work out the volume of the prism.

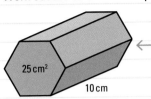

25 cm²

10 cm

> Use volume of prism = area of cross-section × length.
> Here, the area of cross-section = 25 cm² and
> the length = 10 cm.

Volume = 25 × 10 = 250 cm³

> Give the unit with your answer. The unit of area is
> cm², the length is in cm so the unit of volume is cm³.

Example 15

Work out the volume of this prism.

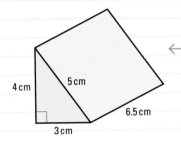

4 cm   5 cm

6.5 cm

3 cm

> The cross-section of the prism is a triangle.
> Remember: area of a triangle = $\frac{1}{2}$ × base × height.
> Here the base = 3 cm and height = 4 cm.

Area of cross-section = $\frac{1}{2}$ × 3 × 4 = 6 cm²
Volume of prism = 6 × 6.5 = 39 cm³

> Use volume of prism = area of cross-section × length.
> Here the area of cross-section = 6 cm² and length = 6.5 cm.

Exercise 10I

1 Work out the volumes of these prisms.

a

12 cm²

6.5 cm

b

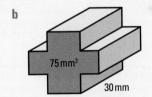

75 mm²

30 mm

c

1.75 m

0.95 m   0.6 m

d

3 cm

6 cm

6 cm

8 cm

2 Work out the volumes of these prisms.

a

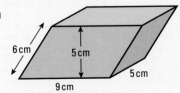

6 cm

5 cm

5 cm

9 cm

b

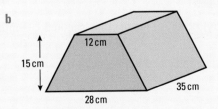

12 cm

15 cm

28 cm

35 cm

C

C

c

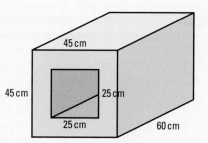

d

3.3 cm
5.9 cm
2.7 cm
3.5 cm
4.6 cm

45 cm
45 cm
25 cm
25 cm
25 cm
60 cm

A02

3 The area of the cross-section of a prism is 45 cm². The volume of the prism is 405 cm³. Work out the length of the prism.

B

4 Here is a prism. Show that the volume of the prism is $8x^3$ cm³.

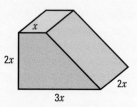

x
$2x$
$2x$
$3x$

5 The diagram shows a triangular prism. The volume of the prism is $45y^3$ cm³. Find an expression for h in terms of y.

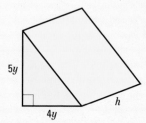

$5y$
$4y$
h

10.8 Volume of a cylinder

◎ Objective

● You can work out the volume of a cylinder.

? Why do this?

You could work out the volume of liquid that your mug can hold if you wanted to boil only that exact amount of water, to save energy.

◈ Get Ready

1. Find the area of these circles:
   a radius 3 cm
   b diameter 5 cm
   c radius 10 cm.

Key Point

◉ Volume of **cylinder** = area of cross-section × length
                    = $\pi r^2 h$
where r is the radius and h is the height.

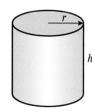

r
h

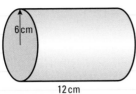 **Example 16** Work out the volume of this cylinder.
Give your answer in terms of π and to 3 significant figures.

> The cross-section of the cylinder is a circle with radius 6 cm. Remember: area of circle = $\pi \times$ radius^2.
> Take π as 3.142.

Area of cross-section = $\pi \times 6^2$
$= 36\pi$

Volume of cylinder = $3.142 \times 6 \times 6 \times 12$ ← Use volume of cylinder = area of cross-section $\times$ length.

$= 1357.344\ cm^3$ ← Do not round your answer at this stage.
Write down all the digits on your calculator display.

$= 1360\ cm^3$ (3 s.f.) ← Give your final answer correct to 3 significant figures.

Exercise 10J

1 Work out the volumes of these cylinders.
Give your answers correct to 3 significant figures.

a 4 cm / 5 cm
b 240 mm / 300 mm
c 30 mm / 5 cm
d 12 cm / 79 cm

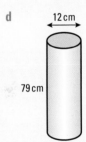

2 Work out the volumes of these cylinders. Give your answers in terms of π.

a 6 cm / 10 cm
b 20 cm / 6.5 cm
c 0.45 m / 0.5 m

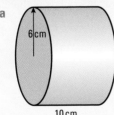

3 An aircraft hangar has a semicircular cross-section of diameter 20 m.
The length of the hangar is 32 m.
Work out the volume of the hangar. Give your answer in terms of π.

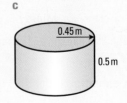
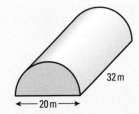

32 m
20 m

C

A03

B · AO3

4 An annulus has an external diameter of 7.8 cm, an internal diameter of 6.2 cm and a length of 6.5 cm. Work out the volume of the annulus. Give your answer correct to 1 decimal place.

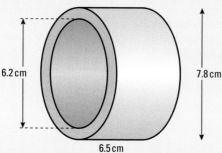

6.2 cm   7.8 cm

6.5 cm

AO3

5 A gold coin has a height of 2.5 mm and a volume of 2000 mm³. Work out the diameter of the gold coin. Give your answer correct to 2 decimal places.

AO2
AO3

6 An oil drum has a radius of 0.9 m and a height of 1.4 m. The oil drum is completely filled with oil. Work out the volume of the oil in the oil drum. Give your answer correct to 3 significant figures.

Chapter review

⦿ **Area** of a triangle $= \frac{1}{2} \times$ base $\times$ height.

$A = \frac{1}{2}bh$

⦿ Area of a parallelogram $=$ base $\times$ height.

$A = bh$

⦿ Area of a trapezium $= \frac{1}{2} \times$ sum of parallel sides $\times$ distance between them.

$A = \frac{1}{2}(a + b)h$

⦿ The perimeter or area of a compound shape can be found by splitting the shape into its simpler parts.

⦿ For all circles, $\dfrac{\text{circumference of circle}}{\text{diameter of circle}} = \pi$ (pi).

⦿ To 3 decimal places, $\pi = 3.142$.

⦿ Circumference of a circle $= \pi d = 2\pi r$ where d is the diameter of the circle, and r is the radius of the circle.

⦿ Area of a circle $= \pi r^2$ where r is the radius of the circle.

⦿ The **net** of a 3D shape is a 2D shape that can be folded to make the 3D shape.

⦿ A 3D shape can have more than one net.

⦿ The **front elevation** is the view from the front.

⦿ The **side elevation** is the view from the side.

⦿ The **plan** is the view from above.

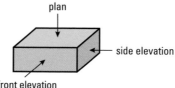

plan

side elevation

front elevation

⦿ Volume of **prism** $=$ area of cross-section $\times$ length.

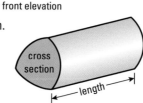

cross section

length

Volume of **cylinder** = area of cross-section × length
= $\pi r^2 h$
where r is the radius and h is the height.

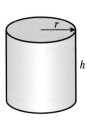

Review exercise

1 The diagram shows some nets and some solid shapes.
An arrow has been drawn from one net to its solid shape.
Draw an arrow from each of the other nets to its solid shape.

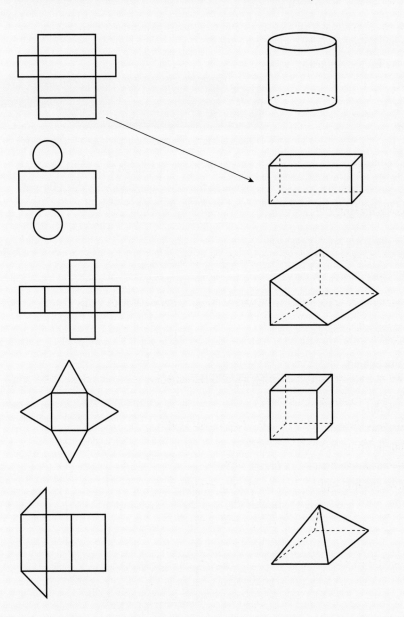

Nov 08

D

2 Find the volume of this prism.

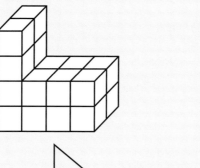

Diagram **NOT** accurately drawn

represents 1 cm³

June 08

3 Work out the area of the shape.

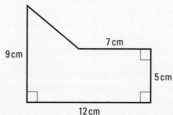

9 cm   7 cm

5 cm

12 cm

Diagram **NOT** accurately drawn

Nov 2008

4 The diagram shows a solid object made of 6 identical cubes.

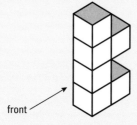

front

ResultsPlus
Exam Question Report

95% of students answered this question poorly because they did not know what the different types of plans and elevations are.

a On a centimetre grid, draw the side elevation of the solid object from the direction of the arrow.

b On a centimetre grid, draw the plan of the solid object.

June 07

5 The diagram shows a cuboid.
The cuboid has:
a volume of 300 cm³
a length of 10 cm
a width of 6 cm.
Work out the height of the cuboid.

height

6 cm

10 cm

Nov 06

6 Boxes are packed into cartons.
A box measures 4 cm by 6 cm by 10 cm.
A carton measures 20 cm by 30 cm by 60 cm.
The carton is completely filled with boxes.
Work out the number of boxes that will completely fill one carton.

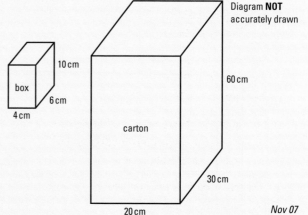

Diagram **NOT** accurately drawn

10 cm

box

6 cm

4 cm

60 cm

carton

30 cm

20 cm

Nov 07

7 Jane makes chocolates.
Each box she puts them in has:

volume = 1000 cm³
length = 20 cm
width = 1000 cm.

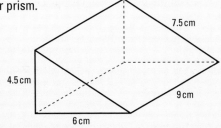

a Work out the height of a box.
Jane makes 350 chocolates.
Each box will hold 18 chocolates.

b Work out: **i** how many boxes Jane can fill completely
ii how many chocolates will be left over.

8 Here is a net of a cuboid. Work out:
a the surface area
b the volume of the cuboid.

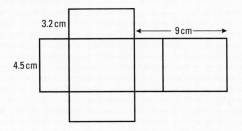

3.2 cm
9 cm
4.5 cm

9 The diagram shows a triangular prism.

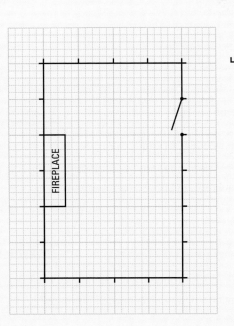

7.5 cm
4.5 cm
9 cm
6 cm

a Draw the elevations and plan for the prism.
b Work out the surface area of the prism.
Give the units with your answer.

* 10 Shelim is replacing the skirting boards and coving in his living room.

Skirting board can be bought in:
4 m lengths at £30.50
3 m lengths at £18.75
2 m lengths at £14.00.

Coving can be bought in:
3 m lengths at £27.50
2.4 m lengths at £22.00.

Coving can be joined together, but skirting board must not be pieced together as the joins will be noticeable.

Find the cost of his materials for both jobs, minimising the waste.

FIREPLACE

= 1 m

A02
A03
D

A03

C

A02
A03

169

C A02 A03

* **11** Amy has saved £600 to spend on carpeting her front room. There are four types she likes:

Natural Twist at £14.50 per m²
Medium Blend at £17.60 per m²
Heavy Weave at £19.00 per m²
Luxury Pile at £24.90 per m².

She also needs to buy underlay, which is available in two types:

Cushion at £2.00 per m²
Super Cushion at £4.00 per m².
Fitting is £50 extra.
What can she afford to buy?

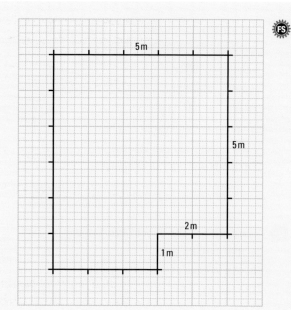

A02 A03

12 A landscape contractor charges:
£40 per square metre for levelling the ground and laying paving stones
£15 per square metre for levelling the ground and sowing grass seed.
Calculate the cost of both paving and seeding the garden shown on the right.

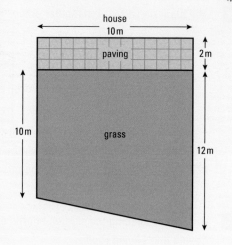

A02 A03

13 A ring-shaped flowerbed is to be created around a circular lawn of radius 2.55 m.

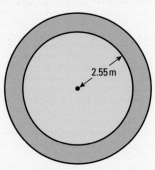

Roses costing £4.20 are to be planted approximately every 50 cm around this flowerbed. How much money will be needed for roses?

14 The diagram shows a garden that includes a lawn, a vegetable patch, a circular pond and a flowerbed. All measurements are shown in metres.
The lawn is going to be relaid with turf costing £4.60 per square metre.
How much will this cost?

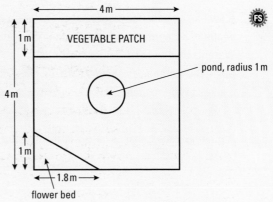

15 You are planning a party for 30 children.
You buy some concentrated orange squash and some plastic cups.

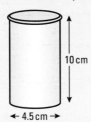

Each plastic cup will have 150 ml of drink in it. (150 ml = 150 cm^3)

a Check that the plastic cup shown can hold 150 ml of drink. Use the formula:

$$\text{volume } 5 \ \pi \times h \times \frac{d^2}{4}$$

Each of the 30 children at the party will have a maximum of three drinks of orange squash.
Each plastic cup is to be filled with 150 ml of drink.
The squash needs to be diluted as shown on the bottle label.
A bottle of concentrated orange squash contains 0.8 litres of squash and costs £1.25.

b How many bottles of concentrated orange squash do you need for the party?

c How much will they cost in total?

16

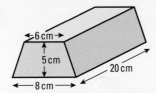

The cross-section of the prism in the diagram is a trapezium.
The lengths of the parallel sides of the trapezium are 8 cm and 6 cm.
The distance between the parallel sides of the trapezium is 5 cm.
The length of the prism is 20 cm.

a Work out the volume of the prism.

The prism is made out of gold.
Gold has a density of 19.3 g/cm^3.

b Work out the mass of the prism. Give your answer in kilograms.

17 A swimming pool has a cross-sectional area in the shape of a trapezium, as shown in the diagram.
Water is pumped in at 2 m³ per minute.
Using the dimensions shown in the diagram, find how long it takes to fill the pool?

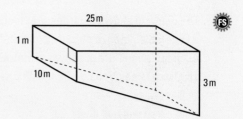

18 A running track consists of two 60 m straights and two semicircular bends of diameter 60 m.

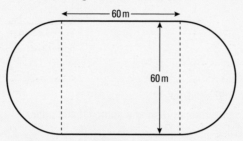

a Find the length of one lap of this running track.

b The owners of the track wish to stage athletics meetings and need it to be exactly 400 m long.
This can be done by just altering the straights or just widening the bends.
Calculate what adjustments would need to be made.

19 Discs of diameter 2 cm are cut from a metal strip that is 2 cm by 100 cm.

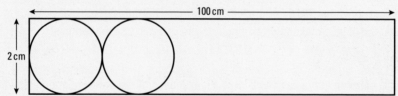

What is the minimum amount of waste material?

20 A cylindrical oil tank has a radius 60 cm and a length of 180 cm.
It is made from reinforced steel and is full of oil.
The oil has a density of 4.3 g/cm³.
The reinforced steel has a mass of 2.8 g/cm².
Find the total mass of the tank and the oil in kg.

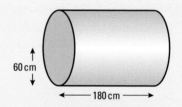

21 The solid shape, shown in the diagram, is made by cutting a hole all the way through a wooden cube.
The cube has edges of length 7 cm.
The hole has a square cross-section of side 2 cm.
a Work out the volume of wood in the solid shape.
The mass of the solid shape is 189 grams.
b Work out the density of the wood.

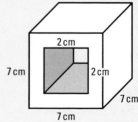

March 2009, adapted

11 AVERAGES AND RANGE

On average, the Rhode Island Red chicken lays 275 eggs per year. Not all chickens of this breed will lay exactly this number of eggs. What do we mean by average and how do we allow for different chickens laying different numbers of eggs?

◎ Objectives

In this chapter you will:
- ◎ learn about mean, mode and median
- ◎ consider the advantages and disadvantages of using each of these three measures of average
- ◎ look at range, quartiles and interquartile range.

◈ Before you start

You should know that:
- ◎ quantitative data are data that can be written as a number
- ◎ frequency is the number of times something occurs.

11.1 Finding the mode and median

⊙ **Objective**

○ You can find the mode and the median of a set of data.

❓ **Why do this?**

You could work out the middle or average value of the students' heights in your class to find out who is taller or shorter than average.

◈ **Get Ready**

1. Arrange the following sets of numbers in order, starting with the smallest number and ending with the largest.

   a 2 5 3 8 12 4 10 5 7 9 1 8   **b** 3.5 6.2 4.5 8.7 12.5 4.6 3.5

🕐 **Key Points**

◉ A quantitative data set is often described by giving a single value that is representative of all the values in the set. We call this value the **average**. For example, we might say, 'The average number of matches in a matchbox is 50'. There are three different measures of average commonly used:

  ◉ the **mode**
  ◉ the **median**
  ◉ the **mean**.

  This section introduces the first two of these measures.

◉ The mode of a set of discrete data is the value that occurs most frequently.

  ◉ It may not exist, if all values occur with exactly the same frequency.
  ◉ When it does exist, it will always be one of the observations.
  ◉ There may be more than one mode.
  ◉ It could be the smallest or largest value.

◉ The median is the middle value when the data are ordered from the smallest to the largest. It splits the data into two.

Lowest value	Median	Highest value
50% of data		50% of data

◉ If there are two middle values in a set of data the median is halfway between them. If there are n observations, add one to n then divide by 2. If this is a whole number the median is the value of this term; if it is not a whole number the median is midway between the values of the two whole numbers either side of it.

📌 **Example 1** Find the mode of each of the following sets of data.

  a The numbers of children in seven families are 0, 3, 4, 2, 6, 7, 2.
  b The numbers of monthly breakdowns recorded by a haulage firm over nine months were 3, 7, 6, 2, 2, 7, 8, 0, 4.
  c The numbers of goals scored by a football team in seven matches were 1, 6, 2, 3, 4, 0, 5.
  d The colours of a sample of five cars were red, green, blue, black, red.

a The mode is 2. ← 2 occurs twice; the other numbers appear only once.

b The modes are 2 and 7. ← Both 2 and 7 occur twice. There are two modes.

c There is no mode. ← All numbers occur with the same frequency.

d The mode is red. ← There are two red cars. There is one each of the other colours.

Example 2 ▶ Eleven people in an examination achieved the following marks. Find the median mark.

26, 32, 48, 37, 33, 32, 29, 41, 42, 36, 35

26 29 32 32 33 35 36 37 41 42 48 ← ┃ Arrange the numbers in order, lowest to highest. ┃

The median is 35. ← ┃ The middle value is 35; five marks are lower than this and five marks are higher. ┃

Example 3 ▶ The temperatures, in °C, at midday on ten consecutive days during a holiday were:

18, 21, 17, 12, 27, 18, 15, 28, 22 and 21.

Find the median temperature.

12 15 17 18 **18 21** 21 22 27 28 ← ┃ Put the numbers in order first. There are two middle values: 18°C and 21°C. ┃

The median is $\dfrac{18 + 21}{2} = 19.5\,°C.$ ← ┃ Halfway between 18 and 21 is 19.5 °C. ┃

⚙ **Exercise 11A**

┃ Questions in this chapter are targeted at the grades indicated. ┃

1 During a football season a school team played 30 matches.
They scored the following numbers of goals in each game.

1	6	7	2	3	3	0	6	3	2
0	1	5	3	0	2	1	4	0	6
2	5	1	0	2	2	4	0	1	2

Find the mode.

2 A salesman buys fuel every day at his local garage.
The following are the quantities he buys in one particular week.

Monday 20 litres   Tuesday 14 litres   Wednesday 16 litres   Thursday 12 litres
Friday 17 litres   Saturday 12 litres   Sunday 30 litres

Find the median amount he buys.

3 Listed below are the ages of the members of a girls' table tennis club.

17	16	18	16	17	15	18	16	18
17	18	17	16	18	17	16	17	17
16	14	16	16	15	15	14		

a Find the mode.   b Find the median age.

4 A train travelling from London to Glasgow makes seven stops.
The following figures show the number of passengers on each stage of the journey.

London to Milton Keynes   240   Milton Keynes to Crewe   250
Crewe to Warrington   220   Warrington to Preston   234
Preston to Lancaster   190   Lancaster to Penrith   170
Penrith to Carlisle   180   Carlisle to Glasgow   245

The train company wants to find the median number of passengers on the train.

11.2 Calculating the mean

Objective

- You can calculate the mean for a set of data.

Why do this?

A swimmer could calculate the mean of their race times to see how well they have done throughout the year.

Get Ready

1. Find the sum of each of the following sets of numbers.

   a 8 7 3 5 9 12

   b 1.6 2.4 8.1 3.1 5.2

Key Points

- The mean of a set of data is the sum of the values divided by the total number of observations. This can be shown as the following formula

$$\text{mean} = \frac{\text{sum of values}}{\text{number of values}}$$

- For a sample of n values of x, the mean $= \frac{\sum x}{n}$.

  $\sum$ is the Greek letter s (sigma) which is short for sum. This, therefore, means the sum of the xs divided by n.

Example 4 Find the mean of the numbers
2, 4, 6, 8, 10, 9, 6, 7.

$2 + 4 + 6 + 8 + 10 + 9 + 6 + 7 = 52$ ← Total the values.

The mean $= \frac{52}{8} = 6.5$ ← Divide by the number of values to get the mean.

Example 5 Over a career lasting 20 seasons, a footballer made a mean number of appearances for the English football team of 1.8 per season.
Work out how many times he played for England.

Total number of appearances $= 20 \times 1.8 = 36$. ← Sum of the values = mean × number of values.

Example 6

The times, to the nearest tenth of a second, run by an athlete in his last ten 400-metre races were: 48.3, 47.2, 49.3, 50.4, 48.6, 60.0, 48.0, 48.2, 51.2, 47.2.

a Find the mode.

b Find the median.

c Find the mean.

a The mode is 47.2 seconds. ← 47.2 occurs twice; each of the other times occurs once.

b 47.2 47.2 48.0 48.2 48.3 48.6 49.3 50.4 51.2 60.0 ← There are two middle values: 48.3 and 48.6.

The median is 48.45 seconds. ← Halfway between is 48.45 seconds.

c The total of all values = 47.2 + 47.2 + 48.0 + 48.2
    + 48.3 + 48.6 + 49.3 + 50.4 + 51.2 + 60.0 = 498.4 ← Find the total sum.

The mean = $\dfrac{498.4}{10}$ = 49.84 seconds. ← Divide by the total number of values.

Exercise 11B

1 A taxi driver has 11 calls during one day.
   The numbers of passengers she carries on each journey are as follows.

   5   4   1   3   6   4   3   1   2   1   3

   Work out the mean number of passengers she carried per journey.

2 An academy has the following numbers of boys and girls in each year.

Year	Boys	Girls
7	134	128
8	138	130
9	160	141
10	162	154
11	156	150
12	110	125
13	92	110

   a Find the mean number of boys per year.
   b Find the mean number of girls per year.

3   During four weeks in July a man earns a mean wage of £323 per week.
Work out how much he earns in total over the four-week period.

4   In a cricket match the eleven players scored the following numbers of runs.

60   23   10   0   12   56   17   10   21   35   20

a   Find the mode.
b   Find the median number of runs.
c   Find the mean number of runs.

11.3 Using the three types of average

◎ Objective

● You can discuss the advantages and disadvantages of the different measures of average.

❓ Why do this?

You would use the mode to work out the most popular meal in the school canteen, but you would use the median or mean to work out an average of how many people eat in the canteen each week.

◈ Get Ready

1. 24   20   28   25   10   19   20
a   Work out the median and mode of these data.
b   Calculate the mean correct to 2 significant figures.

Key Points

● Each of the three measures of average is useful in different situations.

● The following table will help you decide which of the three averages works best in different situations by showing a summary of the advantages and disadvantages of each measure.

Measure	Advantages	Disadvantages
MODE Use the mode when the data are non-numeric or when asked to choose the most popular item.	Extreme values (outliers) do not affect the mode. Can be used with qualitative data.	There may be more than one mode. There may not be a mode, particularly if the data set is small.
MEDIAN Use the median to describe the middle of a set of data that does have an extreme value.	Not influenced by extreme values.	Not as popular as mean. Actual value may not exist.
MEAN Use the mean to describe the middle of a set of data that does not have an extreme value.	Is the most popular measure. Can be used for further calculations. Uses all the data.	Affected by extreme values.

Example 7 A company consists of six workers, and their supervisor. The rates of pay of the six workers are £7, £7, £8, £9, £11 and £12 per hour. The supervisor is paid £25 per hour.

a Find the mode, median and mean rate of pay.

b Write down, giving a reason, which of the three averages you would use in the following situations.

   i When asked the typical wage rate.

   ii When trying to persuade a prospective employee to join the company.

a Mode = £7 per hour ← There are two £7 values.

Median = £9 per hour ← There are three values higher than £9 and three lower.

Total = 7 + 7 + 8 + 9 + 11 + 12 + 25 = 79 ← Total the values and divide by the total sum.

Mean = $\frac{79}{7}$ = £11.29

b i The mode of £7 is the lowest value partly because the number in the sample is small. You therefore would not use this as a 'typical' value. The median is probably the best measure in this case as it is unaffected by the high wage rate of the supervisor.

   ii There are only two values greater than the mean as the high wage rate of the supervisor has pulled the mean value up. As the mean is the highest average it is the one you would use to persuade a prospective employee to join the company.

ResultsPlus

Exam Tip

When comparing averages, look at how well each average represents the numbers as a whole, and give reasons why they would or would not be representative.

Exercise 11C

1 Five friends each buy a new dress for a party. They spend the following amounts of money.

£17   £148   £22   £17   £31

a Work out the mean, the mode and the median values.

b Which average would best describe the amount of money they spent? Give a reason for your answer.

2 Write down one advantage and one disadvantage of using the mean as an average.

3 A restaurant records the number of diners it has every day for a week. The numbers are as follows.

28   40   28   38   110   170   33

a Write down the mode.

b Work out the median number of diners.

c Work out the mean number of diners.

d The manager wishes to sell the restaurant. What average is he likely to use when talking to prospective buyers? Give a reason for your answer.

11.4 Using frequency tables to find averages

Objective

○ You can use a frequency table to find averages.

Why do this?

You can group some data in a frequency table, for example, goals scored in a football season or number of fans in the crowd per match.

Get Ready

1. The following numbers came up when a dice was thrown.

1 5 3 2 4 3 2 4
5 3 2 6 5 2 3 4

Represent these data as a frequency table.

Key Points

◉ When data are given in the form of a **frequency table**, the mode is the number that has the highest frequency.

◉ The median is the number that is the middle value, or halfway between the two middle values.

◉ To work out the mean for discrete data in a frequency table, use the following formula.

Mean $= \dfrac{\sum f \times x}{\sum f}$ where f is the frequency, x is the variable and $\sum$ means 'the sum of'.

◉ You can use the $\boxed{\Sigma x}$ key on a calculator to help you calculate the mean.

Example 8

The following table shows information about the number of goals scored per match over two seasons by a football team.

Number of goals	0	1	2	3	4	5
Frequency	8	15	12	7	3	1

a Write down the mode of these data.

b Find the median of these data.

c Work out the mean of these data.

a The mode is 1 goal. ← 1 has the highest frequency (15).

b

Number of goals (x)	Frequency (f)	Frequency × number of goals (f × x)
0	8	0
1	15	15
2	12	24
3	7	21
4	3	12
5	1	5
Total	46	77

There are 12 occasions when they scored 2 goals so there are 12 × 2 = 24 goals.

The total number of matches is 46.

The total number of goals is the sum of all the $f \times x$ values.

The total frequency is 46 so the median will be the 23.5th value. $\leftarrow$ $\frac{46+1}{2} = 23.5$
The median will be midway between the 23rd and 24th values.

There were 8 games with no goals scored.
There are $15 + 8 = 23$ games with 0 or 1 goals scored.
The 23rd value must be 1 and the 24th value must be 2.
The median is therefore $1\frac{1}{2}$.

c The mean is $\frac{77}{46} = 1.67$ goals. $\leftarrow$ $\text{Mean} = \dfrac{\text{Total number of goals}}{\text{Total frequency}} = \dfrac{\Sigma f \times x}{\Sigma f}$

Exercise 11D

1 The following table shows information about the numbers of siblings (brothers and sisters) a group of children have.

Number of siblings (x)	Frequency (f)	Frequency × number of siblings (f × x)
0	3	
1	8	
2	9	
3	4	
4	3	
5	0	
6	2	
7	1	
Total		

a Copy and complete the table.
b Write down the mode of these data.
c Work out the median number of siblings.
d Work out the mean number of siblings.

2 The table shows information about the numbers of paper clips in each packet of a box containing 60 packets.

Number of paper clips (x)	101	102	103	104	105	106	107
Frequency (f)	6	4	8	20	15	2	5

a Write down the mode of these data.
b Work out the median number of paper clips.
c Work out the mean number of paper clips.

3 Calgom Engineering employs apprentices.
The table shows information about the ages of the apprentices they have in 2008.

Ages of apprentices (x years)	17	18	19	20	21	22
Frequency (f)	10	8	8	2	2	2

a Write down the mode of these data.
b Work out the median age.
c Work out the mean age.

D

11.5 Modal class and median of grouped data

◎ Objectives

○ You can find the modal class for grouped data.
○ You can find a class interval containing the median of grouped data.

⟨?⟩ Why do this?

In order to understand large amounts of data, you can put it into groups. This can be useful when looking at the speeds of cars on a motorway, or the number of albums sold in the UK.

⟨⇧⟩ Get Ready

1. Here are 30 numbers.

1 2 5 16 1 3 18 14 7 12
12 9 4 16 18 10 6 13 2 18
19 4 6 3 11 11 2 19 4 3

Draw up a frequency table for these data using the following class intervals.

1–4   5–9   10–14   15–19

⟨⟩ Key Points

◉ When dealing with continuous data and class intervals, the class interval with the highest frequency is called the **modal class**.

◉ You cannot find the exact value of the median.
You can only write down the class interval in which the median falls.

Example 9

The frequency table below gives information about the number of phone calls made during a day by 21 people.

Number of calls (class interval)	Frequency
3–5	2
6–8	3
9–11	5
12–14	7
15–17	4

> These are discrete data. Each whole number appears in only one class. There will be $3 + 2 = 5$ values less than or equal to 8.

a Find the modal class.
b Find the class into which the median falls.

a The modal class is 12–14. ← Look for the class with the highest frequency.

b There are $3 + 2 = 5$ people that made less than 8 phone calls and $5 + 5 = 10$ people that made less than 12 calls so the median is in the class interval 9–11 calls. ← The median will be the 11th value.

Example 10 The following frequency table gives information about the speed (s), in miles per hour, of 50 cars.

Speed (s mph)	Frequency (f)
$20 \leqslant s < 25$	4
$25 \leqslant s < 30$	10
$30 \leqslant s < 35$	12
$35 \leqslant s < 40$	15
$40 \leqslant s < 45$	9

These are continuous data.

There will be 4 + 10 = 14 values less than 30.

a Find the modal class.

b Find the class into which the median falls.

a The modal class is $35 \leqslant s < 40$.

b The median falls in the class $30 \leqslant s < 35$.

There are 14 cars doing less than 30 mph and 26 doing less than 35 mph.

Exercise 11E

1 A manufacturer produces steel machine parts. The lengths of a sample of 200 parts are shown in the table below.

a Write down the modal class.
b Find the class into which the median falls.

Length (x cm)	Frequency (f)
$69.5 \leqslant x < 69.6$	2
$69.6 \leqslant x < 69.7$	10
$69.7 \leqslant x < 69.8$	30
$69.8 \leqslant x < 69.9$	34
$69.9 \leqslant x < 70.0$	35
$70.0 \leqslant x < 70.1$	56
$70.1 \leqslant x < 70.2$	33

D

2 The weekly wages of the employees in a vehicle repair workshop are shown in the following grouped frequency table.

Weekly wage (£s)	Frequency (f)
£240–£280	4
£281–£320	20
£321–£360	12
£361–£400	14

a Write down the modal class.
b Find the class into which the median falls.

C

C

3 The table shows the weights of silver deposited on an electrode over 30 different experiments.

Weight (x g)	Frequency (f)
$0.30 \leqslant x < 0.35$	3
$0.35 \leqslant x < 0.40$	7
$0.40 \leqslant x < 0.45$	6
$0.45 \leqslant x < 0.50$	14

a Write down the modal class.
b Find the class into which the median falls.

11.6 Estimating the mean of grouped data

Objective

● You can estimate the mean of grouped data.

Why do this?

You could work out the mean amount of time you spend on your mobile phone each month in order to choose the best monthly tariff for you.

Get Ready

1. Which number is halfway between:
 a 56 and 64   b 0.75 and 0.85   c 0.001 and 0.0001

Key Point

● An estimate for the mean of grouped data can be found by using the middle value of the class interval.

Example 11 Work out an estimate for the mean number of phone calls in Example 9.

Number of calls	Frequency (f)	Class midpoint (x)	$f \times x$
3−5	2	4	8
6−8	3	7	21
9−11	5	10	50
12−14	7	13	91
15−17	4	16	64
Totals	21		234

The middle value of the class 3–5 is 4.
The middle value of the class 6–8 is 7.
The 3 people in the class 6–8 might not all have made 7 calls.
This is why it is an estimated mean.

You can now use the formula.

$$\text{Estimated mean} = \frac{\Sigma f \times x}{\Sigma f} = \frac{234}{21} = 11.14 \text{ calls.}$$

Example 12 ▶ Work out an estimate for the mean speed of the cars in Example 10.

Speed (s mph)	Frequency (f)	Class midpoint (x)	f × x
20 ⩽ s < 25	4	22.5	90
25 ⩽ s < 30	10	27.5	275
30 ⩽ s < 35	12	32.5	390
35 ⩽ s < 40	15	37.5	562.5
40 ⩽ s < 45	9	42.5	382.5
Totals	50		1700

Estimated mean = $\dfrac{1700}{50}$ = 34 mph.

Exercise 11F

1 The following group frequency table shows the ages of members of an aerobics class.

Age range (years)	16–25	26–35	36–45	46–55	56–65	66–75
Frequency	4	10	12	4	8	2

Work out an estimate for the mean age of the members.

2 The manager of a supermarket recorded the length of time, in seconds, that customers had to wait in the checkout queue. The results are shown in the grouped frequency table below.

Waiting time (t seconds)	0 ⩽ t < 100	100 ⩽ t < 200	200 ⩽ t < 300	300 ⩽ t < 400
Frequency	10	46	20	8

Work out an estimate for the mean waiting time of the shoppers.

3 A call centre kept a record of the time, in seconds, that callers had to wait to speak to call centre staff over a period of 10 minutes. The results are shown in the grouped frequency table.

Waiting time (t seconds)	0 ⩽ t < 30	30 ⩽ t < 60	60 ⩽ t < 90	90 ⩽ t < 120	120 ⩽ t < 150
Frequency	30	55	26	13	6

Work out an estimate for the mean waiting time of the callers.

C

11.7 Range, quartiles and interquartile range

Objectives

- You can calculate the range of a set of data.
- You can work out the quartiles of a set of data.
- You can find the interquartile range for a set of data.
- You can compare data sets using a measure of average and a measure of range.

Why do this?

To get a better idea of the heights of students in your class, it helps to work out the range and quartiles.

Get Ready

1. Arrange the following numbers in ascending order.
   43   21   18   32   45   16   16   14   23   27   38   49
2. Arrange the following weights in ascending order.
   56.2 kg   43.4 kg   56.2 kg   49.9 kg   43.5 kg   36.0 kg

Key Points

- **Range** = highest value of a data set − lowest value of a data set.
- The **quartiles**, Q_1, Q_2 and Q_3, split the data into four parts.

| Lowest value | Q_1 | Q_2 | Q_3 | Highest value |

25% of data   25% of data   25% of data   25% of data

- For a set of data arranged in ascending order:
  - Q_1, the **lower quartile**, is a quarter of the way through the data
  - Q_2, the second quartile, is halfway through the data (the median)
  - Q_3, the **upper quartile**, is three-quarters of the way through the data.
- Generally for a set of n data values arranged in ascending order:
  - Q_1 is the $\left(\dfrac{n+1}{4}\right)$th value
  - Q_2 (median) is the $\left(\dfrac{n+1}{2}\right)$th value
  - Q_3 is the $\left(\dfrac{3(n+1)}{4}\right)$th value.
- The **interquartile range** (IQR) = $Q_3 - Q_1$.
- The average and the range together give a description of the **distribution** of the data.
- To compare the distributions of sets of data you need to give a measure of average and a measure of spread.

Example 13

Eight students sat two examinations. Their marks, out of 30, are shown below.

Mathematics: 20, 16, 30, 17, 25, 21, 22, 19.

English: 14, 16, 23, 28, 24, 12, 21, 13.

a Work out the range of marks for each exam.

b Which set of marks was the more consistent? Give a reason for your answer.

a Range for mathematics exam = 30 − 16 = 14.
   Range for English exam = 28 − 12 = 16.

b The mathematics marks were more consistent.
   This is because the range was smaller.

Example 14 Find the quartiles and interquartile range of the following set of data:
570, 460, 600, 480, 500, 510, 340, 560, 320, 590, 650.

320 340 **460** 480 500 **510** 560 570 **590** 600 650 ← Write the data in order, starting with the lowest value.

$Q_1 = 460$ $Q_2 = 510$ $Q_3 = 590$ ← Find the quartiles.

$IQR = 590 - 460 = 130$

Example 15 The following data give information about the heights, in metres, of trees commonly found in English hedgerows: 20, 18, 30, 10, 31, 4, 12, 18, 27, 7, 24, 24, 30, 6, 10.
Find the upper and lower quartiles and interquartile range of the heights.

4 6 7 10 10 12 18 18 20 24 24 27 30 30 31 ← Put the data in ascending order.

$n = 15$

$Q_1 = \dfrac{15 + 1}{4} = $ 4th value $= 10\,$m ← Use the formula $\dfrac{n + 1}{4}$.

$Q_3 = \dfrac{3(15 + 1)}{4} = $ 12th value $= 27\,$m ← Use the formula $\dfrac{3(n + 1)}{4}$.

$IQR = 27 - 10 = 17\,$m ← $IQR = Q_3 - Q_1$

Example 16 The heights, in cm, of 11 men and their sons are given below.
Men's heights: 150, 152, 155, 160, 165, 170, 175, 180, 180, 190, 198.
Sons' heights: 163, 166, 168, 170, 170, 173, 175, 178, 183, 183, 185.
a Find the means and interquartile ranges of these data.
b Compare and contrast these results.

A03

a Men's heights:

Mean $= \dfrac{150 + 152 + 155 + 160 + 165 + 170 + 175 + 180 + 180 + 190 + 198}{11}$

$= 170.45\,$cm

$Q_1 = (\dfrac{11 + 1}{4} = $ 3rd value$) = 155\,$cm

$Q_3 = \dfrac{3(11 + 1)}{4} = $ 9th value $= 180\,$cm ← The data are already ordered. Use the formula to find the quartiles.

$IQR = 180 - 155 = 25\,$cm

Sons' heights:

Mean $= \dfrac{163 + 166 + 168 + 170 + 170 + 173 + 175 + 178 + 183 + 183 + 185}{11}$

$= 174\,$cm

$Q_1 = 168\,$cm, $Q_3 = 183\,$cm, $IQR = 183 - 168 = 15\,$cm | Mean $= \dfrac{\text{total of values}}{\text{total frequency}}$

b The mean height of the sons was higher than that of their fathers.
Sons are generally taller than their fathers.
The IQR of the fathers was higher than that of their sons.
The fathers' heights were more spread out.

The mean is used in preference to the median as there are no extreme values.

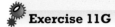 **Exercise 11G**

B

1   Write down another name for Q_2.

2   Eleven college students were asked to record the amount of time they spent on the internet one evening. Their times, in minutes, were:

| 38 | 42 | 50 | 56 | 60 | 62 | 65 | 70 | 70 | 75 | 80 |

   a   Write down Q_1, Q_2 and Q_3 for these data.

   b   Work out the interquartile range.

   c   Work out the range.

3   A lepidopterist set a moth trap for 15 evenings. She recorded the number of moths trapped. They were:

| 5 | 9 | 15 | 12 | 21 | 14 | 19 | 8 |
| 11 | 24 | 16 | 13 | 20 | 7 | 6 | |

   a   Write down Q_1, Q_2 and Q_3 for these data.

   b   Work out the interquartile range.

   c   Work out the range.

4   The number of bags of crisps sold per day in a general shop was recorded over 13 days. The results are shown below.

| 32 | 45 | 36 | 56 | 45 | 68 | 29 | 48 |
| 21 | 45 | 32 | 47 | 59 | | | |

   a   Write down Q_1, Q_2 and Q_3 for these data.

   b   Work out the interquartile range.

   c   Work out the range.

Chapter review

- The **mode** of a set of discrete data is the value that occurs most frequently.
- The **median** is the middle value when the data are ordered from the smallest to the largest.
- If there are two middle values in a set of data, the median is halfway between them.
- The **mean** of a set of data is the sum of the values divided by the total number of observations.
- For a sample of n values of x, mean $= \dfrac{\sum x}{n}$
- For discrete data in a **frequency table**,

  mean $= \dfrac{\sum f \times x}{\sum f}$ where f is the frequency, x is the variable and $\sum$ means 'the sum of'.
- For grouped data:
    - The class interval with the highest frequency is called the **modal class**.
    - You can only write down the class interval in which the median falls.

- An estimate for the mean of grouped data can be found by using the middle value of the class interval.
- **Range** = highest value − lowest value.
- For a set of data arranged in ascending order, the **quartiles**, Q_1, Q_2 and Q_3, split the data into four parts:
  - Q_1, the **lower quartile**, is a quarter of the way through the data
  - Q_2, the second quartile, is halfway through the data
  - Q_3, the **upper quartile**, is three-quarters of the way through the data.
- Generally for a set of n data values arranged in ascending order:
  - Q_1 is the $\left(\dfrac{n+1}{4}\right)$th value
  - Q_2 (median) is the $\left(\dfrac{n+1}{2}\right)$th value
  - Q_3 is the $\left(\dfrac{3(n+1)}{4}\right)$th value.
- The **interquartile range** (IQR) $= Q_3 - Q_1$.
- The average and the range together give a description of the **distribution** of the data.
- To compare the distributions of sets of data you need to give a measure of average and a measure of spread.

Review exercise

1. Peter rolled a 6-sided dice ten times.
    Here are his scores:

    3   2   4   6   3   3   4   2   5   4

    a Work out the median of his scores.

    b Work out the mean of his scores.

    c Work out the range of his scores.

    June 2007

2. Nine friends go to a charity shop. They spend the following amounts of money:

    £4   £6   £4   £38   £10   £4   £3   £7   £5

    a Work out the mode, the median and the mean of the amounts they spent.

    b Which of these three averages best describes the amount they spent? Give a reason for your answer.

3. Write down one advantage and one disadvantage of using each of the following as an average:

    a the mode   b the median   c the mean.

4. The frequency table below shows the number of aeroplanes that took off from a small airport each hour during one day in January 2010.

Number of aeroplanes	Frequency
0	3
1	1
2	2
3	2
4	8
5	5
6	3

    a Work out how many aeroplanes took off in total during the day.

    b Work out the mean number of aeroplanes taking off per hour.

D

D

5 Ali found out the number of rooms in each of 40 houses in a town.
He used the information to complete the frequency table.

Number of rooms	Frequency	
4	4	
5	7	
6	10	
7	12	
8	5	
9	2	

Ali said that the mode is 9.
Ali is wrong.

a Explain why.

b Calculate the mean number of rooms.

Nov 2007

6 Majid carried out a survey of the number of school dinners 32 students had in one week.
The table shows this information.

Number of school dinners	0	1	2	3	4	5
Frequency	0	8	12	6	4	2

Calculate the mean.

Nov 2008

7 The mean of eight numbers is 41.
The mean of two of the numbers is 29.
What is the mean of the other six numbers?

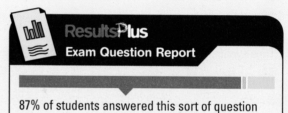

ResultsPlus
Exam Question Report

87% of students answered this sort of question
poorly.

June 2007

8 A man kept a record of the number (x) of junk emails he received each day over a period of 100 days.
Given $\Sigma x = 770$, work out the mean number of junk emails he received each day.

C

9 The wing spans of 30 emperor dragonflies were measured.
The results are shown in the following grouped frequency table.

Wing span (l cm)	Frequency (f)
9.6 – 9.8	3
9.9 – 10.1	4
10.2 – 10.4	9
10.5 – 10.7	14

a Find the modal class.

b Find the class into which the median falls.

10 A vet keeps a record of the weight of all dogs brought into his surgery.
The table shows the adult weights (w kg) of the labradors in his records.

Class interval	Frequency (f)	Class midpoint	$f \times x$
$26 \leqslant w < 29$	4		
$29 \leqslant w < 32$	7		
$32 \leqslant w < 35$	15		
$35 \leqslant w < 38$	12		
$38 \leqslant w < 41$	2		
Totals			

a Copy and complete the table.

b Work out an estimate for the mean weight of the dogs.

11 A researcher was conducting a study into the growth patterns of mice. She recorded the body length of 28 mice. The lengths (l mm) are shown in the following grouped frequency table.

Class interval	Frequency (f)
$70 \leqslant x < 75$	3
$75 \leqslant x < 80$	3
$80 \leqslant x < 85$	5
$85 \leqslant x < 90$	12
$90 \leqslant x < 95$	5

ResultsPlus
Exam Question Report

81% of students answered this sort of question poorly.

Work out an estimate for the mean body length of the mice.

12 Josh asked 30 students how many minutes they each took to get to school.
The table shows some information about his results.

Time (t minutes)	Frequency
$0 < t \leqslant 10$	6
$10 < t \leqslant 20$	11
$20 < t \leqslant 30$	8
$30 < t \leqslant 40$	5

Work out the estimate for the mean number of minutes taken by the 30 students.   *Nov 2008*

13 The table gives information about the times, in minutes, that 106 shoppers spent in a supermarket.

Time (t minutes)	Frequency
$0 < t \leqslant 10$	20
$10 < t \leqslant 20$	17
$20 < t \leqslant 30$	12
$30 < t \leqslant 40$	32
$40 < t \leqslant 50$	25

a Find the class interval that contains the median.

b Calculate an estimate for the mean time that the shoppers spent in the supermarket.
Give your answer correct to 3 significant figures.   *Nov 2007*

C

14 Sethina recorded the times, in minutes, taken to repair 80 car tyres.
Information about these times is shown in the table.

Time (t minutes)	Frequency
$0 < t \leqslant 6$	15
$6 < t \leqslant 12$	25
$12 < t \leqslant 18$	20
$18 < t \leqslant 24$	12
$24 < t \leqslant 30$	8

Calculate an estimate for the mean time taken to repair each car tyre.

June 2009

A02
A03

* **15** There are 50 students in each of the year groups at a school.
A survey was carried out to find how many pets these students owned.
The table shows these results.

Number of pets	0	1	2	3	4
Year 9	1	29	14	5	1
Year 10	5	22	19	4	0
Year 11	32	11	6	1	0

Which year group has the least number of pets?
You must show all your calculations.

B

16 As part of an ongoing research programme, the pups in a small colony of grey seals were weighed, to the nearest kilogram, at the age of four weeks. Their weights were as follows:

42   40   45   47   50   48   39   47   42   50   49

a   Write down Q_1, Q_2 and Q_3 for these data.

b   Work out the interquartile range.

c   Work out the range of the weights.

17 A council is introducing a new traffic management scheme to speed up morning rush-hour traffic. Before the scheme, 11 council workers are asked to record their journey time to work one Wednesday morning. After the scheme was put in place the same 11 workers were asked to record their journey time again one Wednesday morning. The results are shown in the table below.

Worker	A	B	C	D	E	F	G	H	I	J	K
Before (min)	23	30	10	13	15	22	16	19	21	14	15
After (min)	20	25	8	13	12	16	14	17	19	10	11

a   Find the mean time taken before the traffic scheme was introduced.

b   Find the mean time taken after its introduction.

c   Find Q_1 and Q_3 for both sets of data.

d   Find the interquartile ranges.

A03

e   Compare the time taken before the introduction of the traffic scheme with the time taken after it was introduced.

* **18** Ten people work in a small factory. The table shows their salaries.

A03 B

Employees	Salary
1 owner	£180 000
1 manager	£40 000
8 workers	£10 000

The workers want a pay rise, but the owner doesn't want to give them a rise.

Explain how both the owner and the workers could use the word 'average' to justify their case.

* **19** Explain the following sentence:

A03 A

The vast majority of dogs in this country have more than the average number of legs.

12 CONSTRUCTIONS AND LOCI

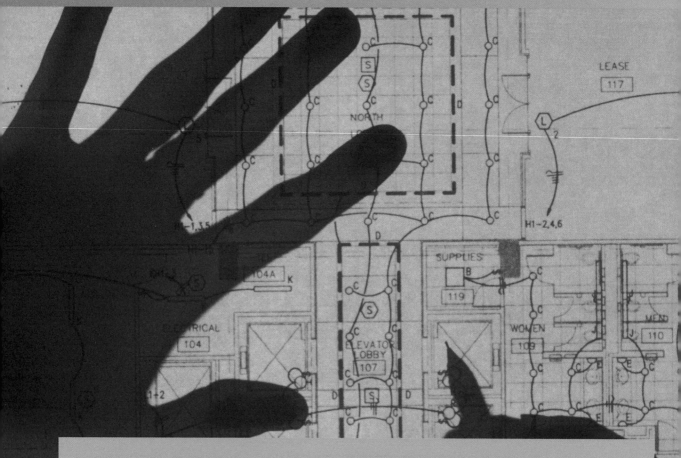

Architects make scale drawings of projects they are working on for both planning and presentation purposes. Originally these were done on paper using ink, and copies had to be made laboriously by hand. Later they were done on tracing paper so that copying was easier. Computer-generated drawings have now largely taken over, but, for many of the top architecture firms, these too have been replaced, by architectural animation.

◉ Objectives

In this chapter you will:
- ◉ use a ruler and a pair of compasses to draw triangles given the lengths of the sides
- ◉ use a straight edge and a pair of compasses to construct perpendiculars and bisectors
- ◉ construct and bisect angles using a pair of compasses
- ◉ draw loci and regions
- ◉ learn how to draw, use and interpret scale drawings.

◈ Before you start

You need to:
- ◉ be able to make accurate drawings of triangles and 2D shapes using a ruler and a protractor
- ◉ be able to draw parallel lines using a protractor and ruler
- ◉ have some understanding of ratio
- ◉ be able to change from one metric unit of length to another.

12.1 Constructing triangles

◉ Objective

◉ You can draw a triangle when given the lengths of its sides.

⟐ Why do this?

If you were redesigning a garden and wanted a triangular border you would need to make a plan first and draw the triangles accurately.

⟐ Get Ready

1. Use a ruler and protractor to make an accurate drawing of this triangle. Measure AC, BC and angle ACB.

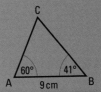

🔑 Key Points

◉ Two triangles are congruent if they have exactly the same shape and size. One of four conditions must be true for two triangles to be congruent: SSS, SAS, ASA and RHS (see Section 8.1).
◉ Constructing a triangle using any one of these sets of information therefore creates a unique triangle.
◉ More than one possible triangle can be created from other sets of information.

⟐ Example 1

Make an accurate drawing of the triangle shown in the sketch.

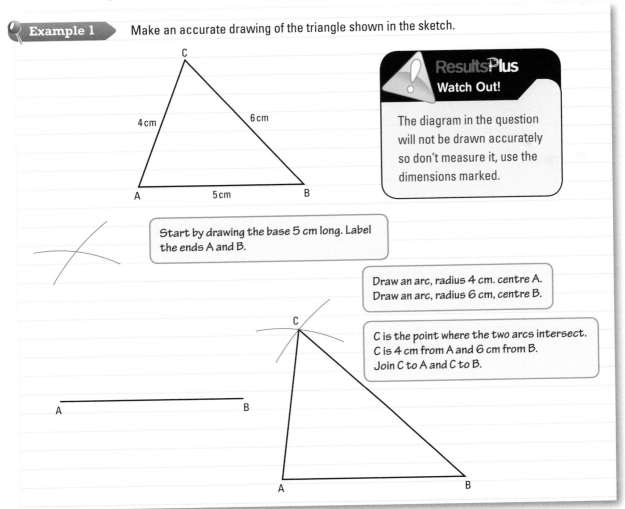

ResultsPlus
Watch Out!

The diagram in the question will not be drawn accurately so don't measure it, use the dimensions marked.

Start by drawing the base 5 cm long. Label the ends A and B.

Draw an arc, radius 4 cm. centre A.
Draw an arc, radius 6 cm, centre B.

C is the point where the two arcs intersect.
C is 4 cm from A and 6 cm from B.
Join C to A and C to B.

Example 2 Show that there are two possible triangles ABC in which AB = 5.6 cm, BC = 3.3 cm and angle A = 31°.

A————————B ← Draw the line AB with length 5.6 cm.

A⟋31°————B ← Using a protractor, draw an angle of 31° at A.

← Draw an arc of 3.3 cm from point B, to locate the possible positions of C.
Triangle ABC₁ and ABC₂ both have the given measurements.

Exercise 12A

Questions in this chapter are targeted at the grades indicated.

D

1   Here is a sketch of triangle XYZ.
     Construct triangle XYZ.

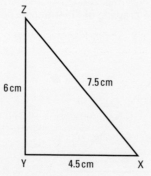

2   Construct an equilateral triangle with sides of length 5 cm.

3   Construct the triangle XYZ with sides XY = 4.2 cm, YZ = 5.8 cm and ZX = 7.5 cm.

4   Here is a sketch of the quadrilateral CDEF.
     Make an accurate drawing of quadrilateral CDEF.

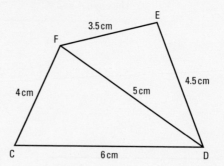

5   The rhombus KLMN has sides of length 5 cm.
     The diagonal KM = 6 cm.
     Make an accurate drawing of the rhombus KLMN.

6   Explain why it is not possible to construct a triangle with sides of length 4 cm, 3 cm and 8 cm.

12.2 Perpendicular lines

◎ Objective

◎ You can construct perpendicular lines using a straight edge and compasses.

❓ Why do this?

Many structures involve lines or planes that are perpendicular, for example the walls and floor of a house are perpendicular.

🔱 Get Ready

1. Draw a circle with a radius of 4 cm.
2. Mark two points A and B 6 cm apart. Mark the points that are 5 cm from A and 5 cm from B.
3. Draw two straight lines which are perpendicular to each other.

🔧 Key Points

◎ A **bisector** cuts something exactly in half.
◎ A **perpendicular bisector** is at right angles to the line it is cutting.
◎ You can use a straight edge and compass in the **construction** of the following:
   ◎ the perpendicular bisector of a **line segment**
   ◎ the perpendicular to a line segment from a point on it
   ◎ the perpendicular to a line segment from a point not on the line.

Example 3 — Construct the perpendicular bisector of the line AB.

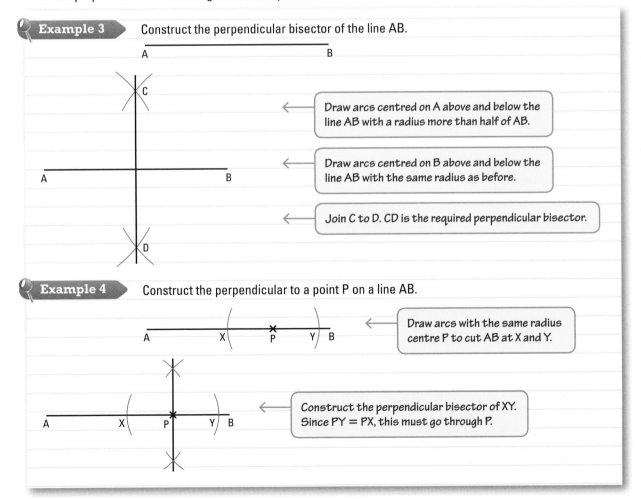

Draw arcs centred on A above and below the line AB with a radius more than half of AB.

Draw arcs centred on B above and below the line AB with the same radius as before.

Join C to D. CD is the required perpendicular bisector.

Example 4 — Construct the perpendicular to a point P on a line AB.

Draw arcs with the same radius centre P to cut AB at X and Y.

Construct the perpendicular bisector of XY. Since PY = PX, this must go through P.

Example 5 Construct the perpendicular to a line AB from a point P not on the line.

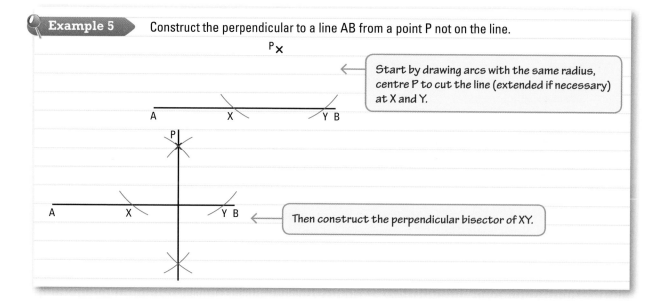

Start by drawing arcs with the same radius, centre P to cut the line (extended if necessary) at X and Y.

Then construct the perpendicular bisector of XY.

Exercise 12B

C

1 Draw line segments of length 10 cm and 8 cm. Using a straight edge and a pair of compasses, construct the perpendicular bisector of each of these line segments.

2 Draw these lines accurately, and then construct the perpendicular from the point P.

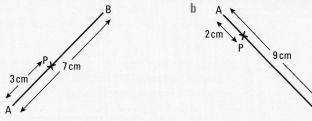

a

b

3 Draw a line segment AB, a point above it, P, and a point below it, Q. Construct the perpendicular from P to AB, and from Q to AB.

12.3 Constructing and bisecting angles

⊚ Objectives

- You can construct certain angles using compasses.
- You can construct the bisector of an angle using a straight edge and compasses.
- You can construct a regular hexagon inside a circle.

⊘ Why do this?

You may need to bisect an angle accurately when cutting a tile to place in an awkward corner.

⊘ Get Ready

1. Draw a circle with a radius of 3 cm.

2. Draw an angle of 60°.

3. Use a protractor to bisect an angle of 60°.

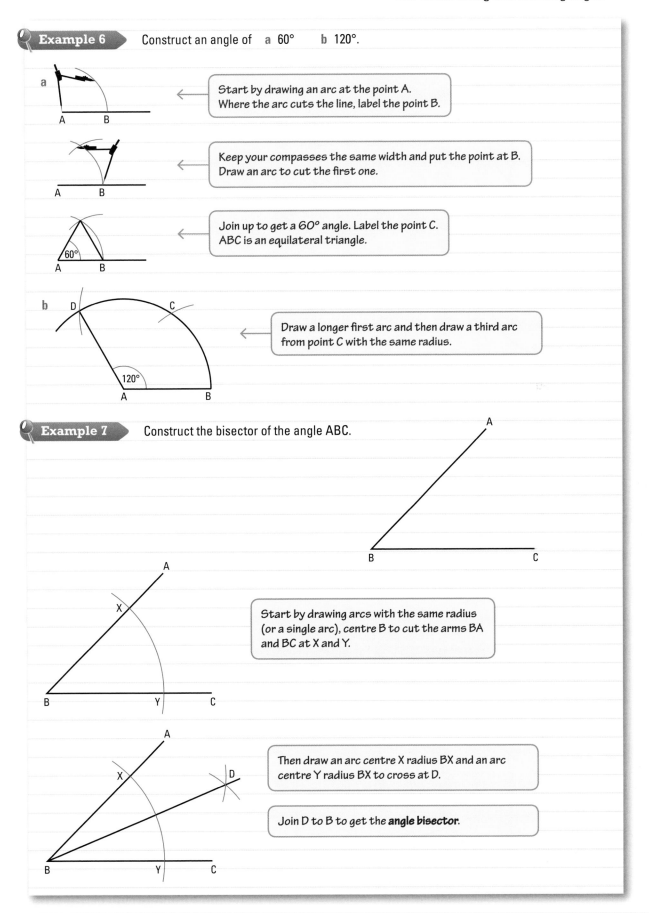

Example 6 Construct an angle of **a** 60° **b** 120°.

a

Start by drawing an arc at the point A.
Where the arc cuts the line, label the point B.

Keep your compasses the same width and put the point at B.
Draw an arc to cut the first one.

Join up to get a 60° angle. Label the point C.
ABC is an equilateral triangle.

b

Draw a longer first arc and then draw a third arc
from point C with the same radius.

Example 7 Construct the bisector of the angle ABC.

Start by drawing arcs with the same radius
(or a single arc), centre B to cut the arms BA
and BC at X and Y.

Then draw an arc centre X radius BX and an arc
centre Y radius BX to cross at D.

Join D to B to get the **angle bisector**.

Example 8   Construct a regular hexagon inside a circle.

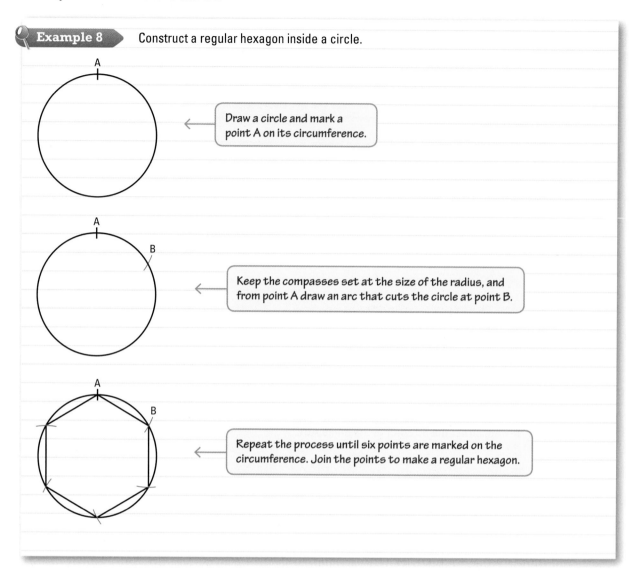

Draw a circle and mark a point A on its circumference.

Keep the compasses set at the size of the radius, and from point A draw an arc that cuts the circle at point B.

Repeat the process until six points are marked on the circumference. Join the points to make a regular hexagon.

Exercise 12C

1   Copy the diagrams and construct the bisector of the angle ABC.

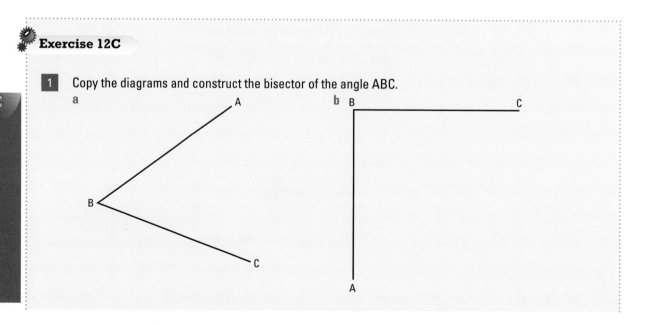

a

b

C

2 Copy the diagrams and construct the bisector of angle Q in the triangle PQR.

a

b

3 Construct each of the following angles.

a 60°   b 120°   c 90°   d 30°   e 45°

4 Draw a regular hexagon in a circle of radius 4 cm.

5 Draw a regular octagon in a circle of radius 4 cm.

12.4 Loci

◉ Objective

○ You can draw the locus
of a point.

◈ Why do this?

Scientists studying interference effects of radio waves need to plot paths
that are equidistant from two or more transmitters. They use loci to do this.

◈ Get Ready

1. Put a cross in your book. Mark some points which are 3 cm from the cross.
2. Put two crosses A and B less than 3 cm apart in your book. Mark points which are 3 cm from each cross.
3. Draw two parallel lines. Mark any points which are the same distance from both lines.

◈ Key Points

◉ A **locus** is a line or curve, formed by points that all satisfy a certain condition.

◉ A locus can be drawn such that:

   ◉ its distance from a fixed point is constant
   ◉ it is **equidistant** from two given points
   ◉ its distance from a given line is constant
   ◉ it is equidistant from two lines.

Example 9

Show the locus of all points which are at a
distance of 3 cm from the fixed point O.

The locus is a circle, radius 3 cm, centre O.
All the points on the circle are 3 cm from O.

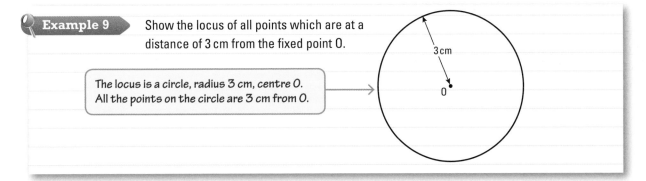

Example 10 Show the locus of all points which are equidistant from the points X and Y.

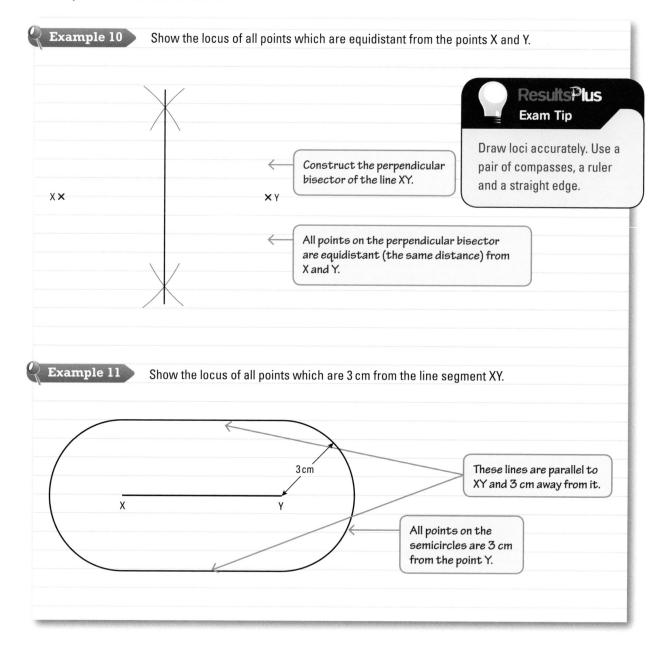

Construct the perpendicular bisector of the line XY.

ResultsPlus
Exam Tip

Draw loci accurately. Use a pair of compasses, a ruler and a straight edge.

X ✕   ✕ Y

All points on the perpendicular bisector are equidistant (the same distance) from X and Y.

Example 11 Show the locus of all points which are 3 cm from the line segment XY.

These lines are parallel to XY and 3 cm away from it.

3 cm

X   Y

All points on the semicircles are 3 cm from the point Y.

Exercise 12D

1 Mark two points A and B approximately 6 cm apart.
Draw the locus of all points that are equidistant from A and B.

2 Draw the locus of all points which are 3.5 cm from a point P.

3 Draw the locus of a point that moves so that it is always 1.5 cm from a line 5 cm long.

4 Draw two lines PQ and QR, so that the angle PQR is acute. Draw the locus of all points that are equidistant between the two lines PQ and QR.

C

12.5 Regions

⊙ Objective

○ You can draw regions.

⊘ Why do this?

If you tether a goat to a point in your garden to eat the grass, you might want to check that the region it can access doesn't include the flowerbed.

⬙ Get Ready

1. Put a cross in your book. Mark some points which are less than 3 cm from the cross.
2. Put two crosses A and B in your book. Mark points which are closer to A than to B.
3. Draw two parallel lines. Mark any points which are further from one line than the other.

⬙ Key Points

◉ A set of points can lie inside a **region** rather than on a line or curve.

◉ The region of points can be drawn such that:
  ◉ the points are greater than or less than a given distance from a fixed point
  ◉ the points are closer to one given point than to another given point
  ◉ the points are closer to one given line than to another given line.

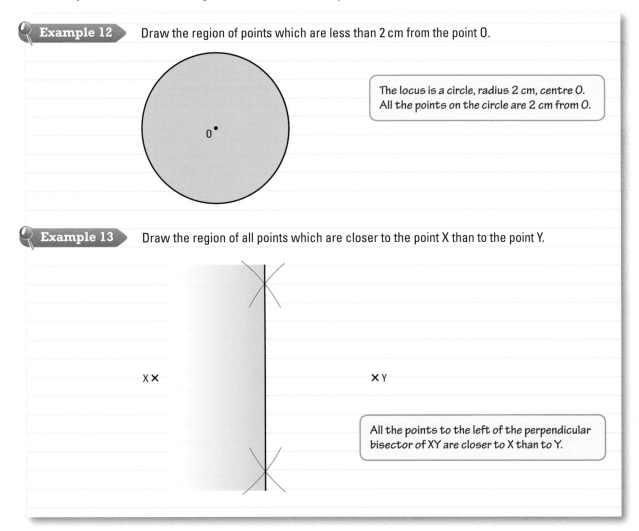

Example 12 ▷ Draw the region of points which are less than 2 cm from the point O.

O•

The locus is a circle, radius 2 cm, centre O.
All the points on the circle are 2 cm from O.

Example 13 ▷ Draw the region of all points which are closer to the point X than to the point Y.

X✕   ✕ Y

All the points to the left of the perpendicular bisector of XY are closer to X than to Y.

Example 14　ABCD is a square of side 4 cm. Draw the region of points inside the rectangle that are both more than 3 cm from point A and more than 2 cm from the line BC.

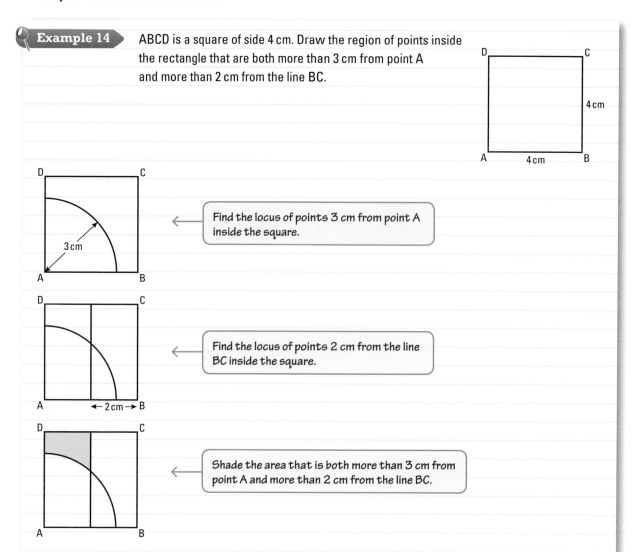

Find the locus of points 3 cm from point A inside the square.

Find the locus of points 2 cm from the line BC inside the square.

Shade the area that is both more than 3 cm from point A and more than 2 cm from the line BC.

Exercise 12E

1　Shade the region of points which are less than 2 cm from a point P.

2　Shade the region of points which are less than 2.6 cm from a line 4 cm long.

3　Mark two points, G and H, roughly 3 cm apart.
Shade the region of points which are closer to G than to H.

4　Draw two lines DE and EF, so that the angle DEF is acute. Shade the region of points which are closer to EF than to DE.

5　Baby Tommy is placed inside a rectangular playpen measuring 1.4 m by 0.8 m. He can reach 25 cm outside the playpen. Show the region of points Tommy can reach beyond the edge of the playpen.

12.6 Scale drawings and maps

Objectives

○ You can read and construct scale drawings.
○ You can draw lines and shapes to scale and estimate lengths on scale drawings.
○ You can work out lengths using a scale factor.

Why do this?

When a new aeroplane is being designed or an extension to a house is planned, accurate scale drawings have to be made.

Get Ready

1. Convert from cm to km:
   a 5 000 000 cm
   b 250 000 cm.

2. Convert from km to cm:
   a 4 km
   b 0.3 km.

Key Points

◉ Here is a picture of a scale model of a Saturn rocket. The model has been built to a scale of 1 : 24. This means that every length on the model is shorter than the length on the real rocket, with a length of 1 cm on the model representing a length of 24 cm on the real rocket.

   ◉ The real rocket is an enlargement of the model with a **scale factor** of 24; the model is a smaller version of the real rocket with a scale factor of $\frac{1}{24}$.
◉ In general, a scale of 1 : n means that:
   ◉ a length on the real object = the length on the **scale diagram** or model × n
   ◉ a length on the scale drawing or model = the length on the real object ÷ n.

Example 15 The Empire State Building is 443 m tall. Bill has a model of the building that is 88.6 cm tall.
   a Calculate the scale of the model. Give your answer in the form 1 : n.
   b The pinnacle at the top of Bill's model is 12.4 cm in length. Work out the actual length of the pinnacle at the top of the Empire State Building. Give your answer in metres.

a Height of building = 443 × 100 = 44 300 cm ⟵ | Both heights have to be in the same units. Change 443 m to cm by multiplying by 100.

Scale factor = $\frac{44\,300}{88.6}$ = 500 ⟵ | Scale factor = $\frac{\text{Height of building}}{\text{Height of model}}$

Scale of model = 1 : 500

b Length of pinnacle on building = 12.4 × 500
   = 6200 cm ⟵ | Length on model = Length on building ÷ 500. Length on building = Length on model × 500.
Length of pinnacle on building = 6200 ÷ 100
   = 62 m ⟵ | Change cm to m by dividing by 100.

Example 16 The scale of a map is 1 : 50 000.

    a On the map, the distance between two churches is 6 cm. Work out the real distance between the churches. Give your answer in kilometres.

    b The real distance between two train stations is 12 km. Work out the distance between the two train stations on the map. Give your answer in centimetres.

Method 1

A scale of 1 : 50 000 means:
real distance = map distance × 50 000.

a Real distance between churches
    = 6 × 50 000 = 300 000 cm

Change cm to m, divide by 100.
Change m to km, divide by 1000.

    = 3000 m
    = 3 km

b 12 km = 12 × 1000 × 100 = 1 200 000 cm

Change km to cm by multiplying by 1000 × 100.

Distance between stations on map
    = 1 200 000 ÷ 50 000 = 24 cm

Map distance = real distance ÷ 50 000

Method 2

Map distance of 1 cm represents real distance of 0.5 km.

1 : 50 000 means 1 cm : 50 000 cm
or 1 cm : 500 m
or 1 cm : 0.5 km

a 6 cm on the map represents real distance of 6 × 0.5 = 3 km.
Distance between the churches = 3 km.

b Real distance of 12 km represents map distance of 12 ÷ 0.5 = 24 cm.
Distance between the stations on map = 24 cm.

Exercise 12F

1 This is an accurate map of a desert island. There is treasure buried on the island at T.
Key to map
P palm trees   R rocks
C cliffs   T treasure
The real distance between the palm trees and the cliffs is 5 km.

    a Find the scale of the map. Give your answer in the form 1 cm represents n km, giving the value of n.

    b Find the real distance of the treasure from: **i** the cliffs **ii** the palm trees **iii** the rocks.

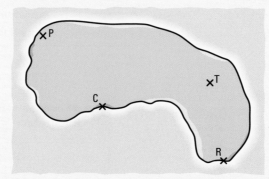

2 On a map of England, 1 cm represents 10 km.
  a The distance between Hull and Manchester is 135 km. Work out the distance between Hull and Manchester on the map.
  b On the map, the distance between London and York is 31.2 cm. Work out the real distance between London and York.

3 Here is part of a map, not accurately drawn, showing three towns: Alphaville (A), Beecombe (B) and Ceeton (C).
  a Using a scale of 1 : 200 000, accurately draw this part of the map.
  b Find the real distance, in km, between Beecombe and Ceeton.
  c Use the scaled drawing to measure the bearing of Ceeton from Beecombe.

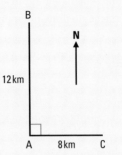

4 This is a sketch of Arfan's bedroom. It is *not* drawn to scale. Draw an accurate scale drawing on cm squared paper of Arfan's bedroom. Use a scale of 1 : 50.

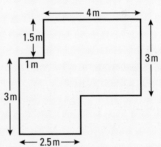

5 A space shuttle has a length of 24 m. A model of the space shuttle has a length of 48 cm.
  a Find, in the form 1 : n, the scale of the model.
  b The height of the space shuttle is 5 m. Work out the height of the model.

6 The distance between Bristol and Hull is 330 km. On a map, the distance between Bristol and Hull is 6.6 cm.
  a Find, as a ratio, the scale of the map.
  b The distance between Bristol and London is 183 km. Work out the distance between Bristol and London on the map. Give your answer in centimetres.

Chapter review

⊙ Two triangles are congruent if they have exactly the same shape and size. One of four conditions must be true for two triangles to be congruent: SSS, SAS, ASA and RHS.

⊙ Constructing a triangle using any one of these sets of information therefore creates a unique triangle.

⊙ More than one possible triangle can be created from other sets of information.

⊙ A **bisector** cuts something exactly in half.

⊙ A **perpendicular bisector** is at right angles to the line it is cutting.

⊙ A **locus** is a line or curve, formed by points that all satisfy a certain condition.

⊙ A locus can be drawn such that
  ⊙ its distance from a fixed point is constant
  ⊙ it is **equidistant** from two given points
  ⊙ its distance from a given line is constant
  ⊙ it is equidistant from two lines.

◉ A set of points can lie inside a **region** rather than on a line or curve.

◉ A region of points can be drawn such that:

  ◉ the points are greater than or less than a given distance from a fixed point

  ◉ the points are closer to one given point than to another given point

  ◉ the points are closer to one given line than to another given line.

◉ A scale of $1 : n$ means that:

  ◉ a length on the real object = the length on the **scale diagram** or model $\times n$

  ◉ a length on the scale drawing or model = the length on the real object $\div n$.

D

✎ Review exercise

1 AB = 8 cm. AC = 6 cm. Angle A = 52°.
Make an accurate drawing of triangle ABC.

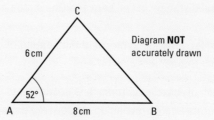

Diagram **NOT** accurately drawn

Nov 2008

2 Make an accurate drawing of the quadrilateral ABCD.

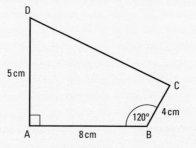

Diagram **NOT** accurately drawn

3 Make an accurate drawing of triangle ABC.

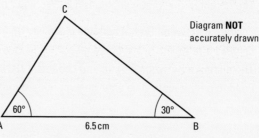

Diagram **NOT** accurately drawn

May 2009

4 Make an accurate drawing of triangle PQR.

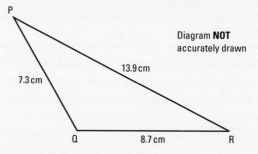

Diagram **NOT** accurately drawn

5 A model of the Eiffel Tower is made to a scale of 2 millimetres to 1 metre.
The width of the base of the real Eiffel Tower is 125 metres.

  a Work out the width of the base of the model. Give your answer in millimetres.

  The height of the model is 648 millimetres.

  b Work out the height of the real Eiffel Tower. Give your answer in metres

June 2008, adapted

6 Beeham is 10 km from Alston.
Corting is 20 km from Beeham.
Deetown is 45 km from Alston.
The diagram below shows the straight road from Alston to Deetown.
This diagram has been drawn accurately using a scale of 1 cm to represent 5 km.

Alston ———————————————————————————— Deetown

On a copy of the diagram, mark accurately with crosses (x), the positions of Beeham and Corting.

Nov 2007

7 ABC is a triangle.
Copy the triangle accurately and shade the region
inside the triangle which is **both** less than
4 centimetres from the point B **and** closer to
the line AC than the line AB.

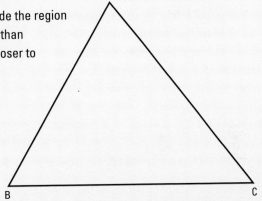

June 2009, adapted

8 On a copy of the diagram, use a ruler and pair of compasses to **construct** an angle of 30° at P.
You **must** show all your construction lines.

P ————————————————————————

ResultsPlus
Exam Question Report

79% of students answered this question
poorly because they did not use two different
constructions.

Nov 2007, adapted

9 **a** Mark the points C and D approximately 8 cm apart. Draw the locus of all points that are equidistant
from C and D.
d Draw the locus of a point that moves so that it is always 3 cm from a line 4.5 cm long.

10 B is 5 km north of A.
C is 4 km from B.
C is 7 km from A.
a Make an accurate scale drawing of triangle ABC.
Use a scale of 1 cm to 1 km.
b From your accurate scale drawing, measure the bearing of C from A.
c Find the bearing of A from C.

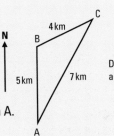

Diagram **NOT**
accurately drawn

Nov 2000

C

11 On an accurate copy of the diagram use a ruler and pair of compasses to construct the bisector of angle ABC.
You must show all your construction lines.

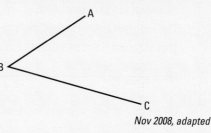

Nov 2008, adapted

A03

12 ABCD is a rectangle.
Make an accurate drawing of ABCD.
Shade the set of points inside the rectangle which are **both**
more than 1.2 centimetres from the point A
and more than 1 centimetre from the line DC.

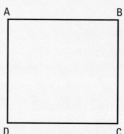

13 Draw a line segment 7 cm long. Construct the perpendicular bisector of the line segment.

14 Draw a line segment ST and a point above it, M. Construct the perpendicular from M to ST.

A03

15 As a bicycle moves along a flat road, draw the locus of:
a the yellow dot
b the green dot.

A03

16

Draw the locus of a man's head as the ladder he is on slips down a wall.

13 LINEAR EQUATIONS

Old-fashioned scales like these are still used in many markets around the world. Just like linear equations, you must "do" the same to both sides to keep them balanced.

◎ Objectives

In this chapter you will:
- solve simple equations
- solve linear equations containing brackets and fractions
- solve linear equations in which the unknown appears on both sides of the equation
- set up and solve simple linear equations
- distinguish between the words 'equation', 'formula', 'identity' and 'expression'.

◈ Before you start

You need to be able to:
- collect like terms in an algebraic expression
- expand or multiply out brackets
- calculate using directed numbers
- apply the rules of BIDMAS.

13.1 Solving simple equations

◎ Objective

○ You can solve simple equations.

⦿ Why do this?

If you have a number of albums on your mp3 player, as well as some individual tracks, you can solve a simple equation to find out how many songs you have in total.

⬥ Get Ready

1. Simplify

   a $4 + 3p - 2q - p + 7q$ b $-3(1 - 2z)$ c $4m(3m + 9)$

🔍 Key Points

◉ In any equation, the value of the left-hand side must always be equal to the value of the right-hand side. So whatever operation is applied to the left-hand side must also be applied to the right-hand side.

◉ Two children sit on a see-saw, equally distant from the middle. The see-saw is level which means that the weight of each child is the same.

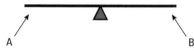

A   B

Weight of child A on the left-hand side = weight of child B on the right-hand side.

The weight of child A is $5x + 12$.

The weight of child B is 57.

So, $5x + 12 = 57$.   This is a **linear equation**.

🔍 Example 1   Solve $5x + 12 = 57$

$$5x + 12 = 57$$
$$5x + 12 - 12 = 57 - 12$$ ← Subtract 12 from both sides of the equation.
$$5x = 45$$
$$\frac{5x}{5} = \frac{45}{5}$$ ← Divide both sides by 5.
$$x = 9$$

ResultsPlus
Watch Out!

Always apply the same operation to **both** sides of the equation.

⚙ Exercise 13A

Questions in this chapter are targeted at the grades indicated.

Solve

| 1 | $2a + 5 = 13$ | 2 | $3b - 4 = 17$ | 3 | $2c + 6 = 11$ |

| 4 | $5d - 1 = 8$ | 5 | $13 + 4e = 7$ | 6 | $1 - 2f = 9$ |

| 7 | $13 = 3g - 5$ | 8 | $5 = 1 + 8h$ | 9 | $2 - 5k = 15$ |

| 10 | $17 = 3 - 4m$ | | | | |

13.2 Solving linear equations containing brackets

⊚ Objective

⊙ You can solve linear equations that require prior simplification of brackets.

⬧ Get Ready

1. Solve

  a $2x + 7 = 19$ **b** $3b + 8 = 20$ **c** $16 = 40 - 3q$

Key Points

⊚ Follow the rules of BIDMAS when solving a linear equation. (See Section 1.3 on BIDMAS).

⊚ **Solutions** can be written as mixed fractions, improper fractions or as decimals.

Example 2   Solve $4(x + 1) = 11$

$$4(x + 1) = 11$$ ⬅   Expand the left-hand side by multiplying out the brackets.
$$4x + 4 = 11$$

$$4x + 4 - 4 = 11 - 4$$ ⬅   Solve the equation as in Section 13.1.
$$4x = 7$$
$$\frac{4x}{4} = \frac{7}{4}$$
$$x = \frac{7}{4}$$
$$x = 1\tfrac{3}{4} \text{ or } 1.75$$

Results Plus
Exam Tip

Your answer can be written as a fraction or as a decimal.

⚙ Exercise 13B

Solve

| 1 | $2(x + 3) = 12$ | 2 | $5(y + 4) = 35$ | 3 | $4(x - 1) = 5$ | 4 | $3(2y - 1) = 9$ |

| 5 | $13 = 4(x + 3)$ | 6 | $2(1 - w) = 10$ | 7 | $5(3 - 4z) = 20$ | 8 | $2 = 3(1 + 3x)$ |

| 9 | $2(2x + 3) + 1 = 11$ | 10 | $17 = 6 - 3(5 - 2y)$ |

D

13.3 Solving linear equations with the unknown on both sides

⊙ Objective

- You can solve linear equations, with integer coefficients, in which the unknown appears on both sides of the equation.

⑦ Why do this?

Knowing how to solve equations helps you solve other problems such as finding one weight when given another.

⬥ Get Ready

1. Solve

   a $4 = 2(x - 6)$ **b** $3(3a + 4) + 3 = 33$ **c** $4 = 8 - 2(8 - 3b)$

◐ Key Point

- Collect terms so that the ones involving the unknown are on one side of the equation.

Example 3   Solve $5x + 5 = 3 - 3x$

$5x + 5 = 3 - 3x$

$5x + 5 + 3x = 3 - 3x + 3x$ ← Add $3x$ to both sides of the equation. Remember $-3x + 3x = 0$.

$8x + 5 = 3$

$8x + 5 - 5 = 3 - 5$ ← Solve the equation as in Section 13.1.

$8x = -2$

$\dfrac{8x}{8} = \dfrac{-2}{8}$

$x = -\dfrac{1}{4}$ or -0.25

ResultsPlus

Watch Out!

Always show each stage of your working.

Example 4   Solve $3 - 6x = 7 - 3x$

Here both terms in x have a negative **coefficient**.

ResultsPlus

Exam Tip

Collect the terms in x on the side of the equation that gives them a positive coefficient.

$3 - 6x = 7 - 3x$

$3 - 6x + 6x = 7 - 3x + 6x$ ← Add $6x$ to both sides of the equation.

$3 = 7 + 3x$

$3 - 7 = 3x$

$-4 = 3x$

$x = -\dfrac{4}{3}$

coefficient

Exercise 13C

Solve

1 $4a + 3 = 8 + 2a$ **2** $5b + 3 = b - 7$ **3** $3c - 2 = 5c - 8$ **4** $d + 7 = 5d + 15$

5 $3 - 2e = 4 - 3e$ **6** $1 - 7f = 3f + 10$ **7** $2(x - 4) = x + 7$

8 $2x + 5 = 1 + 3(2 + x)$ **9** $3(4x + 1) + 2(1 - 5x) = 2 + x$

10 $6 + 2(x - 3) = x - 3(1 - 2x)$

D

C

13.4   Solving linear equations containing fractions

◎ Objective

○ You can solve linear equations containing fractions.

⬦ Get Ready

1. Solve

   a $4x + 6 = 2(4 + x) + 2$ **b** $2 - 8x = 20 + 3(4 + 9x)$ **c** $5 + 4(x - 1) = 2x - 4(2 - 3x)$

Key Point

◎ To solve an algebraic equation involving fractions, eliminate all fractions by multiplying each term by the LCM of the denominators.

Example 5   Solve $\dfrac{12}{p + 2} = 3$

$$\frac{12}{p + 2} = 3$$

$$\frac{12}{p + 2} \times (p + 2) = 3 \times (p + 2)$$

Multiply both sides of the equation by $(p + 2)$.
The terms in $(p + 2)$ on the left-hand side cancel out.

$$12 = 3(p + 2)$$
$$12 = 3p + 6$$
$$12 - 6 = 3p$$
$$3p = 6$$
$$p = 2$$

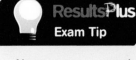
ResultsPlus
Exam Tip

Always try to remove the fraction first.

Example 6   Solve $\dfrac{x+1}{2} - \dfrac{4x-1}{3} = \dfrac{5}{12}$

$$\dfrac{x+1}{2} - \dfrac{4x-1}{3} = \dfrac{5}{12}$$

$$12 \times \dfrac{x+1}{2} - 12 \times \dfrac{4x-1}{3} = 12 \times \dfrac{5}{12}$$

← Multiply each of the three terms by 12.

$${}^{6}\cancel{12} \times \dfrac{x+1}{\cancel{2}_1} - {}^{4}\cancel{12} \times \dfrac{4x-1}{\cancel{3}_1} = {}^{1}\cancel{12} \times \dfrac{5}{\cancel{12}_1}$$

$$6(x+1) - 4(4x-1) = 5$$
$$6x + 6 - 16x + 4 = 5$$
$$-10x + 10 = 5$$
$$-10x = -5$$
$$x = \dfrac{1}{2}$$

Remember: $-4 \times -1 = +4$

Exercise 13D

Solve

B

1 $\dfrac{p}{5} + 3 = 7$

2 $\dfrac{q+2}{3} = 4$

3 $\dfrac{m}{2} + \dfrac{m}{5} = 21$

4 $\dfrac{x}{6} + 1 = \dfrac{x-4}{4}$

5 $3(2y-10) = \dfrac{4y-7}{2}$

6 $2\left(\dfrac{x}{3} - 3\right) = 16$

7 $\dfrac{1}{2n} + \dfrac{1}{3n} = 7$

A

8 $\dfrac{3t+6}{10} + \dfrac{5-2t}{5} = 6$

9 $2 - \dfrac{1-x}{3} = \dfrac{5x+2}{9}$

10 $\dfrac{3y-4}{2} - \dfrac{2y+1}{5} = \dfrac{1-y}{3}$

13.5 Setting up and solving simple linear equations

◎ Objective

● You can set up and solve simple linear equations.

�ⓘ Why do this?

Businesses use linear equations to help them work out how much of a product needs to be produced to make a given profit.

◆ Get Ready

1. Solve

  a $6a + 1 = 2a - 7$

  b $5 - 6b = 10 + 4b$

  c $\left(\dfrac{x}{5}\right) + 2 = \dfrac{(x-3)}{2}$

🔍 Key Points

◉ When setting up an equation, define all the unknowns used that have not already been defined.

◉ Make sure that units are consistent on both sides of the equation.

Example 7

Daniel makes some drinks to sell at the Summer Fair.

From one bottle costing £2.80 he can make 40 drinks.

Daniel wants to make a profit of £2 on each bottle.

a If c is the price of each drink, write an equation in terms of c.

b Solve your equation in **a** to find what the price of each drink should be.

a 40 drinks will cost $40 \times c = 40c$ pence   ← Change all amounts to pence.

One bottle costs £2.80 = 280 pence

Daniel's profit = £2 = 200 pence

Profit = total sales − total costs

$200 = 40c − 280$

b $200 = 40c − 280$

$200 + 280 = 40c$

$40c = 480$

$c = \dfrac{480}{40}$

$c = 12$

So the cost of each drink is 12 pence.

Example 8

By setting up and solving an equation in terms of x, work out the size of the largest angle of this quadrilateral.

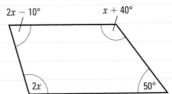

Since the sum of the interior angles of a quadrilateral is 360°, adding the angles gives:

$(2x − 10) + (x + 40) + (2x) + (50) = 360$

$2x − 10 + x + 40 + 2x + 50 = 360$ ← Collect like terms.

$5x + 80 = 360$

$5x = 360 − 80$

$5x = 280$

$x = 56$

ResultsPlus

Watch Out!

Read the question carefully; is the size of the largest angle or the smallest angle required?

Angles are: $2x − 10 = 2 \times 56 − 10 = 112 − 10 = 102°$

$x + 40 = 56 + 40 = 96°$

$2x = 2 \times 56 = 112°$ is the largest angle.

Exercise 13E

1 Viv thinks of a number. She multiplies the number by 5 and then subtracts 2.
Her answer is 23.
If x is the number that Viv was thinking of, work out the value of x.

2 Michelle is 2 years younger than Angela.
If the sum of their ages is 64, work out Michelle's age.

D

A02

C **A02**

3 In an exam, Jessica scored p%.
Mason scored three-quarters of Jessica's score.
Zach scored 10% less than Jessica.
The total of their scores was 210%.
How much did each student score?

A03

4 The length of a rectangle is $(2x + 3)$ cm.
The width of the rectangle is $\frac{1}{2}(x + 3)$ cm.
If the perimeter of the rectangle is 49 cm, find the value of x.

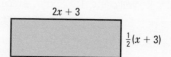
$2x + 3$
$\frac{1}{2}(x + 3)$

A03

5 Joanna works for n hours each week for 5 weeks. In the sixth week she works an extra $4\frac{1}{2}$ hours.
In these six weeks she works a total of 117 hours.
Work out the number of hours Joanna works in the sixth week.

B

6 The diagram shows an isosceles triangle ABC.
The lengths of the sides are given in centimetres and AB = BC.
a Write down an equation in terms of x.
b Work out the lengths of AB and BC.

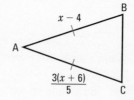
B
$x - 4$
A
$\frac{3(x + 6)}{5}$
C

A03

7 In one quarter Stuart uses 2512 units of electricity. Part of his electricity bill is shown below.
Units used = 2512
Cost of the first x units at 12.86p per unit = £_____
Cost of remaining units at 11.8p per unit = £_____
Total cost of electricity used = £ 298.
Work out the number of units, at 12.86p per unit, that Stuart used. Give your answer to the nearest unit.

13.6 Distinguishing between 'equation', 'formula', 'identity' and 'expression'

⊚ **Objective**

• You can distinguish between the words 'equation', 'formula', 'identity' and 'expression'.

⟡ **Why do this?**

You need to understand mathematical words to be able to understand exam questions.

⬙ **Get Ready**

1. Evaluate these expressions when $x = 4$ and $y = -2$.
a $x - y$ b $x + y$ c $4 + x + 2y$

Key Points

- An example of an algebraic expression is $2p - 10$. It is made up of the terms $2p$ and -10.
- An example of an **equation** is $2p - 10 = 9$. This can be solved to find the value of p.
- $2(p - 5) = 2p - 10$ is called an **identity**. The left-hand side is the same as the right-hand side.
- $C = 2p - 10$ is called a **formula**. If the value of p is known, you can substitute it into the equation to work out the value of C.

Example 9   Write down whether each of the following is an expression, an equation, an identity or a formula.

    a $5ab - 2ab = 3ab$

    b $5p + 2q$

    c $T = 2\pi\sqrt{\dfrac{l}{g}}$

    d $\dfrac{1 - 2x}{3} = 5$

ResultsPlus
Exam Tip

An expression is the only one without an '=' sign.

a   The right-hand side is the same as the left-hand side and so $5ab - 2ab = 3ab$ is an identity.

> Collect like terms of $5ab - 2ab$ to give $3ab$.

b $5p + 2q$ is an expression.

c   If the values of l and g are known, the value of T can be worked out using the formula $T = 2\pi\sqrt{\dfrac{l}{g}}$.

d $\dfrac{1 - 2x}{3} = 5$ can be solved to find the value of x and is therefore an equation.

Exercise 13F

Write down whether each of the following is an expression, an equation, an identity or a formula.

1 $A = 2(l + b)$ **2** $m + m + m + m = 4m$ **3** $2a + 3b$

4 $3y^2 = 243$ **5** $y = mx + c$ **6** $E = mc^2$

7 $x^2 - 5x$ **8** $\dfrac{1}{x} + \dfrac{2}{3x} = 4$ **9** $V = \dfrac{4}{3}\pi r^3$

10 $2(x + 3) = x + 4$ **11** $x^2 - 5x = x(x - 5)$

12 $x^2 \times x^5 = x^7$

D

C

Chapter review

- ⊚ In any equation, the value of the left-hand side is always equal to the value of the right-hand side. So whatever operation is applied to the left-hand side must also be applied to the right-hand side.
- ⊚ Follow the rules of BIDMAS when solving a **linear equation**.
- ⊚ Solutions can be written as mixed fractions, improper fractions or as decimals.
- ⊚ Collect terms so that the ones involving the unknown are on one side of the equation.
- ⊚ To solve an algebraic fraction equation, eliminate all fractions by multiplying each term by the LCM of the denominators.
- ⊚ When setting up an equation, define all unknowns used that have not already been defined.
- ⊚ Make sure that units are consistent on both sides of an equation.
- ⊚ An algebraic expression is made up of terms.
- ⊚ An **equation** can be solved to find the value of a term.
- ⊚ In an **identity** the left-hand side is the same as the right-hand side.
- ⊚ If the value of a term is known, you can substitute it into a **formula** to work out the value of another term.

Review exercise

1 Solve $4t + 1 = 19$

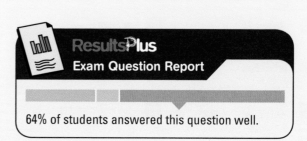

ResultsPlus
Exam Question Report

86% of students answered this question well because they remembered to do the same to both sides of the question.

Nov 2008

2 Solve $4(x - 2) = 10$

3 Solve   **a** $3x + 1 = x + 6$ **b** $4y - 3 = 2y - 8$

4 The sizes of the angles, in degrees, of the quadrilaterals are $x + 10$, $2x$, $x + 90$ and $x + 20$.

Work out the smallest angle of the quadrilateral.

Diagram NOT drawn accurately

Nov 2005

5 The diagram shows a rectangle.

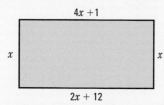

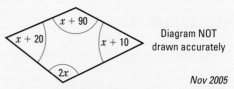

ResultsPlus
Exam Question Report

64% of students answered this question well.

All the measurements are in centimetres.
Work out the perimeter of the rectangle.

June 2009, Adapted

6 Uzma has £x. Hajra has £20 more than Uzma. Mabintou has twice as much as Haira.
The total amount of money they have is £132.
Find how much money they each have.

7

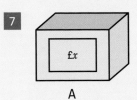
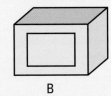
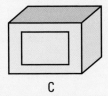

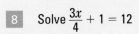

A   B   C

Here are 3 boxes. Box A has £x. Box B has £4 more than box A. Box C has one third of the money in box B.
Altogether there is £24 in the 3 boxes.
Find the amount of money in each box.

8 Solve $\dfrac{3x}{4} + 1 = 12$

9 Solve $\dfrac{x-3}{5} = x - 5$

June 2005

10 Solve $\dfrac{y}{2} - \dfrac{y-1}{3} = 2$

14 PERCENTAGES

One of the most common percentages used to compare schools is the number of students getting five GCSEs grade A*–C. Percentages can also be used to compare results across the sexes; traditionally, girls have outdone boys but the gap between the sexes looked as if it was beginning to close in 2009, with 70.5% of girls getting five GCSEs grade A*–C, compared to 63.6% of boys. This is the smallest gap since 1991.

Objectives

In this chapter you will:
- find a percentage of a quantity
- find quantities after a percentage increase or decrease
- express one quantity as a percentage of another
- find an amount after repeated percentage changes
- find the original amount after a percentage change.

Before you start

You should know:
- how to find a fraction of a quantity
- how to convert between fractions and decimals
- that percent means 'out of 100'
- how to write a percentage as a fraction or a decimal.

14.1 Working out a percentage of a quantity

⊙ Objectives

⊙ You can convert between fractions, decimals and percentages.
⊙ You can find a percentage of a quantity.

⊘ Why do this?

Percentages are a part of our everyday language. Banks pay interest as a percentage of the money in your bank account. Tax is paid as a percentage of money earned.

⊕ Get Ready

1. Write these percentages as   **i** fractions in their simplest form   **ii** decimals.
   a 20%   **b** $12\frac{1}{2}\%$ **c** 60%   **d** $17\frac{1}{2}\%$

Key Points

⊙ There are a number of different methods that can be used to work out a percentage of an amount.
⊙ When using a calculator, first change the percentage to a fraction or decimal then multiply the amount by the fraction or decimal.
⊙ When not using a calculator, first work out either 10% or 1% of the amount then build up the percentage.

Example 1

Colin invests £1850.
The **interest** rate is 3% per year.
How much interest will Colin receive after 1 year?

Method 1

$$3\% = \frac{3}{100}$$ ← Convert the percentage to a fraction.

$$\frac{3}{100} \times 1850 = 55.5$$ ← Multiply the amount by the fraction.

Interest will be £55.50

ResultsPlus
Watch Out!

As the answer is an amount of money, remember to give two decimal places in the answer.

Method 2

$$3\% = 0.03$$ ← Convert the percentage to a decimal.

$$0.03 \times 1850 = 55.5$$ ← Multiply the amount by the decimal.

Interest will be £55.50

ResultsPlus
Exam Tip

Although Method 3 can be used to answer the question, on a calculator paper it is more sensible to use either Method 1 or Method 2 as candidates often make arithmetical errors using the build-up method.

Method 3

$$1\% \text{ of } 1850 = 1850 \div 100$$
$$= 18.5$$ ← Work out 1% of the amount. As $1\% = \frac{1}{100}$, divide by 100.

$$3\% \text{ of } 1850 = 18.5 \times 3$$ ← $3\% = 3 \times 1\%$ so multiply 18.5 by 3.
$$= 55.5$$

Interest will be £55.50

Exercise 14A

Questions in this chapter are targeted at the grades indicated.

1. Work out
   a 30% of £600
   b 15% of 40 kg
   c 5% of £32.40
   d 8% of 62 kg
   e 20% of £30
   f 30% of 150
   g 4.2% of 60 km
   h $17\frac{1}{2}$% of £300

2. There are 150 shop assistants in a large store. 8% of the shop assistants are male.
   How many of the shop assistants are male?

3. Danya invests £250. The interest rate is 4% per year.
   How much interest will she receive after 1 year?

4. A shop has 4600 DVDs. 23% of the DVDs are thrillers.
   How many of the DVDs in the shop are thrillers?

5. There are 154 students in Year 11. 84 of these students are girls.
   50% of the girls and 10% of the boys attend Spanish lessons.
   What fraction of these Year 11 students attend Spanish lessons? Give your fraction in its simplest form.

14.2 Finding the new amount after a percentage increase or decrease

◉ Objective

● You can find quantities after a percentage increase or decrease.

◈ Why do this?

Percentages are often used when a shop has a sale or is offering a discount. A pay rise can be given as a percentage of current earnings.

◈ Get Ready

1. Work out 25% of £300.
2. Work out 20% of £80.
3. Write 5 as a fraction of 20.

◉ Key Points

◉ There are two methods that can be used to increase an amount by a percentage.
   ◉ You can find the percentage of that number and then add this to the starting number.
   ◉ You can use a **multiplier**.

◉ There are two methods that can be used to decrease an amount by a percentage.
   ◉ You can find the percentage of that number and then subtract this from the starting number.
   ◉ You can use a multiplier.

Example 2

Hugh's salary is £25 000 a year.
His salary is increased by 4%.
Work out his new salary.

Method 1

4% of £25 000 = $\frac{4}{100}$ × 25 000

= 1000 ← The increase in his salary is £1000.

25 000 + 1000 = 26 000 ← Add the increase to his original salary.

Hugh's new salary is £26 000.

Method 2

100% + 4% = 104% ← His new salary is 104% of £25 000.

104% = $\frac{104}{100}$ = 1.04 ← 1.04 is the multiplier.

1.04 × 25 000 = 26 000 ← Multiply 25 000 by 1.04. (This increases 25 000 by 4%.)

Hugh's new salary is £26 000.

Example 3

a Write down the single number that you can multiply by to increase an amount by 12.5%.
b Increase £56 by 12.5%.

a 100% + 12.5% = 112.5%

112.5% = $\frac{112.5}{100}$ = 1.125

b 56 × 1.125 = £63 ← Use the multiplier worked out in part a.

Exercise 14B

1 Write down the single number you can multiply by to work out an increase of
  a 64%   b 3%   c 14%   d 40%
  e 13.4%   f 12½%   g 15%   h 2.36%

2 a In order to increase an amount by 40%, what single number should you multiply by?
  b The cost of a theatre ticket is increased by 40% for a special concert.
     What is the new price if the normal price was £15.40?

3 The table shows the salaries of three workers.
  Each worker receives a 4.2% salary increase.
  Work out the new salary of each worker.

Helen	£12 000
Tom	£24 000
Sandeep	£32 000

D

4   Jenny puts £600 into a bank account. At the end of one year 3.5% interest is added.
    How much is in her account at the end of 1 year?

5   a Increase £120 by 20%.   b Increase 56 kg by 25%.   c Increase 2.4 m by 16%.
    d Increase £1240 by 10.5%.   e Increase 126 cm by 2%.

Example 4   The value of a car depreciates by 15% each year.
                The value of a car when new is £14 000.
                Work out the value of the car after 1 year.

Depreciates means that the value of the car decreases.

Method 1
15% of £14 000 = $\frac{15}{100}$ × 14 000   ← The depreciation in 1 year is £2100.

              = 2100

14 000 − 2100 = 11 900   ← Subtract to work out the new value.

Value after 1 year = £11 900.

Method 2
100% − 15% = 85%   ← The final value is 85% of the original value.

85% = $\frac{85}{100}$ = 0.85   ← 0.85 is the multiplier.

0.85 × 14 000 = 11 900   ← Multiply the original amount by 0.85.

Value after 1 year = £11 900.

Exercise 14C

1   Write down the single number you can multiply by to work out a decrease of
    a 7%   b 20%   c 16%   d 27%
    e 5.6%   f $2\frac{1}{2}$%   g $7\frac{1}{4}$%   h 0.8%

D

2   In a sale all prices are reduced by 15%. Work out the sale price of each of the following:
    a a television set that normally costs £300
    b a CD player that normally costs £40
    c a computer that normally costs £1200.

3   Alan weighs 82 kg before going on a diet. He sets himself a target of losing 5% of his original weight.
    What is his target weight?

4   A holiday normally costs £850. It is reduced by 12%. How much will the holiday now cost?   C

5   Ria buys a car for £7300. The value of the car depreciates by 20% each year.
Work out the value of the car at the end of: a   1 year   b   2 years.

14.3 Working out a percentage increase or decrease

Objectives

○ You can express one quantity as a percentage of another.
○ You can find percentage loss or profit.

Why do this?

To work out how well a business is doing, you might want to work out what percentage profit or loss they have made.

Get Ready

1. Work out £24.80 − £12.05.   **2.** Work out £15 000 − £13 700.   **3.** Write 45 out of 200 as a fraction.

Key Points

◉ To write one quantity as a percentage of another quantity:
 ◉ write down the first quantity as a fraction of the second quantity
 ◉ convert the fraction to a percentage.

◉ Percentage problems sometimes involve percentage profit or percentage loss, where:
 ◉ percentage profit (or increase) $= \dfrac{\text{profit (or increase)}}{\text{original amount}} \times 100\%$

 ◉ percentage loss (or decrease) $= \dfrac{\text{loss (or decrease)}}{\text{original amount}} \times 100\%$

Example 5

a   Convert 11 out of 20 to a percentage.
b   Convert 23 cm out of 4 m to a percentage.

a $\dfrac{11}{20}$ ← Write the first number as a fraction of the second number.

$\dfrac{11}{20} \times 100 = 55\%$ ← To convert a fraction to a percentage, multiply by 100.

b $4\,m = 4 \times 100$ ← Multiply by 100 to convert 4 m into centimetres.
  $= 400\,cm$

$\dfrac{23}{400} \times 100 = 5.75\%$ ← Convert the fraction to a percentage.

Watch Out!

When working with quantities in different units, first make sure that all the units are the same.

D

Exercise 14D

1 Write:

a £3 as a percentage of £6

b 2 kg as a percentage of 8 kg

c 4p as a percentage of 10p

d 8 cm as a percentage of 40 cm

e 60p as a percentage of £2.40

f 15 mm as a percentage of 6 cm

g 36 minutes as a percentage of 1 hour

h 50 cm as a percentage of 4 m.

2 Janet scored 36 out of 40 in a German test. Work out her score as a percentage.

3 Jerry took 60 bottles to a bottle bank. 27 of the bottles were green.
What percentage of the bottles were green?

4 A 40 g serving of cereal contains 8 g of protein, 24 g of carbohydrates, 4.5 g of fat and 3.5 g of fibre.
What percentage of the serving is:

a protein

b carbohydrates

c fat

d fibre?

Example 6 Karen bought a car for £1200.
One year later, she sold it for £840.
Work out her percentage loss.

$1200 - 840 = 360$ ← Subtract the selling price from the original price to find her loss.

$\frac{360}{1200}$ ← Write down the fraction $\frac{loss}{original\ price}$

$\frac{360}{1200} \times 100 = 30\%$ ← Multiply $\frac{360}{1200}$ by 100 to change it to a percentage.

Her percentage loss is 30%.

A02
A03

Example 7 Tony bought a box of 24 oranges for £4.
He sold all the oranges for 21p each.
Work out his percentage profit.

$24 \times 21 = 504p$ ← Work out the total amount, in pence, Tony received from selling all the oranges.

$504 - 400 = 104p$ profit ← Subtract the original price from the selling price to find his profit in pence.

$\frac{104}{400}$ ← Write down the fraction $\frac{profit}{original\ price}$

$\frac{104}{400} \times 100 = 26\%$ ← Multiply $\frac{104}{400}$ by 100 to change it to a percentage.

Percentage profit = 26%.

Exercise 14E

1 Calculate the percentage increase or decrease to the nearest 1%:

a £24 to £36   b 12.5 kg to 20 kg   c 45 cm to 39.5 cm

d 2 minutes to 110 seconds.

2 In a sale, the price of a clock is reduced from £32 to £27.20. Work out the percentage reduction.

3 Rob bought a crate of 40 melons for £30. He sold all the melons for £1.05 each.
Work out his percentage profit.

* 4 David owns three shops selling DVDs.

He tells the staff in each of the shops that some of them will receive a bonus.

He will give the bonus to the staff who work in the shop that has the biggest percentage increase in the number of DVDs sold from the first half to the second half of the year.

	DVDs sold Jan–Jun	DVDs sold Jul–Dec
Shop A	12 893	13 562
Shop B	9 875	10 346
Shop C	11 235	11 853

Which shop should receive the bonus? You must show how you decided on your answer.

5 Martin goes to a discount centre.

He buys 10 trays of drinks.

Each tray holds 24 cans of cola and costs £9.50.

Martin sells 150 cans of cola at a fair for 55p each.

He sells the rest of the cans for 35p each the next day at a car boot sale.

Work out the profit or loss percentage that Martin makes.

6 A badminton club has 44 members.

Each member pays £85 per year as a membership fee.

The club has to pay a total of £3700 to the sports centre to hire the badminton courts.

The sports centre decides to increase the cost of hiring courts by 8.5%.

The badminton club will have 46 members next year.

Work out the smallest possible percentage rise in club membership fees so that the club can afford to pay the sports centre. Give your answer correct to 1 decimal place.

14.4 Working out compound interest

◎ Objective

○ You can find an amount after repeated percentage changes.

❓ Why do this?

Percentage changes can happen over a period of time. You may want to work out how much money you will have in two years if you put £100 in your bank account now.

◈ Get Ready

1. Write

a $2 \times 2 \times 2 \times 2 \times 2$ as a power of 2   b 30% as a decimal   c 125% as a decimal

Key Points

◉ Banks and building societies pay **compound interest**.

◉ At the end of the first year, interest is paid on the money in an account. This interest is then added to the account. At the end of the second year, interest is paid on the total amount in the account, that is, the original amount of money plus the interest earned in the first year.

◉ At the end of each year, interest is paid on the total amount in the account at the start of that year.
  For example, if £200 is invested in a bank account and interest is paid at a rate of 5% then

Year	Amount at start of year	Amount plus interest	Total amount at year end
1	£200	200×1.05	£210
2	£210	$210 \times 1.05 = 200 \times 1.05^2$	£220.50
3	£220.50	$220.50 \times 1.05 = 200 \times 1.05^3$	£231.52
4	£231.52	$231.52 \times 1.05 = 200 \times 1.05^4$	£243.10
5	£243.10	$243.10 \times 1.05 = 200 \times 1.05^5$	£255.26
6	£255.26	$255.26 \times 1.05 = 200 \times 1.05^6$	£268.02

◉ To calculate compound interest, find the multiplier:
  ◉ Amount after n years = original amount $\times$ multiplier^n

Example 8   £4000 is invested for 2 years at 5% per annum compound interest.
Work out the **total interest** earned over the 2 years.

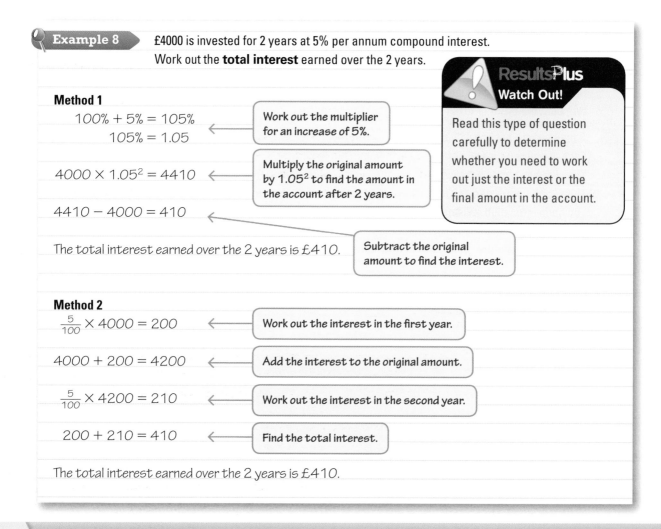

Method 1

$100\% + 5\% = 105\%$
$105\% = 1.05$ ← Work out the multiplier for an increase of 5%.

$4000 \times 1.05^2 = 4410$ ← Multiply the original amount by 1.05^2 to find the amount in the account after 2 years.

$4410 - 4000 = 410$ ←

The total interest earned over the 2 years is £410.   Subtract the original amount to find the interest.

ResultsPlus
Watch Out!

Read this type of question carefully to determine whether you need to work out just the interest or the final amount in the account.

Method 2

$\frac{5}{100} \times 4000 = 200$ ← Work out the interest in the first year.

$4000 + 200 = 4200$ ← Add the interest to the original amount.

$\frac{5}{100} \times 4200 = 210$ ← Work out the interest in the second year.

$200 + 210 = 410$ ← Find the total interest.

The total interest earned over the 2 years is £410.

Example 9　a　Each year the value of a car depreciates by 30%. Find the single number, as a decimal, that the value of the car can be multiplied by to find its value at the end of 4 years.

　b　The value of a house increases by 16% of its value at the beginning of the year. The next year its value decreases by 3% of its value at the start of the second year. Find the single number, as a decimal, that the original value of the house can be multiplied by to find its value at the end of the 2 years.

a　　　$100\% - 30\% = 70\%$　　←　Find the multiplier that represents a decrease of 30%.

　　　　$70\% = \frac{70}{100} = 0.7$

$0.7 \times 0.7 \times 0.7 \times 0.7 = 0.7^4$　←　The depreciation is over 4 years so the single multiplier is 0.7 raised to the power of 4.

　　　　$0.7^4 = 0.2401$

0.2401 is the single number.

b　　　$100\% + 16\% = 116\%$　　←　Find the multiplier for an increase of 16%.

　　　　$116\% = \frac{116}{100} = 1.16$

　　　　$100\% - 3\% = 97\%$　　←　Find the multiplier for a decrease of 3%.

　　　　$97\% = \frac{97}{100} = 0.97$

　　　$1.16 \times 0.97 = 1.1252$　←　The value increases and then decreases so find the product of the two multipliers.

1.1252 is the single number.

Example 10　The value of a machine when new is £8000.

The value of the machine depreciates by 10% each year.

Work out its value after 3 years.

Method 1

$100\% - 10\% = 90\%$　　←　Work out the multiplier for a decrease of 10%.

　　　$90\% = 0.9$

$8000 \times 0.9^3 = £5832$　←　Multiply the value when new by 0.9^3 to find the value after 3 years.

The value of the machine after 3 years is £5832.

Method 2

$\frac{10}{100} \times 8000 = 800$

$8000 - 800 = 7200$

$\frac{10}{100} \times 7200 = 720$

$7200 - 720 = 6480$

$\frac{10}{100} \times 6480 = 648$

$6480 - 648 = 5832$

The value of the machine after 3 years is £5832.

Exercise 14F

D

1 Work out the multiplier as a single decimal number that represents:
   a an increase of 20% for 3 years
   b a decrease of 10% for 4 years
   c an increase of 4% followed by an increase of 2%
   d a decrease of 35% followed by a decrease of 20%.

C

2 £1000 is invested for 2 years at 5% per annum compound interest.
   Work out the total amount in the account after 2 years.

B **A02** **A03**

3 Mrs Bell buys a house for £60 000. In the first year, the value of the house increases by 16%.
   In the second year, the value of the house decreases by 4% of its value at the beginning of that year.
   a Write down the single number, as a decimal, that the original value of the house can be multiplied by to find its value after 2 years.
   b Work out the value of the house after the 2 years.

A03

4 Ben says that an increase of 40% followed by an increase of 20% is the same as an increase of 60%.
   Is Ben correct? You must give a reason for your answer.

5 Jeremy deposits £3000 in a bank account. Compound interest is paid at a rate of 4% per annum.
   Jeremy wants to leave the money in the account until there is at least £4000 in the account.
   Calculate the least number of years Jeremy must leave his money in the bank account.

14.5 Calculating reverse percentages

⊙ Objective

● You know how to find the original amount given the final amount after a percentage increase or decrease.

⊙ Why do this?

If you found a book marked 60% off in a sale, you can use reverse percentages to work out how much it originally cost.

⬆ Get Ready

1. Write down the multiplier for:
   a an increase of 15%   b a decrease of 15%   c an increase of 4%   d a decrease of 4%.

◉ Key Points

◉ There are two methods that can be used to find the original amount if the final amount after a percentage increase or decrease is known using **reverse percentages**.

◉ A flow diagram can be used to represent a percentage change using multipliers.

   original price $\xrightarrow{\times \text{ multiplier}}$ final amount

◉ Drawing a second flow diagram reversing the direction and using the inverse operation shows that to find the original price from the final amount you divide by the multiplier.

   original price $\xleftarrow{\div \text{ multiplier}}$ final amount

Example 11

In a sale, all prices are reduced by 20%.
The sale price of a jacket is £33.60.
Work out the original price of the jacket.

Method 1

$100\% - 20\% = 80\%$ ← Find the multiplier for a decrease of 20%.
$\frac{80}{100} = 0.8$

Original price $= 33.60 \div 0.8$ ← Divide by the multiplier to find the original price.
$= 42$

The original price of the jacket was £42.

Method 2

$100\% - 20\% = 80\%$ ← £33.60 represents 80% of the original price.

$£33.60 = 80\%$
So $1\% = 33.60 \div 80$ ← Divide 33.60 by 80 to find the value of 1%.
$= 0.42$

So, original price $= 0.42 \times 100$ ← The original price is 100% so multiply the
$= 42$ amount that represents 1% by 100.
The original price of the jacket was £42. (Check: $42 \times 0.8 = 33.6$)

Example 12

The price of a new washing machine is £376.
This price includes Value Added Tax (VAT) at $17\frac{1}{2}\%$.
Work out the cost of the washing machine before VAT was added.

Method 1

$100\% + 17\frac{1}{2}\% = 117.5\%$ ← The original cost was increased by 17.5% so
find the multiplier for an increase of 17.5%.
$\frac{117.5}{100} = 1.175$

Original price $= 376 \div 1.175$ ← Divide by the multiplier to find the amount without VAT.
$= 320$

The cost of the washing machine before VAT was added was £320. ← (Check: $320 \times 1.175 = 376$)

Method 2

$100\% + 17.5\% = 117.5\%$ ← £376 represents 117.5% of the original cost.

$£376 = 117.5\%$
So $1\% = 376 \div 117.5$ ← Divide 376 by 117.5 to find the value of 1%.
$= 3.2$

So original price $= 3.2 \times 100$ ← The original cost is 100% so multiply the
$= 320$ amount that represents 1% by 100.
The cost of the washing machine before VAT was added was £320. (Check: $320 \times 117.5 = 376$)

Exercise 14G

B
A03

1. Employees at a firm receive a pay increase of 4%. After the pay increase, Linda earns £24 960. How much did Linda earn before the pay increase?

2. The price of a new television set is £329.
This price includes Value Added Tax (VAT) at $17\frac{1}{2}$%.
Work out the cost of the television set before VAT was added.

A03

3. A holiday is advertised at a price of £403.
This represents a 35% saving on the brochure price. Work out the brochure price of the holiday.

A03

4. Kunal pays tax at a rate of 22%. After he has paid tax, Kunal receives £140.40 per week. How much does Kunal earn per week before he pays tax?

5. In one year, the population of an island increased by 3.2% to 434 472.
Work out the population of the island before the increase.

6. Tasha invests some money in a bank account. Interest is paid at a rate of 8% per annum.
After 1 year, there is £291.60 in the account. How much money did Tasha invest?

Chapter review

- There are a number of different methods that can be used to work out a percentage of an amount.
- When using a calculator, first change the percentage to a fraction or decimal then multiply the amount by the fraction or decimal.
- When not using a calculator, first work out either 10% or 1% of the amount and build up the percentage.
- There are two methods that can be used to increase (or decrease) an amount by a percentage.
  - You can find the percentage of that number and then add this to (or subtract this from) the starting number.
  - You can use a **multiplier**.
- To write one quantity as a percentage of another quantity, write down the first quantity as a fraction of the second quantity, then change the fraction to a percentage.
- Percentage problems sometimes involve percentage profit or percentage loss, where:
  - percentage profit (or loss) $= \dfrac{\text{profit (or loss)}}{\text{original amount}} \times 100\%$
- To calculate **compound interest**, find the multiplier:
  - Amount after n years = original amount $\times$ multiplier^n

- There are two methods that can be used to find the original amount if the final amount after a percentage increase or decrease is known using **reverse percentages**.

- A flow diagram can be used to represent a percentage change using multipliers.

  original price $\xrightarrow{\times \text{ multiplier}}$ final amount

- Drawing a second flow diagram reversing the direction and using the inverse operation shows that to find the original price from the final amount you divide by the multiplier.

  original price $\xleftarrow{\div \text{ multiplier}}$ final amount

⚙ Review exercise

1 Work out: **a** 30% of £800 **b** 25% of 20 kg **c** 15% of 70 kg **d** $17\frac{1}{2}$% of £60.

2 A farmer has a rectangular field. He makes the field 20% longer.
If he wants to keep the same area, what would he have to reduce the width by? **FS**

3 A man invests £20 000 with a guaranteed compound interest rate of 4%.
How long will it be before he has doubled his money? **FS**

*4 The same barbeque set is sold in three different shops. **FS**
Here are the price labels shown on each barbeque set.

Shop A	Shop B	Shop C
£680.00 (inc. VAT)	£640.00 (inc. VAT)	£450.00
Get $\frac{1}{4}$ off when you buy this barbeque set	Now with 20% discount	Plus 15% VAT

Which barbeque is the best buy?

*5 Barry has ben asked to compare the pay for four similar jobs advertised in a newspaper. **FS**

Able Computer Sales	Beta IT Support
Sales Assistant	Sales Consultant
You will spend time in the field, working both from our Manchester headquarters and from home in the North West region.	Full time: 30 hours per week Pay: £15 per hour Tele-sales based in our new offices.
Pay: £23 000 per annum	Daily hours variable.
Compu Systems	Digital Hardware
Sales Agent	Sales Adviser
As a sales agent your pay will be £1800 per month, plus commission of 1% of monthly sales. You can expect to make monthly sales to a minimum value of £22 000.	You will be part of a team with a salary of £20 000 per annum + team bonus. Team bonus last year was 20% of salary.

Which job pays the most?

6 Linda's mark in a Maths test was 36 out of 50.
Find 36 out of 50 as a percentage.

7 Jessica's annual income is £12 000. **FS**
She pays $\frac{1}{4}$ of the £12 000 in rent.
She spends 10% of the £12 000 on clothes.
Work out how much of the £12 000 Jessica has left.

8 A hotel has 56 guests.
35 of the guests are male.
a Work out 35 out of 56 as a percentage.
40% of the 35 male guests wear glasses.
b Write the number of male guests who wear glasses as a fraction of the 56 guests.
Give your answer in its simplest form.

Nov 2007

D

D

9 In April 2004, the population of the European Community was 376 million.
In April 2005, the population of the European Community was 451 million.
Work out the percentage increase in population.
Give your answer correct to 1 decimal place.

Nov 2007

B A02 A03

10 Bill buys a new lawn mower.
The value of the lawn mower depreciates by 20% each year.
a Bill says 'after 5 years the lawn mower will have no value'.
Bill is wrong.
Explain why.
Bill wants to work out the value of the lawn mower after 2 years.
b By what single decimal number should Bill multiply the value of the
lawn mower when new?

Nov 2005

11 In a sale normal prices are reduced by 20%.
Andrew bought a saddle for his horse in the sale.
The sale price of the saddle was £220.
Calculate the normal price of the saddle.

12 The value of a car depreciates by 35% each year.
At the end of 2007, the value of the car was £5460.
Work out the value of the car at the end of 2006.

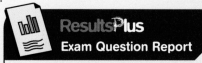

ResultsPlus
Exam Question Report

86% of students answered this question poorly
because they did not find the value of the
reduction first.

Nov 2008

13 Nimer got a pay rise of 5%. His new pay was £1680 per month.
Work out his pay per month before he got this pay rise.

14 Jim is a plumber. He has to work out the VAT on some equipment. VAT is charged at $17\frac{1}{2}$%.
The total cost of the equipment including VAT is £4465. Calculate how much the VAT was.

A02 A03

15 Sophie is offered the following pay deals:
A '5% increase this year, followed by a 4% increase next year'
B '$4\frac{1}{2}$% increase this year, followed by $4\frac{1}{2}$% increase next year'
Which offer should Sophie accept?

The street system of New York is made up of straight lines running at right angles to each other. The city did not have this layout from the start, but in 1811 the mesh of rectangles which characterises Manhattan today was built over the original tangle of city streets. Lanes and paths that didn't fit into the pattern were blocked up and the buildings that lined them torn down. The only street that survived was Broadway – an original Indian track which angled across the island and can still be seen in the city.

◎ Objectives

In this chapter you will:
- draw straight-line graphs
- find the midpoint of a line segment
- find the gradient, y-intercept and equation of a straight line
- understand the relationship between the gradients of parallel and perpendicular lines
- draw and interpret graphs describing real-life situations, including distance-time graphs.

◈ Before you start

You need to be able to:
- draw, label and scale axes
- plot points
- substitute numbers in simple algebraic expressions
- solve simple equations
- draw straight lines by plotting points.

15.1 Drawing straight-line graphs by plotting points

◈ Get Ready

1. Work out the value of $2x + 3$ if
   a $x = 2$ b $x = -1$

2. Solve the equations
   a $2 + y = 3$ b $3 + 2y = 6$

◷ Key Points

- ◉ To draw a straight line by plotting points you need to plot at least two points which fit the equation of the line.
- ◉ The **equation of a straight line** can have several forms:
    - ◉ lines in the form $x = c$ or $y = c$ where c is a number (see Example 1 and Exercise 15A)
    - ◉ lines in the form $y = mx + c$ where m and c are numbers (see Examples 2 and 3 and Exercise 15B)
    - ◉ lines in the form $ax + by = c$ where a, b and c are numbers (see Example 4 and Exercise 15C).

Example 1 Draw the graph of $x = 2$.

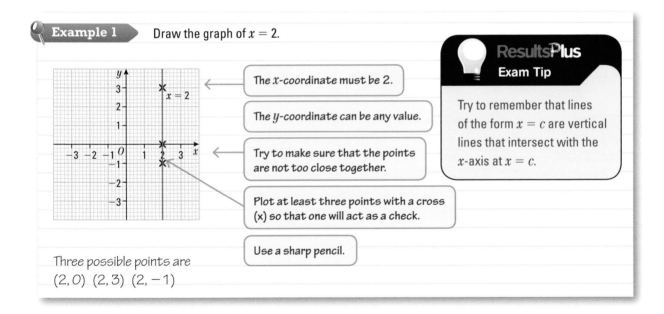

The x-coordinate must be 2.

The y-coordinate can be any value.

Try to make sure that the points are not too close together.

Plot at least three points with a cross (x) so that one will act as a check.

Use a sharp pencil.

ResultsPlus
Exam Tip

Try to remember that lines of the form $x = c$ are vertical lines that intersect with the x-axis at $x = c$.

Three possible points are
$(2, 0)$ $(2, 3)$ $(2, -1)$

✹ Exercise 15A

Questions in this chapter are targeted at the grades indicated.

1 Draw, on separate axes, the graphs of:
     a $x = -3$ b $y = 2$ c $x = 0$
     d $y = 0$ e $x = 1.5$ f $y = -5$

2 Write down the equation of each of these lines.

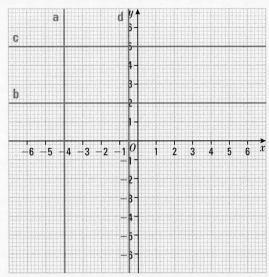

3 Write down the coordinates of the point where the following pairs of lines cross.

   a $x = 1, y = 3$ **b** $x = -4, y = 2$ **c** $y = 3, x = -\frac{1}{2}$

4 Find the perimeter and area of the rectangle formed by the lines $x = -3$, $x = 1$, $y = -2$ and $y = 5$.

A02 **D**

🔍 **Example 2** Draw the graphs of $y = x$ and $y = -x$.

$y = x$
when $x = 0$, $y = 0$ $(0, 0)$
when $x = 3$, $y = 3$ $(3, 3)$
when $x = -3$, $y = -3$ $(-3, -3)$

> Find the **coordinates** of at least two points for which the y value is the same as the x value.

$y = -x$
when $x = 0$, $y = 0$ $(0, 0)$
when $x = 3$, $y = -3$ $(3, -3)$
when $x = -3$, $y = 3$ $(-3, 3)$

> Find the coordinates of at least two points for which the y value is the negative of the x value – that is, it has the opposite sign.

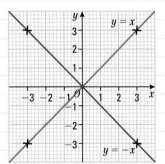

> Plot the three points for each graph and join them with a ruler.

> Label each of the lines with their equation.

ResultsPlus
Exam Tip

These diagonal lines occur frequently so it is useful to be able to draw them from memory.

> **Example 3**　Draw the graph of $y = 3 - 2x$.
>
> Use values of x from -2 to $+5$.

When $x = 0$,　$y = 3 - 2 \times 0 = 3$　$(0, 3)$ ← Use $x = 0$ to make the working out easier.
When $x = 5$,　$y = 3 - 2 \times 5 = -7$　$(5, -7)$
When $x = -2$,　$y = 3 - 2 \times (-2) = 7$　$(-2, 7)$

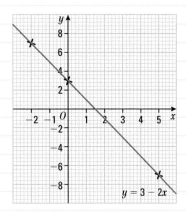

$y = 3 - 2x$

Exercise 15B

1　a　Copy and complete the table of values for $y = 4 - 2x$.

x	-2	-1	0	1	2	3	4
y			4			-2	-4

　　b　Draw the graph of $y = 4 - 2x$. Use values of x from -2 to $+4$.

2　a　Draw the graph of $y = 4x - 8$. Use values of x from -1 to $+4$.
　　b　Write down the coordinates of the point where the graph intersects:
　　　i　the x-axis　　　　　　　ii　the y-axis.
　　c　Use your graph to find:
　　　i　the value of y when $x = 2.5$　　ii　the value of x when $y = 6$.

3　a　On the same axes, draw the graphs of $y = x$, $y = -x$, $y = 2x$ and $y = -3x$.
　　b　What can you say about the graphs of all lines with equations of the form $y = mx + c$ with $c = 0$?

4　a　On the same axes, draw the graphs of $y = 6x - 9$ and $y = 9 - 6x$.
　　b　Write down the coordinates of the point where the two graphs intersect.

5　Work out the area of the triangle formed by the lines $y = 0$, $y = x$ and $y = 2x + 5$.

D

A03

C　A02
A03

Example 4 Draw the graph of $2x + y = 6$.

When $x = 0, 2 \times 0 + y = 6$ $y = 6$ $(0, 6)$ *Use $x = 0$ to find one point.*

When $y = 0, 2x + 0 = 6$ $x = 3$ $(3, 0)$ *Use $y = 0$ to find a second point.*
Solve $2x = 6$.

When $x = 2, 2 \times 2 + y = 6$ $4 + y = 6$ $y = 2$ $(2, 2)$ *Choose a value of x to use as a check point and work out y.*
Solve $4 + y = 6$.

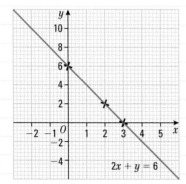

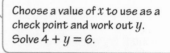

Exam Tip

When you have equations in the form $ax + by = c$, it is easier if you use $x = 0$ and $y = 0$ to find two of your points.

Exercise 15C

1   Draw the graph of $3x + 4y = 12$.

2   **a**   On the same axes, draw the graphs of:
      i $x + y = 2$ **ii** $x + y = 7$ **iii** $x + y = -4$.
   b   What do you notice about the graphs you have drawn?

3   Find the coordinates of the point where the lines $2x + 3y = 12$ and $x - y = 1$ intersect.

4   **a**   On the same axes, draw the graphs of $5x + 2y = 10$ and $5y - 2x = 8$.
   b   What do you notice about the way in which these two graphs intersect?
   c   Write down the equations of two other lines that intersect in this way.

C

A03

A02

A02
A03

15.2 Finding the midpoint of a line segment

◎ Objective

○ You can find the coordinates of the midpoint of a line segment.

⊕ Why do this?

You might do this if you and a friend agree to meet halfway between your house and theirs.

⬆ Get Ready

1. Find the number halfway between:
   a   4 and 6   **b**   5 and 6   **c**   17 and 17.5   **d**   0.20 and 0.25
   e -2 and 4   **f** -6 and -11

Key Points

◉ A line joining two points is called a line segment.
AB is the line segment joining points A and B.

◉ The **midpoint** of a line is halfway along the line.

◉ To find the midpoint you should add the x-coordinates and
divide by 2, and add the y-coordinates and divide by 2.

◉ The midpoint of the line segment AB between

A (x_1, y_1) and B (x_2, y_2) is $\left(\dfrac{x_1 + x_2}{2}, \dfrac{y_1 + y_2}{2}\right)$.

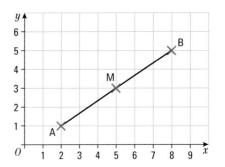

Example 5 Work out the coordinates of the midpoint of the line segment PQ where P is
(2, 3) and Q is (7, 11).

x-coordinate $\qquad 2 + 7 = 9$ $\qquad\qquad\longleftarrow$ $\boxed{\text{Add the } x\text{-coordinates and divide by 2.}}$

$\qquad\qquad\qquad\quad 9 \div 2 = 4\frac{1}{2}$

y-coordinate $\qquad 3 + 11 = 14$ $\qquad\quad\longleftarrow$ $\boxed{\text{Add the } y\text{-coordinates and divide by 2.}}$

$\qquad\qquad\qquad\quad 14 \div 2 = 7$

The midpoint is $(4\frac{1}{2}, 7)$.

Example 6 Find the midpoint of RS.

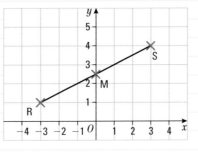

R has coordinates $(-3, 1)$
S has coordinates $(3, 4)$

Add the x-coordinates and divide by 2.
$-3 + 3 = 0 \quad\quad 0 \div 2 = 0$
Add the y-coordinates and divide by 2.
$1 + 4 = 5 \quad\quad 5 \div 2 = 2.5$
M has coordinates $(0, 2.5)$ or $(0, 2\frac{1}{2})$.

Exercise 15D

D

1 Work out the coordinates of the midpoint of each
of the line segments shown on the grid.

    a OA   b BC   c DE

    d FG   e HJ   f KL

    g MN   h PQ   i ST

    j UV

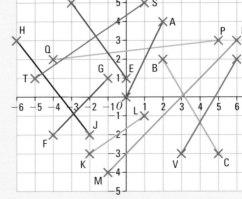

D

2 Work out the coordinates of the midpoint of each of these line segments.

a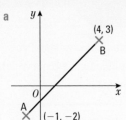

b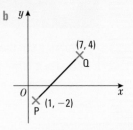

c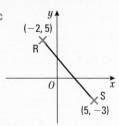

d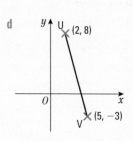

3 Work out the coordinates of the midpoint of each of these line segments.

a AB when A is $(-1, -1)$ and B is $(9, 9)$
b PQ when P is $(2, -4)$ and Q is $(-6, 9)$
c ST when S is $(5, -8)$ and T is $(-2, 1)$
d CD when C is $(1, 7)$ and D is $(-7, 2)$
e UV when U is $(-2, 3)$ and V is $(6, -8)$
f GH when G is $(-2, -6)$ and H is $(7, 3)$

C

15.3 The gradient and y-intercept of a straight line

◎ Objectives

- You can find the gradient of a straight line from its graph.
- You can find the y-intercept of a straight line from its graph.
- You can interpret the gradient of a straight line.

◈ Why do this?

Gradients on graphs often represent important measures in real-life such as speed, acceleration or cost of petrol per litre.

◈ Get Ready

1. Draw, on the same pair of axes, the graphs of $y = x$, $y = 2x$, $y = 3x$
2. Draw, on the same pair of axes, the graphs of $y = -x$, $y = -\frac{1}{2}x$, $y = -2x$
3. Draw, on the same pair of axes, the graphs of $y = 3x - 2$, $y = 3x$, $y = 3x - 3$

◉ Key Points

- The **gradient** of a straight line is a measure of its slope.
- Steeper lines have larger gradients.
- Gently sloping lines have smaller gradients.
- Gradient of a line $= \dfrac{\text{change in } y\text{-direction}}{\text{change in } x\text{-direction}}$
- Lines which slope upwards from left to right have positive gradients.
- Lines which slope downwards from left to right have negative gradients.
- The gradient of the line through the points (x_1, y_1) and (x_2, y_2) is given by $m = \dfrac{y_2 - y_1}{x_2 - x_1}$

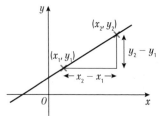

- The **y-intercept** of a line is the value of y when $x = 0$.
  It is shown by the point where the graph crosses the y-axis.

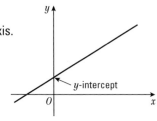

Example 7 Find the gradient and y-intercept of this straight-line graph.

Results Plus

Exam Tip

Working out the gradient is easier if you choose a large triangle with a base which is a whole number of units.

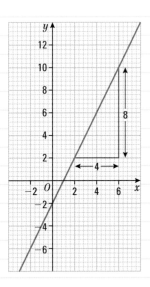

$$\text{Gradient} = \frac{\text{change in } y\text{-direction}}{\text{change in } x\text{-direction}}$$ ← Draw a right-angled triangle on the line.

$$= \frac{10 - 2}{6 - 2}$$ ← Find the difference in the y values and the difference in the x values.

$$= \frac{8}{4}$$ ← Work out the value of the fraction.

$$\text{Gradient} = 2$$

$$\text{Intercept} = -2$$ ← Read off the value where the graph crosses the y-axis.

Example 8 Find the gradient and the coordinates of the y-intercept of this straight-line graph.

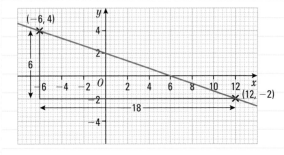

$$\text{Gradient} = \frac{\text{change in } y\text{-direction}}{\text{change in } x\text{-direction}}$$

$$= -\frac{6}{18}$$ ← Put a negative sign in as the gradient slopes down from left to right.

$$= -\frac{1}{3}$$ ← Simplify the fraction if possible.

$$y\text{-intercept} = 2$$
$$y\text{-intercept has coordinates } (0, 2).$$

Results Plus

Exam Tip

Make sure you take into account the scales on the axes when finding the lengths of the sides of the triangle.

Example 9   Find the gradient of the line joining the points A $(-5, -1)$ and B $(4, 5)$.

Method 1

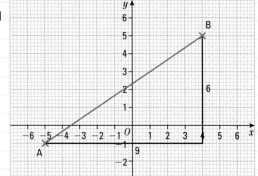

Draw a diagram to help you.

$$\frac{\text{vertical distance}}{\text{horizontal distance}} = \frac{\text{difference in } y\text{-coordinates}}{\text{difference in } x\text{-coordinates}}$$

$$= \frac{5 + 1}{4 + 5}$$

$$= \frac{6}{9}$$

$$\text{Gradient} = \frac{2}{3}$$

Method 2

$$m = \frac{y_2 - y_1}{x_2 - x_1}$$

Use $(x_1, y_1) = (-5, -1)$
and $(x_2, y_2) = (4, 5)$.
Put $x_1 = -5, y_1 = -1$
and $x_2 = 4, y_2 = 5$ into
the formula for m.

$$= \frac{5 - (-1)}{4 - (-5)}$$

$$= \frac{6}{9}$$

$$\text{Gradient} = \frac{2}{3}$$

Exercise 15E

1   Work out the gradient of each line.

a
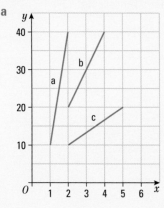

b
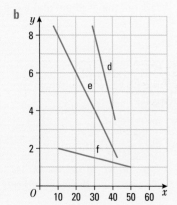

D

245

D

2 Work out the gradient and y-intercept of each straight line.

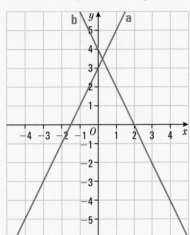 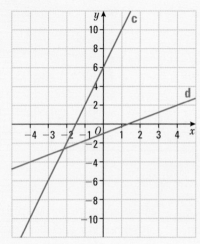

C

3 Using separate axes, using 1 cm to represent 1 unit, draw straight lines with:

a gradient 1, y-intercept 2

b gradient 2, y-intercept -3

c gradient $\frac{1}{2}$, y-intercept 4

d gradient -3, y-intercept 3

e gradient $-\frac{1}{3}$, passing through the point $(0, 2)$.

4 A straight line has gradient 3. The point $(2, -1)$ lies on the line. Find the coordinates of one other point on the line.

5 A is the point $(-4, 6)$. B is the point $(8, 0)$.

a Find the gradient of the line AB.

b Find the coordinates of the y-intercept of the line AB.

Example 10 The graph shows the charge for gas, in £s, supplied to a customer by a gas company.

The cost consists of a standing charge plus a charge for each unit of gas used.

a Write down the standing charge made by the gas company.

b Work out the gradient of the graph and explain what it represents.

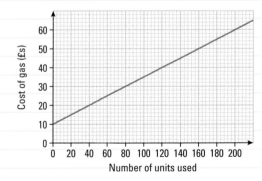

a

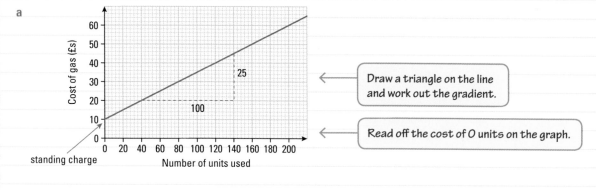

Draw a triangle on the line and work out the gradient.

Read off the cost of 0 units on the graph.

Standing charge = £10

b Gradient = $\dfrac{25}{100}$

= 0.25

The gradient is found by dividing the cost (in £s) by the number of units used. It represents the cost of each unit of gas used.

The cost of gas is 25p per unit used.

Exercise 15F

1 The graph shows the cooking time, *t* minutes, needed for a chicken of weight *w* kilograms.

   a Work out the gradient of the graph and explain what it represents.
   b Describe a rule to give the time needed to cook a chicken of any weight.
   c Why do you think the line doesn't extend down to the *x*-axis?

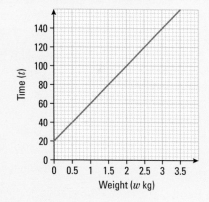

2 This graph can be used to change between temperatures measured in °C and temperatures measured in °F.

   a What temperature in °F is equivalent to 10 °C?
   b Find the gradient of this graph and explain what it represents.

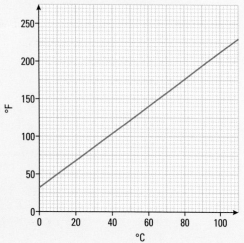

D

C

3 The diagram shows the distance–time graphs of a car, a cycle and a lorry.

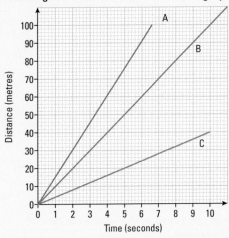

a Work out the gradient of each graph.
b Use your answer to part **a** to match each vehicle to one of the graphs A, B or C.
Give reasons for your answers.

4 Sheila records the depth of water, in cm, in a swimming pool every 30 seconds.
Her results are shown in the table below.

Time (t seconds)	0	30	60	90	120	150	180	210	240
Depth (d cm)	200	175	150	125	100	75	50	25	0

a Draw a straight line graph to show the depth of water for $t = 0$ to $t = 240$.
b Work out the gradient of the straight line.
c Describe what the gradient represents. What do you think is happening?

15.4 The equation $y = mx + c$

⊙ Objectives

- ○ You can find the gradient of a line from its equation.
- ○ You can find the y-intercept of a line from its equation.
- ○ You can find the equation of a straight line.
- ○ You can interpret the equation of a straight line.

⟐ Why do this?

Straight-line graphs often occur in real life and their equations give the link between two quantities. For example, the equation $F = 1.8C + 32$ can be used to convert °C to °F.

⟐ Get Ready

1. Complete the table to show the gradients of these lines. What do you notice?

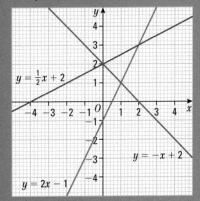

Equation of line	Gradient of line
$y = 2x - 1$	
$y = \frac{1}{2}x + 2$	
$y = -x + 2$	

2. Complete the table to show the y-intercepts of these lines. What do you notice?

Equation of line	y-intercept
$y = 2x - 1$	
$y = \frac{1}{2}x + 2$	
$y = -x + 2$	

Key Points

⊙ The straight line with equation $y = mx + c$ has gradient m.

⊙ The straight line with equation $y = mx + c$ crosses the y-axis at the point $(0, c)$.

⊙ The point $(0, c)$ is known as the y-intercept.

Example 11 For the lines with equations

a $y = 5x + 4$

b $3x + 2y = 6$

find:

i the gradient of the line

ii the y-intercept of the line.

a $\qquad y = 5x + 4 \quad \longleftarrow$ | Compare $y = 5x + 4$ with $y = mx + c$.

i $\qquad$ gradient $= 5 \quad \longleftarrow$ | Write down the value of the gradient m from the term in x.

ii y-intercept $= 4 \quad \longleftarrow$ | Write down the value of the y-intercept c from the constant term.

b i $\qquad 3x + 2y = 6 \quad \longleftarrow$ | Rearrange the equation $3x + 2y = 6$ into the form $y = mx + c$.

$\qquad 2y = 6 - 3x \quad \longleftarrow$

$\qquad y = 3 - 1.5x \quad \longleftarrow$ | Subtract $3x$ from both sides.

$\qquad y = -1.5x + 3 \quad \longleftarrow$ | Divide both sides by 2.

$\qquad$ gradient $= -1.5$

ii $\qquad y$-intercept $= 3$

Example 12 a Draw the straight line which has a gradient of $\frac{1}{2}$ and which crosses the y-axis at the point $(0, 1)$.

b Write down the equation of this line.

a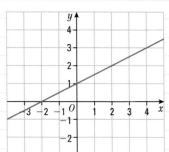

b gradient of line $= \frac{1}{2}$

y-intercept $= 1$

Equation of the line is $y = \frac{1}{2}x + 1 \quad \longleftarrow$ | Substitute $m = \frac{1}{2}, c = 1$ into $y = mx + c$.

Exercise 15G

C

1 A line which passes through the point (0, 5) has gradient 2.
Write down the equation of the line.

2 Find **i** the gradient and **ii** the y-intercept of the lines with the equations
  a $y = 4x + 1$ **b** $y = 3x - 4$ **c** $y = \frac{2}{3}x + 4$
  d $2x + 5y = 20$ **e** $4x - 3y = 12$ **f** $x - 2y = 0$

B

3 Find the equations of the lines shown in the diagram.

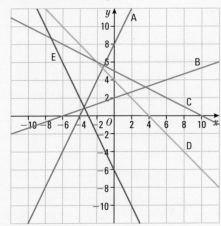

A

4 A line passes through the points with coordinates (1, 3) and (2, 8). Find the equation of the line.

5 The gradient of a line is 3. The point with coordinates (4, 2) lies on the line.
Find the equation of the line.

15.5 Parallel and perpendicular lines

◎ Objectives

- You understand the relationship between the gradients of parallel lines.
- You understand the relationship between the gradients of perpendicular lines.
- You can use the relationship between the gradients to find the equations of parallel and perpendicular lines.

◈ Why do this?

When building a house you need to find the gradients of parallel and perpendicular lines such as struts and beams in ceilings and roofs.

◈ Get Ready

1. Here are three pairs of lines.
Write down the gradient of each line.

a
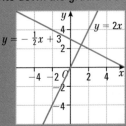

b
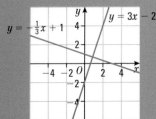

c
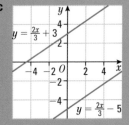

What do you notice?

Key Point

◉ If a line has gradient m then any line drawn parallel to it also has gradient m and any line drawn **perpendicular** to it has gradient $-\frac{1}{m}$ (the negative reciprocal of m).

Example 13   Find the equation of the line parallel to $y = 3x + 7$ and passing through $(0, -2)$.

$y = 3x + 7$, gradient $m = 3$, y-intercept is $(0, 7)$ ← Compare $y = 3x + 7$ with $y = mx + c$.

The gradient of any line parallel to $y = 3x + 7$ is 3 ← Parallel lines have equal gradients.
so the equation of any line parallel to $y = 3x + 7$ is $y = 3x + c$.

The required line has y-intercept $(0, -2)$. ← Write down the value of c from the coordinates of the y-intercept given.
The equation is $y = 3x - 2$.

Example 14   Find the equation of any line which is perpendicular to $y = 2x - 9$.

$y = 2x - 9$, gradient $m = 2$
The gradient of any line perpendicular to this has gradient $-\frac{1}{2}$. ← Find the negative reciprocal of 2.

The equation of any line with gradient $-\frac{1}{2}$ is of the form $y = -\frac{1}{2}x + c$. ← Use $y = mx + c$.

So $y = -\frac{1}{2}x + 1$ is one example of a line ← Pick any value for c.
perpendicular to the line $y = 2x - 9$.

Example 15   Find the equation of the line which is perpendicular to the line joining the points $(-2, 4)$ and $(4, 1)$ and which passes through the point $(1, 5)$.

Method 1

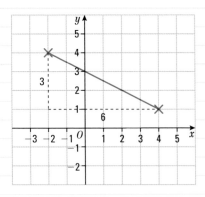

Draw a diagram and find the gradient of the line joining the points $(-2, 4)$ and $(4, 1)$.

Gradient $m = -\frac{3}{6} = -\frac{1}{2}$

The gradient of a perpendicular line is $-\frac{1}{m} = 2$ ← Find the negative reciprocal of $-\frac{1}{2}$.

The equation of any line with gradient 2 is $y = 2x + c$. ← Use $y = mx + c$.

$(1, 5)$ is on this line so

$5 = 2 \times 1 + c$ ← Substitute $x = 1$ and $y = 5$.

$5 = 2 + c$

$c = 3$ ← Find the value of c.

The equation of the line required is $y = 2x + 3$.

Method 2

$$m = \frac{y_2 - y_1}{x_2 - x_1}$$

$$= \frac{1 - 4}{4 - (-2)}$$ ← Use $(-2, 4)$ as (x_1, y_1) and $(4, 1)$ as (x_2, y_2).

$$= \frac{-3}{6}$$

Gradient $= -\frac{1}{2}$

Once the gradient is found, the solution follows **Method 1** above.

Exercise 15H

1 Copy and complete the following table to show the gradients of pairs of lines l_1 and l_2 which are perpendicular to each other.

	a	b	c	d	e
Gradient of line l_1	3	−4	$\frac{1}{5}$		
Gradient of line l_2				3	$-\frac{1}{6}$

2 Write down the equation of a line parallel to the line with the equation
   a $y = 2x + 5$ **b** $y = \frac{1}{3}x - 1$ **c** $y = 4 - x$

3 Write down the equation of a line perpendicular to the line with the equation
   a $y = x - 6$ **b** $y = 3x + 2$ **c** $y = 1 - \frac{1}{2}x$

4 Find the equation of a line which is parallel to the line with the equation $y = 4x - 1$ and which passes through the point $(0, 3)$.

5 Find the equation of a line which is parallel to the line with the equation $2x + y = 4$ and which passes through the origin.

6 Find the equation of a line which is perpendicular to the line with the equation $y = \frac{1}{4}x$ and which passes through the point $(2, -8)$.

7 Find the equation of a line which is perpendicular to the line with the equation $x + y = 10$ and which passes through the point $(-2, -5)$.

15.6 Real-life graphs

◎ Objectives

◉ You can draw and interpret distance–time graphs.

◉ You can draw and interpret other graphs describing real-life situations.

ⓘ Why do this?

Graphs are used to describe a wide range of real-life situations. You may have to plot and interpret the results of your science class experiment.

◈ Get Ready

1. Read off the values shown on the scale.

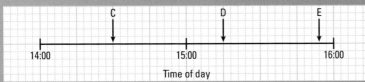

◔ Key Point

◉ Graphs can be used to describe a variety of real-life situations, and show how one variable changes in relation to another – for example, distance against time, the cost of posting a parcel against its weight, or how a liquid fills a container over time.

◉ On a **distance–time graph** (or travel graph):

   ◉ straight lines represent **constant speed**

   ◉ horizontal lines represent no movement

   ◉ the gradient gives the **speed**: average speed $= \dfrac{\text{distance travelled}}{\text{time taken}}$

🔍 Example 16

Steve went for a ride on his bike. He rode from his home to a friend's house and back. The travel graph shows his trip.

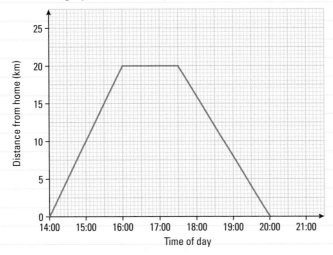

a At what time did Steve reach his friend's house?

b Find Steve's speed on the journey to his friend's house.

c What was Steve doing between 16:00 and 17:30?

d Find Steve's average speed on his journey home.

a Steve reached his friend's house at 16:00 (or 4 pm).

b Speed = $\dfrac{20\text{ km}}{2\text{ hours}}$ ← Work out the gradient of the line representing the first part of his journey.

This represents $\dfrac{\text{distance}}{\text{time}}$ = speed

= 10 km/h ← State the units with your answer.

Steve travels 10 kilometres each hour on the journey to his friend's house.
Steve's speed is 10 km/h.

c The gradient of the line representing Steve's journey between 16:00 and 17:30 is 0 and so his speed is 0 km/h. He is not moving.
Steve stays at his friend's house between 16:00 and 17:30.

d The gradient of the line representing Steve's journey home is

$\dfrac{20\text{ km}}{2.5\text{ hours}}$ ← The distance home is 20 km. Steve arrives home $2\frac{1}{2}$ hours after he leaves.

= 8 km/h

Steve's speed is 8 km/h.

Example 17 Water is poured into a cylindrical container at a constant rate.

a Sketch a graph to describe the relationship between the height (h) of the water and the time taken (t).

b Here is a different-shaped container.
Sketch a graph showing the relationship between h and t in this case.

a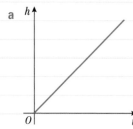

The container is filled at a constant rate, so the height increases by the same amount for each second.

Draw a straight line through $(0, 0)$.

b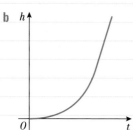

Since the container gets narrower the height of the water increases more rapidly for each second at first. The gradient of the graph increases. The top part of the container is cylindrical so this is represented by a straight line.

Exercise 15I

1 Janine sets off from home to walk to the shopping centre. She does some shopping then gets the bus home. The travel graph shows information about her journey.

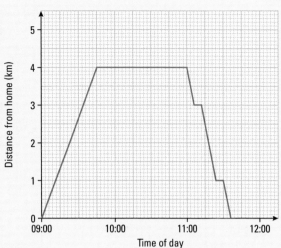

a What distance from Janine's home is the shopping centre?

b For how many minutes is Janine at the shopping centre?

c Work out Janine's walking speed. Give your answer in km/h.

d The bus stops twice on Janine's journey home. For how long does the bus stop each time?

e At what average speed, in km/h, does the bus travel?

2 The graph shows the cost of posting a parcel.

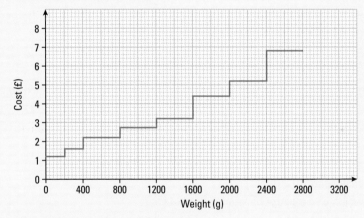

a Find the cost of posting a parcel of weight 1 kg.

b Find the maximum weight of a parcel that can be posted for less than £5.

c Work out the total cost of posting three parcels which have weights 520 g, 1.5 kg and 2.5 kg.

3 Liquid is poured into each of these containers.
Sketch a graph to show the relationship between the depth of water and the volume of water in each container.

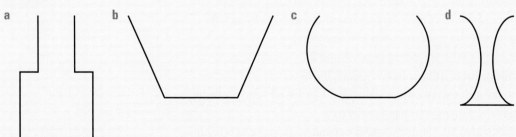

C **A03**

4 Here are the cross-sections of three different swimming pools.
Each pool is to be emptied by pumping out the water. Water is pumped out at a steady rate.

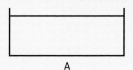

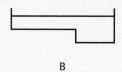

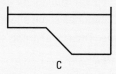

A   B   C

Here are three sketch graphs showing the relationship between the depth of the water left in the pool and the number of minutes since the pump was switched on.

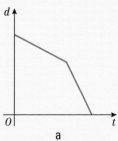

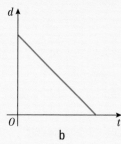

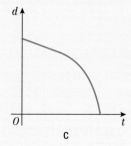

a   b   c

Match each swimming pool with one of the graphs.

A03 **5** The petrol consumption of a car, in kilometres per litre (km/l), depends on the speed of the car.
The table gives some information about the petrol consumption of the car at different speeds.

Speed (km/h)	64	72	80	88	96	104
Petrol consumption (km/l)	12.3	13.8	14.4	14.5	14.1	12.3

Draw axes on graph paper, taking 2 cm to represent 10 km/h on the horizontal axis and 4 cm to represent 1 km/l on the vertical axis.
Start the horizontal axis at 60 and the vertical axis at 12.
Plot the values from the table and join them with a smooth curve.
From your graph estimate:
a the petrol consumption at 70 km/h
b the speeds which give a petrol consumption of 14 km/l.

Chapter review

⦿ To draw a straight line by plotting points you need to plot at least two points which fit the equation of the line.
⦿ The **equation of a straight line** can have several forms:
 ⦿ lines in the form of $x = c$ or $y = c$ where c is a number
 ⦿ lines in the form of $y = mx + c$ where m and c are numbers
 ⦿ lines in the form of $ax + by = c$ where a, b and c are numbers.
⦿ A line joining two points is called a line segment.
 AB is the line segment joining points A and B.
⦿ The **midpoint** of a line is halfway along the line.
⦿ To find the midpoint you should add the x-coordinates and divide by 2, and add the y-coordinates and divide by 2.
⦿ The midpoint of the line segment AB between A (x_1, y_1) and B (x_2, y_2) is $\left(\dfrac{x_1 + x_2}{2}, \dfrac{y_1 + y_2}{2}\right)$.

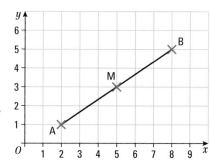

- The **gradient** of a straight line is a measure of its slope.
- Steeper lines have larger gradients.
- Gently sloping lines have smaller gradients.
- Gradient of a line $= \dfrac{\text{change in } y\text{-direction}}{\text{change in } x\text{-direction}}$
- Lines which slope upwards from left to right have positive gradients.
- Lines which slope downwards from left to right have negative gradients.
- The gradient of the line through the points (x_1, y_1) and (x_2, y_2) is given by

$$m = \dfrac{y_2 - y_1}{x_2 - x_1}$$

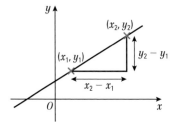

- The **y-intercept** of a line is the value of y when $x = 0$.
  It is shown by the point where the graph crosses the y-axis.

- The straight line with equation $y = mx + c$ has gradient m.
- The straight line with equation $y = mx + c$ crosses the y-axis at the point $(0, c)$.
- The point $(0, c)$ is known as the y-intercept.
- If a line has gradient m then any line drawn parallel to it also has gradient m and any line drawn **perpendicular** to it has gradient $-\frac{1}{m}$ (the negative reciprocal of m).
- On a **distance–time graph** (or travel graph):
  - straight lines represent **constant speed**
  - horizontal lines represent no movement
  - the gradient gives the **speed**: average speed $= \dfrac{\text{distance travelled}}{\text{time taken}}$

Review exercise

1. a On the same axes, draw and label the lines $x = 4$, $y = -2$ and $y = -x$.
   b Work out the area of the triangle formed by the lines $x = 4$, $y = -2$ and $y = -x$.

2. Nicki is going on holiday to the USA.
   She wants to change some pounds (£) to dollars ($). The exchange rate is £1 = $1.65.
   Draw a conversion graph that Nicki could use to change between £ and $.

   A02

3. ABCDE is a pentagon.
   a Find the gradient of:
      i CD
      ii BC
      iii ED.
   b Use your answers to part a to write down the gradients of:
      i AB
      ii AE.

   D

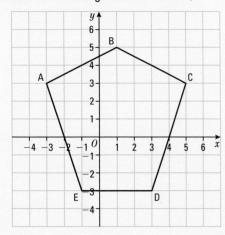

D

4 The formula $F = 2C + 30$ can be used to estimate F given the value of C, where F is the temperature in Fahrenheit and C is the temperature in Celsius.

Copy and complete the table and use it to draw the graph of F against C for values of C from 0 to 100.

C	0	20	40	60	80	100
F			110			

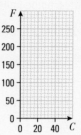

C
A02

5 A straight line has equation $2x + y = 6$.

  a Draw the graph of $2x + y = 6$.

  b Find:

    i the gradient

    ii the y-intercept of the straight line.

6 The distance–time graphs represent the journey made by a bus and a car starting in Swindon, travelling to London and returning to Swindon.

  a How far is it from Swindon to London?

  b How much longer, including stops, did it take the bus to complete the journey from Swindon to London than it did the car?

  c Work out the greatest speed of the car during the journey.

  d The bus stopped at Reading on its journey to London. At what time did the car reach Reading?

  e Work out the average speed of the bus, including stops, on its return journey.

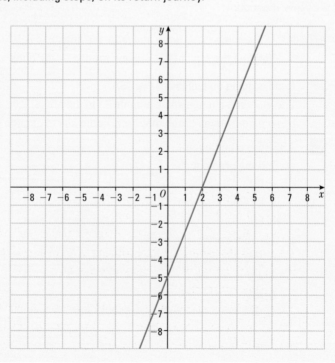

A02
A03

7 Line A has equation $y = 2x + 1$.

Line B passes through the points $(2, -5)$ and $(3, -1)$.

Line C is shown on the diagram.

Which line has the steepest gradient?

Show your working clearly.

8 Here are four containers.

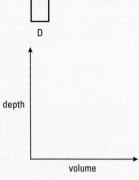

A   B   C   D

Water is poured into each container at a constant rate.
Sketch a graph for each container showing how the depth of water increases with the volume.

depth

volume

9 The equation of a straight line, l, is $y = \frac{1}{2}x + 3$.

a Does the point $(2, -1)$ lie on this line?

b Write down the equation of the line parallel to l which passes through the point $(0, 5)$.

c Find the equation of the line perpendicular to l which passes through the point $(4, 1)$.

10 Use a suitable grid to draw the graph of $y = 7 - 2x$ for values of x from -1 to 4.

11 The straight line L has the equation $y = 4x + 3$.

a State the gradient of this line.

b Find the coordinates of the point where L cuts the x-axis.

The point $(k, 21)$ lies on L.

c Find the value of k.

12 The graph shows the cost of using a mobile phone for one month for three different tariffs.

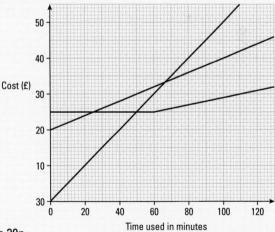

Cost (£)

Time used in minutes

Tariff A   Rental £20   Every minute costs 20p.

Tariff B   Pay as you go   Every minute costs 50p.

Tariff C   Rental £25   First 60 minutes free, then each minute costs 10p.

a Label each line on the graph with the letter of the tariff it represents.

Jim uses tariff A for 100 minutes in one month.

b Find the total cost.

Fiona uses her mobile phone for about 60 minutes each month.

c Explain which tariff would be the cheapest for her to use.

You must give the reasons for your answer.

C AO2 AO3 * **13** Abbie has the option of joining two health clubs.

Hermes has a joining fee of £100 plus a fee of £5 per session.

Atlantis has a joining fee of £200 with a fee of £3 per session.

Which health club should she choose?

You must show all calculations and fully explain your solution.

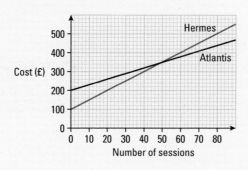

14 P has coordinates (1, 4).

R has coordinates (5, 0).

Find the coordinates of the midpoint of the line PR.

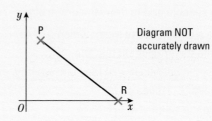

Diagram NOT accurately drawn

June 2008

B **15** Copy and complete the following table.

Equation of line	Gradient	y-intercept
$y = 2x + 5$		
	7	-3
$y = 6 - x$		
	$\frac{2}{3}$	-1
	-4	3

AO3 **16** Here are six temperature/time graphs.

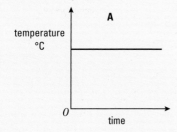

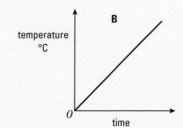

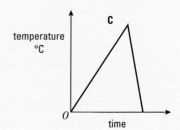

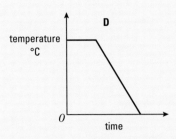

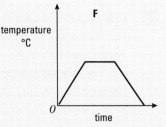

B

Describe the events shown by the graph in each case.

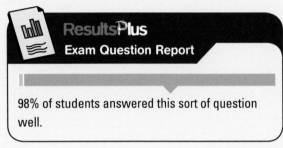

ResultsPlus
Exam Question Report

98% of students answered this sort of question well.

Nov 2008, adapted

17 The diagram shows a rectangle.
All the measurements are in centimetres.
The perimeter of the rectangle is 24 cm.

A03 A

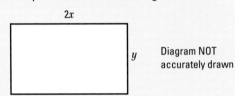

Diagram NOT
accurately drawn

a Explain why $2x + y = 12$.
b Draw the graph of $2x + y = 12$ for values of x from 0 to 6.
c Use your graph to find the value of x which makes the rectangle a square.

18 The point P $(3, k)$ lies on the line with equation $y = 2x + 1$.
Show that P also lies on the line with equation $y = 3x - 2$.

A02

19 The diagram shows three points A $(-1, 5)$, B $(2, -1)$ and C $(0, 5)$.
The line L is parallel to AB and passes through C.
Find the equation of the line L.

A03

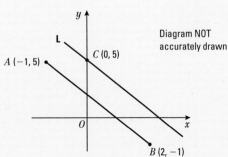

Diagram NOT
accurately drawn

June 2005

A
A02

20 A straight line has equation $y = 2x - 3$.

The point P lies on the straight line.

The y coordinate of P is -4.

a Find the x coordinate of P.

A straight line L is parallel to $y = 2x - 3$ and passes through the point $(3, 4)$.

b Find the equation of the line L.

Nov 2005

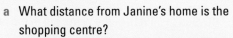

Exercise 15I

1   Janine sets off from home to walk to the shopping
    centre. She does some shopping then gets the
    bus home. The travel graph shows information
    about her journey.

    a What distance from Janine's home is the
       shopping centre?

    b For how many minutes is Janine at the
       shopping centre?

    c Work out Janine's walking speed.
       Give your answer in km/h.

    d The bus stops twice on Janine's journey
       home. For how long does the bus stop
       each time?

    e At what average speed, in km/h, does the bus travel?

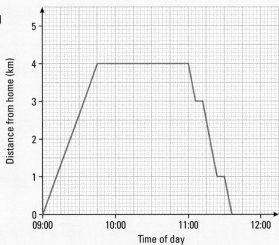

2   The graph shows the cost of posting a parcel.

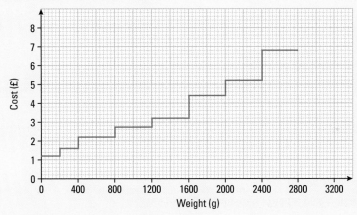

    a Find the cost of posting a parcel of weight 1 kg.

    b Find the maximum weight of a parcel that can be posted for less than £5.

    c Work out the total cost of posting three parcels which have weights 520 g, 1.5 kg and 2.5 kg.

3   Liquid is poured into each of these containers.
    Sketch a graph to show the relationship between the depth of water and the volume of water in each
    container.

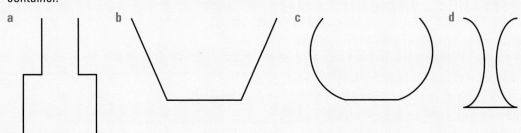

C A03

4 Here are the cross-sections of three different swimming pools.
Each pool is to be emptied by pumping out the water. Water is pumped out at a steady rate.

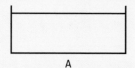 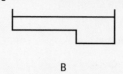 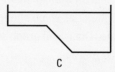

A   B   C

Here are three sketch graphs showing the relationship between the depth of the water left in the pool and the number of minutes since the pump was switched on.

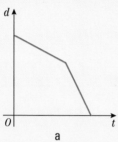 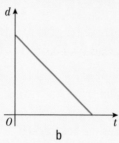 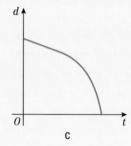

a   b   c

Match each swimming pool with one of the graphs.

A03

5 The petrol consumption of a car, in kilometres per litre (km/*l*), depends on the speed of the car.
The table gives some information about the petrol consumption of the car at different speeds.

Speed (km/h)	64	72	80	88	96	104
Petrol consumption (km/*l*)	12.3	13.8	14.4	14.5	14.1	12.3

Draw axes on graph paper, taking 2 cm to represent 10 km/h on the horizontal axis and 4 cm to represent 1 km/*l* on the vertical axis.
Start the horizontal axis at 60 and the vertical axis at 12.
Plot the values from the table and join them with a smooth curve.
From your graph estimate:

a the petrol consumption at 70 km/h

b the speeds which give a petrol consumption of 14 km/*l*.

Chapter review

- To draw a straight line by plotting points you need to plot at least two points which fit the equation of the line.
- The **equation of a straight line** can have several forms:
  - lines in the form of $x = c$ or $y = c$ where c is a number
  - lines in the form of $y = mx + c$ where m and c are numbers
  - lines in the form of $ax + by = c$ where a, b and c are numbers.
- A line joining two points is called a line segment.
  AB is the line segment joining points A and B.
- The **midpoint** of a line is halfway along the line.
- To find the midpoint you should add the x-coordinates and divide by 2, and add the y-coordinates and divide by 2.
- The midpoint of the line segment AB between A (x_1, y_1) and B (x_2, y_2) is $\left(\dfrac{x_1 + x_2}{2}, \dfrac{y_1 + y_2}{2}\right)$.

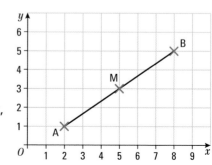

- The **gradient** of a straight line is a measure of its slope.
- Steeper lines have larger gradients.
- Gently sloping lines have smaller gradients.
- Gradient of a line $= \dfrac{\text{change in } y\text{-direction}}{\text{change in } x\text{-direction}}$
- Lines which slope upwards from left to right have positive gradients.
- Lines which slope downwards from left to right have negative gradients.
- The gradient of the line through the points (x_1, y_1) and (x_2, y_2) is given by

  $$m = \dfrac{y_2 - y_1}{x_2 - x_1}$$

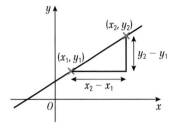

- The **y-intercept** of a line is the value of y when $x = 0$.
  It is shown by the point where the graph crosses the y-axis.

- The straight line with equation $y = mx + c$ has gradient m.
- The straight line with equation $y = mx + c$ crosses the y-axis at the point $(0, c)$.
- The point $(0, c)$ is known as the y-intercept.
- If a line has gradient m then any line drawn parallel to it also has gradient m and any line drawn **perpendicular** to it has gradient $-\frac{1}{m}$ (the negative reciprocal of m).
- On a **distance–time graph** (or travel graph):
  - straight lines represent **constant speed**
  - horizontal lines represent no movement
  - the gradient gives the **speed**: average speed $= \dfrac{\text{distance travelled}}{\text{time taken}}$

Review exercise

1 **a** On the same axes, draw and label the lines $x = 4$, $y = -2$ and $y = -x$.
    b Work out the area of the triangle formed by the lines $x = 4$, $y = -2$ and $y = -x$.

2 Nicki is going on holiday to the USA.
She wants to change some pounds (£) to dollars ($). The exchange rate is £1 = \$1.65.
Draw a conversion graph that Nicki could use to change between £ and \$.

A02

3 ABCDE is a pentagon.
  a Find the gradient of:
    i CD
    ii BC
    iii ED.
  b Use your answers to part **a** to write down the gradients of:
    i AB
    ii AE.

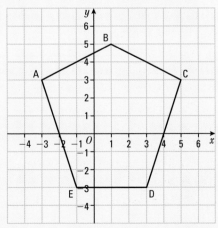

D

D

4 The formula $F = 2C + 30$ can be used to estimate F given the value of C, where F is the temperature in Fahrenheit and C is the temperature in Celsius.

Copy and complete the table and use it to draw the graph of F against C for values of C from 0 to 100.

C	0	20	40	60	80	100
F			110			

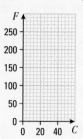

C **A02**

5 A straight line has equation $2x + y = 6$.
  a Draw the graph of $2x + y = 6$.
  b Find:
   i the gradient
   ii the y-intercept of the straight line.

6 The distance–time graphs represent the journey made by a bus and a car starting in Swindon, travelling to London and returning to Swindon.
  a How far is it from Swindon to London?
  b How much longer, including stops, did it take the bus to complete the journey from Swindon to London than it did the car?
  c Work out the greatest speed of the car during the journey.
  d The bus stopped at Reading on its journey to London. At what time did the car reach Reading?
  e Work out the average speed of the bus, including stops, on its return journey.

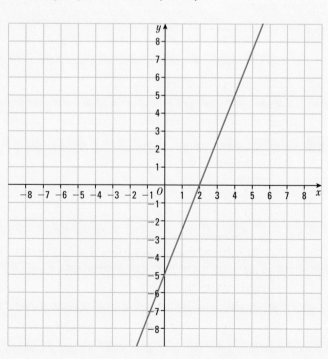

A02 **A03**

7 Line A has equation $y = 2x + 1$.
Line B passes through the points $(2, -5)$ and $(3, -1)$.
Line C is shown on the diagram.
Which line has the steepest gradient?
Show your working clearly.

8 Here are four containers.

  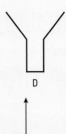

A   B   C   D

Water is poured into each container at a constant rate.
Sketch a graph for each container showing how the depth of water increases with the volume.

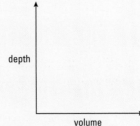

9 The equation of a straight line, l, is $y = \frac{1}{2}x + 3$.
  a Does the point $(2, -1)$ lie on this line?
  b Write down the equation of the line parallel to l which passes through the point $(0, 5)$.
  c Find the equation of the line perpendicular to l which passes through the point $(4, 1)$.

10 Use a suitable grid to draw the graph of $y = 7 - 2x$ for values of x from -1 to 4.

11 The straight line L has the equation $y = 4x + 3$.
  a State the gradient of this line.
  b Find the coordinates of the point where L cuts the x-axis.
  The point $(k, 21)$ lies on L.
  c Find the value of k.

12 The graph shows the cost of using a mobile phone for one month for three different tariffs.

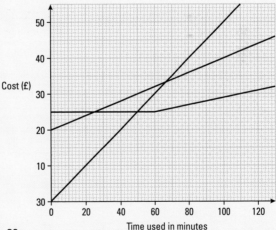

Time used in minutes

Tariff A   Rental £20   Every minute costs 20p.
Tariff B   Pay as you go Every minute costs 50p.
Tariff C   Rental £25   First 60 minutes free, then each minute costs 10p.
  a Label each line on the graph with the letter of the tariff it represents.
Jim uses tariff A for 100 minutes in one month.
  b Find the total cost.
Fiona uses her mobile phone for about 60 minutes each month.
  c Explain which tariff would be the cheapest for her to use.
  You must give the reasons for your answer.

A03 C
A02
A02

C A02 A03 *13 Abbie has the option of joining two health clubs.

Hermes has a joining fee of £100 plus a fee of £5 per session.

Atlantis has a joining fee of £200 with a fee of £3 per session.

Which health club should she choose?

You must show all calculations and fully explain your solution.

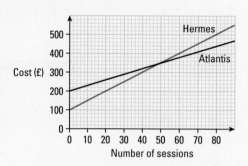

14 P has coordinates (1, 4).

R has coordinates (5, 0).

Find the coordinates of the midpoint of the line PR.

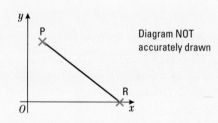

Diagram NOT accurately drawn

June 2008

B 15 Copy and complete the following table.

Equation of line	Gradient	y-intercept
$y = 2x + 5$		
	7	−3
$y = 6 − x$		
	$\frac{2}{3}$	−1
	−4	3

A03 16 Here are six temperature/time graphs.

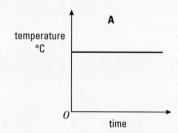

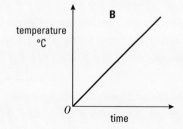

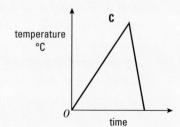

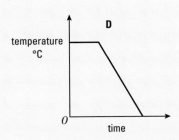

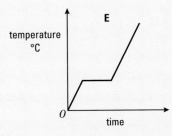

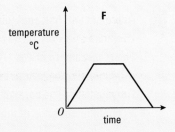

Describe the events shown by the graph in each case.

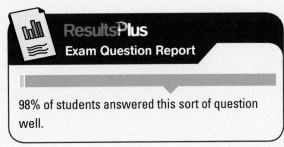

Nov 2008, adapted

17 The diagram shows a rectangle.
All the measurements are in centimetres.
The perimeter of the rectangle is 24 cm.

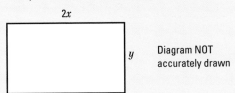

Diagram NOT accurately drawn

a Explain why $2x + y = 12$.

b Draw the graph of $2x + y = 12$ for values of x from 0 to 6.

c Use your graph to find the value of x which makes the rectangle a square.

18 The point P $(3, k)$ lies on the line with equation $y = 2x + 1$.
Show that P also lies on the line with equation $y = 3x - 2$.

19 The diagram shows three points $A (-1, 5)$, $B (2, -1)$ and $C (0, 5)$.
The line L is parallel to AB and passes through C.
Find the equation of the line L.

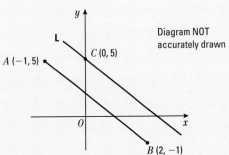

Diagram NOT accurately drawn

June 2005

20 A straight line has equation $y = 2x - 3$.

The point P lies on the straight line.

The y coordinate of P is -4.

a   Find the x coordinate of P.

A straight line L is parallel to $y = 2x - 3$ and passes through the point $(3, 4)$.

b   Find the equation of the line L.

Nov 2005

16 RATIO AND PROPORTION

The ratio of water to land on the Earth's surface is approximately 7 : 3. However, this ratio is likely to change as global warming leads to the melting of the polar icecaps and an increase in the amount of the surface underwater. Countries including Bangladesh, Burma and Egypt could find large parts of their surface area flooded and even parts of the UK such as Lincolnshire and the area around the Thames Estuary may be lost.

◎ Objectives

In this chapter you will:
- simplify ratios
- solve problems using ratios
- share a quantity in a given ratio
- use direct proportion
- use inverse proportion.

◇ Before you start

You should be able to:
- carry out arithmetic using whole numbers
- find the HCF of two numbers
- use ratios in maps and scale drawings
- convert between metric units.

16.1 Introducing ratio

Objectives

- You can simplify ratios.
- You can write down a fraction from a ratio.
- You can write ratios in the unitary form.

Why do this?

Maps use ratio so that the actual distance between two places can be worked out by measuring the distance on the map and then using the ratio.

Get Ready

1. The scale on a road map is 1 : 200 000.
   Sunderland and Newcastle are 9 cm apart on the map.
   Work out the real distance, in km, between Sunderland and Newcastle.

Key Points

- **Ratios** are used to compare quantities.
- The simplest form of a ratio has whole numbers with no common factor.
- Ratios are sometimes given in the form $1 : n$ where n is a number.
  This is called the **unitary** form of a ratio.
  It is most often used for scales in maps and scale drawings (see Section 12.6).
- To write a ratio in the form $1 : n$, divide each number in the ratio by the first number in the ratio.
  For example, $5 : 8 = \frac{5}{5} : \frac{8}{5}$
  $$= 1 : 1.6$$

Example 1

In a library, there are 560 fiction books and 420 non-fiction books.

a Write down the ratio of the number of fiction books to the number of non-fiction books.
  Give your ratio in its simplest form.

b Give your ratio in the form $1 : n$.

a $560 : 420$ ←   Write down the ratio. The number of fiction books goes first.
  $= 56 : 42$ Divide both numbers by 10.
  $= 4 : 3$ Divide both numbers by 14.
  $4 : 3$ is the simplest form.

b $4 : 3 = 1 : \frac{3}{4}$ ←   Divide both numbers by 4 to give the ratio in the form $1 : n$.
  (or $1 : 0.75$)

Exercise 16A

Questions in this chapter are targeted at the grades indicated.

D

1 Write each ratio in the form $1 : n$.
   a $5 : 15$ b $8 : 32$ c $4 : 14$ d $6 : 3$
   e $30 : 9$ f $15 : 9$ g $\frac{1}{2} : 4$ h $\frac{3}{8} : \frac{2}{10}$

2 In a school, there are 120 computers. There are 600 students in the school.
   Advise the Headteacher of the ratio of the number of computers to the number of students.
   Give your ratio in the form $1 : n$.

3 In a cinema, there are 160 children and 200 adults.

   a What fraction of the audience are children?

   b Write down the ratio of the number of children to the number of adults.
Give your ratio in its simplest form.

   c Write your answer to part **b** in the form $1 : n$.

4 The length of a model aeroplane is 16 cm. The length of the real aeroplane is 60 m.
Work out the ratio of the length of the model aeroplane to the length of the real aeroplane.
Write your answer in the form $1 : n$.

5 Write these ratios in the form $1 : n$.

   a 3 hours : $\frac{1}{2}$ hour   **b** £2 : 40p   **c** 2 m : 4 cm   **d** 25 g : 1 kg

16.2 Solving ratio problems

Objective

- You can solve problems using ratio.

Why do this?

A teacher taking some pupils on a school trip knows that the ratio of staff to students must be 1 : 15. Once the number of pupils on the trip is known, the number of staff needed can be calculated.

Get Ready

1. Write the ratio in its simplest form.

   a 10 : 15   **b** 130 : 650   **c** 4 cm : 35 mm   **d** 45 g : 1 kg

Key Point

- If the ratio of two quantities is given and one of the quantities is known, then the other quantity can be found.
This can be done using **equivalent ratios**.

Example 2

To make concrete, 2 parts of cement is used to every 5 parts of sand.

   a Write down the ratio of cement to sand.

   b 4 buckets of cement are used. How many buckets of sand will be needed?

   c 20 buckets of sand are used. How many buckets of cement will be needed?

a 2 : 5

b $4 \div 2 = 2$ ⟵ | The amount of cement has been multiplied by 2. |

    cement : sand

    $\times 2 \left(\begin{array}{ccc} 2 & : & 5 \\ 4 & : & 10 \end{array} \right) \times 2$ ⟵ | Multiply 5 by 2. |

10 buckets of sand will be needed.

c $20 \div 5 = 4$ ← The amount of sand has been multiplied by 4.

cement : sand

$\times 4 \left(\begin{array}{ccc} 2 & : & 5 \\ 8 & : & 20 \end{array} \right) \times 4$ ← Multiply 2 by 4.

8 buckets of cement will be needed.

Exercise 16B

1 In a recipe for pancakes, the ratio of the weight of flour to the weight of sugar is 4 : 1.
Work out the weight of sugar needed for:

a 40 g of flour   b 120 g of flour   c 1 kg of flour.

2 Brass is made from copper and zinc in the ratio 5 : 3 by weight.
a If there are 6 kg of zinc, work out the weight of copper.
b If there are 25 kg of copper, work out the weight of zinc.

3 A map is drawn using a scale of 1 : 500 000. On the map, the distance between two towns is 21.7 cm.
Work out the real distance between the towns. Give your answer in kilometres.

4 George and Henry share some money in the ratio 7 : 9.
If George receives £840, work out how much money Henry gets.

5 The ratio of the widths of two pictures is 6 : 9.
If the width of the first picture is 1.02 m, calculate the width of the second picture.

6 In a school, the ratio of the number of students to the number of computers is $1 : \frac{2}{5}$.
If there are 100 computers in the school, work out the number of students in the school.

16.3 Sharing a quantity in a given ratio

◎ Objective

● You can share a quantity in a given ratio.

◈ Why do this?

If a recipe required 300 g of crumble mixture and you know that the ratio of sugar, fat and flour is 1 : 2 : 3, you could work out how much of each you would need.

◈ Get Ready

Work out

1. $\frac{16.52}{4} \times 5$ 2. $8.4 \times \frac{5}{7}$ 3. $\frac{483}{6} \times 3$

Key Point

◉ There are two methods for sharing a quantity in a given ratio. In one method, you work out how much each share is worth, then multiply by the number of shares each person receives. In the second method, you work out what fraction of the total amount each person receives and multiply the total by these fractions.

Example 3 Anna, Faye and Harriet share £42 in the ratio 1 : 2 : 3.
How much money does each girl get?

$1 + 2 + 3 = 6$ ⟵ Add the numbers in the ratio to get the total number of shares.

$42 \div 6 = 7$ ⟵ Work out what each share is worth.

Anna gets £7. ⟵ Anna gets 1 share.

Faye gets $7 \times 2 = £14.$ ⟵ Faye gets 2 shares so multiply 7 by 2.

Harriet gets $7 \times 3 = £21.$ ⟵ Harriet gets 3 shares so multiply 7 by 3.

 Results**Plus**
Exam Tip

Check your answer is correct by adding up each person's share and check this equals the total number of shares.

Exercise 16C

1 Divide the quantities in the ratios given.
a £14.91 in the ratio 2 : 5
b 600 g in the ratio 3 : 2
c £170.52 in the ratio 1 : 4 : 7
d 34.65 m in the ratio 2 : 4 : 5

2 The angles in a triangle are in the ratio 6 : 5 : 7.
Find the sizes of the three angles.

3 Three boys washed some cars. They earned a total of £87.60.
They shared the money in the ratio of the amount of time that each of them worked.
James worked for 5 hours. Sam worked for $3\frac{1}{2}$ hours and Will also worked for $3\frac{1}{2}$ hours.
Calculate the amount of money James received.

A02
A03

4 Jean and Kevin shared £320 in the ratio 3 : 5.
Jean gave one third of her share to Michael.
Kevin gave half of his share to Michael.
What fraction of the original amount of money did Michael receive?
Give your fraction in its simplest form.

A03

5 Barry bought a box full of fruit. The box contained some apples, oranges and lemons in the ratio 5 : 3 : 1.
Given that there were more than 50 pieces of fruit in the box, work out the minimum number of oranges in the box.

A03

6 Angela and Michelle shared some money in the ratio 4 : 9.
Then Angela gave Daniel half of her share.
Michelle gave Daniel a third of her share.
Daniel was given a total of £20.
Work out how much money was shared originally by Angela and Michelle.

A03

16.4 Using direct proportion

◎ **Objective**

◉ You can use direct proportion.

❓ **Why do this?**

If you know how much you got paid for a 6-hour shift one weekend, you can work out how much you would get paid for an 8-hour shift next weekend.

◆ **Get Ready**

Work out

1. $\frac{5}{12} \times 36$ 2. £4.50 ÷ 9   3. 2 hours $\times \frac{2}{3}$ 4. 2.50 × 1.10

🌀 **Key Points**

◎ Two quantities are in **direct proportion** if their ratio stays the same as the quantities increase or decrease.

◎ There are two methods that can be used to solve problems that involve direct proportion.
   ◉ The first method is called the unitary method because it finds the cost of one item first.
   ◉ The second method (the ratio method) is particularly useful for recipe questions.

Example 4   5 buns cost £2.50.
Work out the cost of 9 of these buns.

$250 \div 5 = 50$ ←

> Work out the cost of 1 bun.
> Divide the cost of five buns by 5.
> It is easier to work in pence.

$50 \times 9 = 450$ ←

> Work out the cost of 9 buns.
> Multiply the cost of one bun by 9.

9 buns cost £4.50. ←

> As the answer is more than £1, give the answer in pounds.

ResultsPlus
Exam Tip

When your answer is more than £1, give your answer in pounds.

Example 5   The weight of card is directly proportional to its area.
A piece of card has an area of 36 cm² and a weight of 15 grams.
A larger piece of the same card has an area of 48 cm².
Calculate the weight of the larger piece of card.

$\frac{15}{36} = \frac{5}{12}$ ←

> Work out the weight of 1 cm² of the card.
> The answer is less than 1, so write this as a fraction in its simplest form.

$\frac{5}{12_1} \times 48^{4} = 20$ ←

> Work out the weight of 48 cm² piece of card.
> Multiply the weight of 1 cm² by 48.

The weight of the larger piece of card is 20 grams.

Exercise 16D

1. A car travels at a steady speed of 50 miles each hour.
   Work out the number of hours it takes to travel:
   a 150 miles   b 350 miles.

2. Four 1-litre tins of paint cost a total of £36.60.
   Work out the cost of seven of the 1-litre tins of paint.

3. Joe is paid £54 for 8 hours' work in a supermarket.
   How much is he paid for 5 hours' work?

4. Prateek buys 17 cakes for £12.75.
   Work out the cost of 6 of these cakes.

5. The cost of ribbon is directly proportional to its length. A 2.5 m piece of ribbon costs £1.35.
   Work out the cost of 6 m of this ribbon.

6. The length of the shadow of an object, at noon, is directly proportional to the height of the object.
   A lamp-post of height 5.4 m has a shadow of length 2.1 m at noon.
   Work out the length of the shadow, at noon, of a man of height 1.8 m.

A03 **D**

A03 **B**

A03 **A**

Example 6

Here is a list of the ingredients needed to make carrot soup for 4 people.

200 g carrots

2 onions

40 g butter

300 ml stock

Work out the amount of each ingredient needed to make carrot soup for 16 people.

$16 \div 4 = 4$ ← *Carrot soup for 16 people needs 4 times as much of each ingredient as carrot soup for 4 people.*

$200 \times 4 = 800$ ← *So multiply each amount by 4.*

$2 \times 4 = 8$
$40 \times 4 = 160$ *The amount of each ingredient is 800 g carrots,*
$300 \times 4 = 1200$ *8 onions, 160 g butter and 1200 ml stock*

Example 7

a Janet went on holiday to France. She changed £200 into euros.
   The exchange rate was £1 = €1.18 euros.
   Work out the number of euros Janet received.

b Janet came home. She had 46 euros left. She changed her 46 euros to pounds.
   The new exchange rate was £1 = €1.15 euros.
   Work out how much Janet received, in pounds, for 46 euros.

a $200 \times 1.18 = 236$ ← *Janet received 1.18 euros for every £1, so multiply the number of pounds by 1.18.*
   Janet received €236.

b $46 \div 1.15 = 40$ ← *Janet received £1 for every 1.15 euros, so divide the number of euros by 1.15.*
   Janet received £40.

Exercise 16E

1 These are the ingredients for 12 cookies.
80 g butter   2 eggs
80 g sugar   100 g flour
   a Work out the amount of butter needed to make 24 cookies.
   b Work out the amount of flour needed to make 18 cookies.

2 A machine fills 720 packets of crisps in 1 hour.
How long will the machine take to fill 1680 packets of crisps? Give your answer in hours and minutes.

3 The exchange rate is £1 = $1.60.
   a Convert £200 to dollars.   b Convert $544 to pounds.

4 Susan bought a coat for €132 in France. The exchange rate was £1 = €1.10.
Work out the cost of the coat in pounds.

*** 5** Angela buys a pair of jeans in England for £45.
She then goes on holiday to America and sees an identical pair of jeans for $55.
The exchange rate is £1 = $1.45.
In which country are the jeans cheaper and by how much?

*** 6** Sian is going on holiday to America. In January she notices that the exchange rate is £1 = $1.74.
When she exchanges £450 for dollars in July the exchange rate has dropped to £1 = $1.61.
How many more dollars would Sian have received if she had exchanged her money in January?

16.5 Using inverse proportion

Objective
● You can use inverse proportion.

Why do this?
A construction company knows that it will take 6 men 10 days to fit the seats in a new athletics stadium. The company can use inverse proportion to work out how many men would be needed to fit the seats in 4 days instead.

Get Ready
1. Work out   a $624 \div 12$ b $24 \div 1\frac{1}{2}$
2. Convert 2.5 days into hours and minutes.

Key Points
● Two quantities are said to be in **inverse proportion** if one quantity increases at the same rate as the other quantity decreases.
● When two quantities are inversely proportional their product is constant.

Example 8   It takes 3 cleaners 6 hours to clean a school.

Work out how long it would take 9 cleaners to clean the school.

3 cleaners take 6 hours.

$9 \div 3 = 3$ ← Divide the new number of cleaners by the original number of cleaners. The original number of cleaners has been multiplied by 3 (3 × 3 = 9).

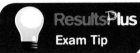
Exam Tip

Think about the problem: if there are more people then the work should take less time.

$6 \div 3 = 2$ ← Divide the number of hours by 3.

9 cleaners would take 2 hours.   ← Check: 3 × 6 = 18 and 9 × 2 = 18.

Example 9   It takes 3 men 2 days to build a wall.

Work out how long it will take 2 men to build the wall.

$3 \times 2 = 6$ ← 1 man would take 6 days to build the wall.

$6 \div 2 = 3$ ← There are now 2 men so divide by 2.

2 men will take 3 days.   ← Check: 3 × 2 = 6 and 2 × 3 = 6.

Exercise 16F

1   It takes 10 men 2 days to cut a hedge.
   Work out how long it will take to cut the hedge if there are:
   a 5 men   b 4 men.

   A03 **D**

2   5 computers process a certain amount of information in 10 hours.
   Work out how long it will take 25 computers to process the same amount of information.

   A02
   A03

3   A factory uses 3 machines to complete a job in 15 hours.
   If 2 extra machines are used, how long will the job take?

   A02
   A03

4   It takes 6 machines 3 days to harvest a crop. How long would it take 2 machines?

   A02
   A03

5   A large ball of wool is used to knit a scarf.
   The scarf is 40 stitches wide and 120 cm long. If the same size ball of wool is used to knit a scarf 25 stitches wide, work out the length of the new scarf.

   A02
   A03

6   A document will fit onto exactly 32 pages if there are 500 words on a page.
   If the number of words on each page is reduced to 400, how many more pages will there be in the document?

   A02
   A03

Chapter review

- **Ratios** are used to compare quantities.
- The simplest form of a ratio has whole numbers with no common factor.
- Ratios are sometimes given in the form 1 : n where n is a number. This is called the **unitary** form of a ratio.
- To write a ratio in the form 1 : n, divide each number in the ratio by the first number in the ratio.
- If the ratio of two quantities is given and one of the quantities is known, then the other quantity can be found. This can be done using **equivalent ratios**.
- There are two methods for sharing a quantity in a given ratio. Either work out how much each share is worth and multiply by the number of shares each person receives, or work out what fraction of the total amount each person receives and multiply the total by these fractions.
- Two quantities are in **direct proportion** if their ratio stays the same as the quantities increase or decrease.
- There are two methods that can be used to solve problems that involve direct proportion: the unitary method and the ratio method.
- Two quantities are said to be in **inverse proportion** if one quantity increases at the same rate as the other quantity decreases.
- When two quantities are inversely proportional their product is constant.

Review exercise

1. There are some sweets in a bag.
   18 of the sweets are toffees.
   12 of the sweets are mints.
   Write down the ratio of the number of toffees to the number of mints.
   Give your ratio in its simplest form.

   ResultsPlus
   Exam Question Report

   79% of students answered this question well because they displayed their answers in the form asked for in the question.

   June 2009

2. A coin is made from copper and nickel.
   84% of its weight is copper.
   16% of its weight is nickel.
   Find the ratio of the weight of copper to the weight of nickel.
   Give your answer in its simplest form.

   June 2008

3. The distance from Ailing to Beeford is 2 km. The distance from Ceetown to Deeton is 800 metres.
   Write the following as a ratio.
   Distance from Ailing to Beeford : Distance from Ceetown to Deeton
   Give your answer in its simplest form.

4. Alice builds a model of a house. She uses a scale of 1 : 20.
   The height of the real house is 10 metres.
   a Work out the height of the model.
   The width of the model is 80 cm.
   b Work out the width of the real house.

D

5 Mr Brown makes some compost.

He mixes soil, manure and leaf mould in the ratio 3 : 1 : 1.

Mr Brown makes 75 litres of compost.

How many litres of soil does he use?

Nov 2006

6 A garage sells British cars and foreign cars.

The ratio of the number of British cars sold to the number of foreign cars sold is 2 : 7.

The garage sells 45 cars in one week.

Work out the number of British cars the garage sold that week.

June 2008

7 There are 600 counters in a bag.

90 of the 600 are yellow. 180 of the 600 are red.

The rest of the counters in the bag are blue or green.

There are twice as many blue counters as green counters.

Work out the number of green counters in the bag.

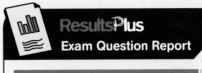

ResultsPlus
Exam Question Report

72% of students answered this sort of question poorly because they did not work out the total number of shares first.

May 2009

* **8** Robert wants to buy some new golf clubs.

He is considering buying them from the USA over the internet instead of from his local golf professional.

Use the prices quoted below to find which option is cheaper. Use the exchange rate £1 = $1.50.

Show all of your working.

Local professional	*Imported from USA*
£435	*$570 plus taxes and duties of 20%*

* **9** Which bottle of tomato ketchup gives better value for money?

Show all your calculations.

720 g £1.79

460 g £1.00

10 Bob lays 200 bricks in one hour. He always works at the same speed.

Bob takes 15 minutes morning break and 30 minutes lunch break.

Bob has to lay 960 bricks. He starts work at 9 am.

Work out the time at which he will finish laying bricks.

June 2006, adapted

11 A map is drawn to a scale of 1 : 50 000. A field is in the shape of a rectangle on the map. The area on the map is 6 cm². Work out the true area of the field. Give your answer in km².

12 The points A, B, C and D represent four jars of jam.

a Which jar costs the most?

b Which is the heaviest jar?

c Which two jars represent the same value for money?

d Which jar is the best value for money?

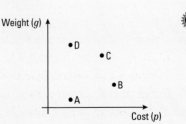

Weight (*g*)

• D
• C
• B
• A

Cost (*p*)

273

MULTIPLICATION

The following question helps you to develop both your ability to select and apply a method (AO2) and your ability to solve problems using your skills of interpretation (AO3). Your AO3 skills are particularly required as you will need to work through several steps to solve this problem. There are also some functional elements as this is a real-life situation and there is a problem to solve.

Example

Adam runs a coach company. He has 6 small coaches, 4 medium coaches, 3 large coaches and 1 double-decker coach.

The table gives information on how many passengers each coach can seat, the cost of hiring the coach and a driver for a day, and how many of these coaches Adam owns.

Adam's Coach Company			
Coach type	**Number of seats**	**Cost of hire**	**Number owned**
Small	25	£100	6
Medium	38	£110	4
Large	54	£120	3
Double-decker	78	£140	1

Rachel wants to hire some coaches from Adam to take 222 people out for the day. What is the cheapest way for Rachel to do this?

Solution

> As the number of seats increases, the cost goes down proportionally. Therefore you need to use the largest coach, the double-decker, first.

1 double-decker	£140	78
3 large	+ £360	+ 162
	£500	240 seats

> This leaves 144 people to fit in. This could be done with three large coaches but would leave 12 empty seats.

1 double-decker	£140	78
2 large	£240	108
1 medium	+ £110	+ 38
	£490	224 seats

> If two large coaches are used then this would leave 36 people to fit in, so a medium coach would be needed as well

The cheapest way costs £490 with two spare seats.

Now try these

1 Sam is a salesman. He is paid expenses when he drives his car on company business.

He is paid 45p for each mile he drives.

He is also paid a meal allowance.

Here is Sam's time and mileage sheet for one week.

Meal Allowance
Lunch £8.50
Dinner £22

*Only paid if Sam arrives home after 8 pm

Day	Miles driven	Lunch claimed	Time arrived home
Monday	180	Yes	9 pm
Tuesday	48		5 pm
Wednesday	64	Yes	8.30 pm
Thursday	33		5 pm
Friday	75	Yes	7.30 pm

Work out Sam's total expenses for the week.

2 Lynsey took part in a sponsored swim. Her target was to raise £100 for charity. Her nan promised her that she would make up the £100 if Lynsey did not raise enough.

Here is Lynsey's sponsor form.

Lynsey swam 32 lengths in a pool of length 40 m.

Will her nan have to give her any money?

You must explain your answer.

Sponsor	Amount
Ali	£5
Rob	25p for each length
Will	30p for each length
Mum	50p for each length
Jade	2p for each metre

3 Here are the rates charged for Mr Pitkin's telephone.

Line rental	£29.36
Daytime cost	4p for each minute
Evening and weekend	3p for each minute
To mobiles	11p for each minute
International rate (anytime)	8p for each minute

Here are the details of calls made by Mr Pitkin in one quarter.

Type of call	Minutes
Daytime	78
Evening	312
To mobiles	42
International rate	25

Calculate Mr Pitkin's telephone bill for that quarter.

The following question helps you develop your ability to select and apply a method (AO2) and your ability to analyse and interpret problems (AO3).

Example

The side of a shed is the shape of a trapezium as shown in the diagram.

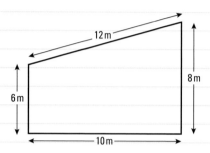

The side is to be given two coats of paint. The paint is sold in 1 litre cans costing £3 each.
1 litre of paint covers 15 square metres.
How much will it cost to paint the wall?

Solution

Using the formula

Area (trapezium) = $\frac{1}{2}(a + b)h$

$= \frac{1}{2}(6 + 8)10$

$= 70\,\text{m}^2$

> You need to find the area of the side of the shed.
> You may choose to use the formula, or divide the shape into a rectangle and a triangle.
> The formula is given on the formulae sheet.

Dividing the shape

Area of rectangle = $6 \times 10 = 60\,\text{m}^2$

Area of triangle = $\frac{1}{2} \times 10 \times 2 = 10\,\text{m}^2$

Total area = $70\,\text{m}^2$

The number of tins of paint required
for one coat is $70 \div 15$.

> Total area ÷ area covered by 1 tin

The number of tins needed for two coats is
$140 \div 15 = 9.3$.
So 10 tins will be needed.
The cost of the paint will be £30.

1   This shape is made by joining six squares.

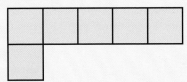

Find two shapes which have the same area but different perimeters.

2   The diagram shows a wall which is to be built with bricks.
The bricks measure 200 mm × 100 mm.
They are sold in packs of 100. One pack costs £35.
Find the cost of the bricks.

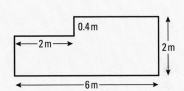

 A02 A03 **D**

3   The diagram shows a rectangular path around a lawn. The path is 1 m wide.

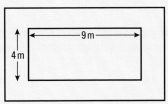

 A02 A03

Gravel costs £124 per tonne.
1 tonne of gravel covers 15 m².
Work out the cost of covering the path with gravel.

4   Find the perimeter of three different rectangles which each have an area of 36 cm².

A03

5   The diagram shows a bathroom wall in the shape of a trapezium. The wall is to be painted.

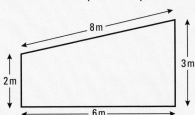

 A03 **C**

The paint chosen is sold in 1 litre cans costing £4 each. 1 litre covers 12 square metres.
How much will it cost to paint the wall?

A03

The following question helps you to develop both your ability to select and apply a method (AO2) and your ability to solve problems using your skills of interpretation and proof (AO3). The AO2 skills are developed as there is more than one way of working out the number of average size people that should be allowed in the lift. You could choose to work with the mean and range or the median and interquartile range. Your AO3 skills are needed in this question as you will need to give a reasoned explanation for your answer.

Example

One hundred people, selected at random, were weighed.
The results were put in a frequency table.

The maximum weight limit for the lift is 700 kg.
How many people of average size should be allowed in the lift?
Give a reason for your answer.

Weight (w kg)	Frequency (f)
$50 \leqslant w < 60$	6
$60 \leqslant w < 70$	29
$70 \leqslant w < 80$	45
$80 \leqslant w < 90$	19
$90 \leqslant w < 110$	1
Total	**100**

Solution

Student could choose to use

Weight (w kg)	Frequency (f)	mid-value	mid-value $\times f$	Cumulative frequency
$50 \leqslant w < 60$	6	55	330	6
$60 \leqslant w < 70$	29	65	1885	35
$70 \leqslant w < 80$	45	75	3375	90
$80 \leqslant w < 90$	19	85	1615	99
$90 \leqslant w < 110$	1	95	95	100
Total	**100**		7300	

Mean – best answer for safety reasons as it uses all of the values and allows for any extremes.
Mean is $\frac{7300}{100} = 73$ kg $\quad \frac{700}{73} = 9.59$
Allow 9 people to use the lift with a 0.59 kg safety margin.

Mode – this is the most frequently occurring group.
Modal group is $70 \leqslant w < 80$ using the upper limit $\frac{700}{80} = 8.75$
Allow 8 people to use the lift with a 0.75 kg safety margin.

Median – this is the middle person and occurs in the $70 \leqslant w < 80$ group.
Estimate about 73 kg $\quad \frac{700}{73} = 9.59$
Allowing 9 people in the lift with a 0.59 kg safety margin.

Now try these

1   A is the set of data 1, 2, 4, 5, 8, 10.
    B is the set of data 3, 4, 6, 7, 10, 12.
    C is the set of data 2, 4, 8, 10, 16, 20.
    a   Compare the mean of A and the mean of B.
    b   Compare the range of A and the range of B.
    c   Compare the mean of A and the mean of C.
    d   Compare the range of A and the range of C.
    e   Write down a data set which has the same mean as B but twice the range.

    D A03

2   The table below shows the marks given in two Maths tests.

    A02 A03

Boys	62	75	67	81	79	91	69	73	85	81
Girls	74	69	83	83	78	68	88	81	68	

    Compare the distributions for boys and girls.
    Farida was absent. What is the minimum mark she needs to score when she takes the test to keep the girls average better than that of the boys?

3   The mean mark for 10 pupils in an English Language Examination was x.
    5 of the students were awarded an extra 4 marks for the quality of their written communication.
    What difference does this make to the average for the group?

    A03

4   The table below shows the Maths and English marks for a group of 50 pupils.

    B A02 A03

Mark	21 – 30	31 – 40	41 – 50	51 – 60	61 – 70	71 – 80	81 – 90	91 – 100
Maths	1	2	8	11	14	8	4	2
English	1	1	3	20	14	9	1	1

    Choose a suitable diagram to display these results.
    Compare the distribution of the two sets of marks.

5   Kay noted the number of hours of sunshine in May and June.
    The results of her survey are given below.

    A02 A03

Hours of sunshine per day	May	June
Mean (hrs)	9.6	8.4
Range	12	16

    a   Compare the number of hours of sunshine for May and June.
    b   Kay chose to use the mean and range instead of using the median and interquartile range.
        i   Give one reason why this might be a good choice.
        ii Give one reason why it might be better to use the median and interquartile range.

PRICE COMPARISONS

This question tests selecting and applying a method (AO2) as there are a number of different ways of approaching percentages and ensuring both prices are compared over the same time scale. As you can see below, it can be solved using algebra or by drawing a graph. You could also use trial and improvement – try this method on one of the questions that follows.

Example

Elsa is comparing electricity prices.
Energee has a standing charge of £65 per annum and a unit rate of 8.5p.
Powero has a standing charge of £116 per annum and charge 7.5p per unit.
Discuss which firm Elsa should use.

Solution

The best deal will depend upon how much electricity Elsa uses.
For a small number of units Energee is obviously cheaper, but for large amounts Powero will cost less.
For your answer you will need to find out after how many units Powero becomes the best option.

To decide you could use algebra, graphs or trial and improvement.

Using algebra
Write down expressions for the charges.

$$\text{Energee} = 65 + 0.085u$$
$$\text{Powero} = 116 + 0.075u$$
$$65 + 0.085u = 116 + 0.075u$$
$$u = 5100$$

> Find the value of u for which the prices are the same.

A customer using more than 5100 units is better off with Powero.

Using a graph
The lines meet at 5100 units,
so if Elsa uses more than 5100 units she
should use Powero.

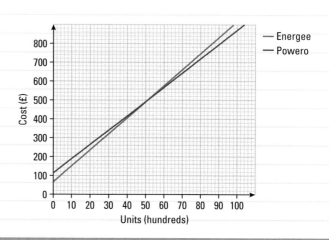

Now try these

*1 Ahmed is trying to find the cheapest provider for gas in his area.
Cogas has a standing charge of £63 per annum with a charge of 3.5p per unit.
Ourgas has a standing charge of £120 per annum and charges 2p per unit.
Discuss which company Ahmed should use.

2 The cost of installing cable broadband is £30. The monthly cost of the contract is £5.
The monthly cost of broadband using a wireless router is £6.50 for a minimum period of 18 months.
The router is free and there is no installation charge.

a Work out an expression in m for each company for the cost of broadband over m months.

b Investigate which of the two broadband deals is cheaper.

3 The cost of hiring a car from ACars and BMotors is shown in the advertisements below.

a Work out an expression in x for the cost of hiring a car for a day and travelling x miles.

b Investigate which of the two firms is cheaper.

ACARS
£60 per day
32p per mile

BMotors
£50 per day
40p per mile

*4 Anna works in a small business and has to decide which courier her
company should use when sending parcels.
Parcels Fly delivers parcels at a cost of £5.50 if they weigh less
than 2 kg. For heavier parcels, it charges 85p per 250 g.
Quick Delivery charges £3 for the first kg then £1.90 per 500 g.
Investigate which company Anna should use.

*5 Tom works for a builder and has to order the concrete for drives and paths.
His firm uses two different suppliers.
Pete's Mix sells concrete for £70 per cubic metre with a delivery charge of £80.
Concrete Sue sells concrete at £85 per cubic metre with a delivery charge of £30.
Investigate which firm he should use.

C

INTERPRETING AND DISPLAYING DATA

This question helps students develop their ability to choose and apply a method (AO2) and their ability to find an estimate by interpreting a graph (AO3). The AO2 skills are developed throughout the question by deciding what type of sampling to use, the choice of diagram to display the data and in the method of estimating between class intervals. The AO3 skills occur when estimating the number of boys who were between 152 cm and 156 cm tall from a cumulative frequency graph or a frequency density graph.

Example

Alan is doing a survey of the heights of 50 boys from his school.
He wants his sample to be fair and unbiased.

a Suggest a method he can use to take his sample.

b Explain how this method avoids bias.

The table shows information from his survey.

c Use a suitable diagram to display this data.

d Estimate how many boys were between 152 cm and 156 cm tall.

Height of boys in cm	Frequency
$140 \leqslant h < 145$	8
$145 \leqslant h < 148$	11
$148 \leqslant h < 150$	20
$150 \leqslant h < 154$	9
$154 \leqslant h < 160$	2

Solution

a Random stratified sample taking the year groups as strata within the sample.

b This method avoids bias by ensuring the number of boys selected from each year group is proportional to the size of the group. This ensures that one year group is not over- or under-represented.

c

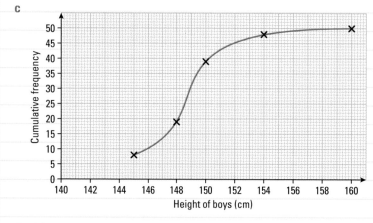

d 4 boys

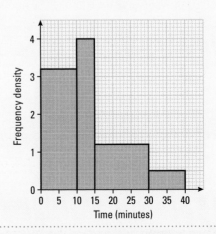

Now try these

1 A geologist has taken a sample of pebbles from an area of interest.
The table below shows some information about the weights of the pebbles.

Weight (w grams)	Frequency (f)
$0 \leqslant w < 20$	1
$20 \leqslant w < 30$	14
$30 \leqslant w < 40$	21
$40 \leqslant w < 50$	29
$50 \leqslant w < 60$	19
$60 \leqslant w < 70$	10
$70 \leqslant w < 80$	6

Estimate the number of pebbles that weigh between 54 g and 62 g.

2 The cumulative frequency curve shows the time
taken for some workers to complete a task.
Estimate the number of workers who took
20–30 minutes to complete the task.

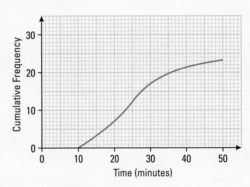

3 Within a radius of 5 miles from Sue's home there are 5 state secondary schools.
Avon has 850 students, Moorside has 986 students, Heaton has 1296 students, Moortop has 1138
students and Brambell has 1450 students.
Explain how a sample of 60 students could be taken to give a fair representation of all of the schools.

4 The histogram below shows the waiting time for patients to
be seen one morning by a doctor in a health centre.

20 people waited between 10 and 15 minutes
for an appointment.
No one waited more than 40 minutes.
How many patients were seen by the doctor that morning?

A02
A03

PROBABILITY

The following question helps you develop both your ability to select and apply a method (AO2) and your ability to analyse and solve problems (AO3). The AO2 skills are used in the first part of the question. Setting up an equation is the most efficient way to work out the probability of choosing a mint but some students may choose to estimate with trial and improvement. Your AO3 skills will be developed as you will have to work out how to solve the problem.

> **Example**
>
> A bag of sweets contains mints, toffees and creams.
> The probability of choosing a toffee is $\frac{1}{3}$.
> The probability of choosing a cream is x.
> The probability of choosing a mint is $2x$.
> What is the probability of choosing a mint?
> Give your answer as a numerical value.

> **Solution 1**

One strategy is to set up an equation.

$$\frac{1}{3} + x + 2x = 1$$
$$3x = \frac{2}{3}$$
$$x = \frac{2}{9}$$

Probability of choosing a mint is

$$2x = \frac{4}{9}$$

> This question requires you to work with algebra and probability.

> **Solution 2**

Alternatively we can estimate the number of sweets. As thirds are involved choose a multiple of 3. Select a total number of sweets which is a multiple of 3 e.g. 36

$$\text{toffees} = 12 \text{ sweets}$$
$$\text{mints} + \text{creams} = 24 \text{ sweets}$$

$$P(\text{mint}) = 2P(\text{cream}) \longleftarrow \boxed{\text{We can compare quantities.}}$$
So 16 mints to 8 creams

$$\text{So } P(\text{mint}) = \frac{16}{36} = \frac{4}{9}$$

Now try these

1   In a box of chocolates, there are 3 types of chocolate.
    The probability of a cream is $2x$.
    The probability of a toffee is $3x$.
    The probability of a hard centre is $4x$. Calculate the probability of choosing a toffee.

    AO2 C

2   A bag contains 20 sweets. x of the sweets are chocolates, the
    rest are toffees.
    Mona takes a toffee from the bag and eats it.
    She then offers the bag to Sam who eats a sweet.
    Explain why the probability of Sam eating a chocolate is not $\frac{x}{20}$.

    AO3

3   A drawer contains a number of black and grey socks.
    The probability that the first sock Ali pulls from the drawer will be black is x.
    Explain why the probability of pulling a second black sock from the drawer is not x.

    AO3

4   A sandwich shop has 3 types of sandwiches; ham, cheese and prawn.
    The probability of a customer choosing ham is $\frac{1}{2}$.
    The probability of a customer choosing cheese is x.
    The probability of a customer choosing prawn is $3x$.
    Mark chooses a sandwich.
    What is the probability of him choosing prawn?
    Give your answer as a numerical value.

    **AO2 A
    AO3**

5   The probability that a train arrives on time is 70%.
    The probability that it arrives early is x.
    The probability that it arrives late is $2x$.
    Calculate the numerical value of it arriving late.

    **AO2
    AO3**

6   The top floor of a block of flats can be reached by two lifts.
    The probability of both of the lifts working is 0.85.
    The probability of only one lift working is $2x$.
    The probability of none of the lifts working is x.
    Calculate the numerical probability of only one lift working.

    AO2

7   In an athletics competition Dan enters for two events.
    The probability of Dan winning one event and losing one is $2x$.
    The probability of him winning two events is x.
    The probability of him not winning any events is $5x$.
    Calculate the probability that he does not win any event.

    AO2

The following question tests your ability to analyse and interpret problems (AO3).

Example

Here is a diagram of a perfume bottle.
The bottle is in the shape of a square-based pyramid.

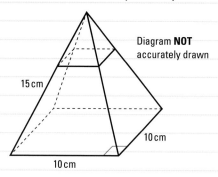

Diagram **NOT** accurately drawn

15 cm

10 cm

10 cm

The lengths of the edges of the base are 10 cm.
The lengths of all the four slant edges of the pyramid are 15 cm.
The bottle is to be sold in a box in the shape of a cuboid.
Find the height of the smallest box that could be used.
Give your answer to 3 significant figures.

Solution

The pyramid-shaped bottle must fit into a box. The minimum height of the box will be the height of the pyramid. You need to find the vertical height of a square-based pyramid given the base lengths and the slant heights.

Diagonal of base $= \sqrt{10^2 + 10^2} = 14.142$ cm

> The only sensible method of solution is to use Pythagoras' theorem.

> Firstly, find the diagonal of the square base.

Half the diagonal $= 7.071$ cm

> Then work out the height on a vertical right-angled triangle which has the slant edge of the pyramid as the hypotenuse.

Height $= \sqrt{15^2 - 7.071^2} = 13.2$ cm

Now try these

1   Jenny has a pencil tin in the shape of a cuboid.

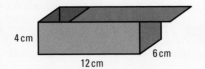

The dimensions for the inside of the box are 12 cm, 6 cm and 4 cm.
What is the length of the longest pencil that Jenny can fit into her tin?

2   Miriam has a stick that is 30 cm long.
She uses the stick to stir paint. She leaves the stick in the paint tin with some of it sticking out at the top.
The tin has a diameter of 10 cm and height of 15 cm.
What is the shortest length of stick that could stick out of the tin?

3   Dave has a garden shed that is 6 ft long by 4 ft wide. Its walls are 6 ft high.
The tallest point of the roof is 7 ft from the ground.

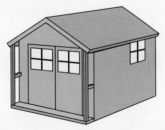

Dave wants to store some bean poles in the shed.
The poles are 9 ft long.
Explain, by showing your working, that it is possible to store the bean poles in the shed.

S nowdon is the highest mountain in Wales. To get to the summit of Snowdon you can walk or go by train. Read the information below then answer the questions opposite.

Here are 5 possible routes up Snowdon.

Route	Distance (km)	Ascent (m)	Average time taken (hrs)	Level of difficulty
Llwybyr Llanberis Path	9	1020	3.5	
Snowdon Ranger	7.5	1100	3	
Miner's Track	7	930	2.5	
South Ridge	7	890	2.5	
Lliwedd	6	970	4	

Fact 1: 1 foot = 0.3048 metres.

Fact 2: The height of Snowdon in feet is 3560 ft.

Here are the train fares.

TRAVEL FARES

	Llanberis – Clogwyn Return fare	Llanberis – Clogwyn Single fare	Llanberis – Summit Return fare	Llanberis – Summit Single fare
Adult	£17	£12	£22	£16
Senior / Student	£12	£8	£17	£13
Child	£14	£9	£15	£10
Early Bird Adult*	£9	£7	£13	£12
Early Bird Child*	£6	£5	£6.50	£5.50

*The Early Bird train departs at 9 am.

1. Jon, Sarah and their two children, Poppy and Mark, are going to buy single fare tickets from Llanberis to the summit of Snowdon. They can go by the 9 am train or the 11.30 am train. Which train should they catch to get the cheapest fares and by how much?

2. Rashid and Chelsea each walk to the summit of Snowdon. Rashid takes the South Ridge route and starts at 09:45. Chelsea takes the Llwybyr Llanberis Path route and starts at 09:30. They both arrive at the summit at 12:30. Rashid says that he is the faster walker. Is he right? Give a reason for your answer.

3. a Draw a chart or graph to convert feet into metres.
   b The highest mountain in Scotland is Ben Nevis. The height of Ben Nevis is 1344 m. Which is higher, Ben Nevis or Snowdon?

Exercise is not based on real data.

LINKS

◉ You need to be able to read data from tables in **Question 1**. You learnt this skill in **Chapter 6**.

◉ You need to work out time intervals for **Question 2**. You learnt this in **Chapter 6**.

◉ **Question 3** asks you to draw a conversion graph. You learnt this in **Chapter 24**.

MUSIC SALES

Listening to music is one of the most popular pastimes for people of all ages. In the twenty-first century, music can be purchased in a number of formats. Read the information below, then answer the questions.

QUESTION

1. Here is some information about music sales between 2006 and 2009 in the UK. Explore the trend in sales for CDs and downloads between 2006 and 2009. You must show your working.

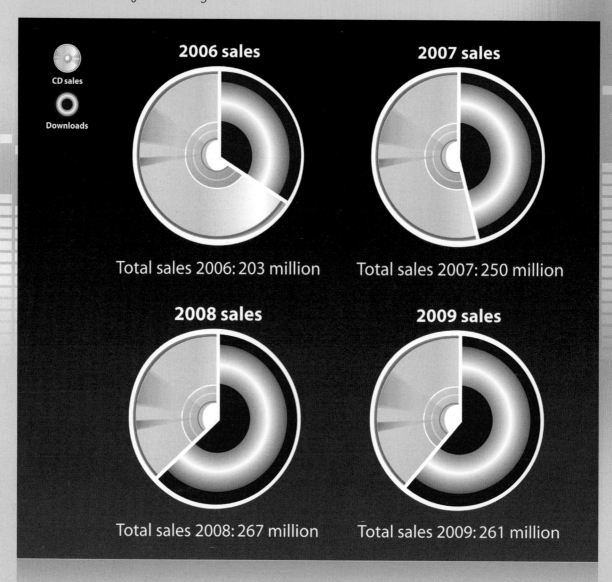

CD sales

Downloads

2006 sales

Total sales 2006: 203 million

2007 sales

Total sales 2007: 250 million

2008 sales

Total sales 2008: 267 million

2009 sales

Total sales 2009: 261 million

QUESTION

2. The tables show a breakdown by age for the sales of CDs and downloads in 2008..

Simon wants to show this information in a suitable diagram or graph that will allow people to compare the information visually. Produce an example of how Simon could do this.

CD Sales

Age (years)	CDs (millions)
$0 \leqslant x < 15$	13
$15 \leqslant x < 30$	22
$30 \leqslant x < 45$	37
$45 \leqslant x < 60$	13
$60 \leqslant x < 75$	8
$75 \leqslant x < 90$	5

Downloads

Age (years)	Downloads (millions)
$0 \leqslant x < 15$	42
$15 \leqslant x < 30$	63
$30 \leqslant x < 45$	29
$45 \leqslant x < 60$	19
$60 \leqslant x < 75$	10
$75 \leqslant x < 90$	6

QUESTION

3. Prepare a short questionnaire to obtain information about how much people spend on music and what format they buy. You need information about their age and gender as part of the information you collect.

Design a suitable data collection sheet for collating the information.

LINKS

⊙ For **Question 1** you need to be able to interpret pie charts. You learnt how to do this in **Chapter 18**.

⊙ For **Question 2** you need to find the best way of visually representing data. You learnt about the various ways of doing this in **Chapter 18**.

⊙ You learnt how to design questionnaires in **Chapter 1**. You will need to do this in **Question 18**.

Exercise is not based on real data.

COMMUNICATION

In the twenty-first century people communicate using a wide range of technologies. The cost, speed and quality of communication can help determine how and when they are used.

1. When buying a mobile phone the two most popular options are a monthly contract or 'pay as you go'. You estimate that you send about 85 texts and make 5 hours of calls per month. Compare the costs of these two options for a variety of time periods up to 2 years.

MONTHLY CONTRACT

Free phone
100 texts/month
400 min/month
£30/month

PAY-AS-YOU-GO

Phone £90
Texts = 5p each
Calls 7p/min

Using this estimate, compare the monthly costs of these two options. How does this comparison change if you try different estimates for texts sent and calls made?

2. Broadband is the internet access choice for most households but the speed varies greatly. The speed is measured in megabytes per second using the formula:

$$\text{Connection speed (Mb/sec)} = \frac{\text{Data received (Mb)}}{\text{Time (sec)}}$$

Paul's current broadband provides a top speed of 8 megabytes per second but he is thinking of upgrading to 50 megabytes per second. Paul regularly downloads TV programmes from the internet. These typically consist of 2 gigabytes (1 gigabyte = 1024 megabytes) and he does this on average once per week. Estimate how much time per month he will save with the upgraded broadband.

3. Photos contain thousands of little dots called pixels. These pixels are arranged in rows and columns to create the picture. The resolution of the photo when printed or displayed on a screen is stated as the number of pixels per inch (ppi).

A photograph when taken has a fixed number of pixels and as the image is enlarged the resolution gets poorer.

Ranji has a photo of his trip go-karting with 2106 pixels in each row and 1443 pixels in each column. He has three sizes of photo frame to choose from: 13 cm by 19 cm, 15 cm by 23 cm and 20 cm by 30 cm.

Which size should he print the photo at to give him a resolution of at least 200 ppi and the smallest amount of distortion to the photo?

1 cm = 0.39 inches

1443ppi

2106 ppi

LINKS

◉ For **Question 1** you need to be able to use decimals in calculations. You learnt about this in **Chapter** ⸴

◉ You learnt about using formulae in **Chapter 19**. You will need to be able to put figures into a formula f⸴ **Question 2**.

◉ For **Question 3** you need to be able to convert between metric and imperial units of measure. You learnt how to do this in **Chapter 7**.

ENERGY EFFICIENCY

Fuel bills are one of the largest expenses for homeowners. People can reduce their fuel bills by making their homes more energy efficient.

QUESTION

1. The diagram shows part of the loft of a bungalow. The floor of the loft measures 8 m by 7 m. Gaps of 370 mm are separated by joists which are 30 mm across. Estimate the best cost for insulating the whole loft to a thickness of 100 mm, 150 mm and 200 mm.

8m

7m

370mm ⤴ ⤴ 30mm

Here are some prices at a DIY store.

Economy Roll	
7m x 370mm x 50mm	£5 per roll
Easy Roll	
4m x 370mm x 100mm	£5 per roll
Space Blanket (thick)	
4m x 370mm x 200mm	£6 per roll
Space Blanket (medium)	
5m x 370mm x 150mm	£6 per roll
Space Combi loft roll	
100mm covers 14m²	£10 per roll

QUESTION

2. Energy-saving lightbulbs can save significant amounts of money.
They are best used where lights are left on for more than an hour at a time.
Mary decides to replace the old bulbs in her hall light, which has two 60 watt bulbs, and her living room light, which has three 100 watt bulbs, with energy efficient bulbs.

20 watt energy efficient bulbs are the equivalent of 100 watt ordinary bulbs.
11 watt energy efficient bulbs are the equivalent of 60 watt bulbs. The cost of electricity is worked out from the number of units, E, used. The formula for E is:

$$E = \frac{p\,t}{1000} \text{ where } p \text{ is the power in watts and } t \text{ is the time in hours}$$

In the winter, her hall light is usually on from 5pm until 11pm and her living room light is on for an average of 5 hours a night. Her electricity provider bills her every 13 weeks for a quarterly period, and charges her 11.47p per unit. How much money can she expect to save on her winter bill?

All new appliances come with energy labels that provide you with information on the efficiency of the product.

3. Isaac uses his washing machine five times a week. He is currently being charged 14p per unit for his electricity. The energy consumption per cycle is the number of units used when one complete cycle is done. Isaac is considering replacing his current machine with the new Spinner Max. How long would it take for the running cost savings to exceed the initial purchase price?

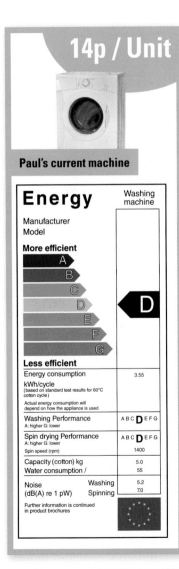

Paul's current machine

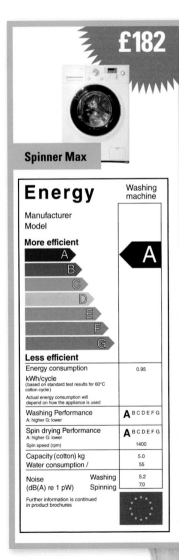

Spinner Max

LINKS

- For **Question 1** you will need to convert between different units of measure. You learnt how to do this in **Chapter 7**.

- You learnt how to use formulae in **Chapter 19**. You will need to be able to do this in **Question 2**.

- For **Question 3** you will need to be able to use decimals in your calculation. You learnt how to do this in **Chapter 4**.

GOING ON HOLIDAY

FS

British people take more than 60 million holidays abroad each year. Of these, 75% are taken during the months of July and August, when many people travel to southern Europe and the Mediterranean. Some reward card companies allow people to collect points when they spend money in particular stores. You can use the points to pay for flights abroad.

QUESTION

1. Jared wants to book a return flight from London to Valetta in Malta. The number of points he needs is calculated using the formula:

$$p = \frac{d}{c^2}$$

p = points required
d = distance of flight (km)
c = class

Class

First class: $c=1$
Business class: $c=2$
Economy class: $c=3$

The distance of the journey is calculated in a straight line for the purpose of the reward points. Approximately how many points does he need to save up to travel in economy class?

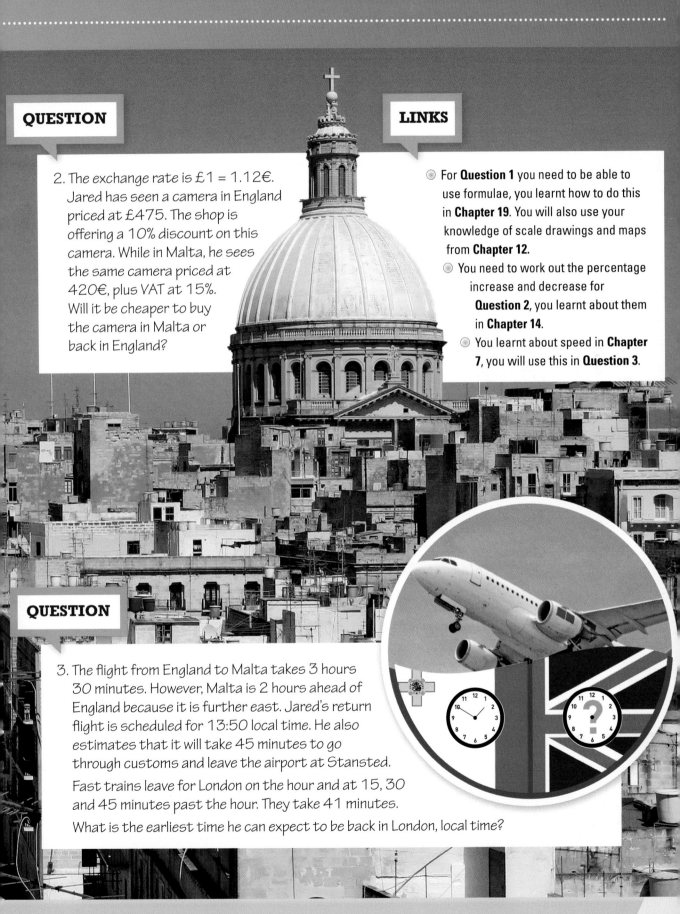

2. The exchange rate is £1 = 1.12€.
Jared has seen a camera in England
priced at £475. The shop is
offering a 10% discount on this
camera. While in Malta, he sees
the same camera priced at
420€, plus VAT at 15%.
Will it be cheaper to buy
the camera in Malta or
back in England?

- For **Question 1** you need to be able to
  use formulae, you learnt how to do this
  in **Chapter 19**. You will also use your
  knowledge of scale drawings and maps
  from **Chapter 12**.
- You need to work out the percentage
  increase and decrease for
  Question 2, you learnt about them
  in **Chapter 14**.
- You learnt about speed in **Chapter
  7**, you will use this in **Question 3**.

QUESTION

3. The flight from England to Malta takes 3 hours
30 minutes. However, Malta is 2 hours ahead of
England because it is further east. Jared's return
flight is scheduled for 13:50 local time. He also
estimates that it will take 45 minutes to go
through customs and leave the airport at Stansted.

Fast trains leave for London on the hour and at 15, 30
and 45 minutes past the hour. They take 41 minutes.

What is the earliest time he can expect to be back in London, local time?

FS ALL AT SEA

Ships navigate using bearings. They can also calculate their position according to their bearings from two known points. Trigonometry, loci and circle theorems can all be helpful tools when understanding and solving bearings problems.

QUESTION

1. A ship sails around an island from a port on the west coast to a harbour on the north shore. The harbour is 11 046 metres away on a bearing on 065°. There is a lighthouse on an outcrop of rocks on the north-west tip of the island, which is on a bearing of 042° from the port and at a distance of 6000 metres. To avoid the rocks, the ship must sail no closer than 720 metres from the lighthouse, passing to the north. The harbour is also 6000 metres from the lighthouse. Work out the total distance sailed.

QUESTION

2. Two coastguard stations are 8000 metres apart with one due east of the other. Simultaneously, they receive a call for help from a ship a sea. They are able to identify the direction from which the call is made but this is subject to a possible error of ±5° due to the fact that the ship is still moving and the accuracy of their equipment.

One coastguard station estimates the bearing from which the call was made to be 065° whilst the other estimates the bearing to be 310°. Draw a scale diagram to identify the search area to which helicopters and lifeboats should be sent.

3. Distances at sea are normally given in nautical miles. A nautical mile is slightly longer than a mile. Ship's speeds are measured in knots: the number of nautical miles per hour.

At midday a ship's captain sees a radio mast that he knows is 8 nautical miles away on a bearing of 020°. The ship is sailing on a bearing of 045° at a speed of 15 knots. At what time will the ship be nearest to the radio mast?

LINKS

◉ For **Question 1** you need to be able to use bearings and Pythagoras' Theorem in your calculations. You learnt how to do this in **Chapter 5** and **Chapter 29**.

◉ You learnt how to draw scale diagrams in **Chapter 12**. You will need to use this for **Question 2**.

◉ For **Question 3** you need to understand how to do calculations involving speed. You learnt this in **Chapter 7**.

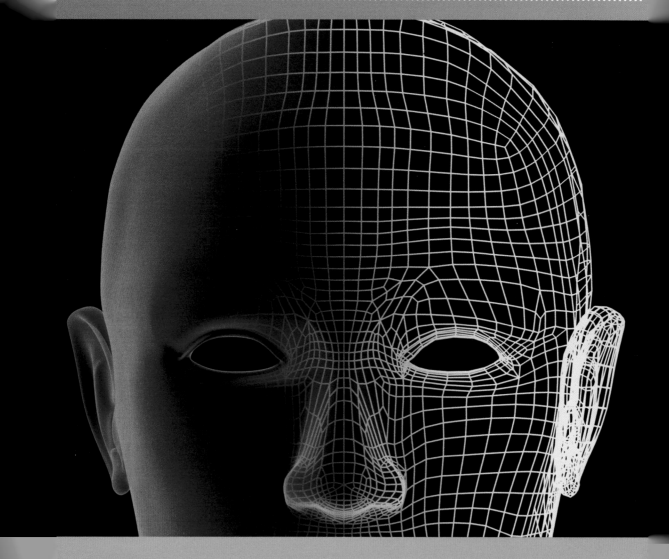

Three-dimensional head models have many uses, including predicting the impact of car crashes. The original computerised head models were made using a mass of 3D points which were then moved around to change the shape of the head and features.

◎ Objectives

In this chapter you will:
- ◎ translate, rotate, reflect and enlarge a 2D shape
- ◎ describe the translation, rotation, reflection and enlargement of a 2D shape
- ◎ combine transformations and describe the single transformation that has the same effect as a combination of transformations.

◈ Before you start

You need to:
- ◉ be able to use coordinates in all four quadrants
- ◉ understand what it means to move (translate), flip (reflect), turn (rotate) and change the size of a 2D shape
- ◉ recognise and be able to draw lines with equations $x = a, y = b, y = \pm x$
- ◉ be able to draw scale diagrams.

17.1 **Using translations**

- ◎ You know that in a translation all points of a shape move the same distance in the same direction.
- ◎ You understand that translations are described by the distance and the direction moved.
- ◎ You can use a vector to describe a translation.
- ◎ You know that when shape A is mapped to shape B by a translation, shape A and shape B are congruent.

⊕ **Get Ready**

1. Here is a letter square.

U	V	W	X	Y
T	S	R	Q	P
K	L	M	N	O
J	I	H	G	F
A	B	C	D	E

Start at square A, go 2 to the right, then 3 up and stop. Then go 3 down and stop. What mathematical word is this?

Start at square M, go 2 to the right and 2 down and stop. Then go 4 to the left and stop. Then go 3 to the right and 2 up and stop. What mathematical word is this?

Give instructions to get **a** KITE, **b** CIRCLE, **c** ADD.

d Make up some examples of your own.

🔍 **Key Points**

- ◉ In the diagram, shape A has been mapped onto shape B by a **translation**.
- ◉ All points of shape A move 3 squares to the right and 6 squares up.
  This can be written as $\binom{3}{6}$.
- ◉ In a translation, all points of the shape move the same distance in the same direction.
- ◉ In a translation:
  - ◉ the lengths of the sides of the shape do not change
  - ◉ the angles of the shape do not change
  - ◉ the shape does not turn.
- ◉ In a translation, any shape is congruent to its image because the lengths of the sides and angles of the shape are preserved by the translation.

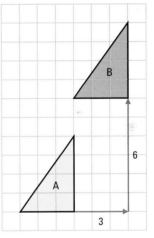

- ◉ $\binom{3}{6}$ is a **vector**. Vectors can be used to describe translations.
  The top number shows the number of squares moved parallel to the x-axis, to the right or left.
  The bottom number shows the number of squares moved parallel to the y-axis, up or down.
  To the right and up are positive.
  To the left and down are negative.
- ◉ Some translations of the yellow shape to the red shape and their **column vectors** are shown on the grid.

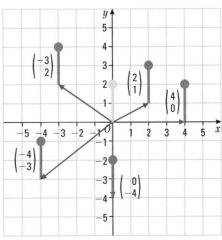

Example 1

Describe the translation that maps triangle P onto triangle Q.

Choose one corner of triangle P.

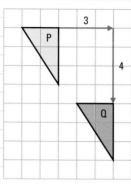

Count the number of squares to the right and the number of squares down from this corner on triangle P to the same corner on triangle Q.

The translation from triangle P to triangle Q is 3 squares to the right and 4 squares down.

This translation can also be written as $\begin{pmatrix} 3 \\ -4 \end{pmatrix}$.

Example 2

a Describe the transformation that maps shape A onto **i** shape B **ii** shape C.

b Translate shape A by the vector $\begin{pmatrix} -3 \\ -5 \end{pmatrix}$.

Label this new shape D.

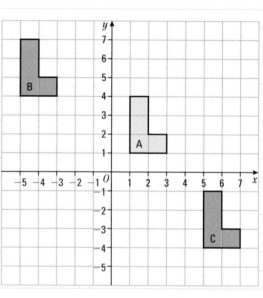

a

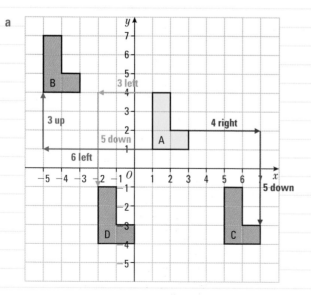

ResultsPlus
Exam Tip

The question asks for the transformation so as well as the vector, you must say it is a translation.

i From A to B is the translation 6 to the left and 3 up, or the translation with vector $\begin{pmatrix} -6 \\ 3 \end{pmatrix}$.

> Count the number of squares moved to the left (negative) and up (positive) from any corner in A to the same corner in B.

> $\begin{pmatrix} -3 \\ -5 \end{pmatrix}$ means 3 to the left and 5 down.

ii From A to C is the translation with vector $\begin{pmatrix} 4 \\ -5 \end{pmatrix}$.

> Choose one corner of shape A.
> Count from this corner 3 squares to the left and then count 5 squares down to find where this corner has moved to.
> The new shape is the same as shape A.
> Draw the new shape and label it D.

b D is marked on the diagram.

Exercise 17A

> Questions in this chapter are targeted at the grades indicated.

C

1 Describe, with a vector, the translation that maps triangle A onto:
  a triangle B
  b triangle C
  c triangle D
  d triangle E
  e triangle F
  f triangle G.

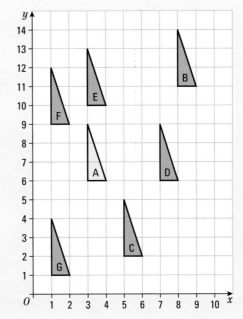

2 On a copy of the diagram translate triangle A:
  a 5 to the right and 4 up.
  Label your new triangle B.
  b 4 to the right and 6 down.
    Label your new triangle C.
  c 7 to the left. Label your new triangle D.
  d by the vector $\begin{pmatrix} 3 \\ 2 \end{pmatrix}$.
    Label your new triangle E.
  e by the vector $\begin{pmatrix} -6 \\ -4 \end{pmatrix}$.
    Label your new triangle F.

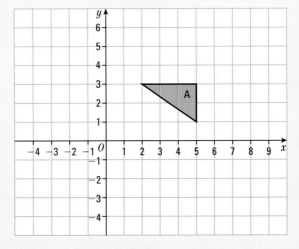

C

3   The coordinates of point A of this kite are (−2, 1).
     The kite is translated so that the point A is mapped
     onto the point (3, 4).
   a   On a copy of the diagram draw the image of the
        kite after this translation.
   b   Describe this translation with a vector.

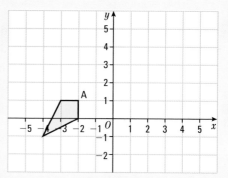

4   Draw the following translations on a copy of the diagram.
   a   Translate kite A by the vector $\begin{pmatrix} 4 \\ 7 \end{pmatrix}$.
        Label this new kite B.

   b   Translate kite B by the vector $\begin{pmatrix} -6 \\ -3 \end{pmatrix}$.
        Label this new kite C.

   c   Describe, with a vector, the translation that
        maps kite A onto kite C.
   d   Describe, with a vector, the translation that
        maps kite C onto kite A.

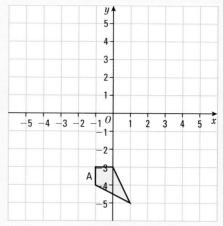

17.2 Transforming shapes using reflections

⊚ **Objectives**

○ You know that in a reflection the image is as far behind the mirror line as
   the object is in front of the mirror line.
○ You understand that reflections are described by the mirror line.
○ You can find an equation of a mirror line.
○ You know that when shape A is mapped to shape B by a reflection,
   shape A and shape B are congruent.

⊘ **Why do this?**

Many interesting patterns can
be produced using reflections.
The patterns in a kaleidoscope
are caused by light being
reflected many times.

◈ **Get Ready**

1. Write down the equation of each of the lines A, B, C and D.

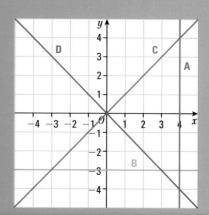

🔍 Key Points

◉ When you look in a mirror, you see your **reflection**.

The diagram below shows triangle P reflected in the mirror line to triangle Q.

The reflection of point A is point A′ so A and A′ are corresponding points.

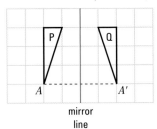

mirror
line

Point A′ is the same distance behind the mirror line as point A is in front.

The line joining points A and A′ is perpendicular to the mirror line.

Triangle Q is the reflection of triangle P in the mirror line. Each corner of Q is the reflection in the mirror line of the corresponding corner of P.

Triangle Q is as far behind the mirror line as triangle P is in front.

In mathematics all mirror lines are two-way mirrors so triangle P is also the reflection of triangle Q in the mirror line.

◉ To describe a reflection, give the mirror line.

◉ In a reflection:
  ◉ the lengths of the sides of the shape do not change
  ◉ the angles of the shape do not change
  ◉ the reflection of a shape (the **image**) is as far behind the mirror line as the shape is in front.

◉ In a reflection, any shape is congruent to its image because the lengths of the sides and angles of the shape are preserved by the reflection.

◉ The mirror line is the line of symmetry.

🔍 Example 3

Reflect trapezium T in the mirror line.
Label the new trapezium U.

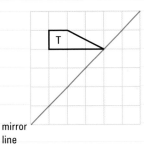

mirror
line

Method 1

> Reflect each corner of T in the mirror line so that its reflection is the same distance behind the mirror line as the corner is in front.
> Notice that:
> the line joining each corner to its image is perpendicular to the mirror line
> the image of the corner which is on the mirror line is also on the mirror line.

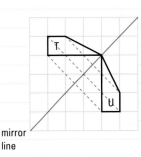

mirror
line

Method 2

> Put the edge of a sheet of tracing paper on the mirror line and make a tracing of the trapezium.
> Turn the tracing paper over and put the edge of the tracing paper back on the mirror line.
> Mark the images of the corners with a pencil or compass point.
> Method 2 is particularly useful when the shape is not a polygon or not drawn on a grid.

Example 4

Triangle T is a reflection of triangle S.
Draw the mirror line of the reflection.

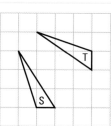

Join each corner of triangle S to its image on triangle T.
The mirror line passes through the mid-points (marked
with crosses) of these lines.

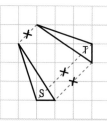

Draw the mirror line by joining the crosses.

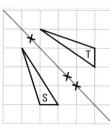

Example 5

Describe fully the transformation which
maps triangle P onto triangle Q.

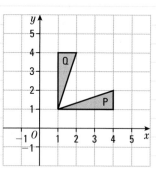

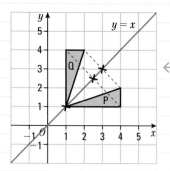

The transformation is a
reflection as triangle P has
been 'flipped over' to triangle Q.
Notice that the point on the
mirror line does not move.

ResultsPlus
Exam Tip

Make sure that you can recognise the lines
with equations $x = a$, $y = b$, $y = \pm x$

The transformation is a reflection in the line with equation $y = x$.

Exercise 17B

1   Make a copy of the diagram and complete the following
    reflections.
    a Reflect triangle P in the line $x = 1$.
       Label this new triangle Q.
    b Reflect triangle P in the line $y = 2$.
       Label this new triangle R.
    c Describe the reflection that maps
       triangle Q onto triangle T.

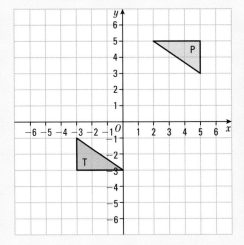

A03 D

2   On a copy of the diagram, complete the following
    reflections.
    a Reflect triangle A in the line $y = x$.
       Label this new triangle B.
    b Reflect triangle A in the line $y = -x$.
       Label this new triangle C.
    c Describe fully the transformation that
       maps triangle B onto triangle A.

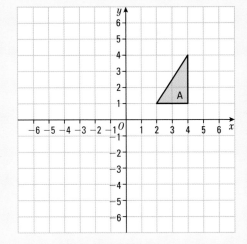

A03 C

3   a Give the equation of the mirror line of
       the reflection that maps:
       i shape P onto shape Q
       ii shape P onto shape R.
    b Describe fully the transformation that maps
       shape Q onto shape P.

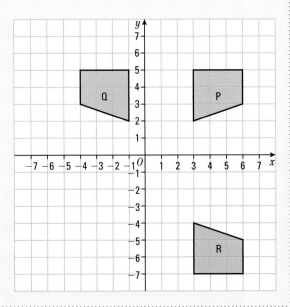

17.3 Transforming shapes using rotations

◎ Objectives

○ You know that in a rotation all points of a shape move around circles with the same centre.

○ You understand that rotations are described by a centre and an angle of turn.

○ You can find a centre of rotation.

○ You know that when shape A is mapped to shape B by a rotation, shape A and shape B are congruent.

⟐ Why do this?

Many everyday objects turn or rotate, for example, cycle wheels and the hands of a clock. It is often necessary to describe the rotation.

◈ Get Ready

1. Here is a clock face with only one hand. The hand is pointing to 12.

   a The hand is turned 90° anticlockwise. What number is the hand pointing to now?

   b Describe as fully as you can how the hand can turn to point to:

   i 3 **ii** 6 **iii** 5.

2. Imagine that the hand is pointing to 5. Describe as fully as you can how the hand can turn to point to:

   a 8 **b** 2 **c** 11 **d** 12.

🔍 Key Points

◉ To rotate means to turn. This face on a stick has rotated 60° clockwise about the point O. The size of the face has not changed.

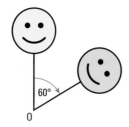

ResultsPlus
Exam Tip

A common mistake when describing a rotation is to call it a turn instead of a rotation and forgetting to say where the centre of rotation is.

◉ To describe a **rotation** you need to give:

  ◎ the angle of turn

  ◎ the direction of turn (clockwise or anticlockwise)

  ◎ the point the shape turns about (the **centre of rotation**).

◉ In a rotation:

  ◎ the lengths of the sides of the shape do not change

  ◎ the angles of the shape do not change

  ◎ the shape turns

  ◎ the centre of rotation does not move.

◉ In a rotation, any shape is congruent to its image because the lengths of the sides and angles of the shape are preserved by the rotation.

Example 6   Rotate the triangle a quarter turn clockwise about the point A.

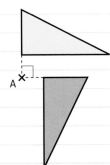

> Tracing paper can be used to rotate the shape. Trace the triangle and mark the point A.
> Fix the point A with a pencil or a compass point so that the point A does not move. Turn the tracing paper about A, clockwise through a quarter turn (90°).
> Now the position of the image of the triangle can be seen.
> Notice that each line of the triangle has turned through a quarter turn clockwise.

Example 7   Describe the transformation that maps triangle A onto triangle B.

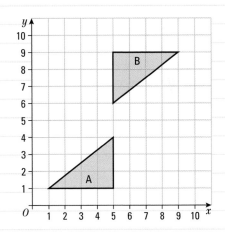

Triangle A is mapped onto triangle B by a rotation of 180° (a half turn) about the point (5, 5).

Tracing paper can be used to check that the transformation is a rotation of 180° with the centre of rotation the point (5, 5).

Exercise 17C

1   On a copy of the diagram, complete the following rotations.
   a   Rotate trapezium A a half turn about the origin O. Label the new trapezium B.
   b   Rotate trapezium A a quarter turn clockwise about the origin O. Label the new trapezium C.
   c   Rotate trapezium A a quarter turn anticlockwise about the origin O. Label the new trapezium D.

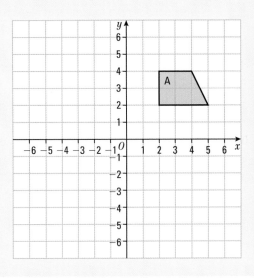

D

C

2 Make three copies of this diagram showing trapezium P.

a On copy 1 of the diagram, rotate trapezium P 180° about the point (2, 0). Label the new trapezium Q.

b On copy 2 of the diagram, rotate trapezium P 90° clockwise about the point (−2, 2). Label the new trapezium R.

c On copy 3 of the diagram, rotate trapezium P 90° anticlockwise about the point (−1, −1). Label the new trapezium S.

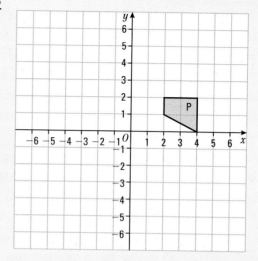

3 a Describe fully the rotation that maps shape A onto: i shape B ii shape C iii shape D.

b Describe fully the rotation that maps shape B onto shape A.

c Describe fully the rotation that maps shape B onto shape D.

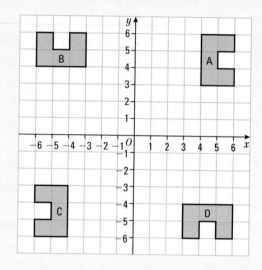

4 a Describe fully the rotation that maps triangle A onto:
   i triangle B ii triangle C iii triangle D iv triangle E
   v triangle F.

b Describe the transformation that maps triangle B onto triangle E.

c Describe the transformation that maps:
   i triangle D onto triangle B ii triangle F onto triangle E.

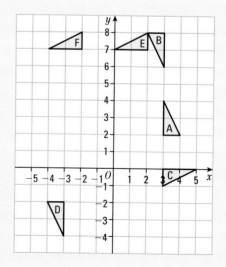

17.4 Enlargements and scale factors

Objectives

- You can enlarge a shape given the scale factor.
- You know that enlargements preserve angles but change lengths.
- You understand that enlargements are described by a centre and a scale factor.
- You can find the centre of an enlargement.
- You can use positive and negative scale factors.

Why do this?

If you have holiday photos blown up for a poster, you are making an enlarged version of the original photo.

Get Ready

1. Plot the following points on graph paper and join them up.
   a (0, 1)   **b** (1, 1)   **c** (1, 0)   **d** (0, 0)
2. Then plot the following points on the same graph and join them up.
   a (0, 2)   **b** (2, 2)   **c** (2, 0)   **d** (0, 0)
3. What can you say about these two shapes?

Scale factors

Key Points

- Here is a photograph of a shark.

Here is an **enlargement** of the photograph.

The sharks in the two photographs are the same but each length in the enlargement is 2 times the corresponding length in the original photograph.

For example, the length of the shark's fin in the enlargement is 2 times the length of the fin in the original photograph.

- The scale factor of an enlargement is the number of times by which each original length has been multiplied. So the larger photograph is an enlargement with scale factor 2 of the smaller photograph as

$$\text{scale factor} = \frac{\text{length of side in image}}{\text{length of corresponding side in object}}$$

- The scale factor can be found from the ratio of the lengths of two corresponding sides; in this case the ratio is 1 : 2.

- An enlargement changes the size of an object but not the shape of the object.
- Notice that each angle in the original photograph has the same size as the corresponding angle in the enlargement.
- So in an enlargement:
  - the lengths of the sides of the shape change
  - the angles of the shape do not change.

Example 8 Triangle B is an enlargement of triangle A.

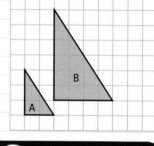

a Work out the scale factor of the enlargement that maps triangle A onto triangle B.

b Work out the scale factor of the enlargement that maps triangle B onto triangle A.

a Scale factor of the enlargement that maps triangle A onto triangle B = $\frac{4}{2}$.
Scale factor = 2

> The base of triangle A is 2 squares. The base of triangle B is 4 squares. Notice that pairs of corresponding sides are parallel.

b Scale factor of the enlargement that maps triangle B onto triangle A = $\frac{2}{4}$.
Scale factor = $\frac{1}{2}$

> This answer means that the length of each side of triangle A is $\frac{1}{2}$ the length of the corresponding side of triangle B.

ResultsPlus
Exam Tip

The transformation is still called an enlargement when the scale factor is a positive fraction less than 1 so that the image is smaller than the object.

Exercise 17D

1 Here is a right-angled triangle.
The triangle is enlarged with a scale factor of 4.
a Work out the length of each side of the enlarged triangle.
b Compare the perimeter of the enlarged triangle with the perimeter of the original triangle.

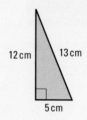

2 Copy the shape on squared paper and draw:
a an enlargement of shape A with scale factor 3.
Label this enlargement shape B.
b an enlargement of shape A with scale factor $\frac{1}{2}$.
Label this enlargement shape C.
c Shape B is an enlargement of shape C.
Work out the scale factor of the enlargement.

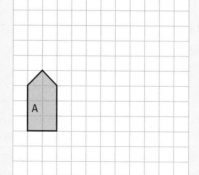

3 Rectangle P has a base of 4 cm and a height of 2 cm.

Rectangle Q is an enlargement of rectangle P with a scale factor of 2.

Rectangle R is an enlargement of rectangle P with a scale factor of 3.

a On squared paper, draw rectangles P, Q and R.

b Find the perimeter of: **i** rectangle P **ii** rectangle Q **iii** rectangle R.

c Find the area of: **i** rectangle P **ii** rectangle Q **iii** rectangle R.

d Work out the value of: **i** $\dfrac{\text{Perimeter of Q}}{\text{Perimeter of P}}$ **ii** $\dfrac{\text{Perimeter of R}}{\text{Perimeter of P}}$.

Write down anything that you notice about these values.

e Work out the value of: **i** $\dfrac{\text{Area of Q}}{\text{Area of P}}$ **ii** $\dfrac{\text{Area of R}}{\text{Area of P}}$.

Write down anything that you notice about these values.

f Rectangle S is an enlargement of rectangle P with a scale factor of 8.

What is the perimeter of rectangle S?

Centre of enlargement

> **Key Points**

◉ In the diagram, triangle P has been enlarged by a scale factor of 2 to give triangle Q.

The corner A of triangle P is mapped onto the corner A′ of triangle Q. A line has been drawn joining A and A′.

Lines have also been drawn joining the other pairs of corresponding points of triangles P and Q.

The lines meet at a point C called the **centre of enlargement**.

C to A is 2 squares across and 3 squares up.

C to A′ is 4 squares across and 6 squares up.

So $\dfrac{CA'}{CA} = 2$, the scale factor of the enlargement.

◉ To describe an enlargement you need to give:

◉ the scale factor

◉ the centre of enlargement.

◉ In general, when shape P is mapped onto shape Q by an enlargement with centre C and scale factor k, $CA' = k \times CA$ for any point A of shape P and the corresponding point A′ of shape Q.

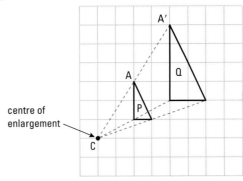

centre of enlargement

> **Example 9** Describe fully the transformation which maps triangle A onto triangle B.

The lengths of the sides of triangle B are twice those of triangle A.

This means that the transformation is an enlargement.

To find the centre of enlargement, join each corner (vertex) of triangle A to the corresponding vertex of triangle B.

The centre of enlargement C is the point where these lines cross.

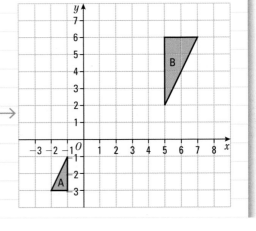

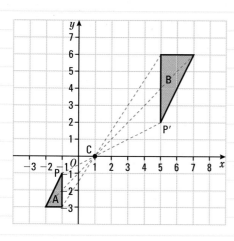

Notice that point C is between the object A and the image B. From C to P' is twice the distance from C to P but in the opposite direction.
The scale factor of the enlargement is −2.

ResultsPlus
Watch Out!

When a shape is enlarged by a negative scale factor, the image is on the opposite side of the centre of enlargement to the object.

The transformation is an enlargement with scale factor −2, centre (1, 0).

Example 10

a Enlarge triangle PQR by a scale factor of $-\frac{1}{3}$ with centre of enlargement C (3, 5).

b Describe fully the transformation that maps triangle P'Q'R' onto triangle PQR.

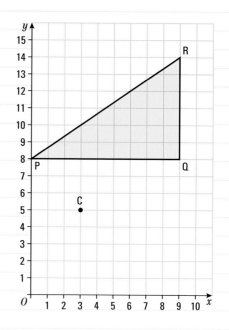

a

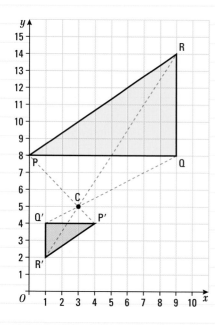

From C to P is 3 squares to the left and 3 squares up. So from C to P' is $-\frac{1}{3} \times 3 = -1$ square to the left, or 1 square to the right, and $-\frac{1}{3} \times 3 = -1$ square up, or 1 square down.

In the same way, from C to Q' is 2 squares to the left and 1 square down, from C to R' is 2 squares to the left and 3 squares down

b The transformation that maps triangle P'Q'R' onto triangle PQR is an enlargement with scale factor −3, centre (3, 5).

The lengths of the sides of triangle PQR are 3 times those of triangle P'Q'R' and the centre of enlargement is between the two triangles.

ResultsPlus
Exam Tip

The word 'enlargement' is used even when the new shape is smaller than the original shape.

ResultsPlus
Exam Tip

In an enlargement, corresponding sides in the object and the image are parallel.

Exercise 17E

1 Copy the shape on squared paper and draw the enlargement of the shape with the given scale factor and centre of enlargement marked with a dot (•).

a Scale factor 3.

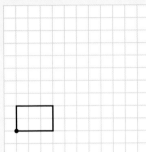

b **i** Scale factor 3.

ii Scale factor 2.

iii Scale factor $\frac{1}{2}$.

Draw all three enlargements on the same diagram.

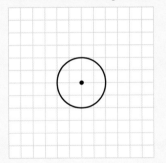

2 On a copy of the diagram complete the following enlargements.

a Enlarge triangle A with a scale factor of −2, centre (0, 0). Label this new triangle B.

b Enlarge triangle A with a scale factor of −$\frac{1}{3}$, centre (1, 6). Label this new triangle C.

c Find the scale factor of the enlargement that maps triangle C onto triangle B.

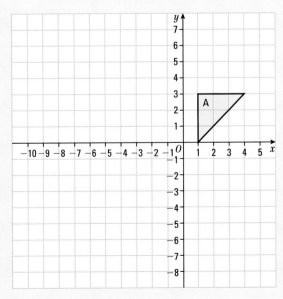

C

A03

3　**a** On a copy of the diagram, enlarge shape P with a scale factor of −1, centre (1, 2). Label this new shape Q.
The mapping of shape P onto shape Q is also a rotation.
　b Describe fully the rotation that maps shape P onto shape Q.

A03

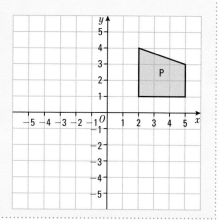

17.5 Combinations of transformations

Objectives

- You can transform a shape using combined translations, rotations, reflections or enlargements.
- You can find a single transformation which has the same effect as a combination of transformations.

Why do this?

Many designs for wallpaper and fabric are based on combinations of transformations.

Key Points

- A combination of transformations is when shape P is transformed to shape Q and then shape Q is transformed to shape R. It may be possible to find a single transformation which maps shape P onto shape R.
For example, a reflection in the y-axis has the same effect as a reflection in the x-axis followed by a rotation of 180° about the origin.

Example 11　**a** Reflect triangle P in the x-axis. Label the new triangle Q.
　b Rotate triangle Q 180° about the origin O. Label the new triangle R.
　c Describe fully the single transformation which maps triangle P onto triangle R.

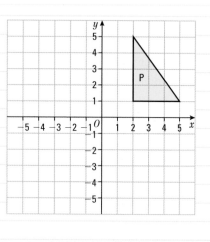

a, b

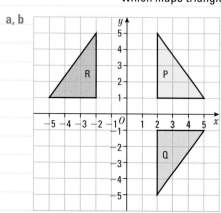

　c The single transformation which maps triangle P onto triangle R is a reflection in the y-axis.

Example 12

a Enlarge triangle P with scale factor 3 and centre of enlargement (2, 1). Label the new triangle Q.

b Enlarge triangle Q with scale factor $\frac{1}{3}$ and centre of enlargement (8, 10). Label the new triangle R.

c Describe fully the single transformation which maps triangle P onto triangle R.

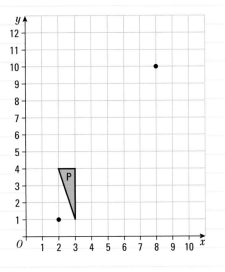

a, b

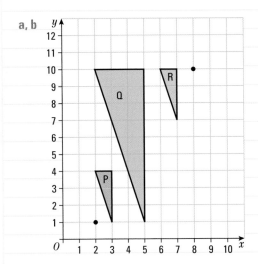

c From P to R is 4 to the right and 6 up.
The single transformation which maps triangle P onto triangle R is the translation with vector $\begin{pmatrix} 4 \\ 6 \end{pmatrix}$.

Exercise 17F

For each question, make a copy of the diagram.

1 Complete the following translations.

a Translate flag F by the vector $\begin{pmatrix} 3 \\ 8 \end{pmatrix}$.
Label the new flag G.

b Translate flag G by the vector $\begin{pmatrix} 6 \\ -4 \end{pmatrix}$.
Label the new flag H.

c Describe fully the single transformation which maps flag F onto flag H.

d Describe fully the single transformation which maps flag H onto flag F.

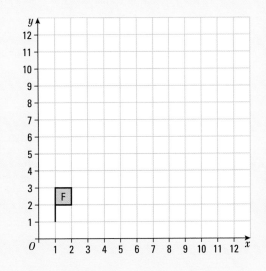

C

A03

A03

B

2 Complete the following transformations.
  a Rotate triangle T 180° about (2, 1).
    Label the new triangle U.
  b Translate triangle U by the vector $\begin{pmatrix} 4 \\ 4 \end{pmatrix}$.
    Label the new triangle V.
  A03
  c Describe fully the single transformation which maps
    triangle T onto triangle V.

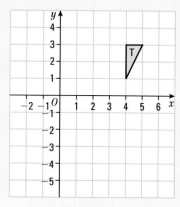

3 Complete the following transformations.
  a Rotate triangle T 90° clockwise about the origin O.
    Label the new triangle U.
  b Reflect triangle U in the line $y = -x$. Label the new triangle V.
  A03
  c Describe fully the single transformation which has the same
    effect as a rotation of 90° clockwise about the origin O followed
    by a reflection in the line $y = -x$.

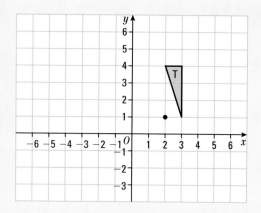

A02
A03
* 4 Use your copy of the graph paper to find and describe fully the single
  transformation which has the same effect as a
  translation with vector $\begin{pmatrix} 4 \\ 0 \end{pmatrix}$ followed by a reflection
  in the line $x = 7$.

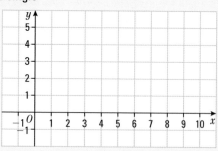

A02
A03
* 5

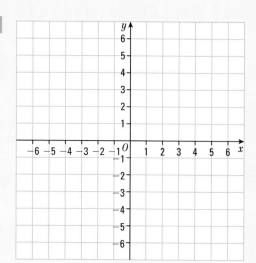

Use your copy of the graph paper to find and describe
fully the single transformation which has the same
effect as a rotation of 180° about (0, 0) followed by a
reflection in the y-axis.

Chapter review

⦿ In a **translation**, all points of the shape move the same distance in the same direction.
⦿ In a translation
   ⦿ the lengths of the sides of the shape do not change
   ⦿ the angles of the shape do not change
   ⦿ the shape does not turn.
⦿ In a translation, any shape is congruent to its image because the lengths of the sides and angles of the shape are preserved by the translation.
⦿ **Vectors** can be used to describe translations.
⦿ The top number shows the number of squares moved parallel to the x-axis, to the right or left.
⦿ The bottom number shows the number of squares moved parallel to the y-axis, up or down.
⦿ To the right and up are positive.
⦿ To the left and down are negative.
⦿ In a **reflection**:
   ⦿ the lengths of the sides of the shape do not change
   ⦿ the angles of the shape do not change
   ⦿ the **image** is as far behind the mirror line as the shape is in front.
⦿ To describe a reflection, give the mirror line. The mirror line is the line of symmetry.
⦿ In a reflection, any shape is congruent to its image because the lengths of the sides and angles of the shape are preserved by the reflection.
⦿ To describe a **rotation**, give:
   ⦿ the angle of turn
   ⦿ the direction of turn (clockwise or anticlockwise)
   ⦿ the point the shape turns about (the **centre of rotation**).
⦿ In a rotation:
   ⦿ the lengths of the sides of the shape do not change
   ⦿ the angles of the shape do not change
   ⦿ the shape turns
   ⦿ the centre of rotation does not move.
⦿ In a rotation, any shape is congruent to its image because the lengths of the sides and angles of the shape are preserved by the rotation.
⦿ In an **enlargement**:
   ⦿ the lengths of the sides of the shape change
   ⦿ the angles of the shape do not change.
⦿ To describe an enlargement, give:
   ⦿ the scale factor
   ⦿ the centre of enlargement.
⦿ If each vertex of shape P is joined to the corresponding vertex of shape Q, the joining lines intersect at the **centre of enlargement**.
⦿ In general, when shape P is mapped onto shape Q by an enlargement with centre C and scale factor k, $CA' = k \times CA$ for any point A of shape P and the corresponding point A′ of shape Q.
⦿ A combination of transformations is when shape P is transformed to shape Q and then shape Q is transformed to shape R. It may be possible to find a single transformation which maps shape P onto shape R.

Review exercise

1   On a copy of the grid, draw an enlargement of the shaded shape with a scale factor of 3.

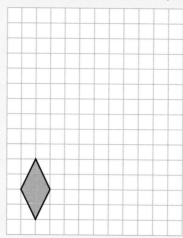

Nov 2006

D

2   **a** On a copy of the grid, rotate the shaded shape 90° clockwise about the point O.

    b Describe fully the single transformation that will map shape P onto shape Q.

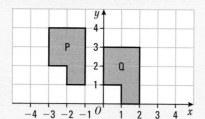

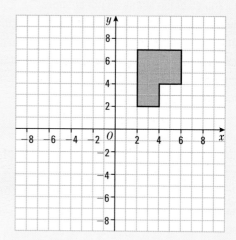

May 2009

3

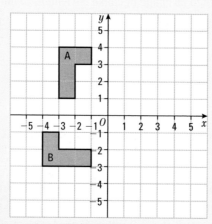

   a On a copy of the grid, reflect shape A in the y-axis.

   b Describe fully the **single** transformation which takes shape A to shape B.

Nov 2008

D

4

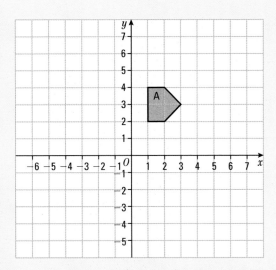

On a copy of the grid:

a reflect shape A in the y-axis. Label your new shape B.

b translate shape A by 3 squares right and 2 squares down. Label your new shape C.

Nov 2007

5 You have been asked to design a bathroom tile with reflective symmetry.

Draw a design in the top left 4 by 4 corner.

Then reflect your design in the vertical and horizontal lines to create the full pattern.

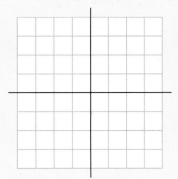

6 Describe fully the single transformation that will map shape P onto shape Q.

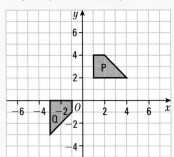

A03 C

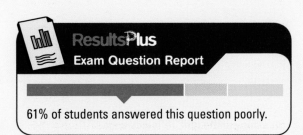

ResultsPlus

Exam Question Report

61% of students answered this question poorly.

Nov 2007

C

7

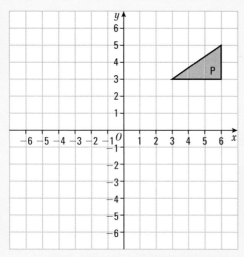

On a copy of the grid, enlarge triangle A by scale factor $-\frac{1}{2}$, centre $(-1, -2)$.
Label your triangle B.

Nov 2005

8 **a** Describe fully the single transformation that maps triangle A onto triangle B.

    b On a copy of the grid, rotate triangle A 90° anticlockwise about the point $(-1, 1)$. Label your new triangle C.

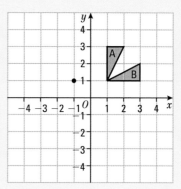

Nov 2006

9

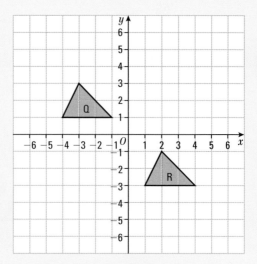

    a On a copy of the grid, reflect triangle P in the line $x = 2$.

    b Describe fully the **single** transformation that takes triangle Q to triangle R.

Nov 2006

10

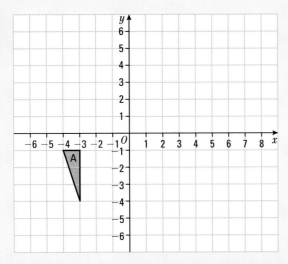

On a copy of the grid, enlarge triangle A by scale factor −2, centre (0, −1).

11 Triangle A is reflected in the x-axis to give triangle B.
Triangle B is reflected in the line $x = 1$ to give triangle C.
Describe the **single** transformation that takes
triangle A to triangle C.

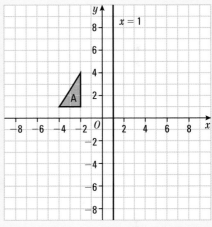

June 2008

12 Triangle A is reflected in the y-axis to give triangle B.
Triangle B is then reflected in the x-axis to give triangle C.
Describe the single transformation that takes triangle A
to triangle C.

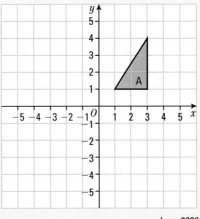

June 2006

18 PROCESSING, REPRESENTING AND INTERPRETING DATA

When you buy food, the packaging gives you information about the nutritional value of that food, but you will need to interpret it to understand what it means for your health. For example, a grilled salmon fillet gives you 30g of protein but unless you know that a woman needs approximately 46g of protein a day and a man approximately 56g, this is of little use. Now you can work out that for a woman, the salmon fillet gives her about 65% of her daily protein intake and for a man about 54%.

◎ Objectives

In this chapter you will be able to produce and interpret the following, for various types of data:
- pie charts
- stem and leaf diagrams
- bar charts and composite bar charts
- frequency diagrams
- histograms for continuous data
- frequency polygons
- cumulative frequency graphs
- box plots.

◆ Before you start

You need to be able to:
- measure and draw angles to the nearest degree
- measure and draw lines to the nearest mm
- understand grouped data.

18.1 Producing pie charts

◎ Objective

○ You can represent the proportions of different categories of data using a pie chart.

? Why do this?

When a council collects council tax they like to show the taxpayers how they are spending their money. They might use a pie chart to show the proportions spent on different things.

⟡ Get Ready

1. How many degrees are there in a circle?
2. How many degrees are there in a quarter-circle?
3. What is **a** $\frac{1}{3}$ of 360   **b** $\frac{2}{8}$ of 180   **c** $\frac{4}{6}$ of 90?

Key Points

◉ A **pie chart** is often used to show data. It shows how the total is split up between the different categories.

◉ In a pie chart the area of the whole circle represents the total number of items.

◉ The area of a **sector** represents the number of items in the category represented by that sector.
This pie chart shows how the population of the United Kingdom is split between the different countries.
It shows that the lowest number of people live in Northern Ireland.
The greatest number of people live in England.

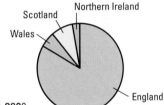

◉ The angles at the centre must add up to 360°.

◉ The angle for a particular sector is found as follows: sector angle $= \dfrac{\text{frequency} \times 360°}{\text{total frequency}}$

Example 1

The table shows the number of theatre-goers who attended each type of performance at least once in a 1-year period.

Performance	Musical	Play	Entertainment	Dance	Opera
Number	38	27	14	17	24

Draw a pie chart to represent this information.

Musical $\dfrac{38}{120} \times 360° = 114°$

Play $\dfrac{27}{120} \times 360° = 81°$

Entertainment $\dfrac{14}{120} \times 360° = 42°$

Dance $\dfrac{17}{120} \times 360° = 51°$

Opera $\dfrac{24}{120} \times 360° = 72°$

> Total frequency = 38 + 27 + 14 + 17 + 24 = 120
> Use the formula to find each angle.

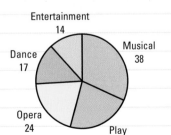

Results Plus
Exam Tip

Add the angles together to make sure they add up to 360°.

Check: 114 + 81 + 42 + 51 + 72 = 360

Exercise 18A

Questions in this chapter are targeted at the grades indicated.

1 The numbers of drinks dispensed by a vending machine in one day are shown in the table.

Type of drink	Tea	Black coffee	Chocolate	Orange	Coke	Latte
Number of drinks	54	42	18	30	12	24

Draw a pie chart to represent these data. Use a radius of 4 cm.

2 The snack bar at a bus station sold 120 sandwiches one lunch time.
The table shows the number of each type of sandwich sold.

Type of sandwich	Cheese	BLT	Tuna	Prawn	Ham	Chicken
Number of sandwiches	10	35	20	30	15	10

Draw a pie chart to represent these data. Use a radius of 4 cm.

3 A factory manager asks the employees how they travel to work.
The table shows these data.

Method of getting to work	Walk	Car	Cycle	Train	Motorbike
Number of employees	14	32	12	10	4

Draw a pie chart to represent these data.

18.2 Interpreting pie charts

Objective

○ You can interpret a pie chart.

Why do this?

Election results are often shown in a pie chart. You can interpret these graphs to see how many people voted for each party.

Key Points

◉ To read frequencies from a pie chart use the formula

$$\text{Frequency} = \frac{\text{sector angle} \times \text{total frequency}}{360°}$$

◉ The frequency represented by corresponding sectors in two pie charts is dependant upon the total populations represented by each of the pie charts.

Example 2

The pie chart shows the number of Bronze Age finds made with a metal detector and the outcomes when they were submitted to a local museum.

There were 36 finds altogether.

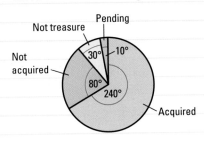

Watch Out!

In an exam 'work out' means calculate the frequency, so don't just measure the angle.

a Which type of outcome was most frequent?

b Work out the frequency for each outcome.

a 'Acquired' was the most common outcome.

b Acquired frequency $= \dfrac{\text{sector angle} \times \text{total frequency}}{360°} = \dfrac{240° \times 36}{360°} = 24$

ResultsPlus
Exam Tip

Always add up the frequencies for each sector to make sure they total to the right number.

Not acquired frequency $= \dfrac{80° \times 36}{360°} = 8$

Not treasure frequency $= \dfrac{30° \times 36}{360} = 3$

Pending frequency $= \dfrac{10° \times 36}{360} = 1$

Check: $24 + 8 + 3 + 1 = 36$

Exercise 18B

1 The pie chart shows how the 180 boys in Year 11 at Windup Academy chose from five sports options.

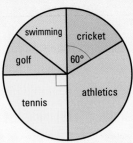

a Write down the least popular option.

b Write down the most popular option.

c Work out how many boys chose tennis.

d Work out how many boys chose cricket.

* **2** A company owns two coffee shops in Twyfield.
  They do a survey to find the number of each type of coffee they sell between 9 am and 10 am on one
  particular day.

Coffee type	Frequency	
	Shop A	**Shop B**
Espresso	5	5
Americano	15	12
Latte	10	24
Mocha	40	20
Cappuccino	20	11

Compare and contrast the information by drawing two pie charts.

18.3 Representing and interpreting data in a stem and leaf diagram

◉ Objectives

● You can represent data as a stem and leaf diagram.
● You can use a stem and leaf diagram to find the
mode, median, range and quartiles of a set of data.

❓ Why do this?

If you surveyed the number of DVDs that your
friends have, you could use a stem and leaf
diagram to show the pattern of the results.

◈ Get Ready

1. Write the numbers in each set in order of size, with smallest number first.

   a 65, 54, 72, 50   **b** 4.3, 4.6, 4.0, 4.4   **c** 0.11, 0.1, 0.01, 0.12

🔑 Key Points

● A **stem and leaf diagram** is a way of presenting data that makes it easy to see the pattern without losing the
actual data.

● A stem and leaf diagram should always have a key.

● From a stem and leaf diagram you can find statistics about the data. The lower quartile (Q_1) is the value a quarter
of the way through the data, the second quartile (Q_2) or median is halfway through, and the upper quartile (Q_3) is
three-quarters of the way through.

● The interquartile range (IQR) is the difference between the upper and lower quartiles = $Q_3 - Q_1$.

Example 3

Here are the numbers of cigarettes smoked per day by 15 people who are going to attempt to give up smoking:

20, 35, 40, 42, 32, 15, 22, 30, 28, 34, 40, 43, 28, 41, 25

a Write these data as an **ordered stem and leaf diagram**.
b Write down the mode of these data.
c Find the median of these data.
d Work out the range of these data.
e Find the lower and upper quartiles and interquartile range.

> The digit that each number begins with is called the stem.

a

Stem	Leaf				
1	5				
2	0	2	8	8	5
3	5	2	0	4	
4	0	2	0	3	1

Key 1|5 stands for 15

> The following digit is called the leaf.

> Under stem, write the numbers 1 to 4.

> Opposite each stem, write the leaves. Don't worry about the order. This gives you an unordered stem and leaf diagram.

1	5				
2	0	2	5	8	8
3	0	2	4	5	
4	0	0	1	2	3

Key 1|5 stands for 15

> Next draw a stem and leaf with the leaves in order, starting with the smallest. This is an ordered stem and leaf diagram as asked for in the question.

b There are two modes: 28 and 40.

> Each appears twice, the others only once.

c The median is 32.

> 32 is the middle value.

d The range is $43 - 15 = 28$

> The range is the difference between the largest and smallest values. The largest and smallest values are the first leaf and the last leaf.

e $Q_1 = 25$ $Q_3 = 40$
$IQR = 40 - 25 = 15$

> $Q_1 = \frac{16}{4}$th value = 4th value
> $Q_3 = 3 \times \frac{16}{4}$th value = 12th value
> You can find the values by counting in from each end.

Exercise 18C

1   Nassim records the number of emails he receives every day for 35 days.
    The data he collects are shown in the stem and leaf diagram.

0	6	7	9	9				
1	4	7	7	8	8	9	9	
2	2	3	5	5	6	7	8	9	9	9
3	1	5	6	6	6	6	7	
4	3	6	8	9				
5	2	3	3					

Key 3 | 1 stands for 31

a   Write down the mode of these data.

b   Find the median of these data.

c   Work out the range of these data.

d   Find Q_1 and Q_3 of these data.

e   Work out the interquartile range for these data.

2   Here are the number of minutes a sample of 19 people had to wait to see a dentist.

10	12	8	9	21	24	17	4	28	30
5	7	9	15	7	9	14	9	6	

a   Draw an ordered stem and leaf diagram for these data.

b   Use your stem and leaf diagram to find the mode of these data.

c   Use your stem and leaf diagram to find the median of these data.

d   Work out the range of these data.

e   Use your stem and leaf diagram to find Q_1 and Q_3 of these data.

f   Work out the interquartile range for these data.

3   A delivery driver does a journey on 23 days every month.
    Here are the distances, in kilometres, that he travelled in March.

56	74	83	74	65	92	52	59
64	68	72	94	82	63	74	65
88	69	68	85	68	74	63	

a   Draw an ordered stem and leaf diagram for these data.

b   Use your stem and leaf diagram to find the mode of these data.

c   Use your stem and leaf diagram to find the median of these data.

d   Work out the range of these data.

e   Use your stem and leaf diagram to find Q_1 and Q_3 of these data.

f   Work out the interquartile range for these data.

18.4 Interpreting comparative and composite bar charts

Objectives

- You can interpret comparative bar charts.
- You can interpret composite bar charts.

Why do this?

You may want to compare the sales of various categories of music in two shops. Composite bar charts would allow you to do this.

Get Ready

1. What can you say about the data in these two charts?

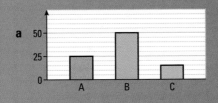

Key Points

- Composite bar charts (sometimes called compound or **component bar charts**) can be drawn to compare data.
- A composite bar chart shows the size of individual categories split into their separate parts.
- A **dual bar chart** is a type of comparative bar chart.
  In a comparative bar chart, two (or more) bars are drawn side-by-side for each category, and the heights of the bars can be compared category-by-category.

Example 4

The **dual bar chart** shows the number of houses sold by two agents in four months.

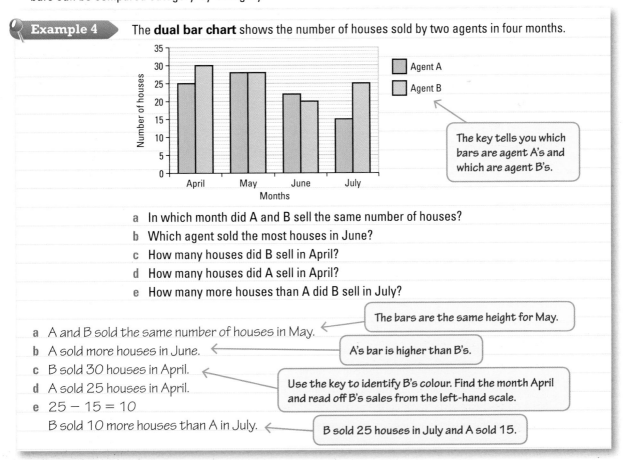

The key tells you which bars are agent A's and which are agent B's.

a In which month did A and B sell the same number of houses?

b Which agent sold the most houses in June?

c How many houses did B sell in April?

d How many houses did A sell in April?

e How many more houses than A did B sell in July?

a A and B sold the same number of houses in May. ← The bars are the same height for May.

b A sold more houses in June. ← A's bar is higher than B's.

c B sold 30 houses in April. ← Use the key to identify B's colour. Find the month April and read off B's sales from the left-hand scale.

d A sold 25 houses in April.

e 25 − 15 = 10

B sold 10 more houses than A in July. ← B sold 25 houses in July and A sold 15.

Example 5

This composite bar chart shows the amounts of protein, carbohydrate, fat and fibre in 100 g of white and wholemeal flour.

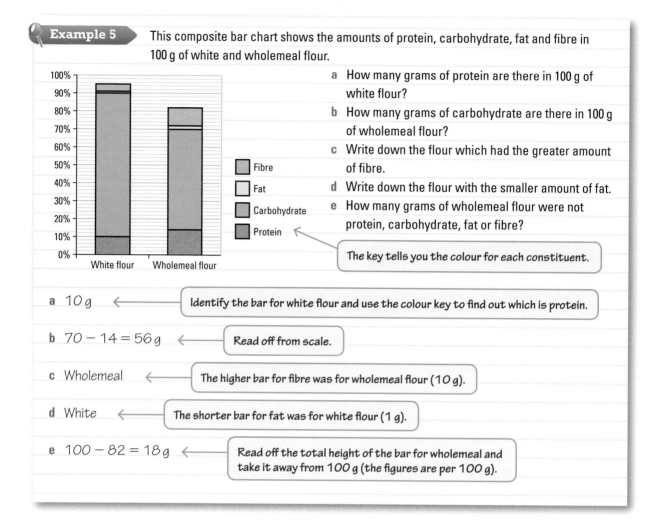

a How many grams of protein are there in 100 g of white flour?

b How many grams of carbohydrate are there in 100 g of wholemeal flour?

c Write down the flour which had the greater amount of fibre.

d Write down the flour with the smaller amount of fat.

e How many grams of wholemeal flour were not protein, carbohydrate, fat or fibre?

The key tells you the colour for each constituent.

a 10 g ← *Identify the bar for white flour and use the colour key to find out which is protein.*

b 70 − 14 = 56 g ← *Read off from scale.*

c Wholemeal ← *The higher bar for fibre was for wholemeal flour (10 g).*

d White ← *The shorter bar for fat was for white flour (1 g).*

e 100 − 82 = 18 g ← *Read off the total height of the bar for wholemeal and take it away from 100 g (the figures are per 100 g).*

Exercise 18D

1 The dual bar chart shows the temperature in a number of resorts in April and October.

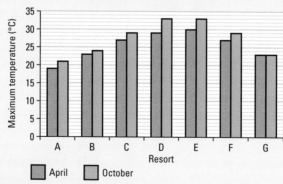

a Write down the maximum temperature in April.

b Write down the maximum temperature in October.

c Write down the resort that had the same maximum temperature in both months.

d Write down the resorts in which the maximum temperature in October was 29°C.

e Write down the resort in which the maximum temperature in April was 19°C.

2 The composite bar chart shows how David spends his money.

a What did David spend most on?

b What did David spend least on?

c What percentage of his income did he spend on housing?

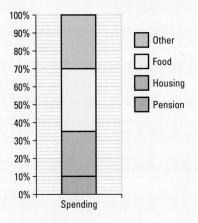

3 The composite bar charts show the make-up of 100 grams of each of two cereals: Wheatees and Fruitbix.

a How many grams of carbohydrate are there in 100 g of Wheatees?

b Estimate the number of grams of fat in 100 g of Fruitbix.

c Write down the name of the cereal that has more fibre.

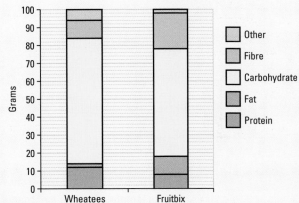

4 Market days in Ulvston are on Thursday and Saturday. Mattie runs a market stall that sells jumpers on each of these days. The composite bar chart shows his sales in one week in November.

a On which day were most jumpers sold overall?

b On which day were most green jumpers sold?

c How many red jumpers were sold on Saturday?

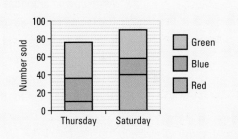

18.5 Drawing and interpreting frequency diagrams and histograms

Objectives

- You can draw a frequency diagram for grouped discrete data.
- You can draw a histogram for continuous data.

Why do this?

An exam board may choose to illustrate how the candidates in a particular year performed in one of its exams by creating a frequency diagram or histogram with the data.

Get Ready

1. What is the width of each class interval?

 a $0 \leqslant h < 3$ b $8 \leqslant h < 24$ c $75 \leqslant h < 100$

> **Key Points**

- A **frequency diagram** can be drawn from grouped discrete data.
- A frequency diagram for grouped discrete data looks the same as a bar chart except that the label underneath each bar represents a group.
- A **histogram** can be drawn from grouped continuous data.
- A histogram is similar to a bar chart but represents continuous data so there is no gap between the bars.
- You can find information from a histogram, such as the median or the number of people in a given interval.

> **Example 6**

The table shows the number of pizzas ordered in a restaurant from 7 pm to 8 pm on consecutive nights. Draw a frequency diagram for this information.

Number ordered	Frequency
1–5	2
6–10	4
11–15	8
16–20	6

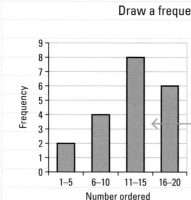

There is a gap between the bars because, for example, there is no whole number between 15 and 16.

> **Example 7**

The **grouped frequency table** shows information about the lengths of a series of roadworks.

a Write down the modal class interval.

b The length of one set of roadworks is 177.2 m. In which class interval is this recorded?

c The length of another set is exactly 180 m. In which class interval is this length recorded?

d Draw a histogram for these data.

Length (l metres)	Frequency
$160 \leqslant l < 165$	10
$165 \leqslant l < 170$	14
$170 \leqslant l < 175$	8
$175 \leqslant l < 180$	5
$180 \leqslant l < 185$	3
$185 \leqslant l < 190$	2

a The modal class is $165 \leqslant l < 170$. ← This class interval has the highest frequency, 14.

b This set is in the class interval $175 \leqslant l < 180$. ← 177.2 m is greater than 175 m but less than 180 m.

c This set is in the class interval $180 \leqslant l < 185$. ← 180 is shown at the end of one class interval and at the beginning of another. The sign for 'less than or equal to' ($\leqslant$) shows that 180 m should go in the class interval $180 \leqslant l < 185$.

d

In this histogram the area of the bars is proportional to the frequency. In the class interval 160 to 165 there are 50 little squares representing a frequency of 10. Each little square is equal to a frequency of $\frac{1}{5}$.

In the class interval 165 to 170 there are 70 little squares so it represents a frequency of $70 \times \frac{1}{5} = 14$.

Exercise 18E

D

1 The grouped frequency table shows information about the number of computer games owned by each of 35 college students.

Draw a frequency diagram for this information.

Number of games	Frequency
0 to 2	2
3 to 5	5
6 to 8	9
9 to 11	12
12 to 14	7

2 The grouped frequency table shows information about the wingspans of 36 snowy owls.

a Write down the modal class.

b The first snowy owl measured had a wingspan of 140 cm. In which class interval is this recorded?

c Draw a histogram for these data.

Wingspan (w cm)	Frequency
$125 \leqslant w < 130$	2
$130 \leqslant w < 135$	10
$135 \leqslant w < 140$	14
$140 \leqslant w < 145$	7
$145 \leqslant w < 150$	3

3 In a research project 40 young otters were weighed. Some information about their weights is shown in the table.

a Write down the modal class.

b In which class interval does the weight of 137 g fall?

c Draw a histogram for these data.

Weight (w g)	Frequency
$135 \leqslant w < 137$	3
$137 \leqslant w < 139$	10
$139 \leqslant w < 141$	14
$141 \leqslant w < 143$	8
$143 \leqslant w < 145$	5

18.6 Drawing and using frequency polygons

Objectives

- You can draw frequency polygons.
- You can recognise simple trends from a frequency polygon.
- You can use two polygons to make comparisons between two sets of data.

Why do this?

If you take a sample of your classmates' long-jump results, a frequency polygon would give you a good idea of how the lengths are distributed.

Get Ready

1. Which number is halfway between:

a 3 and 7   b 15 and 20   c 112 and 119?

Key Points

⊙ A **frequency polygon** is another graph which shows data.

⊙ When drawing a frequency polygon you draw a histogram then mark the midpoints of the tops of the bars and join these with straight lines.

⊙ More than one frequency polygon can be drawn on the same grid to compare data.

Example 8   Draw a frequency polygon for the data in Example 7.

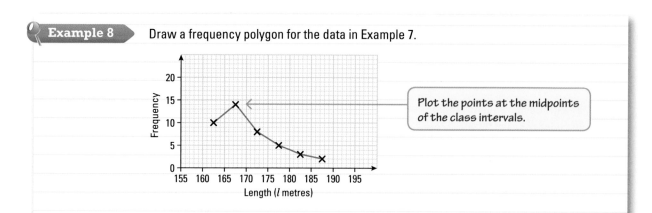

Plot the points at the midpoints of the class intervals.

Example 9   The frequency table gives information about the time waited, in seconds, at a set of traffic lights.

a Write down the modal class.

b Use the information to draw a histogram.

c Draw a frequency polygon to represent the information.

Time waited (t seconds)	Frequency
$90 \leqslant t < 95$	6
$95 \leqslant t < 100$	6
$100 \leqslant t < 105$	7
$105 \leqslant t < 110$	4
$110 \leqslant t < 115$	5
$115 \leqslant t < 120$	2

a The modal class is $100 \leqslant t < 105$.

b, c

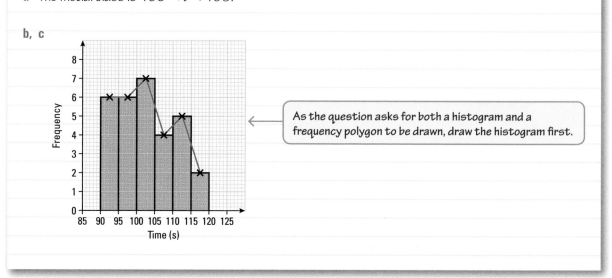

As the question asks for both a histogram and a frequency polygon to be drawn, draw the histogram first.

Example 10 These two frequency polygons show the heights of seedlings growing in two different composts.
Compare the heights of the two groups.
Give reasons for your answers.

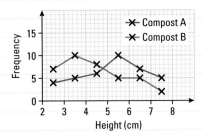

Compost A gives taller seedlings overall. ⟵ Above 5 cm, the line showing the heights with compost A is above the line for compost B.

There are more very tall seedlings with compost A. ⟵ There are five seedlings in the 7–8 cm class interval which were grown in compost A compared to two for compost B.

There are more very short seedlings with compost B. ⟵ There are seven seedlings grown in compost B but only four for compost A in the 2–3 cm class interval.

Exercise 18F

1 A seed producer wants to know the numbers of peas in pods of a new variety of peas.
He records the number of peas in 60 pods. The table shows this information.

Number of peas	3	4	5	6	7	8
Frequency	2	4	7	10	22	15

Draw a frequency polygon for these data.

2 The noise levels at 40 locations near an airport were measured in decibels.
The data collected are shown in the grouped frequency table.

Noise level (d decibels)	$60 \leqslant d < 70$	$70 \leqslant d < 80$	$80 \leqslant d < 90$	$90 \leqslant d < 100$
Frequency	15	16	7	2

a Write down the modal class.

b Use the information in the table to draw a histogram.

c Use your answer to part b to draw a frequency polygon.

C

3 In a fishing competition the lengths, in centimetres, of all the trout caught were measured.
The information collected is shown in the table.

Trout length (l cm)	Frequency
$24 \leqslant l < 25$	4
$25 \leqslant l < 26$	14
$26 \leqslant l < 27$	6
$27 \leqslant l < 28$	10
$28 \leqslant l < 29$	6

Draw a frequency polygon for these data.

A03
* 4 The two frequency polygons show the amount of time it took a
group of boys and a group of girls to do a crossword puzzle.
Who were better at doing the puzzle, boys or girls?
Give a reason for your answer.

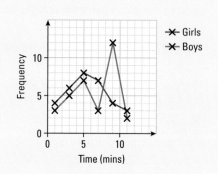

18.7 Drawing and using histograms with unequal class intervals

◎ Objectives

● You can draw a histogram with unequal class intervals.
● You understand frequency density.
● You can find the number of people in a given interval.

❓ Why do this?

If you measure the heights of a number of people, they will cluster around a middle value. Adjusting the size of the class intervals makes these irregularities less noticeable.

Key Points

● In histograms, when there are unequal class intervals in a bar you adjust the height by using a scale of **frequency density** rather than width, where:

$$\text{frequency density} = \frac{\text{frequency}}{\text{class width}}$$

or frequency = frequency density × class width.

● The area of each bar gives its frequency.

Example 11 The table gives information about the times taken, in seconds, by a number of workers to complete an operation in a factory.

Time taken (t seconds)	Frequency
$10 < t \leqslant 30$	5
$30 < t \leqslant 35$	4
$35 < t \leqslant 40$	8
$40 < t \leqslant 50$	27
$50 < t \leqslant 70$	24

Draw a histogram for these data.

Time taken (t seconds)	Frequency	Class width	Frequency density $= \dfrac{\text{frequency}}{\text{class width}}$
$10 < t \leqslant 30$	5	20	$\frac{5}{20} = 0.25$
$30 < t \leqslant 35$	4	5	$\frac{4}{5} = 0.8$
$35 < t \leqslant 40$	8	5	$\frac{8}{5} = 1.6$
$40 < t \leqslant 50$	27	10	$\frac{27}{10} = 2.7$
$50 < t \leqslant 70$	24	20	$\frac{24}{20} = 1.2$

Work out the width of each class interval (the class width).
Divide the frequency by the class width to find the frequency density which gives the height of each bar.

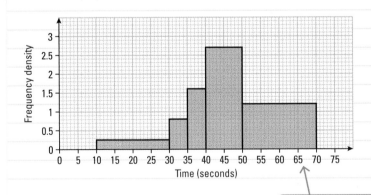

On a grid label the horizontal axis 'Time (seconds)' and the vertical axis 'Frequency density'.
Scale the horizontal axis from 0 to 75 and the vertical axis from 0 to 3.
Draw the bars with no gaps between them.
The first bar goes from 10 to 30 and has a height of 0.25

Example 12 The histogram gives information about the time, in seconds, taken by students to solve a puzzle.

Time taken (t seconds)	Frequency
$0 < t \leqslant 20$	
$20 < t \leqslant 30$	8
$30 < t \leqslant 40$	
$40 < t \leqslant 50$	
$50 < t \leqslant 70$	

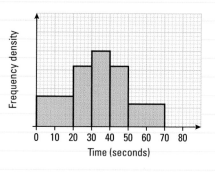

a Complete the frequency table.

b Use the histogram to estimate the number of people who took between 10 and 36 seconds to solve the puzzle.

a Frequency density for $20 < t \leqslant 30$ seconds $= \frac{8}{10} = 0.8$. ← Frequency density $= \dfrac{\text{frequency}}{\text{class width}}$

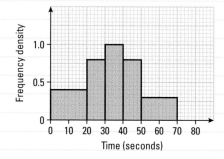

Now put a scale on the histogram.

Frequency = frequency density × class width →

Time taken (t seconds)	Frequency
$0 < t \leqslant 20$	$20 \times 0.4 = 8$
$20 < t \leqslant 30$	8
$30 < t \leqslant 40$	$10 \times 1.0 = 10$
$40 < t \leqslant 50$	$10 \times 0.8 = 8$
$50 < t \leqslant 70$	$20 \times 0.3 = 3$

b

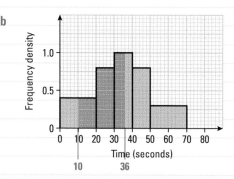

Frequency $= (10 \times 0.4) + (10 \times 0.8) + (6 \times 1.0)$
$= 4 + 8 + 6$
$= 18$ people

Work out the area between time = 10 and 36 seconds using frequency = frequency density × class width.

Exercise 18G

1 The table gives information about the lifetime of a certain make of torch battery.

Lifetime (l hours)	Frequency	Class width	Frequency density
$10 \leqslant l < 15$	4		
$15 \leqslant l < 20$	10		
$20 \leqslant l < 25$	20		
$25 \leqslant l < 30$	15		
$30 \leqslant l < 40$	6		

  a Copy and complete the table.

  b Draw a histogram for these data.

* 2 The table gives information about the distances a group of workers have to travel to work.

Distance (d kilometres)	Frequency
$0 < d \leqslant 5$	8
$5 < d \leqslant 10$	16
$10 < d \leqslant 20$	30
$20 < d \leqslant 30$	20
$30 < d \leqslant 40$	6

Draw a histogram for these data and find an estimate of the number of workers who travel between 15 and 25 minutes.

3 The table gives information about the age of people visiting a theme park one April morning.

  a Copy and complete the table and histogram. Add a scale where necessary.

Age (y years)	Frequency
$0 < y \leqslant 5$	10
$5 < y \leqslant 10$	28
$10 < y \leqslant 20$	
$20 < y \leqslant 40$	
$40 < y \leqslant 70$	

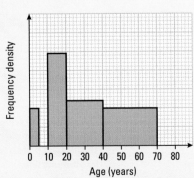

  b Find an estimate of how many people between 5 years and 30 years visited the theme park that morning.

18.8 Drawing and using cumulative frequency graphs

Objectives

- You can construct a cumulative frequency table.
- You can draw a cumulative frequency graph.

Why do this?

Data are sometimes displayed in a cumulative frequency curve, for example, weights of babies as they get older.

Key Points

- The **cumulative frequency** of a value is the total number of observations that are less than or equal to that value.
- **Cumulative frequency diagrams (graphs)** can be used to find estimates for the number of items up to a certain value.

Example 13 The grouped frequency table shows information about the time, in minutes, taken by 40 runners who had competed in a cross-country race.

Time (t minutes)	Frequency
$t \leqslant 60$	0
$60 < t \leqslant 65$	2
$65 < t \leqslant 70$	12
$70 < t \leqslant 75$	21
$75 < t \leqslant 80$	5

a Draw up a **cumulative frequency table**.
b Draw a cumulative frequency graph.

a

Time (t minutes)	Frequency	Cumulative frequency
$t \leqslant 60$	0	0
$60 < t \leqslant 65$	2	$0 + 2 = 2$
$65 < t \leqslant 70$	12	$2 + 12 = 14$
$70 < t \leqslant 75$	21	$14 + 21 = 35$
$75 < t \leqslant 80$	5	$35 + 5 = 40$

> Each time add the frequency to the previous cumulative frequency. The previous frequency was 2 so add the frequency 12 to get the new cumulative frequency 14.

b

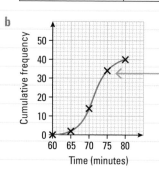

> The cumulative frequency 35 for the interval $70 < t \leqslant 75$ is plotted at $(75, 35)$.
> The plotted points may be joined by a curve or by straight lines.

Example 14

Forty students took a test. The cumulative frequency graph gives information about their marks.

a Use the graph to estimate the number of students who had marks less than or equal to 26.

b Use the graph to work out an estimate for the number of students whose mark was greater than 44.

c 26 students passed the test. Work out the pass mark for the test.

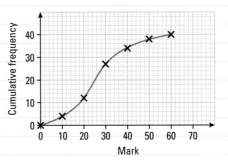

a There are 20 students with a mark less than 26.

b There are 36 students with a mark less than or equal to 44 so there are 40 − 36 = 4 with a mark greater than 44.

c If 26 pass there will be 40 − 26 = 14 that fail. From the graph the pass mark was 22.

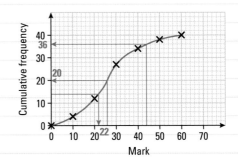

Exercise 18H

1   The table shows the ages of people using a bowling alley.

B

Age (x years)	Frequency	Cumulative frequency
$x \leqslant 10$	3	
$10 < x \leqslant 15$	7	
$15 < x \leqslant 20$	10	
$20 < x \leqslant 25$	15	
$25 < x \leqslant 30$	8	
$30 < x \leqslant 35$	5	
$35 < x \leqslant 40$	2	

a Copy and complete the table.

b Draw a cumulative frequency graph for these data.

B

2 The cumulative frequency graph shows the time a group of
girls spent on school computers.
   a Use the cumulative frequency graph to estimate the
      number of girls who spent up to 4 hours on the computer.
   b Use the cumulative frequency graph to estimate the
      number of girls who spent more than 6 hours on the
      computer.
   c Use the cumulative frequency graph to estimate
      the number of girls who spent between $3\frac{1}{2}$ and
      $6\frac{1}{2}$ hours on the computer.

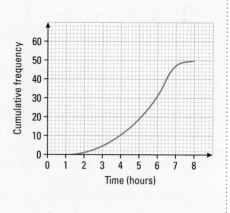

3 The cumulative frequency graph shows the speeds of cars
on a motorway.
   a Use the cumulative frequency graph to find an estimate
      for the number of motorists
      i driving at 45 mph or less
      ii driving at between 40 mph and 70 mph.
   b How many motorists' speeds were recorded altogether?
   c The speed limit on a motorway is 70 mph.
      Estimate the percentage of cars with a speed greater
      than 70 mph.

A03

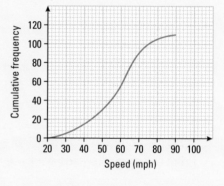

18.9 Finding quartiles from a cumulative frequency graph

◉ Objective

● You can estimate the median and quartiles
from a cumulative frequency graph.

❓ Why do this?

Looking at the age of Britain's population in a frequency
table, it is difficult to estimate the median and range. A
cumulative frequency graph makes it easy to find the values.

◈ Get Ready

1. Look at this list of numbers: 5, 5, 6, 7, 9, 9, 12, 13, 18, 20, 22, 23.

   Which numbers are: **a** halfway along the list   **b** three-quarters along the list?

Key Points

● The quartiles divide the frequency into four equal parts.
● If there are n values then the quartiles can be estimated from the cumulative frequency graph.
● The estimate for the lower quartile is the $\frac{n}{4}$th value.
● The estimate for the median is the $\frac{n}{2}$th value.
● The estimate for the upper quartile is the $\frac{3n}{4}$th value.
● You can compare measures of spread for two cumulative frequency graphs.

Example 15

The cumulative frequency graph shows information about the times, in minutes, taken by 40 runners who competed in a cross-country race.

    a Find estimates for the median and quartiles.

    b Find estimates for the range and interquartile range.

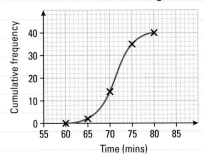

a
$$Q_1 = 69 \text{ min}$$
$$\text{median} = Q_2 = 71.5 \text{ min}$$
$$Q_3 = 73.5 \text{ min}$$

> Q_1 is the $\frac{40}{4} = 10$th value.
> Q_2 is the $\frac{40}{2} = 20$th value.
> Q_3 is the $3 \times \frac{40}{4} = 30$th value.

b $\text{Range} = 80 - 60 = 20 \text{ min}$
$\text{IQR} = 73.5 - 69 = 4.5 \text{ min}$

> $\text{Range} = \text{highest} - \text{lowest values}$
> $\text{IQR} = Q_3 - Q_1$

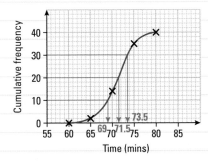

Exercise 18I

1 The cumulative frequency graph shows the scores a group of 100 apprentices got in an engineering examination.

    a Find an estimate for the median (Q_2).

    b Find an estimate for Q_1 and Q_3.

    c Work out the interquartile range.

    d Work out the range.

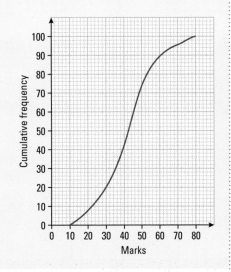

B

345

B

2 The cumulative frequency graph shows the prices of second-hand cars at a garage.

a Find an estimate for the median (Q_2).

b Find an estimate for Q_1 and Q_3.

c Work out the interquartile range.

3 The cumulative frequency graph shows the prices of detached houses on an estate agent's website.

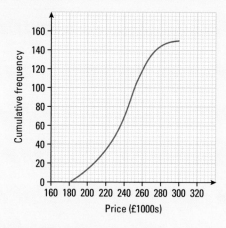

a Find estimates for the median and quartiles.

b Find estimates for the range and the interquartile range.

18.10 Drawing and interpreting box plots

⊙ Objectives

- You can construct a box plot given the raw data.
- You can find the median, quartiles and interquartile range given a box plot.

⊘ Why do this?

You can easily show the median and range of data with a box plot. For example, speeds of cars on a section of motorway.

⬦ Get Ready

1. What are the median, lower and upper quartiles, and interquartile range of this list of numbers?

5   6   6   8   11   13   13   15   17   20   22   25   26   26   29

Key Points

- Box plots (sometimes called **box and whisker plots**) are diagrams that show the median, upper and lower quartiles and the maximum and minimum values of a set of data and are often used to compare distributions.

Example 16 The times run by an athlete had a maximum of 52.1 seconds, a minimum of 47.2 seconds, a median of 48.8 seconds and upper and lower quartiles of 49.3 seconds and 48.2 seconds. Draw a box plot for these data.

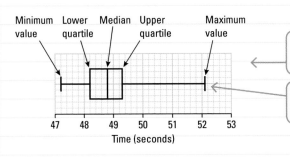

The box shows the spread over the middle 50% of the data (the interquartile range).

The whiskers show the lower 25% and the upper 25% of the data.

Example 17 The numbers of downloads from a music site during 15 time periods were as follows.

5	5	7	12	16	20	21	23
26	26	27	27	28	29	31	

Draw a box plot for these data.

5   5   7   (12)   16   20   21   (23)   26   26
27   (27)   28   29   31

The lowest value is 5 and the highest is 31.

The lower quartile is the $\frac{1}{4}(15 + 1)$th
= 4th value = 12.

Lower quartile is the $\frac{1}{4}(n + 1)$th value.

The median is the $\frac{1}{2}(15 + 1)$th
= 8th value = 23.

Median is the $\frac{1}{2}(n + 1)$th value.

Upper quartile is the $\frac{3}{4} \times (15 + 1)$th
= 12th value = 27.

Upper quartile is the $\frac{3}{4}(n + 1)$th value.

ResultsPlus
Exam Tip

These formulae are only for discrete data.

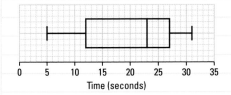

Time (seconds)

Example 18 The cumulative frequency graphs give information about the number of sales of mobile phones at two shops over 100 days.

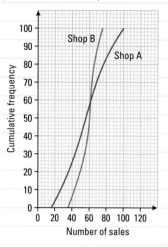

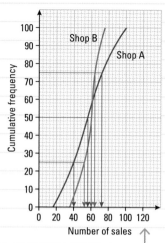

a Draw comparative box plots for these data.

b Compare the sales of the two shops.

> Find the maximum and minimum values and the median and quartiles from your graph.

a

	Shop A	Shop B
Least number	16	36
Lower quartile	40	52
Median	56	60
Upper quartile	72	64
Greatest number	100	75

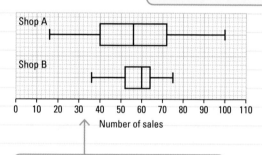

> Draw your box plots to the same scale.

b Shop B had a higher median so their sales are generally greater.

Both the range and interquartile range of shop A were greater than those of shop B.

The sales of shop A are more variable from day to day.

Exercise 18J

1 A wildlife park ranger estimated the heights of all the adult giraffes in the park. The tallest was 5.8 metres tall and the shortest was 4.2 metres. The median height was 5 metres, the lower quartile 4.6 metres and the upper quartile 5.6 metres. Draw a box plot for these data.

2 The heights of the trees in a small piece of mature woodland were measured in metres. They were as follows.

29 29.2 30.1 32 32.5 34.5 34.5 36.7 38 39.2 39.5 40.0 40.3 40.3 40.4

Draw a box plot for these data.

3 The cumulative frequency graph gives information about the ages of the male and female members of a cycling club.

    a Use the cumulative frequency diagram to find the quartiles and the maximum and minimum values.

*   **b** Draw two box plots on the same scale using these data and compare and contrast the data.

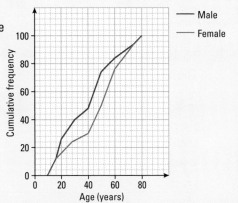

A03

Chapter review

- In a **pie chart** the area of the whole circle represents the total number of items.
- The area of each **sector** represents the number of items in that category.
- The angles at the centre must add up to 360°.

$$\text{sector angle} = \frac{\text{frequency} \times 360°}{\text{total frequency}} \quad \text{or} \quad \text{frequency} = \frac{\text{sector angle} \times \text{total frequency}}{360°}$$

- The frequency represented by corresponding sectors in two pie charts is dependant upon the total populations represented by each of the pie charts.
- A **stem and leaf diagram** is a way of presenting data that makes it easy to see the pattern without losing the actual data.
- A stem and leaf diagram should always have a key.
- From a stem and leaf diagram you can find statistics about the data. The lower quartile (Q_1) is the value a quarter of the way through the data, the second quartile (Q_2) or median is halfway through, and the upper quartile (Q_3) is three-quarters of the way through.
- The interquartile range (IQR) is the difference between the upper and lower quartiles $= Q_3 - Q_1$.
- A composite bar chart shows the size of individual categories split into their separate parts.
- A comparative bar chart shows two or more bars side-by-side for each category.
- A **frequency diagram** for grouped discrete data looks the same as a bar chart except that the label underneath each bar represents a group.
- A **histogram** is similar to a bar chart but because it represents continuous data, no gap is left between the bars.
- You can find information from a histogram, such as the median or the number of people in a given interval.
- When drawing a **frequency polygon** you draw a histogram then mark the midpoints of the tops of the bars and join these with straight lines.
- More than one frequency polygon can be drawn on the same grid to compare data.
- In histograms the area of each bar is proportional to the frequency it represents.

$$\textbf{Frequency density} = \frac{\text{frequency}}{\text{class width}} \quad \text{or} \quad \text{frequency} = \text{frequency density} \times \text{class width}.$$

- The **cumulative frequency** of a value is the total number of observations that are less than or equal to that value.
- The quartiles divide the frequency into four equal parts and can be estimated from the **cumulative frequency graph**.
- If there are n values, the estimates are:

    lower quartile $= \frac{n}{4}$th value, median $= \frac{n}{2}$th value, upper quartile $= \frac{3n}{4}$th value.

- You can compare measures of spread for two cumulative frequency graphs.
- Box plots (sometimes called **box and whisker plots**) are diagrams that show the median, upper and lower quartiles and the maximum and minimum values of a set of data and are often used to compare distributions.

Review exercise

1. 60 students were asked to choose one of four subjects.
   The table gives information about their choices.

Subject	Number of students	Angle
Art	12	72°
French	10	
History	20	
Music	18	

   Copy and complete the pie chart to show this information.

   Nov 2008

2. The table gives information about the drinks sold in a café one day.

Drink	Frequency	Size of angle
Hot chocolate	20	80°
Soup	15	
Coffee	25	
Tea	30	

   Copy and complete the pie chart to show this information.

   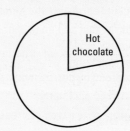

   Nov 2008

3. Mr White recorded the number of students absent one week.
   The dual bar chart shows this information for the first four days.

   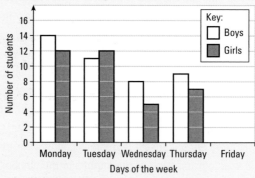

   a. How many boys were absent on Monday?

   b. How many girls were absent on Wednesday?

   On Friday, 9 boys were absent and 6 girls were absent.

   c. Use this information to complete the bar chart.

   On only one day more girls were absent than boys.

   d. Which day?

   March 2008

4. Zoe recorded the weights, in kilograms, of 15 people. Here are her results.

   87   51   46   77   74   58   68   78   48   63   52   64   79   60   66

   a. Draw a diagram to show these results.

   b. Write down the number of people with a weight of more than 70 kg.

   c. Work out the range of the weights.

   March 2009, amended

5 Jason collected some information about the heights of 19 plants.
This information is shown in the stem and leaf diagram.

D

```
1 | 1   2   3   3
2 | 3   3   5   9   9
3 | 0   2   2   6   6   7
4 | 1   1   4   8
```

Key 4|8 means 48 mm

Find the median.

Nov 2008

6 The table shows some information about the weights (w grams) of 60 apples.

On a copy of the grid, draw a frequency polygon
to show this information.

C

Weight (w grams)	Frequency
$100 \leqslant w < 110$	5
$110 \leqslant w < 120$	9
$120 \leqslant w < 130$	14
$130 \leqslant w < 140$	24
$140 \leqslant w < 150$	8

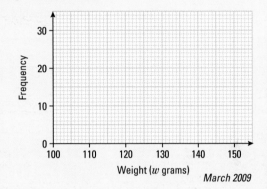

March 2009

7 60 students take a science test.
The test is marked out of 50.
This table shows information about the students' marks.

Science mark	0–10	11–20	21–30	31–40	41–50
Frequency	4	13	17	19	7

On a copy of the grid, draw a frequency
polygon to show this information.

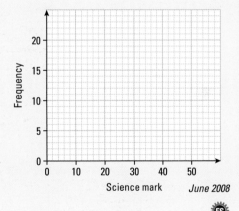

June 2008

*8

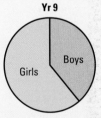

Pie chart showing proportion
of boys and girls in Year 9

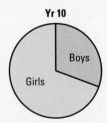

Pie chart showing proportion
of boys and girls in Year 10

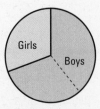

Pie chart showing proportion of boys
and girls in Year 9 and Year 10

To draw the pie chart for boys and girls in Years 9 and
10 combined, Kimberly drew the pie chart on the right:

James said that this could not be correct.
Explain who is right.

A03

351

B A03

* 9 John and Peter each own a garage. They both sell used cars.
The box plots show some information about the prices of cars at their garages.

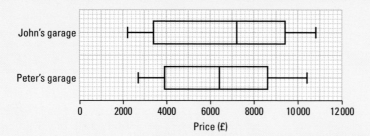

Compare the distribution of the prices of cars in these two garages.
Give **two** comparisons.

Nov 2008

10 The cumulative frequency graph shows some information about the ages of 100 people.

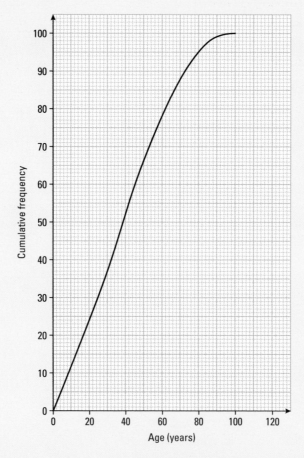

a Use the graph to find an estimate for the number of these people less than 70 years of age.

b Use the graph to find an estimate for the median age.

c Use the graph to find an estimate for the interquartile range of the ages.

Nov 2008

11 Verity records the heights of the girls in her class.
The height of the shortest girl is 1.38 m.
The height of the tallest girl is 1.81 m.
The median height is 1.63 m.
The lower quartile is 1.54 m.
The interquartile range is 0.14 m.

a Using this scale, draw a box plot for this information.

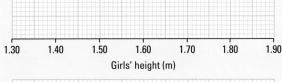

1.30   1.40   1.50   1.60   1.70   1.80   1.90
Girls' height (m)

The box plot shows information about the heights of the boys in Verity's class.

b Compare the distributions of the boys' heights and the girls' heights.

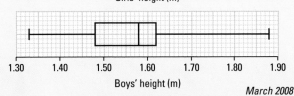

1.30   1.40   1.50   1.60   1.70   1.80   1.90
Boys' height (m)

March 2008

* 12 Lucy did a survey about the amounts of money spent by 120 men during their summer holidays. The cumulative frequency table gives some information about the amounts of money spent by the 120 men.

A survey of the amounts of money spent by 200 women during their summer holidays gave a median of £205. Compare the amounts of money spent by the women with the amounts of money spent by the men.

Amount (£A) spent	Cumulative frequency
$0 < A \leqslant 100$	13
$0 < A \leqslant 150$	25
$0 < A \leqslant 200$	42
$0 < A \leqslant 250$	64
$0 < A \leqslant 300$	93
$0 < A \leqslant 350$	110
$0 < A \leqslant 400$	120

May 2009

13 The box plot gives information about the distribution of the weights of bags on a plane.

a Jean says the heaviest bag weighs 23 kg.
She is **wrong**. Explain why.

b Write down the median weight.

c Work out the interquartile range of the weights.
There are 240 bags on the plane.

d Work out the number of bags with a weight of 10 kg or less.

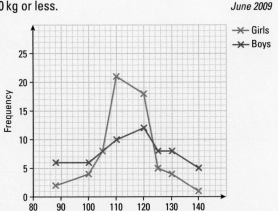

0   5   10   15   20   25   30
Weight (kg)

June 2009

14 The frequency polygons show information about the IQs of a group of boys and a group of girls.

a Write down an estimate for the number of girls with an IQ of 110.

b Write down an estimate for the number of boys with an IQ of 110.

c Use the frequency polygon to compare the overall IQs of the boys and the girls.

A

15 The table gives some information about the lengths of time some boys took to run a race.

Draw a histogram for the information in the table.

Time (t minutes)	Frequency
$40 \leqslant t < 50$	16
$50 \leqslant t < 55$	18
$55 \leqslant t < 65$	32
$65 \leqslant t < 80$	30
$80 \leqslant t < 100$	24

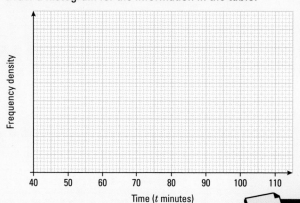

ResultsPlus
Exam Question Report

73% of students answered this sort of question poorly.

March 2009

16 On Friday, Peter went to the airport.
He recorded the number of minutes that each plane was delayed.
He used his results to work out the information in this table.

	Minutes
Shortest delay	0
Lower quartile	2
Median	8
Upper quartile	18
Longest delay	41

a Using this scale, draw a box plot to show the information in the table.

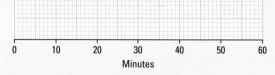

Peter also went to the airport on Saturday.
He recorded the number of minutes that each plane was delayed.
The box plot below was drawn using this information.

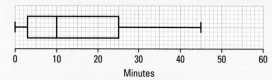

A03

b Comment on the plane delays.

March 2009, adapted

17 The speeds of 100 cars on a motorway were recorded.

The grouped frequency table shows some information about the speeds of these cars.

Speed (s mph)	Frequency
$40 < s \leqslant 50$	4
$50 < s \leqslant 60$	19
$60 < s \leqslant 70$	34
$70 < s \leqslant 80$	27
$80 < s \leqslant 90$	14
$90 < s \leqslant 100$	2

a On a copy of the grid, draw an appropriate graph for your table.

b Find an estimate for the median speed.

c Find an estimate for the interquartile range.

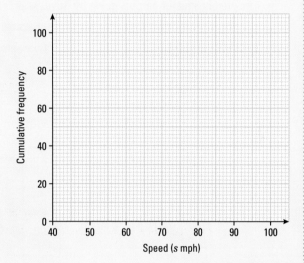

June 2008, adapted

18 The incomplete histogram and table give some information about the distances some teachers travel to school.

a Use the information in the histogram to complete the frequency table.

Distance (d km)	Frequency
$0 < d \leqslant 5$	15
$5 < d \leqslant 10$	20
$10 < d \leqslant 20$	
$20 < d \leqslant 40$	
$40 < d \leqslant 60$	10

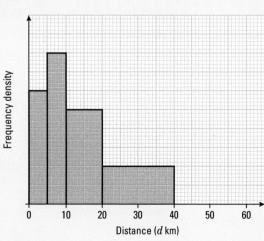

b Use the information in the table to complete the histogram. *Nov 2008*

A A02 A03 | 19 The table gives information about parcel sizes and their frequency.

Weight (w kg)	Frequency	Frequency density
$0 < w \leqslant 5$	20	
$5 < w \leqslant 15$	30	
$15 < w \leqslant 25$	15	
$25 < w \leqslant 35$	10	
$35 < w \leqslant 40$	5	

a Copy and complete the table.

b Draw a histogram for these data.

The weight limit for parcels going by Royal Mail is 20 kg.

c Work out an estimate for the number of parcels which will weigh 20 kg or less.

d Work out an estimate for the number of parcels weighing between 10 and 30 kg.

A☆ | 20 The histogram shows information about the lifetime of some batteries.

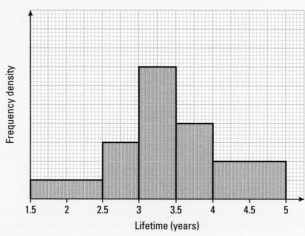

Two of the batteries had a lifetime of between 1.5 and 2.5 years.

Find the total number of batteries.

June 2008

INEQUALITIES AND FORMULAE

The coldest ever recorded temperature in Antarctica was −89°C on 21 July 1983. The warmest ever temperature was 15°C on 5 January 1974. So using inequalities, you could say that the temperature in Antarctica is $\geq$ −89°C and $\leq$ 15°C. On top of these extremes of temperature, Antarctica has winds of up to 320 km/hour, and average precipitation of less than 5 cm per year.

◎ Objectives

In this chapter you will:
- represent inequalities on a number line using the correct notation
- solve simple linear inequalities in one variable
- solve graphically several inequalities in two variables and find the solution set
- use and derive algebraic formulae
- change the subject of simple and more complex formulae.

◈ Before you start

You should already know how to:
- calculate using directed numbers
- set up and solve linear equations
- plot and read points on coordinate axes
- draw lines with the equation $y = mx + c$
- collect like terms in an algebraic expression
- know how to substitute into algebraic expressions.

19.1 Representing inequalities on a number line

◉ Objective

● You can represent inequalities on a number line using the correct notation.

❓ Why do this?

One way to check minimum and maximum temperatures in your greenhouse would be to represent them as an inequality on a number line.

⬆ Get Ready

1. What are the values of a, b, c, and d on the number line?

d   c b   a

−2 −1 0 1 2 3

🔍 Key Points

● In this chapter the symbols $<$, $\leq$, $>$ and $\geq$ are used.

  ◉ $<$ means 'less than'.

  ◉ $\leq$ means 'less than or equal to'.

  ◉ $>$ means 'greater than'.

  ◉ $\geq$ means 'greater than or equal to'.

Note: If you think of the sign '$>$' as an arrow, it always 'points' to the smaller value.

$2 < 7$ $(2 \leftarrow 7)$ and $7 > 2$ $(7 \rightarrow 2)$

● **Inequalities** have more than one value in their solution set. The solution set can be shown on a **number line**.

● When showing inequalities on a number line, an open circle shows the number is not included and a closed circle shows the number is included.

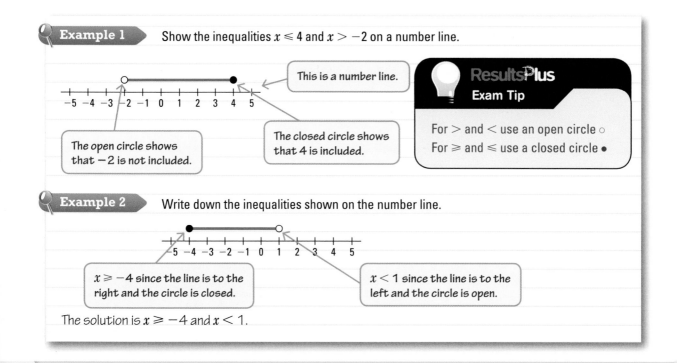

Example 1   Show the inequalities $x \leq 4$ and $x > -2$ on a number line.

This is a number line.

The closed circle shows that 4 is included.

The open circle shows that −2 is not included.

ResultsPlus

Exam Tip

For $>$ and $<$ use an open circle ○
For $\geq$ and $\leq$ use a closed circle ●

Example 2   Write down the inequalities shown on the number line.

$x \geq -4$ since the line is to the right and the circle is closed.

$x < 1$ since the line is to the left and the circle is open.

The solution is $x \geq -4$ and $x < 1$.

C

Exercise 19A

Questions in this chapter are targeted at the grades indicated.

1 Show these inequalities on a number line.
  a $x \leqslant 3$ b $x > 0$ c $x < 3$ and $x > -4$
  d $x \leqslant 1$ and $x \geqslant -3$ e $x < 0$ and $x \geqslant -5$

2 Write down the inequalities shown on these number lines.

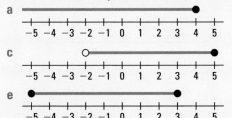

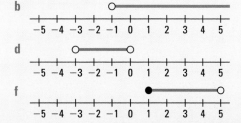

19.2 Solving simple linear inequalities in one variable

⊙ Objective

⊙ You can solve simple linear inequalities in one variable, and represent the solution set on a number line.

? Why do this?

All lifts have a maximum weight capacity. If there are 8 people in a lift, you can solve a simple inequality to find the maximum average weight that each person can be.

✦ Get Ready

Solve these equations.
1. $x + 10 = 4(x - 5)$ 2. $2x + 1 = 6x - 1$
Show these on a number line.
3. $x \leqslant 7$ 4. $x > 10$

Key Points

⊙ You solve a linear inequality using a similar method to the one you use for solving a linear equation (see Section 13.5).
⊙ If you multiply both sides of an inequality by a negative number, then you must reverse the inequality sign.

🔍 Example 3   Solve $3(x + 2) > 5 - x$ and show your answer on a number line.

$3(x + 2) > 5 - x$ ⟵ Expand the brackets.
$3x + 6 > 5 - x$

$3x + 6 + x > 5$ ⟵ Add x to both sides.
$4x + 6 > 5$

$4x > 5 - 6$ ⟵ Subtract 6 from both sides.
$4x > -1$

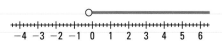

$x > -0.25$ ⟵ Divide both sides by 4.

359

Example 4 Solve $-3x \leqslant 12$.

$$-3x \leqslant 12$$

$$-12 \leqslant 3x \leftarrow$$
$$3x \geqslant -12$$

> Add $3x$ to both sides and subtract 12 from both sides.

$$x \geqslant -4 \leftarrow$$

> Divide both sides by 3.

ResultsPlus
Exam Tip

Check your answer by putting a value that satisfies your answer into the inequality in the question.

Exercise 19B

C

1 Solve these inequalities and show each answer on a number line.

   a $x + 1 > 5$ **b** $x - 3 \leqslant -2$ **c** $2x + 5 \leqslant 1$ **d** $10x - 7 > 9$

B

2 Solve these inequalities.

   a $3x < x + 9$ **b** $5x - 3 > 2x + 9$ **c** $2(x + 3) \leqslant 11$ **d** $5x - 7 > 3(x + 2)$

A☆

3 Solve these inequalities.

   a $x + 3 \geqslant 5(x - 2)$ **b** $3(x + 1) < 4(x - 5)$

   c $\dfrac{2 - 3x}{5} \leqslant 1 + \dfrac{x}{2}$ **d** $\dfrac{5x - 3}{4} + 1 \geqslant \dfrac{1 - 2x}{6}$

19.3 Finding integer solutions to inequalities in one variable

◉ Objective

● You can find integer solutions to inequalities in one variable.

◈ Why do this?

You could find out the grade boundary to get a grade A in a test, and then find each individual exam mark that you could get that would give you an A.

◈ Get Ready

Solve these inequalities.

1. $5x - 1 \geqslant 19$ **2.** $3x + 4 > 16$ **3.** $2x + 5 \leqslant 9$ **4.** $10x - 7 > 23$

Key Point

◉ When a value has a **lower limit** and an **upper limit**, you need to seperate the two inequalities, and solve them separately.

Example 5 $-3 \leqslant n < 4$

n is an integer. Find all the possible values of n.

> Show the inequality $-3 \leqslant n < 4$ on a number line.

$n = -3, -2, -1, 0, 1, 2, 3.$

> Write down the integers from the number line.

ResultsPlus

Watch Out!

Remember that 0 is an integer.

Example 6 $-3 \leqslant 2p - 1 < 8$

p is an integer. Find all the possible values of p.

$-3 \leqslant 2p - 1$ and $2p - 1 < 8$

> Write the two inequalities seperately.

$1 - 3 \leqslant 2p$	$2p < 8 + 1$
$-2 \leqslant 2p$	$2p < 9$
$-1 \leqslant p$	$p < 4.5$

> Solve each inequality.

So $-1 \leqslant p < 4.5$
$p = -1, 0, 1, 2, 3, 4.$

> Write down the integer values satisfying the inequality.

Exercise 19C

Find the possible integer values of x in these inequalities.

1 **a** $-2 < x \leqslant 5$ **b** $-5 < x < 2$ **c** $0 < x \leqslant 3$ **d** $-5 \leqslant x \leqslant 4$

2 **a** $-8 < 2x \leqslant 6$ **b** $-21 < 5x < 36$ **c** $-5 < 10x \leqslant 42$ **d** $-11 \leqslant 3x \leqslant 28$

3 **a** $-5 < 2x + 1 < 9$ **b** $-7 < 3x - 2 \leqslant 11$ **c** $-12 < 4x - 7 \leqslant 10$ **d** $-9 \leqslant 2x + 5 \leqslant 13$

4 **a** $-1 < \dfrac{x}{3} \leqslant 2$ **b** $-2 < \dfrac{2x}{5} \leqslant 3$ **c** $-1 < \dfrac{3x - 2}{4} \leqslant 2$ **d** $-3 < \dfrac{2 - x}{3} \leqslant 2$

C

B

A☆

19.4 Solving graphically several linear inequalities in two variables

Objective

● You can solve graphically several linear inequalities in two variables and find the solution set.

Why do this?

If you're planning a party you could solve inequalities to work out who to invite. For example, if you wanted at least twice as many girls as boys and 60 people maximum you can express this as $G \geqslant 2B$ and $G + B \leqslant 60$.

Get Ready

Draw these lines on a coordinate grid.

1. $y = 4x + 1$ **2.** $y = 3x - 4$ **3.** $2y = 3x + 4$

Key Points

◉ You can show the points that satisfy an inequality on a graph.

◉ Lines which are boundaries for regions that do include values on the line are shown as solid lines. Lines which are boundaries for regions that do not include values on the line are shown as dotted lines.

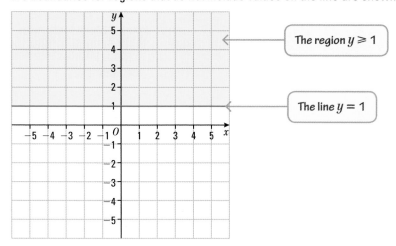

All points in the region $y \geqslant 1$ satisfy the inequality $y \geqslant 1$.
This includes points on the solid line.

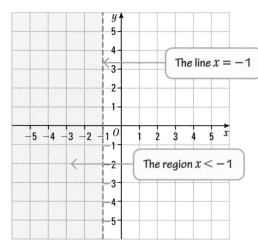

All points in the region $x < -1$ satisfy the inequality $x < -1$.
This does not include points on the dotted line.

Example 7 Write down the three inequalities satisfied by the coordinates of all points in the shaded region.
Write down the equations of each of the lines.

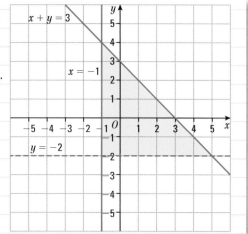

$x \geqslant -1$, since the shaded region is to the **right** of the solid line $x = -1$.
$x + y \leqslant 3$, since the shaded region is **below** the solid line $x + y = 3$.
$y > -2$, since the shaded region is **above** the dotted line $y = -2$.

Example 8

a On the grid, shade the region of points whose
coordinates satisfy these inequalities.
$y < 3$, $x < 2$, $y \geqslant 2x - 3$, and $x \geqslant -1$

b x and y are integers. Write down the coordinates
of all points (x, y) which satisfy the inequalities in **a**.

a $y < 3$, $x < 2$, $y \geqslant 2x - 3$, and $x \geqslant -1$
Draw the dotted lines $y = 3$ and $x = 2$.
Draw the solid lines $y = 2x - 3$ and $x = -1$.
Shade the region:
$y < 3$ (points below the **dotted line**)
$x < 2$ (points to the left of the **dotted line**)
$y \geqslant 2x - 3$ (points above the **solid line**)
$x \geqslant -1$ (points to the right of the **solid line**).

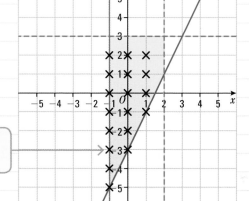

b The points are:
$(-1, 2), (0, 2), (1, 2), (-1, 1), (0, 1),$
$(1, 1), (-1, 0), (0, 0), (1, 0), (-1, -1),$
$(0, -1), (1, -1), (-1, -2), (0, -2),$
$(-1, -3), (0, -3), (-1, -4), (-1, -5)$

Mark the points
with a cross (x).

Exercise 19D

1 x and y are integers.
$-2 < x \leqslant 1$ $y > -2$ $y < x + 1$
On a copy of the grid, mark with a cross (x), each of the
six points which satisfy all of these three inequalities.

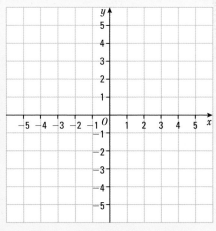

2 On a grid scaled from -6 to 6 on each axis, shade the region of points whose coordinates satisfy the
following inequalities.
a $x \leqslant 4$ b $y < 1$ c $-2 < x \leqslant 5$ d $-4 \leqslant y < 2$

3 On a grid scaled from -6 to 6 on each axis, shade the region of points whose coordinates satisfy the
following inequalities.
a $-1 < x \leqslant 3$ and $-5 \leqslant y < 1$ b $-3 \leqslant x \leqslant 2$ and $-1 \leqslant y < 4$
c $-2 \leqslant x \leqslant 3$, $y > -4$ and $y < x$ d $x \geqslant 0$, $y + x < 4$ and $y > 3x - 2$

D

B

A

4 The diagram shows a shaded region bounded by three lines.

  a Write down the equation of each of these lines.

  b Write down the three inequalities satisfied by the coordinates of the points in this shaded region.

  c If x and y are integers, write down the coordinates of the points in this shaded region.

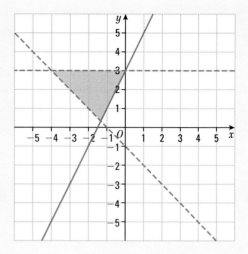

5 The diagram shows a shaded region bounded by four lines.

  a Write down the equation of each of these four lines.

  b Write down the four inequalities satisfied by the coordinates of the points in this shaded region.

  c If x and y are integers, write down the least value of y in this shaded region.

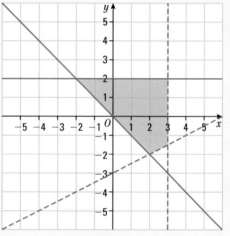

19.5 Using formulae

Objective

● You can use formulae from mathematics and other subjects.

Why do this?

Mobile phone bills are calculated using a formula which is based on the number and length of calls made and number of texts sent.

Get Ready

1. Work out

a 64.9×82.4 **b** $\sqrt{7\frac{1}{9}}$

Key Points

● A formula is a way of describing a relationship between two or more sets of values.
  Area = length × width is an example of a **word formula**.
● A formula is written using words or algebraic expressions.
  Einstein's theory of relativity is described by the **algebraic formula** $\boxed{E = mc^2}$

word formula   algebraic formula

Example 9

The area of a rectangle is given by the formula.

Area of a rectangle = length × width.

   a Find the area of a rectangle of length 8 cm and width 5 cm.

   b Find the length of a rectangle with an area of 27 mm² and width of 10 mm.

a Length = 8 cm, width = 5 cm   ←   *Substitute the values for the length and the width.*

   Area = length × width = 8 cm × 5 cm

   Area = 40 cm²

b Area = length × width   ←   *Divide both sides of the equation by 10.*

   27 = length × 10

   Length = 27 ÷ 10

           = 2.7 mm

ResultsPlus

Exam Tip

The first step to finding a value from a formula is to simply replace each word by its value.

Example 10

Use the formula $E = mc^2$ to work out the value of E when $m = 4$ and $c = 3$.

$E = mc^2$ ←   *Substitute $m = 4$ and $c = 3$ into the formula.*

$E = 4 \times 3^2$

$ = 4 \times 9$

$E = 36$

Exercise 19E

1 $v = u + at$

   Work out the value of v when

   a $u = 80$, $a = 10$ and $t = 4$ **b** $u = 35$, $a = -5$ and $t = 12$

2 $T = 3p^2 - 2p$

   Work out the value of T when   **a** $p = 5$ **b** $p = -1$

3 $V = \frac{1}{3}\pi r^2 h$

   Work out the value of V when

   a $\pi = 3.14$, $r = 10$ and $h = 15$ **b** $\pi = 3.14$, $r = 2.4$ and $h = 20$

4   Use the formula **distance = speed × time** to work out:

   a   the distance travelled by a car travelling for $2\frac{1}{2}$ hours at an average speed of 48 mph

   b   the average speed of an athlete running 100 metres in 12.5 seconds.

5   The cooking time, in minutes, of a turkey is given by the following formula:

   cooking time = weight of turkey in lb × 25 + 30.

   Use this formula to work out:

   a   the cooking time for a turkey weighing 11 lb, giving your answer in hours and minutes

   b   the weight, in lb, of a turkey taking 8 hours to cook.

6 $a = \sqrt{c^2 - b^2}$

   Work out the value of a when

   a $c = 13$ and $b = 5$ **b** $c = 41$ and $b = 40$

D

C

19.6 Deriving an algebraic formula

◎ Objective

○ You can derive an algebraic formulae from information given.

⊘ Why do this?

Chemists need to derive algebraic formulae when producing the correct balance of ingredients in medicines.

⬆ Get Ready

1. $V = I^2R$
   Work out V when
   a $I = 4, R = 200$ **b** $I = 0.3, R = 100$

◉ Key Points

◉ You can use information given to form an algebraic expression. You give each variable a letter.

◉ You must define each variable.

Example 11

A farmer sells sheep and cows at the local market. Each sheep is sold for £s and each cow is sold for £c.

     a Write down a formula for his total sales £T, in terms of s and c, if he sells 120 sheep and 45 cows.

     b Find the value of T if $s = 150$ and $c = 480$.

a Income from sale of 120 sheep is $120 \times s = 120s$.
Income from sale of 45 cows is $45 \times c = 45c$.
Total sales is $120s + 45c$.   ←   The total is required, so add the expressions.
The formula is therefore $T = 120s + 45c$.

b $s = 150$ and $c = 480$ ←   Substitute given values into the formula.
Substituting into $T = 120s + 45c$
$$T = 120 \times 150 + 45 \times 480$$
$$T = 18\,000 + 21\,600$$
$$T = 39\,600$$

ResultsPlus
Watch Out!

Do not try to combine the terms if the variables are different.

⚙ Exercise 19F

C

1 Duncan hires a car whilst on holiday in Spain. The cost of hiring a car is €90 plus €50 for each day that the car is hired for.
     a Write down a formula that could be used to find the total cost, €C, to hire a car for d days.
     b Use your formula to work out the cost of hiring a car for 14 days.

2 In some games, 5 points are awarded for a win, 3 points are awarded for a draw and 1 point is awarded for a loss.
In one evening, Caroline wins x games, draws y games and loses z games.
Write down a formula, in terms of x, y and z, for the total points (P) scored by Caroline.

3 David owns a hairdressing salon. The average length of time spent on a male client is 45 minutes. The average length of time spent on a female client is 75 minutes.

In one week David had m male clients and f female clients.

Write down a formula, in terms of m and f, for the total time, T hours, that David spent on his clients during this week.

A02 **C**

4 The diagram shows the plan of an L-shaped room. The dimensions of the room are given in metres. Write down a formula, in terms of x and y, for the perimeter, P metres, of the room.

A02

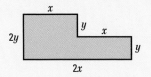

5 Adult cinema tickets cost £a and child cinema tickets cost £c.

Mr Brown buys 2 adult tickets and 5 child tickets.

a Write down a formula, in terms of a and c, for the total cost, £T, of these tickets.

The following week the cinema has a special offer.

'For each adult ticket bought, one child ticket is free.'

Mr Brown again buys 2 adult tickets and takes the same 5 children.

b Write down a formula, in terms of a and c, for the new total cost, £P, of these tickets.

A02

19.7 Changing the subject of a formula

◎ Objective

● You can change the subject of a simple formula.

⊘ Why do this?

You can change the subject of the formula for density, which is useful if you knew the density and volume of something like a ball-bearing in a pinball machine, but still needed to know the mass.

◈ Get Ready

1. Use the formula distance = speed × time to find how far a car travelling at an average speed of 40 mph travels in 1.5 hours.

2. Use the formula $E = mc^2$ to find E when $m = 2$ and $c = 3 \times 10^8$.

◉ Key Point

● You can use the techniques you learnt in Chapter 13 to change the **subject** of a formula by isolating the terms involving the new subject.

Example 12 — Make t the subject of the following formula

$$v = u + at. \leftarrow \boxed{\text{The aim is to isolate the term in } t.}$$

$v = u + at$
$v - u = u + at - u$ ← $\boxed{\begin{array}{l}\text{Just like you did when solving linear equations}\\ \text{in Chapter 13, subtract } u \text{ from both sides.}\end{array}}$

$v - u = at$ ← $\boxed{\text{Divide both sides by } a.}$

$\dfrac{v - u}{a} = \dfrac{at}{a}$ ← $\boxed{t \text{ is now the subject of the formula.}}$

$t = \dfrac{v - u}{a}$

Example 13 — Make p the subject of the following formula

$$T = \dfrac{2(p - 1)}{q}$$

Exam Tip

Only apply one operation in each stage of your working.

$T = \dfrac{2(p - 1)}{q}$

$T = \dfrac{2p - 2}{q}$ ← $\boxed{\text{Following the rules of BIDMAS (see section 1.3), first expand the brackets.}}$

$T \times q = \dfrac{2p - 2}{q} \times q$ ← $\boxed{\text{Multiply both sides by } q \text{ to remove the fraction.}}$

$Tq = 2P - 2$

$Tq + 2 = 2p - 2 + 2$ ← $\boxed{\text{Add 2 to both sides.}}$

$Tq + 2 = 2p$

$p = \dfrac{Tq + 2}{2}$ ← $\boxed{\text{Dividing both sides by 2 gives } p \text{ as the subject of the formula.}}$

Exercise 19G

In each of the following formulae, change the subject to the letter given in brackets.

1. $y = 5x + 3$ (x)
2. $c = 5d - 2$ (d)
3. $v = u + at$ (a)

4. $P = 5xy + y$ (x)
5. $E = 4 - 3m$ (m)
6. $f = 3(g - 10)$ (g)

7. $T = \dfrac{x + 2}{7}$ (x)
8. $Y = \dfrac{3(n - m)}{2}$ (n)

9. $W = \dfrac{2y}{3}(1 + p)$ (p)
10. $A = 1 + \dfrac{1 - w}{3}$ (w)

19.8 Changing the subject in complex formulae

◎ Objective

○ You can change the subject of a formula where the subject appears twice, or where a power of the subject appears.

⬙ Why do this?

Physicists and engineers rearrange many complex formulae in order to find important measures.

⬥ Get Ready

1. Make u the subject of the formula $v = u + at$.
2. Make R the subject of the formula $V = IR$.
3. Make m the subject of the formula $E = mc^2$.

Example 14   Make a the subject of the following formula.

$$T = 2\pi\sqrt{a^2 - 4}.$$

$T = 2\pi\sqrt{a^2 - 4}$ ←   Divide both sides by 2π.

$\dfrac{T}{2\pi} = \dfrac{2\pi\sqrt{a^2 - 4}}{2\pi}$

$\dfrac{T}{2\pi} = \sqrt{a^2 - 4}$

$\dfrac{T^2}{2^2 \times \pi^2} = a^2 - 4$ ←   Square both sides using the result that $\sqrt{x} \times \sqrt{x} = x$.

$\dfrac{T^2}{4\pi^2} + 4 = a^2$ ←   Add 4 to both sides.

$a = \sqrt{\left(\dfrac{T^2}{4\pi^2} + 4\right)}$ ←   Finally take the square root to give a as the subject of the formula.

Example 15   Make x the subject of the formula.
$$P = qx + 2x + 2a$$

Notice that x appears twice in this formula.

ResultsPlus
Exam Tip

Collect all terms in the 'new' subject together and then factorise.

$P = qx + 2x + 2a$

$P - 2a = qx + 2x$ ←   Subtract $2a$ from both sides.

$P - 2a = x(q + 2)$ ←   Factorising the right-hand side leaves x appearing once only.

$x = \dfrac{P - 2a}{q + 2}$ ←   Divide both sides by $(q + 2)$.

Exercise 19H

In each of the following formulae, change the subject to the letter given in brackets.

1	$A = \pi R^2$	(R)		**2**	$P = \sqrt{x - y}$	(x)

3 $v^2 = u^2 + 2as$ (u) **4** $T = \sqrt{\dfrac{2s}{g}}$ (g)

5 $y = 5\sqrt{3 - 2x^2}$ (x) **6** $f = 3g - 4 + g$ (g)

7 $T = \dfrac{1 + 3m}{m}$ (m) **8** Rearrange $4(p + 3) = q(1 - p)$ to make p the subject.

9 Make c the subject of $a - bc = 3 + 7c$. **10** Make T the subject of the formula $W = \sqrt{\dfrac{3T + 7}{2T}}$

Chapter review

- ◉ $<$ means 'less than'.
- ◉ $\leq$ means 'less than or equal to'.
- ◉ $>$ means 'greater than'.
- ◉ $\geq$ means 'greater than or equal to'.
- ◉ **Inequalities** have more than one value in their solution set. The solution set can be shown on **a number line**.
- ◉ When showing inequalities on a number line, an open circle shows that the number is not included and a closed circle shows that the number is included.
- ◉ You solve a linear inequality using a similar method to the one you use for solving a linear equation.
- ◉ If you multiply both sides of an inequality by a negative number, then you must reverse the inequality sign.
- ◉ When a value has a **lower limit** and an **upper limit**, you need to split the two inequalities and solve them separately.
- ◉ You can show the points that satisfy an inequality on a graph.
- ◉ Lines which are boundaries for regions that do include values on the line are shown as solid lines. Lines which are boundaries for regions that do not include values on the line are shown as dotted lines.
- ◉ A formula is a way of describing a relationship between two or more sets of values.
- ◉ A **word formula** is written in words.
- ◉ An **algebraic formula** is written using algebraic expressions.
- ◉ You can use information given to form an algebraic expression. You give each variable a letter.
- ◉ You must define each variable.
- ◉ You can change the **subject** of a formula by isolating the terms involving the new subject.

Review exercise

1 $v^2 = u^2 + 2as$ $u = 6, a = 3, s = 7.5$
Work out the value of v.

Nov 2008

2 Adam, Barry and Charlie each buy some stamps.

Barry buys three times as many stamps as Adam.

a Write down an expression for the number of stamps Barry buys.

Charlie buys 5 more stamps than Adam.

b Write down an expression for the number of stamps Charlie buys.

3 The cost of hiring a car can be worked out using this rule.

$$\text{Cost} = \text{£}90 + 50\text{p per mile}$$

Bill hires a car and drives 80 miles.

a Work out the cost.

The cost of hiring a car and driving m miles is C pounds.

b Write the formula for C in terms of m.

Nov 2007

4 $y = \dfrac{a^2 - c^2}{a^2 + c^2}$ $a = 3.2,$ $c = 1.6$

Work out the value of y.

5 $S = 4p + 3q$

a $p = 5, q = -4$

Work out the value of S.

b Make P the subject of the formula $S = 4p + 3q$.

6 Write down the inequalities represented on the number lines.

a 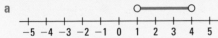   **b**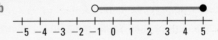

c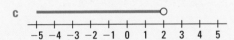

7 Draw a separate number line from -5 to 5 for each part. Show the inequality given in each case.

a $-3 < x < 2$ **b** $x \geqslant 1$ **c** $x < -2$ **d** $-1 \leqslant x \leqslant 3$

8 $-3 \leqslant n < 2$

n is an integer.

Write down all the possible values of n.

Nov 2006

9 **a** Solve the inequality

$3t + 1 < t + 12$

b t is a whole number.

Write down the largest value of t that satisfies

$3t + 1 < t + 12$

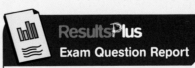

ResultsPlus
Exam Question Report

43% of students answered this question poorly. Remember to keep the inequality sign in throughout your working.

May 2009

C

10 $-6 \leqslant 2y < 5$ y is an integer. Write down all the possible values of y.

Nov 2005

B

11 Solve the inequality $4p - 8 < 7 - p$.

June 2006

A02

12

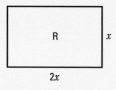

The perimeter of the rectangle R is less than the perimeter of the square S.

Write down the range of values of x.

A02

13

Diagram NOT
accurately drawn

A	B	C

←x→

Here are 3 rods.

The length of rod A is x cm.

Rod B is 4 cm longer than rod A.

The length of rod C is twice the length of rod B.

The total length of all 3 rods is L cm

a   Show that $L = 4x + 12$.

The total length of all 3 rods must be less than 50 cm.

b   Write down the inequality that must be satisfied.

c   Work out the range of possible values of x.

A02

14 The region **R** satisfies the inequalities

$x \geqslant 2, y \geqslant -1, x + y \leqslant 6$

Draw a suitable graph and use shading to show the region **R**.

A

15 $\dfrac{1}{u} + \dfrac{1}{v} = \dfrac{1}{f}$ $u = 2\tfrac{1}{2}, v = 3\tfrac{1}{3}$

a   Find the value of f.

b   Rearrange $\dfrac{1}{u} + \dfrac{1}{v} = \dfrac{1}{f}$ to make u the subject of the formula.

Give your answer in its simplest form.

ResultsPlus
Exam Question Report

93% of students answered this question poorly because they were not accurate in their calculations.

May 2009

16 Make x the subject of $5(x - 3) = y(4 - 3x)$.

Nov 2005

A

17 $P = \pi r + 2r + 2a$ $P = 84, r = 6.7$

   a Work out the value of a. Give your answer correct to 3 significant figures.

   b Make r the subject of the formula $P = \pi r + 2r + 2a$.

June 2005

18 **a** $4x + 3y < 12$

     x and y are both integers.

     Write down two possible pairs of values
     that satisfy this inequality.

         $4x + 3y < 12, \ y < 3x, \ y > 0, \ x > 0$

   b On the grid mark with a cross ($\times$) each of
     the three points which satisfy all these four inequalities.

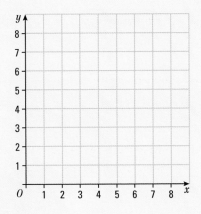

* 19 The cost of sweets is £2 per kg. The cost of chocolate is £5 per kg.

   Jim buys x kg of sweets and y kg of chocolate.

   He buys at least 2 kg of sweets.

   He buys at least 3 kg of chocolate.

   He spends at most £20.

   a Write down 3 inequalities in x and/or y.

   b Draw a suitable graph and show, by shading, the region that satisfies all 3 inequalities.

A03

20 PYTHAGORAS' THEOREM AND TRIGONOMETRY 1

Pythagoras lived in Greece in about 500 BC, and though one of the most famous results in mathematics was named after him, it is thought possible that the Egyptians used the result of the theorem to build their pyramids.

◉ Objectives

In this chapter you will:

- use Pythagoras' Theorem to find the length of sides in right-angled triangles
- find the length of a line segment
- use trigonometry to find the sizes of angles and the lengths of sides in right-angled triangles.

◈ Before you start

You need to:

- know how to find the square and square root of a number
- be able to recognise a right-angled triangle.

20.1 Pythagoras' Theorem

Objectives

- You can use Pythagoras' Theorem to find the length of the hypotenuse in a right-angled triangle.
- You can use Pythagoras' Theorem to find the length of a shorter side in a right-angled triangle.

Why do this?

The size of a TV is given as the length of the diagonal across the screen. You can use Pythagoras' Theorem to work out the length and width of the screen.

Get Ready

1. Find the value of **a** 12^2 **b** 7.4^2
2. Work out **a** $9^2 + 40^2$ **b** $13.8^2 + 9.3^2$
3. Work out **a** $17^2 - 8^2$ **b** $16.4^2 - 12.9^2$
4. Work out **a** $\sqrt{24^2 + 7^2}$ **b** $\sqrt{15^2 - 9^2}$
5. Work out **a** $\sqrt{3.8^2 + 5.2^2}$ **b** $\sqrt{13.1^2 - 9.6^2}$

Give your answers correct to 3 significant figures.

Key Points

- The right-angled triangle in this diagram has sides of length 3 cm, 4 cm and 5 cm. Squares have been drawn on each side of the triangle and each square has been divided up into squares of side 1 cm.

  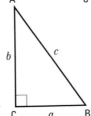

  The area of the square on the side of length 3 cm is 9 cm².
  The area of the square on the side of length 4 cm is 16 cm².
  The area of the square on the side of length 5 cm (the **hypotenuse**) is 25 cm².
  Notice that $25 = 9 + 16$, that is, $5^2 = 3^2 + 4^2$.
  In other words, 5^2 (the area of the square on the hypotenuse) is equal to the sum of 3^2 and 4^2 (the areas of the squares on the other two sides added together).
  This is an example of **Pythagoras' Theorem**. It is only true for right-angled triangles.

- Here is a right-angled triangle, ABC.
  The angle at C is the right angle.
  The side, AB, opposite the right angle is called the hypotenuse.
  It is the longest side in the triangle.

- Pythagoras' Theorem states that:
  In a right-angled triangle, the square of the hypotenuse is equal to the sum of the squares of the other two sides.

$$c^2 = a^2 + b^2$$

or

$$AB^2 = BC^2 + CA^2$$

Where AB^2 means the length of the side AB squared.

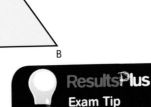

Results Plus

Exam Tip

You will need to learn Pythagoras' Theorem for your exam.

- You can also use Pythagoras' Theorem to work out the length of one of the shorter sides in a right-angled triangle when you know the lengths of the other two sides.

Example 1 Work out the length of the hypotenuse in this triangle.

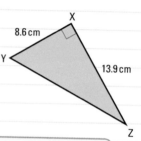

15 cm c cm

8 cm

$c^2 = a^2 + b^2$ ← Write down Pythagoras' Theorem.

$c^2 = 8^2 + 15^2$ ← Put in the known values.

$c^2 = 64 + 225 = 289$ ← Work out the value of c^2.

$c = \sqrt{289} = 17$ ← Square root to find the value of c.
Length of hypotenuse = 17 cm.

Example 2 In triangle XYZ, angle X = 90°, XY = 8.6 cm and
XZ = 13.9 cm. Work out the length of YZ.
Give your answer correct to 3 significant figures.

X

8.6 cm

Y

13.9 cm

Z

$YZ^2 = XY^2 + XZ^2$ ← YZ is the hypotenuse as it is opposite the right angle.

$YZ^2 = 8.6^2 + 13.9^2$ ← Put in the values.

$YZ^2 = 73.96 + 193.21$

$YZ^2 = 267.17$ ← Work out the value of YZ^2.

$YZ = \sqrt{267.17} = 16.34...$ ← Square root.

$YZ = 16.3$ cm (to 3 s.f.)

ResultsPlus
Exam Tip

When you are showing your calculations on your exam paper, write down at least four figures of the calculator display.

Exercise 20A

Questions in this chapter are targeted at the grades indicated.

C

1   Work out the length of each hypotenuse marked with letters in these triangles.
    Where appropriate, give each answer correct to 3 significant figures.

a

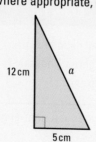

12 cm a

5 cm

b

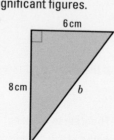

6 cm

8 cm b

c

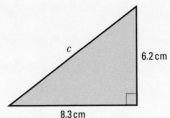

6.2 cm

c

8.3 cm

d

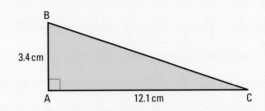

d

10.6 cm

4.8 cm

2　**a** In triangle ABC, angle A = 90°, AB = 3.4 cm and
AC = 12.1 cm. Work out the length of BC.
Give your answer correct to 3 significant figures.

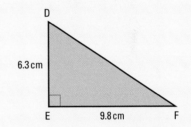

B

3.4 cm

A　　　12.1 cm　　　C

b In triangle DEF, angle E = 90°, DE = 6.3 cm and
EF = 9.8 cm. Work out the length of DF.
Give your answer correct to 3 significant figures.

D

6.3 cm

E　　9.8 cm　　F

c In triangle PQR, angle R = 90°, PR = 5.9 cm and QR = 13.1 cm.
Work out the length of PQ. Give your answer correct to 3 significant figures.

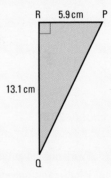

R　5.9 cm　P

13.1 cm

Q

d

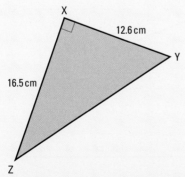

X

12.6 cm

Y

16.5 cm

Z

In triangle XYZ, angle X = 90°, XY = 12.6 cm and XZ = 16.5 cm.
Work out the length of YZ. Give your answer correct to 3 significant figures.

Example 3

In triangle ABC, angle A = 90°, BC = 17.4 cm and AC = 5.8 cm. Work out the length of AB. Give your answer correct to 3 significant figures.

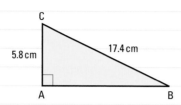

$BC^2 = AC^2 + AB^2$ ← Angle A is the right angle so the hypotenuse is BC.

$17.4^2 = 5.8^2 + AB^2$ ← Write down Pythagoras' Theorem and put in the values.

$302.76 = 33.64 + AB^2$

$302.76 - 33.64 = AB^2$
$269.12 = AB^2$

← Work out the value of AB^2 by subtracting 33.64 from both sides.

ResultsPlus
Exam Tip

Check that the hypotenuse is the longest side of the triangle.

$AB = \sqrt{269.12} = 16.4...$

$AB = 16.4$ cm (to 3 s.f.) ← Give the length of AB correct to 3 s.f.

Exercise 20B

C

1 Work out the lengths of the sides marked with letters in these triangles.
Where appropriate give each answer correct to 3 significant figures.

a
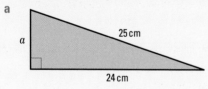

b
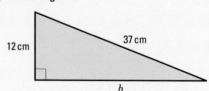

c
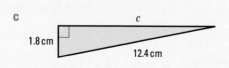

d
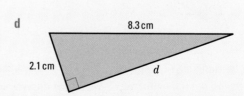

2 **a** In triangle ABC, angle A = 90°, AB = 5.9 cm and BC = 16.3 cm. Work out the length of AC.
Give your answer correct to 3 significant figures.

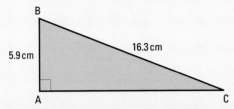

b In triangle PQR, angle R = 90°, PQ = 11.2 cm and QR = 9.6 cm.
Work out the length of RP. Give your answer correct to 3 significant figures.

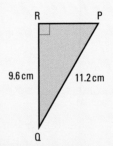

c In triangle DEF, angle E = 90°, DF = 10.1 cm and EF = 7.8 cm.
i Draw a sketch of the right-angled triangle DEF and label sides DF and EF with their lengths.
ii Work out the length of DE. Give your answer correct to 3 significant figures.

20.2 Applying Pythagoras' Theorem

◉ Objective

◉ You can solve problems using Pythagoras' Theorem.

? Why do this?

If you travel 30 kilometres north and then 20 kilometres east, you can use Pythagoras' Theorem to work out the direct distance 'as the crow flies'.

◈ Get Ready

1. Draw a sketch of an A4 page with sides of lengths 21.2 cm and 29.7 cm. Round the lengths to the nearest centimetre and work out an estimate of the length of the diagonal of the page.

Key Points

◉ Pythagoras' Theorem can be used to solve problems.
◉ Isosceles triangles can be split into two right-angled triangles. You can then use Pythagoras' Theorem.

Example 4 A boat travels due North for 5.7 km. The boat then turns and travels due East for 7.2 km.
Work out the distance between the boat's finishing point and its starting point.
Give your answer in km correct to 3 significant figures.

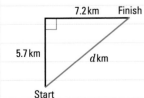

$d^2 = 5.7^2 + 7.2^2$
$d^2 = 32.49 + 51.84$
$d^2 = 84.33$
$d = \sqrt{84.33} = 9.183...$
Distance = 9.18 km (to 3 s.f.)

> Draw a sketch of the boat's journey. A reminder about bearings can be found in Section 5.5.

> The triangle is right-angled and the distance between the start and the finish is the hypotenuse, so you can use Pythagoras' Theorem.

Example 5

The diagram shows an isosceles triangle ABC.
The midpoint of BC is the point M. In the triangle,
AB = AC = 8 cm and BC = 6 cm.
Work out the height, AM, of the triangle.
Give your answer correct to 3 significant figures

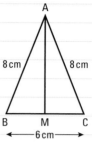

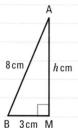

By Pythagoras
$AB^2 = AM^2 + BM^2$
$8^2 = h^2 + 3^2$
$64 = h^2 + 9$
$64 - 9 = h^2$
$55 = h^2$
$h = \sqrt{55} = 7.416...$
$h = 7.42$
Height of triangle = 7.42 cm (to 3 s.f.)

Pythagoras' Theorem cannot be used in triangle ABC as it is not a right-angled triangle.
As M is the midpoint of the base of the isosceles triangle, the line AM is the line of symmetry of triangle ABC.
So AM is perpendicular to the base and angle AMB = 90°.
BM = 3 cm as M is the midpoint of BC.

Draw a sketch of triangle ABM.

Triangle ABM is right-angled with hypotenuse AB. The height, AM, of the triangle is marked h cm on the sketch.

Exercise 20C

1 Find the lengths of the sides marked with letters in each of these triangles.
Give each answer correct to 3 significant figures.

a

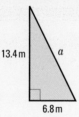

13.4 m
a
6.8 m

b

9.7 cm
b
6.8 cm

c

11.4 cm
c
14.8 cm

2 The diagram shows a ladder leaning against a vertical wall.
The foot of the ladder is on horizontal ground, 3.6 m from the wall. The length of the ladder is 5 m.
Work out how far up the wall the ladder reaches.
Give your answer correct to 3 significant figures.

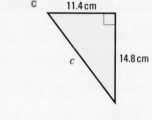

5 m
3.6 m

3

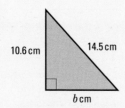

10.6 cm
14.5 cm
b cm

Work out the area of the triangle.
Give your answer correct to 3 significant figures.

4

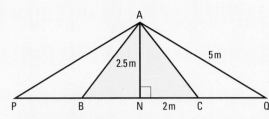

The diagram represents the end view of a tent, triangle ABC; two guy-ropes, AP and AQ; and a vertical tent pole, AN. The tent is on horizontal ground so that PBNCQ is a straight horizontal line. Triangles ABC and APQ are both isosceles triangles. BN = NC = 2 m, AN = 2.5 m and AP = AQ = 5 m

A02
A03
B

a Work out the length of the side AC of the tent. Give your answer correct to 3 significant figures.

b Work the length of **i** NQ, **ii** CQ.   Give your answers correct to 3 significant figures.

There is a tent peg at P and a tent peg at Q.

c Work out the distance between the two tent pegs at P and Q.

Give your answer correct to 3 significant figures.

5 Here are the lengths of sides of six triangles.

A03

Triangle 1   5 cm, 12 cm and 13 cm   Triangle 2   9 cm, 40 cm and 41 cm
Triangle 3   10 cm, 17 cm and 18 cm   Triangle 4   20 cm, 21 cm and 29 cm
Triangle 5   8 cm, 17 cm and 20 cm   Triangle 6   33 cm, 56 cm and 65 cm

Which of these triangles are right-angled triangles?

6 The diagram shows two right-angled triangles.
Work out the length of the side marked a.
Give your answer correct to 3 significant figures.

A02
A03

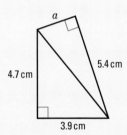

7 The size of a computer monitor or TV screen is determined by the length of the diagonal across the screen.

A02
A03

a A computer monitor has a screen that is a rectangle with dimensions 13.2 inches by 10.8 inches. Work out the size of this monitor by working out, correct to the nearest inch, the length of the diagonal of the rectangle.

b A widescreen TV has a 37 inch screen which is a rectangle with sides of lengths in the ratio 16 : 9. Work out the height and width of the TV screen.

20.3 Finding the length of a line segment

◎ Objective

○ You can calculate the length of a line segment.

◈ Get Ready

1. Calculate   **a** $\sqrt{10^2 + 24^2}$ **b** $\sqrt{18^2 + 7^2}$ **c** $\sqrt{4.8^2 + 9.1^2}$

Key Points

- By creating a right-angled triangle, Pythagoras' Theorem can be used to find the length between two points on a line.

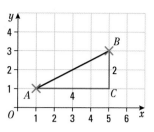

- The length of the line segment AB between $A\,(x_1, y_1)$ and $B\,(x_2, y_2)$ is $\sqrt{(x_2 - x_1)^2 + (y_2 - y_1)^2}$

Example 6   Find the length of the line joining   **a** A (3, 2) and B (15, 7)   **b** P (−9, 4) and Q (7, −5)

a

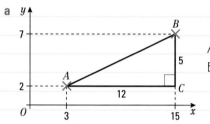

$AC = 15 - 3 = 12$
$BC = 7 - 2 = 5$

> Draw a sketch showing A and B and complete the right-angled triangle ABC.

$AB^2 = 12^2 + 5^2$
$AB^2 = 144 + 25 = 169$
$AB = \sqrt{169} = 13$

> Use Pythagoras' Theorem to find the length of AB.

b   P (−9, 4)

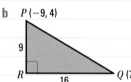

$QR = 7 - -9 = 7 + 9 = 16$
$PR = 4 - -5 = 4 + 5 = 9$

> Draw a sketch showing P and Q and complete the right-angled triangle PQR.

$PQ^2 = 16^2 + 9^2$
$PQ^2 = 256 + 81 = 337$
$PQ = \sqrt{337} = 18.4 \text{ (to 3 s.f.)}$

> Use Pythagoras' Theorem to find the length of PQ.

Exercise 20D

1   Work out the length of the line joining each of these pairs of points.

   a   (3, 1) and (11, 7)   **b**   (2, 5) and (12, 29)

   c   (−6, 9) and (8, 13)   **d**   (−4, −6) and (6, 12)

   e   (9, −15) and (−11, 6)   **f**   (0, −5) and (9, −11)

2   The point A has coordinates (5, 2), the point B has coordinates (8, 6) and the point C has coordinates (1, 5).

   a   Work out the length of   **i** AB   **ii** BC   **iii** AC

   b   What does your answer to part **a** tell you about triangle ABC?

3   A circle has centre point O (4, 2). The point A (9, 14) lies on the circle.
  a   Work out the radius of the circle.
  b   Determine by calculation which of the following points also lie on the circle.
     i   B (16, 7)   ii   C (−1, −10)   iii   D (7, 16)   iv   E (4, 15)

20.4 Trigonometry in right-angled triangles

◉ Objective

○ You can use trigonometry to find the sizes of angles in right-angled triangles.

❓ Why do this?

Trigonometry is used in science and engineering. It is used to plan the movements of the robotic arm on the International Space Station.

◈ Get Ready

Look at these right-angled triangles. For each one, name the side that is
a the hypotenuse b the side opposite c the side adjacent to the angle marked $x°$.

1. 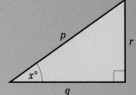   **2.** 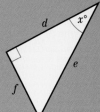   **3.**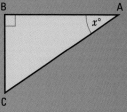

🔑 Key Points

○ The hypotenuse (hyp) of a right-angled triangle is the longest side of the triangle and is opposite the right angle. The other two sides of the triangle are named adjacent and opposite. The side opposite an angle is called the opposite side (opp).
The side next to this angle is called the adjacent side (adj).

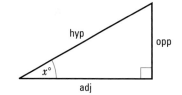

○ Here is a right-angled triangle with its hypotenuse of length 1.
The length of the opposite side (opp) in this triangle is known accurately and is called the **sine** of 70° and is written sin 70°.

Its value can be found on any scientific calculator. Not all calculators are the same but the key sequence to find sin 70° applies to many calculators.

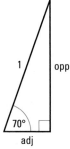

○ Make sure that the angle mode of your calculator is degrees, usually shown by 'D' on the calculator screen.

Press ⌈sin⌉   Key in ⌈7⌉⌈0⌉   Press ⌈=⌉

The number ⌈ 0.93969262 ⌋ should appear on your calculator screen.

So, correct to four decimal places, sin 70° = 0.9397

- The length of the adjacent side (adj) is called the **cosine** of 70° and is written cos 70°. Using a similar sequence to the one above, but using the cos key, correct to four decimal places, cos 70° = 0.3420

- The terms sine and cosine are called trigonometric ratios, or trig ratios.

- There is another trig ratio called the **tangent** of 70° and written tan 70°. As above, but using the tan key, correct to four decimal places, tan 70° = 2.7475

- You can find the sine, cosine and tangent of any angle.

- Here are three right-angled triangles.

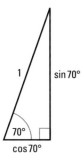 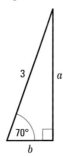 

The second triangle is an enlargement of the first triangle with a scale factor of 3.

This means that $a = 3 \times \sin 70° $ or $3\sin 70°$ and $b = 3 \times \cos 70°$ or $3\cos 70°$

The third triangle is an enlargement of the first triangle with a scale factor of hyp.

This means that opp = hyp $\times \sin 70°$ and adj = hyp $\times \cos 70°$

These results can also be written as $\sin 70° = \dfrac{opp}{hyp}$ and $\cos 70° = \dfrac{adj}{hyp}$.

- Results like these are true for all right-angled triangles so that

$$\sin x° = \frac{\textbf{opp}}{\textbf{hyp}} \qquad \cos x° = \frac{\textbf{adj}}{\textbf{hyp}}$$

When the opposite side and the adjacent side are involved

$$\tan x° = \frac{\textbf{opp}}{\textbf{adj}}$$

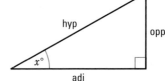

SOHCAHTOA might help you remember these results.

Sin Opp Hyp Cos Adj Hyp Tan Opp Adj

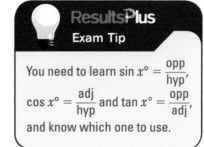

ResultsPlus
Exam Tip

You need to learn $\sin x° = \dfrac{opp}{hyp}$,

$\cos x° = \dfrac{adj}{hyp}$ and $\tan x° = \dfrac{opp}{adj}$,

and know which one to use.

Example 7   Use a calculator to write down, correct to 4 decimal places, the value of cos 74.6°.

Make sure that your calculator is in degree mode.

$cos\ 74.6° = 0.265\,556\,117$ ←   Use the COS key and key in 74.6.

$cos\ 74.6° = 0.2656$ correct to 4 decimal places

Example 8 Find the value of x when $\tan x° = 2.7$.
Give your answer correct to 1 decimal place.

When the value of sine, cosine or tangent is given, a calculator can be used to find the size of the angle. To do this, use the SHIFT or INV key on your calculator.

ResultsPlus
Watch Out!

Not all calculators are the same. It is important that you know how to work your calculator. Make sure that your calculator is in degree mode.

Press $\boxed{\text{SHIFT}}$ Press $\boxed{\tan^{-1}}$ ← The display will show $\tan^{-1}$

Key in $\boxed{2}\boxed{\cdot}\boxed{7}$ ← The display will show $\tan^{-1} 2.7$

Press $\boxed{=}$ ← The display will show **69.676 863 17**

$x = 69.7$ correct to 1 decimal place.

Exercise 20E

1 Use a calculator to find the value of
  a $\sin 20°$ b $\sin 72.6°$ c $\cos 60°$ d $\cos 18.9°$
  e $\tan 45°$ f $\tan 86.4°$ g $\cos 137.8°$ h $\tan 4°$
  i $\sin 127.2°$ j $\sin 14.7°$ k $\tan 159.5°$ l $\cos 87.3°$
  Give each answer correct to four decimal places, where necessary.

2 Use a calculator to find the value of x when
  a $\cos x° = 0.6$ b $\sin x° = 0.43$ c $\cos x° = 0.5$
  d $\tan x° = 0.96$ e $\sin x° = 0.8516$ f $\tan x° = 2.03$
  g $\sin x° = 0.047$ h $\tan x° = \sqrt{3}$ i $\cos x° = \dfrac{\sqrt{2}}{2}$
  Give each answer correct to 1 decimal place where necessary.

C

Example 9 Write down the trigonometric ratio needed to calculate:
  a the size of the angle marked $x°$ b the length of the side marked p.

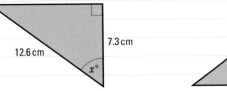

12.6 cm   7.3 cm $x°$

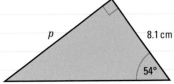

p 8.1 cm   54°

  a The given sides are the hypotenuse and the side adjacent to the angle $x°$ so cosine is needed.
  b The hypotenuse is not involved, p is opposite the given angle and 8.1 cm is adjacent so tangent is needed.

B

Exercise 20F

1   Write down which trigonometric ratio is needed to calculate either the length of the side marked p or the size of the angle marked x in each of these triangles. You do not have to calculate anything.

a
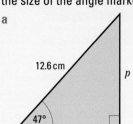

b
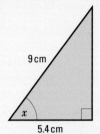

c
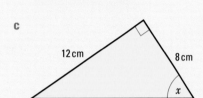

d
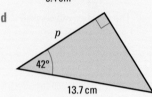

Example 10   Work out the size of each of the marked angles.
Give each answer correct to one decimal place.

a

b
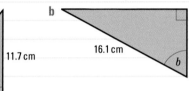

c

a $\sin a = \dfrac{11.7}{15.9} = 0.7358...$ ⟵

$a = 47.379...°$

$a = 47.4°$ correct to 1 d.p.

> 15.9 cm is the hypotenuse.
> 11.7 cm is opposite angle a.
> $\sin = \dfrac{\text{opp}}{\text{hyp}}$
> Use your calculator to find:
> $\sin^{-1} 0.7358...$ which is $47.379...°$.

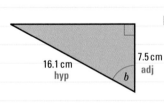

b $\cos b = \dfrac{7.5}{16.1} = 0.4658...$ ⟵

$b = 62.235...°$

$b = 62.2°$ correct to 1 d.p.

> 16.1 cm is the hypotenuse.
> 7.5 cm is adjacent to angle b.
> $\cos = \dfrac{\text{adj}}{\text{hyp}}$
> Use your calculator to find:
> $\cos^{-1} 0.4658...$ which is $62.235...°$.

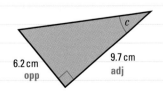

c $\tan c = \dfrac{6.2}{9.7} = 0.6391...$ ⟵

$c = 32.585...°$

$c = 32.6°$ correct to 1 d.p.

> 6.2 cm is opposite angle c.
> 9.7 cm is adjacent to angle c.
> $\tan = \dfrac{\text{opp}}{\text{adj}}$
> Use your calculator to find:
> $\tan^{-1} 0.6391...$ which is $32.585...°$.

B

Exercise 20G

1   Work out the size of each of the lettered angles. Give each answer correct to one decimal place.

a   b   c

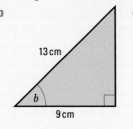

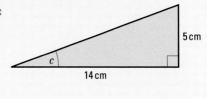

18 cm   11 cm   13 cm   5 cm

a   b   c

                                         9 cm   14 cm

d   e   f

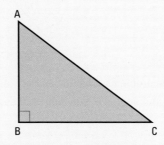

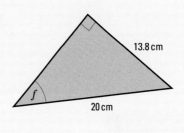

18.3 cm   17 cm   13.8 cm

d   15.8 cm   f

14 cm   e   20 cm

2   Triangle ABC is right-angled at B.

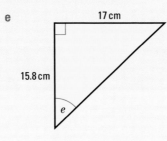

A

B   C

a   AB = 8.9 cm and BC = 12.1 cm. Calculate the size of angle ACB.
    Give your answer correct to 0.1°.

b   BC = 15.5 cm and AC = 24.7 cm. Calculate the size of angle BAC.
    Give your answer correct to 0.1°.

c   AB = 6.3 cm and AC = 11.8 cm. Calculate the size of angle ACB.
    Give your answer correct to 0.1°.

3   In triangle ACD, the point B lies on AD so that
    CB and AD are perpendicular.

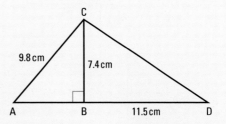

    C

    9.8 cm   7.4 cm

    A   B   11.5 cm   D

a   Using triangle ABC, calculate the size of angle ACB.
    Give your answer correct to one decimal place.

b   Using triangle BCD, calculate the size of angle BCD.
    Give your answer correct to one decimal place.

c   Hence calculate the size of angle ACD.
    Give your answer to the nearest degree.

20.5 Working out lengths of sides using trigonometry

◎ Objectives

- ◉ You can use trigonometry to find the lengths of sides in a right-angled triangle.
- ◉ You can solve problems using Pythagoras' Theorem and trigonometry in right-angled triangles.

❓ Why do this?

Astronomers use trigonometry to work out how far away stars are from Earth.

⬆ Get Ready

1. Draw sketch diagrams to show these situations. (You do not need to do any working out.)
   a A ship is 20 km from lighthouse X on a bearing of 055°.
   b A ladder 8 m in length leans against a wall, with the foot of the ladder 3 m from the wall.
2. For question **1b**, which trig ratio would you use to find the angle the ladder makes with the ground?

🕒 Key Points

◉ The results used in the last section can be written as

$opp = hyp \times \sin x°$
$adj = hyp \times \cos x°$
$opp = adj \times \tan x°$

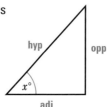

◉ Trigonometry can be used to solve problems. Sometimes Pythagoras' Theorem is needed as well. Some questions involve bearings and angles of elevation and depression (see Sections 5.4–5.5).

Example 11 ▶ Work out the length of each of the lettered sides.
Give each answer correct to 3 significant figures.

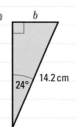

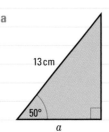

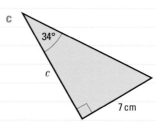

a

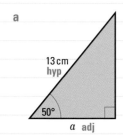

$a = 13 \times \cos 50° = 13 \cos 50°$
$a = 13 \times 0.6427\ldots$
$a = 8.356\ldots$
$a = 8.36 \, cm$ (to 3 s.f.)

> 13 cm is the hypotenuse.
> a is adjacent to the 50° angle.

> adj and hyp are involved so use cos.
> adj = hyp × cos

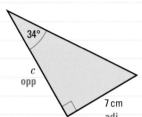

$b = 14.2 \times \sin 24° = 14.2 \sin 24°$
$b = 14.2 \times 0.4067\ldots$
$b = 5.7756\ldots$
$b = 5.78\,cm$ (to 3 s.f.)

> 14.2 cm is the hypotenuse.
> b is opposite to the 24° angle.

> opp and hyp are involved so use sin.
> opp = hyp × sin

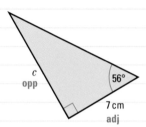

$c = 7 \times \tan 56° = 7 \tan 56°$
$c = 7 \times 1.4825\ldots$
$c = 10.3779\ldots$

$c = 10.4\,cm$ (to 3 s.f.)

> The hypotenuse is not given and opposite and adjacent are involved so use tan.
> opp = adj × tan

> When using tan to find a length, it is easiest to find the opposite side. Relative to the angle of 34°, 7 cm is the opposite side and c is the adjacent side.
> The third angle in the triangle is $(180 − 90 − 34)° = 56°$
> Relative to the angle of 56°, c is the opposite side and 7 cm is the adjacent side.

Example 12 The diagram shows a lighthouse 30 m in height, standing on horizontal ground.

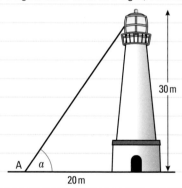

a Work out the **angle of elevation** of the top of the lighthouse from point A on the ground.

b Work out the angle of depression of point A from the top of the lighthouse.

a $\tan a = \dfrac{30}{20}$ ← $\tan a = \dfrac{opp}{adj}$

$\tan a = 1.5$
$a = 56.3°$ (to 3 s.f.)

b Angle of depression = 56.3° (to 3 s.f.) ← The angles of elevation and depression are the same; alternate angles.

Exercise 20H

B

1   Work out the length of each lettered side.
    Give each answer correct to three significant figures.

a

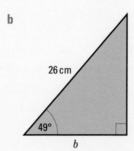

16 cm

a

20°

b

26 cm

49°

b

c

c

37°

15.4 cm

d

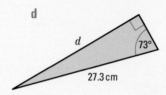

d

73°

27.3 cm

e

e

14.9 cm

55°

f

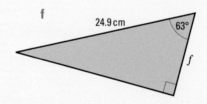

24.9 cm

63°

f

2   Triangle PQR is right-angled at Q.

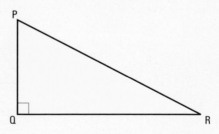

In each part, calculate the length of QR.
Give each answer correct to three significant figures.

a   PQ = 7.3 cm, angle QPR = 68°
b   PR = 17.2 m, angle QRP = 39°
c   PR = 12.6 cm, angle QPR = 59°

A03

3   In triangle ABD, the point C lies on AD so that BC and AD are perpendicular.

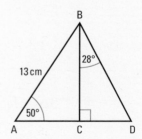

a   Using triangle ABC, work out the length of **i** BC **ii** AC.
    Give each answer correct to three significant figures.
b   Using triangle BCD, work out the length of CD,
    correct to three significant figures.
c   Hence calculate the length of AD, correct to three significant figures.
d   Calculate the area of triangle ABD.
    Give your answer correct to the nearest cm².

Example 13

Two towns, Aytown and Beeville, are 40 km apart.

The bearing of Beeville from Aytown is 067°.

   a Calculate how far east and how far north Beeville is from Aytown.

      Give your answers to 3 significant figures.

Ceeham is 60 km east of Beeville.

   b Calculate the distance between Aytown and Ceeham.

      Give your answer to the nearest km.

   c Calculate the bearing of Ceeham from Aytown.

      Give your answer to the nearest degree.

a

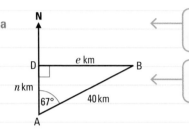

First draw a diagram showing the positions of Aytown (A) and Beeville (B). Draw a line from B 'west' to meet the 'north' line from A at D.

In the right-angled triangle ABD, the length of AD gives how far B is north of A (n km). The length of DB gives how far B is east of A (e km).

In triangle ABD, the 40 km is the hypotenuse and e km is opposite the 67° angle.

opp = hyp × sin

$e = 40\sin 67° = 36.82\ldots$

Distance east = 36.8 km (to 3 s.f.)

n km is adjacent to the 67° angle.

adj = hyp × cos

$n = 40\cos 67° = 15.629\ldots$

Distance north = 15.6 km (to 3 s.f.)

Having worked out the value of e, the value of n could be found using Pythagoras' Theorem.

b

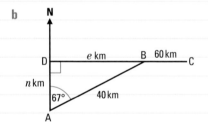

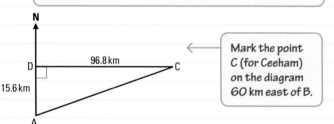

Mark the point C (for Ceeham) on the diagram 60 km east of B.

$AC^2 = AD^2 + DC^2$

$ = 15.6^2 + 96.8^2$

$AC^2 = 9613.6$

$AC = 98.04\ldots$ km

Ceeham is 15.6 km north of Aytown.
Ceeham is 60 + 36.8 = 96.8 km east of Aytown.
Draw triangle ADC.

The distance between Aytown and Ceeham is the length of AC. Find the length of AC using Pythagoras' Theorem.

Distance between Aytown and Ceeham is 98 km (to nearest km).

c $\tan DAC = \dfrac{96.8}{15.6} = 6.205\ldots$

$DAC = 80.8\ldots°$

$ = 81°$ (to nearest degree)

To find the bearing of C from A, calculate the size of angle DAC. Any of the trigonometric ratios can be used as all three sides of triangle ACD are known.

Using $\tan = \dfrac{opp}{adj}$

Bearing of Ceeham from Aytown is 081° (to nearest degree).

The bearing should be given as a three-figure bearing.

Exercise 20I

Where necessary give lengths correct to 3 significant figures and angles correct to 1 decimal place.

1 The diagram shows the plans for the sails of a boat.

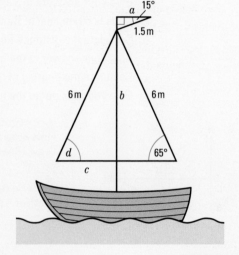

  a Work out the length of the side marked

   i a **ii** b **iii** c

  b Work out the size of the angle marked d.

2 The diagram shows a vertical building standing on horizontal ground.
  The points A, B and C are in a straight line on the ground.
  The point T is at the top of the building so that TC is vertical.
  The angle of elevation of T from A is 40°, as shown in the diagram.

  a Work out the height, TC, of the building.

  b Work out the size of the angle of elevation of T from B.

  c Work out the size of angle ATB.

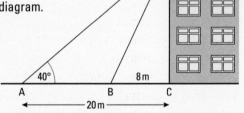

3 The points P and Q are marked on a horizontal field. The distance from P to Q is 100 m. The bearing of Q from P is 062°. Work out how far:

  a Q is north of P

  b Q is east of P.

4 The diagram shows a circle centre O.
  The line ABC is the tangent to the circle at B so that
  the size of angle OBA is 90°.

  a Work out the radius of the circle.

  b Work out the size of angle OCB.

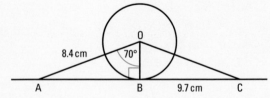

5 A, B and C are three buoys marking the course of a yacht race.

  a Calculate how far B is: **i** north of A **ii** east of A.

  b Calculate how far C is: **i** north of B **ii** east of B.

  c Hence calculate how far C is: **i** north of A **ii** east of A.

  d Calculate the distance and bearing of C from A.

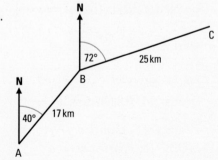

6 The diagram shows an isosceles trapezium.

a Work out the distance, h cm, between the two parallel sides of the trapezium.

The length of the shorter parallel side of the trapezium is 5.8 cm, as shown in the diagram.

b Work out the length of the longer parallel side of the trapezium.

c Calculate the area of the trapezium. Give your answer to the nearest cm².

A02
A03
A

Chapter review

◉ In a right-angled triangle, the side opposite the right angle is called the **hypotenuse**. It is the longest side in the triangle.

◉ **Pythagoras' Theorem** states that:

In a right-angled triangle, the square of the hypotenuse is equal to the sum of the squares of the other two sides.

That is $c^2 = a^2 + b^2$

or

$AB^2 = BC^2 + CA^2$

where AB^2 means the length of the side AB squared.

◉ You can also use Pythagoras' Theorem to work out the length of one of the shorter sides in a right-angled triangle when you know the lengths of the other two sides.

◉ Pythagoras' Theorem can be used to solve problems.

◉ Isosceles triangles can be split into two right-angled triangles. You can then use Pythagoras' Theorem.

◉ By creating a right-angled triangle, Pythagoras' Theorem can be used to find the length between two points on a line.

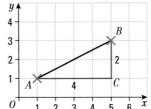

◉ The length of the line segment AB between $A\ (x_1, y_1)$ and $B\ (x_2, y_2)$ is $\sqrt{(x_2 - x_1)^2 + (y_2 - y_1)^2}$

◉ The three sides of a triangle are named hypotenuse, adjacent and opposite. The side opposite an angle is called the opposite side (opp). The side next to this angle is called the adjacent side (adj).

◉ The terms **sine** (sin), **cosine** (cos) and **tangent** (tan) are called trigonometric ratios, or trig ratios.

$$\sin x° = \frac{\text{opp}}{\text{hyp}} \quad \cos x° = \frac{\text{adj}}{\text{hyp}} \quad \tan x° = \frac{\text{opp}}{\text{adj}}$$

SOHCAHTOA might help you remember these results.

◉ These trig rations can also be written as

opp = hyp × sin $x°$

adj = hyp × cos $x°$

opp = adj × tan $x°$

✦ Review exercise

C

1 AC = 12 cm. Angle ABC = 90°. Angle ACB = 32°.
Calculate the length of AB.
Give your answer correct to 3 significant figures.

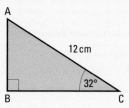

June 2007

2 ABC is a right-angled triangle.
AC = 6 cm.
BC = 9 cm.
Work out the length of AB.
Give your answer correct to 3 significant figures.

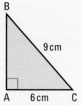

June 2009

B

3 The diagram shows three cities.
Norwich is 168 km due east of Leicester.
York is 157 km due north of Leicester.
Calculate the distance between Norwich and York.
Give your answer correct to the nearest kilometre.

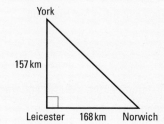

Nov 2006

4 In triangle ABC
Angle ABC = 90°
BC = 8 cm
AC = 21 cm
Work out the length of AB.
Give your answer correct to 3 significant figures.

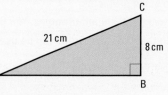

March 2007

5 PQR is a right-angled triangle. PR = 12 cm. QR = 4.5 cm. Angle PRQ = 90°.
Work out the value of x.
Give your answer correct to one decimal place.

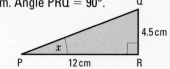

Nov 2007

6 Here is a right-angled triangle.

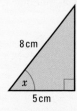

Here is another right-angled triangle.

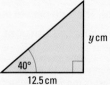

a Calculate the size of the angle marked x.
Give your answer correct to 1 decimal place.

b Calculate the value of y.
Give your answer correct to 1 decimal place.

June 2009

7 The diagram shows a vertical tower DC on horizontal ground ABC.
ABC is a straight line.
The angle of elevation of D from A is 28°.
The angle of elevation of D from B is 54°.
AB = 25 m
Calculate the height of the tower.
Give your answer to 3 significant figures.

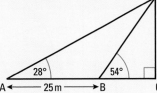

June 2006

8 Paul flies his helicopter from Ashwell to Birton.
He flies due west from Ashwell for 4.8 km. He then flies due south for 7.4 km to Birton.
Calculate the bearing from Birton to Ashwell.

9

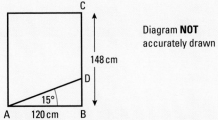

Diagram **NOT** accurately drawn

AB and BC are two sides of a rectangle.
AB = 120 cm and BC = 148 cm.
D is a point on BC.
Angle BAD = 15°.
Work out the length of CD.
Give your answer correct to the nearest centimetre.

10 The diagram shows a field 80 metres by 60 metres.
Alan runs around the field from point A to C (via point B) at 5 m/s.
Bhavana sets off at the same time, but runs directly across the diagonal of the field (shown by the dotted line) from A to C at 3 m/s.

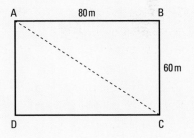

a Who will reach point C first?
b How long will it be before the second person arrives?

11 a Calculate the length of the side of the largest square that fits inside a 10 cm diameter circle.
b Work out the length of the side of the smallest square that surrounds a 10 cm diameter circle.

12 Hamish wants to refelt the roof of his lean-to shed.
Felt is sold in 5 m rolls that are 1 m wide. They cost £12 each.
a How many rolls will he need to buy?
The felt is stuck on with an adhesive which costs £6.99 for a 2.5 litre tin.
It will cover 6 m².
b How much will Hamish have to pay for the materials to do the job?

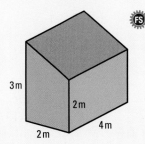

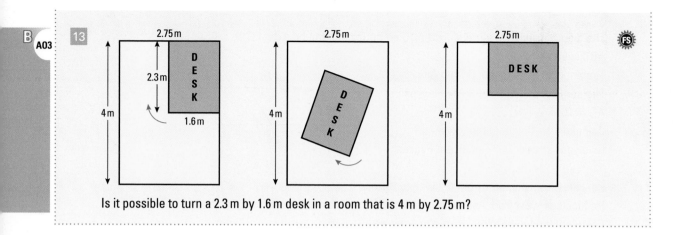

Is it possible to turn a 2.3 m by 1.6 m desk in a room that is 4 m by 2.75 m?

21 MORE GRAPHS AND EQUATIONS

Different-shaped curves are seen in many areas of mathematics, science and engineering. Galileo showed that if an object is thrown it traces out a type of curve called a parabola. This bridge is the Butterfly Bridge in Bedford and it has two parabolic arches.

Objectives

In this chapter you will:
- recognise and draw graphs of quadratic, cubic, reciprocal and exponential functions
- use graphs to solve quadratic equations
- find approximate solutions of equations by using a trial and improvement method.

Before you start

You should be able to:
- draw straight-line graphs
- work out the value of a given expression by substituting values.

21.1 Graphs of quadratic functions

◉ Objectives

- ○ You can recognise and draw graphs of quadratic functions.
- ○ You can use graphs to solve quadratic equations.

◇ Why do this?

You can use quadratic functions to represent the path of projectiles, such as the trajectory of a cannonball or of a drop-goal in rugby.

◈ Get Ready

1. Draw the graph of $y = 2x + 3$ for values of x from -3 to $+3$.
2. Work out the value of $x^2 - 5$ when   **a** $x = 1$ **b** $x = -3$
3. Work out the value of $2x^2$ when   **a** $x = 1$ **b** $x = -3$

◉ Key Points

- ◉ A **quadratic function** (or expression) is one in which the highest power of x is x^2.

- ◉ All quadratic functions can be written in the form $ax^2 + bx + c$ where a, b and c represent numbers. Examples of quadratic functions include $x^2 + 1$, $x^2 - 2x + 3$, $3x^2 + x - 2$, and $3 - x^2$.

- ◉ The graph of a quadratic function is called a **parabola**. It has a smooth $\smile$ or $\frown$ shape according to whether $a > 0$ or $a < 0$.

- ◉ The lowest point of a quadratic graph is where the graph turns, and is called the **minimum point**.

- ◉ The highest point of a quadratic graph is where the graph turns, and is called the **maximum point**.

- ◉ All quadratic graphs have a line of symmetry.

- ◉ You can solve **quadratic equations** of the form $ax^2 + bx + c = 0$ by reading off the x-coordinate where the graph
  $y = ax^2 + bx + c$ crosses the x-axis.

- ◉ You can solve quadratic equations of the form $ax^2 + bx + c = mx + k$ by reading off the x-coordinate at the point of intersection of the graph $y = ax^2 + bx + c$ with the straight-line graph $y = mx + k$.

⬤ Example 1   Draw the graph of $y = x^2 + 1$ taking values of x from -3 to $+3$.

When $x = 3, y = 3 \times 3 + 1 = 10$ ← Work out the value of y for each value of x.
When $x = 2, y = 2 \times 2 + 1 = 5$
When $x = 1, y = 1 \times 1 + 1 = 2$
When $x = 0, y = 0 \times 0 + 1 = 1$
When $x = -1, y = (-1) \times (-1) + 1 = 2$ ← Put negative values of x in brackets when substituting them.
When $x = -2, y = (-2) \times (-2) + 1 = 5$
When $x = -3, y = (-3) \times (-3) + 1 = 10$

x	-3	-2	-1	0	1	2	3
y	10	5	2	1	2	5	10

← These results can be shown in a table of values.

---

quadratic function   parabola   minimum point   maximum point   quadratic equations

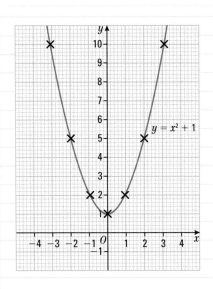

Be careful to plot the points accurately.
Use a sharp pencil cross for each point.
Join your points with a smooth curve.

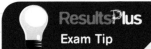

Exam Tip

Make sure your curve passes through all the points.

Example 2

a Complete the **table of values** for $y = x^2 - 2x - 4$.

x	-2	-1	0	1	2	3	4
y		-1	-4			-1	

b Draw the graph of $y = x^2 - 2x - 4$ for $x = -2$ to $x = 4$.

c Write down the equation of the line of symmetry of this curve.

d Write down the values of x where the graph crosses the x-axis.

a When $x = 4, y = 4^2 - 2 \times 4 - 4$ $\quad = 16 - 8 - 4 = 4$
When $x = 2, y = 2^2 - 2 \times 2 - 4$ $\quad = 4 - 4 - 4 \ = -4$
When $x = 1, y = 1^2 - 2 \times 1 - 4$ $\quad = 1 - 2 - 4 \ = -5$
When $x = -2, y = (-2)^2 - 2 \times (-2) - 4 = 4 + 4 - 4 \ = 4$

Work out the value of y for each value of x in turn.

x	-2	-1	0	1	2	3	4
y	4	-1	-4	-5	-4	-1	4

b

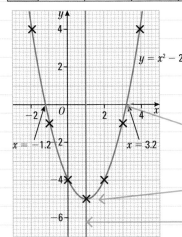

Look at the values of y in the table to determine the extent of the y-axis. Use values of -6 to $+5$.

Draw a smooth symmetrical curve through all the plotted points.

Read off the values where the curve crosses the x-axis.

This is the minimum point.

The curve has a line of symmetry.

c The line of symmetry has equation $x = 1$.

d $x = -1.2$ and $x = 3.2$

Questions in this chapter are targeted at the grades indicated.

Exercise 21A

B

1   Here is the table of values for $y = x^2 - 3$.

x	-3	-2	-1	0	1	2	3
y	6		-2	-3			6

   a   Copy and complete the table of values.
   b   Draw the graph of $y = x^2 - 3$ for $x = -3$ to $x = 3$.
   c   Write down the equation of the line of symmetry of your graph.
   d   Write down the coordinates of the minimum point.

2   a   Copy and complete the table of values for $y = 4 - x^2$.

x	-3	-2	-1	0	1	2	3
y		0	3	4			-5

   b   Draw the graph of $y = 4 - x^2$ for $x = -3$ to $x = 3$.
   c   Write down the coordinates of the maximum point.
   d   Write down the values of x where the graph crosses the x-axis.

3   a   Copy and complete the table of values for $y = 2x^2 + 2$.

x	-3	-2	-1	0	1	2	3
y	20		4	2			20

   b   Draw the graph of $y = 2x^2 + 2$ for $x = -3$ to $x = 3$.
   c   Use your graph to find:
      i   the value of y when $x = 1.5$
      ii the two values of x when $y = 11$.

4   Draw the graph for each of the following equations:
   a $y = x^2 - 4x - 1$ for values of x from -2 to 6
   b $y = 2x^2 - 4x - 3$ for values of x from -2 to 4
   c $y = (x + 2)^2$ for values of x from -6 to 2
   d $y = 5 + 3x - 2x^2$ for values of x from -2 to 4.

   For each case use your graphs to:
      i   write down the values of x when the graph crosses the x-axis
      ii draw in and write down the equation of the line of symmetry.

Example 3

Here is the graph of $y = x^2 - 2x - 2$.

a Use the graph to solve the equation
$x^2 - 2x - 2 = 0$.
Give your answers correct to 1 decimal place.

b Use the graph to solve the equation
$x^2 - 2x - 5 = 0$.
Give your answers correct to 1 decimal place.

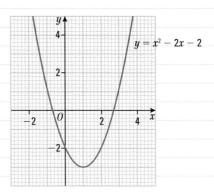

a

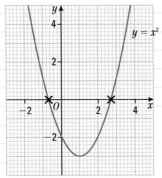

$x^2 - 2x - 2 = 0$ ← Find where the graph crosses the x-axis – that is where $y = 0$.

$x = -0.7$ and $x = 2.7$ ← Read off the values.

b

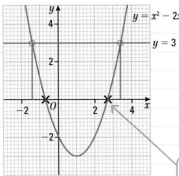

$x^2 - 2x - 5 = 0$ ← Rearrange the equation so that one side is $x^2 - 2x - 2$.

$x^2 - 2x - 2 = 3$ ← Add 3 to each side of the equation. Find where $x^2 - 2x - 2$ intersects $y = 3$.

$x = -1.4$ and $x = 3.4$

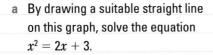
Read off the x values.

Example 4

The diagram shows the graph of $y = x^2$.

a By drawing a suitable straight line on this graph, solve the equation
$x^2 = 2x + 3$.

b By drawing another suitable straight line on this graph, solve the equation
$x^2 + x - 3 = 0$.

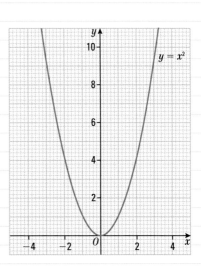

a

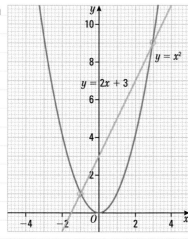

$x^2 = 2x + 3$ ← Draw the graph of $y = 2x + 3$.

$x = -1$ and ← Read off the x values
$x = 3$ at the points of intersection.

ResultsPlus
Exam Tip

If the equation in the question is in terms of x remember to give only the value of the x-coordinate.

b

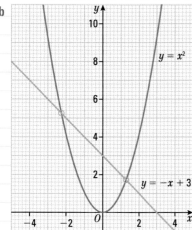

$x^2 + x - 3 = 0$ ← Rearrange the equation so that one side is x^2.

$x^2 = -x + 3$ ← Draw the line $y = -x + 3$.

$x = -2.3$ and ← Read off the x values
$x = 1.3$ at the points of intersection.

⚙ **Exercise 21B**

B

1   Here are four graphs.

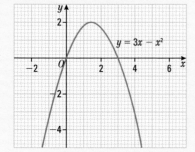

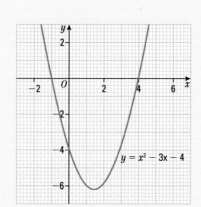

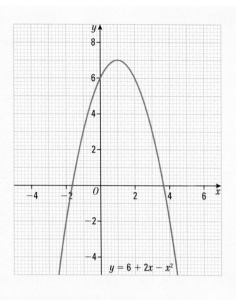

Use these graphs to solve the equations

a $3x - x^2 = 0$

b $x^2 - 3x - 4 = 0$

c $2x^2 - 3x - 7 = 0$

d $6 + 2x - x^2 = 0$

2 Use the graphs in question 1 to solve the equations

a $3x - x^2 = 1$

b $x^2 - 3x - 4 = -5$

c $2x^2 - 3x - 10 = 0$

d $4 + 2x - x^2 = 0$

A02

3 Here is a table of values for $y = 1 + 2x - x^2$.

x	-2	-1	0	1	2	3	4
y		-2		2	1		-7

a Copy and complete the table.

b Draw the graph of $y = 1 + 2x - x^2$.

c By drawing a suitable line on your graph, solve the equation $2 + 4x - 2x^2 = 2 - 2x$.

A03

4 a Make a table of values for $y = 3x^2 - x + 2$, taking values of x from -3 to $+3$.

b Draw the graph of $y = 3x^2 - x + 2$.

c By drawing a suitable line on your graph, solve the equation $3x^2 - 3x - 2 = 0$.

A03

21.2 Graphs of cubic functions

◎ Objective

○ You can recognise and draw graphs of cubic functions.

⊘ Why do this?

Engineers use cubic models, for example, when testing the strength of rubber in car tyres.

◈ Get Ready

1. Write down the first five cube numbers.

2. Work out the value of x^3 when

a $x = 100$ b $x = -10$

> **Key Points**

- A **cubic function** (or expression) is one in which the highest power of x is x^3.
- All cubic functions can be written in the form $ax^3 + bx^2 + cx + d$ where a, b, c and d represent numbers. Examples of cubic functions include $4 - x^3$ and $x^3 - 2x^2 + 3$.
- The graph of a cubic function has one of the following shapes.

for $a > 0$

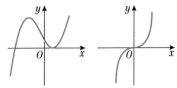

for $a < 0$

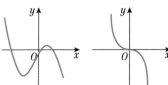

- To draw the graph of a cubic function, make a table of values, then plot the points from your table and join them with a smooth curve.

> **Example 5**

a Draw the graph of $y = x^3 - 4x^2 + 5$ for $-2 \le x \le 4$.

b Use your graph to solve the equation $x^3 - 4x^2 - x + 5 = 0$.

a When $x = 4$, $y = 4^3 - 4 \times 4^2 + 5$ $= 5$ ←   Work out y for each value of x.

When $x = 3$, $y = 3^3 - 4 \times 3^2 + 5$ $= -4$

When $x = 2$, $y = 2^3 - 4 \times 2^2 + 5$ $= -3$

When $x = 1$, $y = 1^3 - 4 \times 1^2 + 5$ $= 2$

When $x = 0$, $y = 0^3 - 4 \times 0^2 + 5$ $= 5$

When $x = -1$, $y = (-1)^3 - 4 \times (-1)^2 + 5 = 0$

When $x = -2$, $y = (-2)^3 - 4 \times (-2)^2 + 5 = -19$

Make a table of values.

x	-2	-1	0	1	2	3	4
y	-19	0	5	2	-3	-4	5

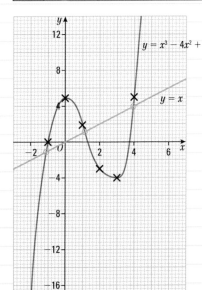

Plot the points and draw a smooth curve through all the points.

b $x^3 - 4x^2 - x + 5 = 0$ ←

$x^3 - 4x^2 + 5 = x$

Compare the expression to $y = x^3 - 4x^2 + 5$.
Add x to each side of the equation.

> **ResultsPlus**
> **Exam Tip**
>
> Take care drawing the graph of $y = x$ if there are different scales on the axes.

The solutions are $x = -1.1$ ←

$x = 1.2$ and $x = 3.9$.

Draw $y = x$ on the graph and find where it intersects with $y = x^3 - 4x^2 + 5$.

Exercise 21C

1 a Copy and complete the table of values for $y = x^3 + 2$.

x	-3	-2	-1	0	1	2	3
y							

   b Draw the graph of $y = x^3 + 2$ for $-3 \leqslant x \leqslant 3$.
   c Use your graph to find the value of y when $x = 2.5$.

2 Here is a table of values for $y = x^3 - 9x$.

x	-4	-3	-2	-1	0	1	2	3	4
y	-28	0		8	0		-10		28

   a Copy and complete the table.
   b Draw the graph of $y = x^3 - 9x$ for $-4 \leqslant x \leqslant 4$.
   c Use your graph to find the solutions to the equation $x^3 - 9x = 0$.

3 a Copy and complete the table of values for $y = 12x + 3x^2 - 2x^3$.

x	-3	-2	-1	0	1	2	3	4
y	$+45$		-7		$+13$	$+20$		-32

   b Draw the graph of $y = 12x + 3x^2 - 2x^3$ for $-3 \leqslant x \leqslant 4$.
   c By drawing a suitable line on your diagram, solve the equation $12x + 3x^2 - 2x^3 = 2x - 1$.

4 Here are four graphs.

A

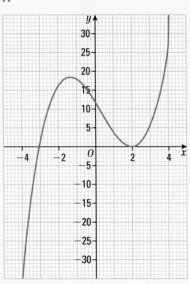

B

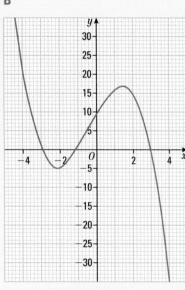

A

C

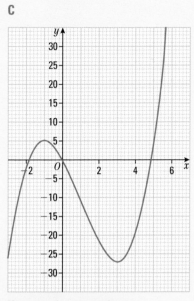

D

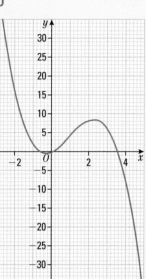

Here are four equations

    i $y = x^3 - 3x^2 - 9x$ ii $y = x^3 - x^2 - 8x + 12$

   iii $y = 2x + 3x^2 - x^3$ iv $y = 9 + 9x - x^2 - x^3$

Match each equation to one of the graphs.

Give reasons for your answers.

21.3 Graphs of reciprocal functions

⊙ Objective

● You can recognise and draw graphs of reciprocal functions.

⑦ Why do this?

You might see a reciprocal graph if you're doing an experiment on volume and pressure. If you compress gas in a container, the volume will decrease but the pressure will increase, and vice-versa.

⬦ Get Ready

1. Work out the value of $\dfrac{1}{x}$ when

   a $x = 4$ **b** $x = \dfrac{1}{4}$ **c** $x = 2.5$ **d** $x = 0.4$

2. Explain what happens to the value of $\dfrac{1}{x}$ as x gets bigger.

⬡ Key Points

◎ The reciprocal of x is $\dfrac{1}{x} = 1 \div x$.

◎ Expressions of the form $\dfrac{k}{x}$, where k is a number are called reciprocal functions.

◎ The reciprocal of 0 is not defined since division by 0 is not possible.

◎ This means that the graph of $y = \dfrac{1}{x}$ does not have a point on the y-axis where $x = 0$.

● The graphs of reciprocal functions have similar shapes.
● They are **discontinuous** and have two parts.
● They do not cross or touch the x-axis or the y-axis, but get nearer and nearer to them. We say that the axes are asymptotes to the graphs.
● Here are the general shapes of reciprocal functions of the form $y = \dfrac{k}{x}$.

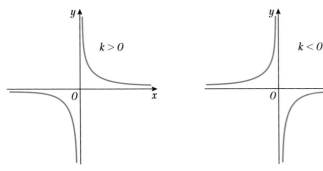

Example 6

a Draw the graph of $y = \dfrac{1}{x}$ where $x \neq 0$.

b Write down the equations of any lines of symmetry of the graph.

a

x	-4	-3	-2	-1	$-\frac{1}{2}$	$-\frac{1}{4}$	$\frac{1}{4}$	$\frac{1}{2}$	1	2	3	4
y	$-\frac{1}{4}$	$-\frac{1}{3}$	$-\frac{1}{2}$	-1	-2	-4	4	2	1	$\frac{1}{2}$	$\frac{1}{3}$	$\frac{1}{4}$

Complete a table of values.

$y = \frac{1}{x}$

Plot the points and join the two parts with smooth curves.

b

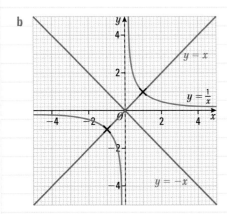

$y = x$

$y = \frac{1}{x}$

$y = -x$

The equations of the lines of symmetry of the graph are $y = x$ and $y = -x$.

Exercise 21D

B
A03

1 **a** Copy and complete the table of values for $y = \dfrac{5}{x}$ for $0 < x \leqslant 20$.

x	0.2	0.4	0.5	1	2	4	5	10	20
y	25	12.5		5	2.5			0.5	

b Using your answer to part **a**, copy and complete the following table of values for
$y = \dfrac{5}{x}$ for $-20 \leqslant x < 0$.

x	-20	-10	-5	-4	-2	-1	-0.5	-0.4	-0.2
y									

c Draw the graph of $y = \dfrac{5}{x}$ for $-20 \leqslant x \leqslant 20$.

A

2 Draw the graph of $y = -\dfrac{2}{x}$ for $-10 \leqslant x \leqslant 10$.

A☆
A03

3 **a** Draw the graph of $y = \dfrac{12}{x+1}$ for $-5 \leqslant x \leqslant 3$.

b Write down the value of x for which $y = \dfrac{12}{x+1}$ is not defined.

21.4 Graphs of exponential functions

◎ Objective

● You can recognise and draw graphs of exponential functions.

⧉ Why do this?

Scientists work out how quickly radioactive materials will break down using a graph of their radioactive half-life, which is an exponential graph.

◆ Get Ready

1. Work out the values of
   a 3^4 **b** 3^0 **c** 3^{-2}
2. Find the value of x in each of these equations.
   a $2^x = 16$ **b** $5^x = 25$ **c** $10^x = 1000$

◗ Key Points

● Expressions of the form a^x, where a is a positive number are called **exponential functions**. Examples are 2^x, 10^x, $\left(\frac{1}{2}\right)^x$ and $(1.05)^x$.
● The graphs of exponential functions have similar shapes.
● They are continuous and always lie above the x-axis.
● They increase very quickly at one end and get nearer and nearer to the x-axis at the other end.
● They cross the y-axis at (0, 1) since $a^0 = 1$ for all values of a.
● Here are the general shapes of exponential functions of the form $y = a^x$ and $y = a^{-x}$.

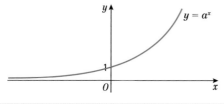
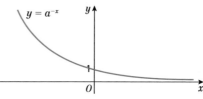

Example 7 **a** Draw the graph of $y = 2^x$, for values of x from -3 to $+3$.
b Use your graph to find an estimate for the solution of the equation $2^x = 6$.

a When $x = 0$, $y = 2^0 = 1$ ← | Substitute $x = 0$ into $y = 2^x$ and work out the value.
When $x = 1$, $y = 2^1 = 2$
When $x = 2$, $y = 2^2 = 4$ ← | Repeat the process for other integer values of x.
When $x = 3$, $y = 2^3 = 8$
When $x = -1$, $y = 2^{-1} = \frac{1}{2^1} = 0.5$
When $x = -2$, $y = 2^{-2} = \frac{1}{2^2} = 0.25$ ← | Use the result $a^{-n} = \frac{1}{a^n}$ to work out the values for negative values of x (see Section 25.1).
When $x = -3$, $y = 2^{-3} = \frac{1}{2^3} = 0.125$

Complete a table of values.

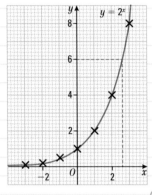

x	-3	-2	-1	0	1	2	3
y	0.125	0.25	0.5	1	2	4	8

Draw the graph from the table of values.

ResultsPlus
Exam Tip

Make sure your curve gets nearer and nearer to the x-axis without touching it.

b $x = 2.6$ ← | Use your graph to find the value of x when $y = 6$.

Example 8 The sketch shows part of the graph of $y = pq^x$.
The points with coordinates (0, 5) and (2, 45) lie on the graph.
a Work out the value of p and of q.
b Find the value of y when $x = 3$.

a $y = pq^x$
The point (0, 5) lies on the graph so
$5 = p \times q^0$
$5 = p \times 1$ ← | Substitute $x = 0$, $y = 5$ into $y = pq^x$. Use the result $q^0 = 1$.
$p = 5$

So the equation of the curve is $y = 5q^x$.
The point (2, 45) also lies on the graph so ← | Substitute $x = 2$, $y = 45$. Solve the equation. Work out the positive value of q.
$45 = 5 \times q^2$
$9 = q^2$
$q = 3$
$p = 5, q = 3$

b $y = pq^x$ so $y = 5 \times 3^x$ ← | Substitute $p = 5$, $q = 3$ into $y = pq^x$.
When $x = 3$ $y = 5 \times 3^3$
$= 5 \times 27$ ← | Put $x = 3$ into $y = 5 \times 3^x$.
$= 135$
$y = 135$

A

Exercise 21E

1 **a** Copy and complete the table of values for $y = 3^x$. Give the values correct to 2 decimal places.

x	-3	-2	-1	0	1	2	3
y	0.04		0.33		3		27

   b Draw the graph of $y = 3^x$ for $-3 \leqslant x \leqslant 3$.
   c Use your graph to find an estimate for:
      i the value of y when $x = 1.5$
      ii the value of x when $y = 15$.

A03

2 The diagram shows the graphs of $y = 2^x$, $y = 5^x$, $y = \left(\frac{1}{2}\right)^x$ and $y = 3^{-x}$.

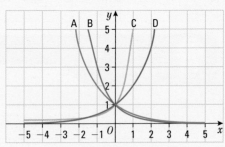

Match each graph to its equation.

A03

3 The number of bacteria, n, after time t minutes is modelled by the equation $n = 10 \times 2^t$.
   a Work out the number of bacteria initially (when $t = 0$).
   b Work out the number of bacteria after 5 minutes.
   c Find the time taken for the number of bacteria to increase to one million. Give your answer to the nearest minute.

A03

4 The points with coordinates (1, 10) and (3, 2560) lie on the graph with equation $y = pq^x$ where p and q are constants. Work out the values of p and q.

21.5 Solving equations by the trial and improvement method

◉ Objectives

● You can use a systematic method to solve an equation to any degree of accuracy.

? Why do this?

Computers can be programmed to solve complex equations using the trial and improvement method.

◈ Get Ready

1. Write these numbers correct to 1 decimal place (1 d.p.).
   a 4.613   **b** 2.157   **c** 1.498
2. Show that $x = 1$ is a solution of the equation $x + \dfrac{1}{x} = 2$.
3. Work out the value of $x^3 - 3x$ when   **a** $x = 1$ **b** $x = 2$

Key Points

- A **trial and improvement** method is a systematic way of finding solutions of equations to any degree of accuracy.

- A first **approximation** is found for the solution then the **method of trial and improvement** is used to obtain a more accurate answer. The process can be repeated in order to get closer to the correct value.

- The root(s) of an equation are found between the two points where the value of the equation changes sign.

Example 9

a   Show that the equation $x^3 - 2x = 15$ has a solution between 2 and 3.

b   Use a trial and improvement method to find this solution correct to 1 decimal place.

a

x	$x^3 - 2x$	Too high or too low	Comment
2	$2^3 - 2 \times 2 = 4$	too low since 4 is less than 15	x is greater than 2
3	$3^3 - 2 \times 3 = 21$	too high since 21 is more than 15	x is between 2 and 3

> Substitute $x = 2$ into the left-hand side of the equation and compare your answer with 15.

> Substitute $x = 3$ into the left-hand side of the equation and compare your answer with 15.

So the solution is between $x = 2$ and $x = 3$ since when
$x = 2$, $x^3 - 2x$ is too low and when $x = 3$, $x^3 - 2x$ is too high.

b

x	$x^3 - 2x$	Too high or too low	Comment
2.5	$2.5^3 - 2.5 \times 2$ $= 10.625$	too low	x is between 2.5 and 3
2.6	$2.6^3 - 2 \times 2.6$ $= 12.376$	too low	x is between 2.6 and 3
2.7	$2.7^3 - 2 \times 2.7$ $= 14.283$	too low	x is between 2.7 and 3
2.8	$2.8^3 - 2 \times 2.8$ $= 16.352$	too high	x is between 2.7 and 2.8
2.75	$2.75^3 - 2 \times 2.75$ $= 15.297$	too high	x is between 2.7 and 2.75

> Substitute $x = 2.5$ into the left-hand side of the equation and decide whether your answer is too high or too low. Record the interval in which it lies.

> Choose a value between 2.5 and 3 and decide on a new interval.

x lies between 2.7 and 2.75.
So the solution is $x = 2.7$ correct to 1 decimal place.
We write $x = 2.7$ (1 d.p.).

A02

Example 10 Use a trial and improvement method to find a solution of the equation $x^2 - \dfrac{3}{x} = 1$ correct to 2 decimal places.

x	$x^2 - \dfrac{3}{x} = 1$	H or L	Comment
1	$1^2 - \dfrac{3}{1} = -2$	L	$x > 1$
2	$2^2 - \dfrac{3}{2} = 2.5$	H	$1 < x < 2$
1.5	$1.5^2 - \dfrac{3}{1.5} = 0.25$	L	$1.5 < x < 2$
1.7	$1.7^2 - \dfrac{3}{1.7} = 1.125\ldots$	H	$1.5 < x < 1.7$
1.6	$1.6^2 - \dfrac{3}{1.6} = 0.685$	L	$1.6 < x < 1.7$
1.65	$1.65^2 - \dfrac{3}{1.65} = 0.904\ldots$	L	$1.65 < x < 1.7$
1.67	$1.67^2 - \dfrac{3}{1.67} = 0.992\ldots$	L	$1.67 < x < 1.7$
1.68	$1.68^2 - \dfrac{3}{1.68} = 1.036\ldots$	H	$1.67 < x < 1.68$
1.675	$1.675^2 - \dfrac{3}{1.675} = 1.014\ldots$	H	$1.67 < x < 1.675$

> Try substituting whole-number values until you find two consecutive integers between which the solution lies: in this case $x = 1$ and $x = 2$.

> Substitute values until you find two consecutive numbers with a difference of 0.1 between which the solution lies: in this case $x = 1.6$ and $x = 1.7$.

> Substitute values until you find two consecutive numbers with a difference of 0.01 between which the solution lies: in this case $x = 1.67$ and $x = 1.68$.

> Substitute the value halfway between 1.67 and 1.68 to find out whether the solution is nearer 1.67 or nearer to 1.68.

x lies between 1.670 and 1.675
so $x = 1.67$ (2 d.p.).

Exercise 21F

C
A02

1 For each of the following equations find two consecutive whole numbers between which a solution lies.

   a $x^3 + 2x = 4$ b $x^3 + x^2 = 1$ c $x + \dfrac{1}{x^2} = 5$ d $x^2 - \dfrac{10}{x} = 0$

2 Use a trial and improvement method to find one solution of these equations correct to 1 decimal place.

   a $x^3 + x = 7$ b $x^3 - x^2 + 4 = 0$ c $x + \dfrac{1}{x} = 5$

3 Use a trial and improvement method to find a positive solution of these equations correct to 2 decimal places.

   a $x^3 - x = 25$ b $2x^2 - \dfrac{1}{x} = 9$ c $x^2(x + 2) = 150$

B **A02** **A03**

4 A cuboid has height x cm.

The length of the cuboid is 2 cm more than its height.

The width of the cuboid is 2 cm less than its height.

The volume of the cuboid is 600 cm³.

   a Show that x satisfies the equation $x^3 - 4x = 600$.

   b Use a trial and improvement method to solve the equation $x^3 - 4x = 600$ correct to 1 decimal place.

   c Write down the length, width and height of the cuboid.

Chapter review

- A **quadratic function** is an expression of the form $ax^2 + bx + c$ where the highest power of x is x^2.
- The graph of a quadratic function is called a **parabola**. It has one of the following shapes.

$$y = ax^2 + bx + c, a > 0 \qquad y = ax^2 + bx + c, a < 0$$

- The graph of a quadratic function has one line of symmetry.
- The lowest point of a quadratic graph is where the graph turns, and is called the **minimum point**.
- The highest point of a quadratic graph is where the graph turns, and is called the **maximum point**.
- You can solve **quadratic equations** of the form $ax^2 + bx + c = 0$ by reading off the x-coordinate where the graph $y = ax^2 + bx + c$ crosses the x-axis.
- You can solve quadratic equations of the form $ax^2 + bx + c = mx + k$ by reading off the x-coordinate at the point of intersection of the graph $y = ax^2 + bx + c$ with the straight-line graph $y = mx + k$.
- A **cubic function** is an expression of the form $ax^3 + bx^2 + cx + d$ where the highest power of x is x^3.
- The graph of a cubic function has one of the following shapes.

$$y = ax^3 + bx^2 + cx + d, a > 0 \qquad y = ax^3 + bx^2 + cx + d, a < 0$$

- A reciprocal function is an expression of the form $\dfrac{k}{x}$.
- The graph of a reciprocal function has one of the following shapes.

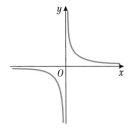 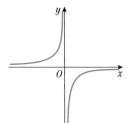

$$y = \frac{k}{x}, k > 0 \qquad y = \frac{k}{x}, k < 0$$

- An **exponential function** is an expression of the form a^x or a^{-x}, where $a > 0$.
- The graph of an exponential function has one of the following shapes.

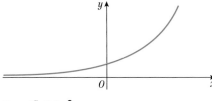 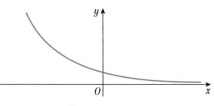

$$y = a^x, a > 0 \qquad y = a^{-x}, a > 0$$

- The graphs cross the y-axis at (0, 1) since $a^0 = 1$ for all values of a.
- You can find approximate solutions to equations which cannot be solved exactly by using a **trial and improvement** method.

Review exercise

C

1 A load, fitted with a parachute, is dropped from an aeroplane.
The parachute opens when the load is 400 m above the ground.
The distance fallen, in metres, t seconds after the parachute has opened, is given by the equation $s = 20t + 2.25t^2$.

  a Copy and complete the table of values for $s = 20t + 2.25t^2$.

t	0	2	3	4	5	10
s						

  b Draw the graph of $s = 20t + 2.25t^2$ for $t = 0$ to $t = 10$.
  c Use your graph to find out
  i how far the load falls in the first three seconds immediately after the parachute opens
  ii when the load hits the ground.

B

2 **a** Copy and complete the table of values for $y = x^2 + 2x$.

x	-4	-3	-2	-1	0	1	2
y	8		0	-1			8

  b Draw the graph of $y = x^2 + 2x$ for $x = -4$ to $x = 2$.
  c Write down the equation for the line of symmetry of this curve.
  d Use your graph to find:
  i the value of y when $x = 0.5$
  ii the values of x when $y = 6$.

A02

3 **a** Make a table of values for $y = 2 + x - x^2$ for $-3 \leqslant x \leqslant 3$.
  b Draw the graph of $y = 2 + x - x^2$ for $-3 \leqslant x \leqslant 3$.
  c Solve the equations
  i $2 + x - x^2 = 0$
  ii $5 + x - x^2 = 0$
  d Write down the coordinates of the maximum point of the graph of $y = 2 + x - x^2$

A02
A03

4 **a** Show that the equation $x^2 - 3x - 2 = x - 2$ can be rewritten as $x^2 - 4x = 0$.
  b Solve the equation $x^2 - 4x = 0$.
  c The equation $x^2 - 2x - 4 = 0$ can be solved by finding the intersection of the graph of $y = x^2 - 3x - 2$ with the graph of a suitable straight line. Find the equation of this straight line.

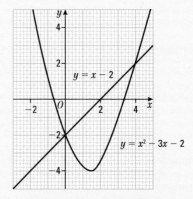

5 **a** Show that there is a solution of the equation $x^2 + \dfrac{2}{x} + 3 = 0$ between $x = -0.5$ and $x = -1$.
  b Use a trial and improvement method to find this solution correct to 2 decimal places.

6 **a** Copy and complete the table of values for $y = x^3 - 2x^2 - 4x$.

x	-2	-1	0	1	2	3	4
y		1	0	-5		-3	

b Draw the graph of $y = x^3 - 2x^2 - 4x$ for $x = -2$ to $x = 4$.

c Solve the equations:

    i $x^3 - 2x^2 - 4x = 0$

    ii $x^3 - 2x^2 - 4x + 5 = 0$

7 **a** Copy and complete the table of values for $y = 3 - \dfrac{2}{x}$ $x \neq 0$

x	-3	-2	-1	-0.5	-0.1	0.1	1	2	3
y	3.7		5	7		-17		2	

b Draw the graph of $y = 3 - \dfrac{2}{x}$ for $-3 \leqslant x \leqslant 3$.

c This graph approaches two lines without touching them. These lines are called asymptotes. Write down the equation of each of these two lines.

8 **a** Copy and complete this table of values for $y = \dfrac{4}{x}$

x	0.5	1	2	3	4
y		4			

b Draw the graph of $y = \dfrac{4}{x}$ for values of x from 0.5 to 4

c Use the graph to find an estimate for the solution of $\dfrac{4}{x} = 6 - x$

d Use the method of trial and improvement to find this value correct to 2 decimal places.

9 The equation $h = 15t - 5t^2$ gives the height, in metres, of a ball moving through the air t seconds after it was projected from ground level by a machine.

a Draw the graph of $h = 15t - 5t^2$ for $t = 0$ to $t = 3$, using values of t every 0.5 seconds.

b Use your graph to find out how long it takes the ball to reach its maximum height.

c The ball is caught at a height of 2 m as it falls back to the ground.

How long has the ball been in the air?

10 A boy throws a ball through the air.

The equation $y = 2 + x - \dfrac{1}{80}x^2$ describes the path of the ball where x represents the horizontal distance, in metres, of the ball from the boy and y represents the height, in metres, of the ball above the ground.

a Draw a graph to show the path of the ball for values of x from 0 to 80.

b Use the graph to find

    i the maximum height reached by the ball

    ii the horizontal distances of the ball from the boy when it is at a height of 10 m above the ground

    iii the horizontal distance of the ball from the boy when it hits the ground.

11 The diagram shows a rectangle.

All the measurements are in cm.

The width is x and the length is 3 cm more than the width.

The area of the rectangle is 20 cm^2.

a Draw a suitable graph

b Find an estimate for the value of x

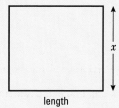

length

415

A **A03**

12 The diagram shows a cuboid.
The base of the cuboid is a square of side x cm.
The height of the cuboid is $(x + 4)$ cm.
The volume of the cuboid is 100 cm³.
Find the height of the cuboid.

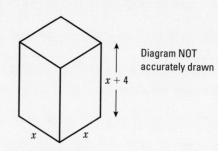

Diagram NOT
accurately drawn

$x + 4$

x x

June 2005

A☆ **A03**

13 Match each of the equations with its graph.

i $y = \dfrac{1}{x}$ ii $y = 3^x$ iii $y = x^2 - 4$

iv $y = x^3$ v $y = x^3 - x^2 - 6x$

A

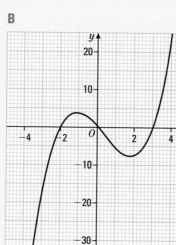

B

C

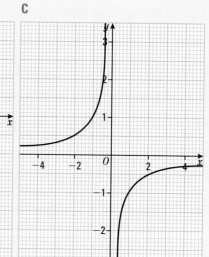

D

E

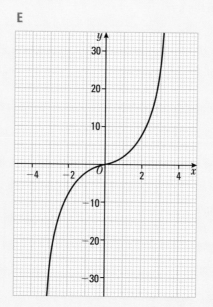

14 The diagram shows a sketch of the graph of $y = ab^x$
The curve passes through the points A (0.5, 1) and B (2, 8).
The point $C\,(-0.5, k)$ lies on the curve.
Find the value of k.

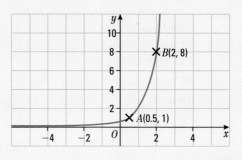

June 2006

A02

22 QUADRATIC AND SIMULTANEOUS EQUATIONS

The graph of a quadratic function is a curve called a parabola. A javelin would follow the path of a perfect parabola if the effects of air resistance, wind and rotation didn't affect it. In the Ancient Olympics, athletes had to throw the javelin from the back of a galloping horse.

◎ Objectives

In this chapter you will:

- set up and solve a pair of simultaneous equations in two unknowns
- solve a pair of simultaneous equations using a graphical approach
- solve quadratic equations by factorisation, completing the square, and using the formula
- solve algebraic fraction equations and quadratic equations
- construct graphs of simple loci
- solve a pair of simultaneous equations in two unknowns when one equation is linear and the other is quadratic
- solve a pair of simultaneous equations in two unknowns when one equation is linear and the other is a circle.

◇ Before you start

You should already know how to:

- factorise algebraic expressions
- solve a simple linear equation
- construct linear and quadratic graphs
- substitute into a quadratic expression
- use surds.

22.1 Solving simultaneous equations

⊙ Objective

○ You can solve a pair of simultaneous equations in two unknowns.

❔ Why do this?

Economists use simultaneous equations when considering the changes in the price of a commodity, and the resulting change in market demand.

⬆ Get Ready

1. Solve
   a $2x + 7 = 4$
   b $4y - 3 = 9 + y$
2. Work out the value of $6 - 5m$ when
   a $m = 2$
   b $m = -3$

🌀 Key Points

◉ When there are two unknowns you need two equations. These are called **simultaneous equations**.

◉ Simultaneous equations can be solved using elimination or substitution.

◉ To **eliminate** an unknown, multiply the equations so that the coefficients of that unknown are the same.
Add or subtract the equations to eliminate the chosen unknown.
Sometimes the equations have to be multiplied by numbers before an unknown can be eliminated.

◉ To substitute an unknown, rearrange one of the equations to make the unknown the subject, then substitute its value (in terms of the second unknown) into the other equation.

◉ Once you know one unknown, you can use substitution to find the other.

Example 1 ▶ Solve the simultaneous equations
$$4x - y = 3$$
$$x + y = 7$$

Method 1

$4x - y = 3$ (1) ← Label the equations (1) and (2).
$x + y = 7$ (2)
$5x + 0 = 10$ ← Since $-y$ and $+y$ are of different sign, add equations (1) and (2) to eliminate terms in y.

$5x = 10$ so $x = 2$ ← Divide both sides by 5.

When $x = 2, 2 + y = 7$ ← Substitute $x = 2$ into equation (2) and solve to find the value of y.
$\quad y = 7 - 2 = 5$
So the solution is $x = 2, y = 5$.
Check: $4 \times 2 - 5 = 8 - 5 = 3$ ✓ ← Check your solution by substituting into equation (1).

ResultsPlus
Exam Tip

When deciding which unknown to eliminate, if possible choose the unknown where the signs are different. You can then eliminate the unknown by adding the equations.

Method 2

$4x - y = 3$ (1) ← Label the equations (1) and (2).

$x + y = 7$ (2)

$y = 7 - x$ ← Rearrange equation (2) to make y the subject.

> **ResultsPlus**
> **Exam Tip**
>
> If a fraction is introduced when making y the subject of an equation, use an alternative method since the fraction will complicate your working.

$4y - (7 - x) = 3$ ← Substitute $y = 7 - x$ into equation (1).

$4x - 7 + x = 3$

$5x - 7 = 3$ ← Expand the bracket and solve by the **balance method**.

$5x = 3 + 7 = 10$

$5x = 10$ so $x = 2$ ← Divide both sides by 5.

When $x = 2$, $2 + y = 7$ ← Substitute $x = 2$ into equation (2) and solve to find the value of y.

$y = 7 - 2 = 5$

So the solution is $x = 2$, $y = 5$. ← Check your solution by substituting into equation (1).

Check: $4 \times 2 - 5 = 8 - 5 = 3$ ✓

Example 2   Solve the simultaneous equations

$$5x - 6y = 13$$
$$3x - 4y = 8$$

$5x - 6y = 13$ (1)

$3x - 4y = 8$ (2)

$15x - 18y = 39$ (3) ← Multiply (1) by 3 and (2) by 5 to make the coefficients of x equal. Label the new equations (3) and (4).

$15x - 20y = 40$ (4)

$0 + 2y = -1$ ← Subtract equation (4) from equation (3) to eliminate the terms in x. $-18y - (-20y) = -18y + 20y = +2y$

$y = -\frac{1}{2}$

$5x - (6 \times -\frac{1}{2}) = 13$ ← Substitute $y = -\frac{1}{2}$ into equation (1).

$5x - (-3) = 13$

$5x + 3 = 13$

$5x = 10$

$x = 2$

So the solution is $x = 2$, $y = -\frac{1}{2}$.

Check: $3 \times 2 - (4 \times -\frac{1}{2}) = 6 + 2 = 8$ ← Check your solution by substituting into equation (2).

Exercise 22A

Questions in this chapter are targeted at the grades indicated.

Solve these simultaneous equations.

1 $2x + y = 9$
$x + y = 5$

2 $3x - y = 12$
$2x + y = 13$

3 $5x - 2y = 9$
$3x - 2y = 7$

4 $x + 4y = 6$
$3x - 2y = 4$

5 $x + 2y = 9$
$y = x + 3$

6 $2x + 5y = 12$
$y = 3 - x$

7 $5x - y = -4$
$y = 2x + 1$

8 $3x - 4y = -2$
$y = x + 1$

9 $8x - 3y = -2$
$y = 3 - 2x$

10 $4x - 3y = 14$
$2x + 2y = -7$

11 $3x + 2y = 11$
$2x - 5y = 1$

12 $4x + 6y = 5$
$3x + 4y = 4$

13 $5x + 4y = 5$
$3x - 5y = -34$

14 $7x - 2y = 13$
$4x - 3y = 13$

15 $4x - 3y = 5$
$2x + 2y = -1$

B

A

22.2 Setting up equations in two unknowns

Objective

○ You can set up and solve a pair of simultaneous equations in two unknowns.

Why do this?

You can set up equations in two unknowns to explain practical situations. For example, 2 adult cinema tickets and 1 child ticket costs £23.50, but 1 adult and 3 children would cost £25.50.

Get Ready

Solve these simultaneous equations.

1. $x + 2y = 8$
$4y - x = 10$

2. $3x + y = 5$
$5y + 4x = 14$

3. $4x + y = 13$
$3x + 2y = 11$

Key Point

○ When setting up your simultaneous equations, clearly define the unknowns used.

Example 3

Zach has some ten pence coins and some twenty pence coins in his piggy bank.
In his piggy bank he has a total of 18 coins which amounts to £2.30. Work out the number of ten pence coins and the number of twenty pence coins in Zach's piggy bank.

Let x be the number of 10p coins in the piggy bank. ← Define the unknowns.
Let y be the number of 20p coins in the piggy bank.

$10x + 20y = 230$ ⠀⠀(1) ← The sum of the 10p and 20p coins is £2.30.

$x + y = 18$ ⠀⠀(2) ← There are 18 coins altogether.

Results**Plus**
Watch Out!

Make sure that both sides of each equation have consistent units.

$$10x + 20y = 230 \quad (1)$$
$$10x + 10y = 180 \quad (3)$$

Multiply equation (2) by 10 to give equation (3).

$$10y = 50$$

Subtract to eliminate x.

$$y = 5$$
$$x + 5 = 18$$

Substitute $y = 5$ in (2) to find the value of x.

$$x = 13$$

So Zach has 13 ten pence coins and 5 twenty pence coins.

Check: $13 \times 10p + 5 \times 20p = £2.30$

Exercise 22B

A

1. The sum of two numbers is 19 and their difference is 5. Find the value of each of the numbers.

A02 A03

2. The total cost of a meal and a bottle of wine is £28.10.
   The meal cost £8.90 more than the bottle of wine. Find the cost of the meal.

A03

3. A taxi company charges a fixed amount of £f plus x pence for each mile of a journey.
   A journey of 10 miles costs £10.20.
   A journey of 6 miles costs £7.40.
   Work out the cost of a journey of 8 miles.

A✩ A02 A03

4. Cinema tickets for 1 adult and 3 children cost £13.50.
   The cost for 2 adults and 5 children is £24. Find the cost of one adult ticket and the cost of one child ticket.

A02 A03

5. Three nuts and six bolts have a combined mass of 72 g.
   Four nuts and five bolts have a combined mass of 66 g.
   Find the combined mass of one nut and one bolt.

6. The diagram shows a rectangle.
   All sides are measured in centimetres.
   a Write down a pair of simultaneous equations in a and b.
   b Solve your pair of simultaneous equations to find a and b.

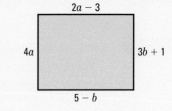

A02 A03

7. Mary and Ann together receive a total of £306 for baby-sitting.
   Ann is paid for 14 days' work and Mary is paid for 15 days' work.
   Ann's pay for 6 days' work is £6 more than Mary gets for 4 days.
   Work out how much they each earn per day.

A02 A03

8. Atif walks for x hours at 5 km/h and runs for y hours at 10 km/h. He travels a total of 35 km and his average speed is 7 km/h. Find the value of x and y.

22.3 Using graphs to solve simultaneous equations

⊙ Objective

○ You can solve a pair of simultaneous equations by using a graph.

⦾ Why do this?

It can be easier to find solutions to some simultaneous equations if they are plotted on a graph.

⟡ Get Ready

The sides of a rectangle are $4a$, $24 - 3b$, $3b + 4$ and $3a$.

1. Set up two simultaneous equations in a and b.
2. Find a and b.
3. Find the perimeter of the rectangle.

🔍 Key Point

◉ Simultaneous equations can be solved graphically, by drawing the graphs of the two equations and finding the coordinates of their point of **intersection**.

Example 4

Solve the simultaneous equations

$$2x - y = 5$$
$$x + y = 4$$

$2x - y = 5$ ⟵ Rearrange $2x - y = 5$ to make y the subject.
$\quad 2x = y + 5$
$\quad\quad y = 2x - 5$

For $y = 2x - 5$, ⟵ Find and plot any three points on $y = 2x - 5$.
when $x = 0$, $y = -5$
when $x = 2$, $y = -1$
when $x = 4$, $y = 3$.

For $x + y = 4$, ⟵ Find and plot the points where $x + y = 4$ crosses the axes and one other point.
when $x = 0$, $y = 4$
when $y = 0$, $x = 4$
when $x = 2$, $y = 2$.

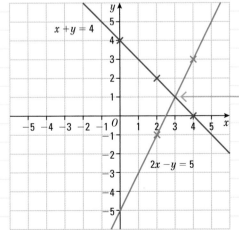

Draw the lines by joining the plots.

Find the coordinates of the point where the lines cross. This gives the solution of the simultaneous equations.

ResultsPlus
Exam Tip

Check your solution by substitution into both equations.

So the solution is $x = 3$, $y = 1$.
Check: $2 \times 3 - 1 = 5$
and $3 + 1 = 4$.

Exercise 22C

1 The diagram shows three lines **A**, **B** and **C**.

a Match the three lines to these equations.

$y = 2x$

$x + y = 3$

$x - 2y = 3$

b Use the diagram to solve these simultaneous equations.

  i $y = 2x$ **ii** $x + y = 3$ **iii** $y = 2x$

    $x + y = 3$ $x - 2y = 3$ $x - 2y = 3$

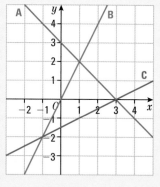

2 For each of these pairs of simultaneous equations, draw two linear graphs on the same grid and use them to solve the equations. Use a scale of -10 to $+10$ on each axis.

a $4x + y = 8$ b $2x + 3y = 12$ c $x + y = 10$

   $x - y = 2$ $y = 2x - 3$ $y = 3x + 2$

22.4 Solving quadratic equations by factorisation

◎ Objective

● You can solve quadratic equations by factorising, when possible.

❔ Why do this?

Quadratic equations can be used to work out car stopping distances.

⬆ Get Ready

Factorise

1. a $x^2 - 4x$ b $2y^2 + 5y$ c $2x^2 + 5x - 3$ d $6y^2 - 7y - 20$

2. Work out the value of $b^2 - 4ac$, when

   a $a = 1, b = 4, c = 2$ b $a = 2, b = -5, c = 3$ c $a = 5, b = -7, c = -5$

🔍 Key Points

◉ A quadratic equation can always be written in the form $ax^2 + bx + c = 0$ where a ($\neq 0$), b and c represent numbers.

◉ To factorise a quadratic equation you need to find two numbers whose sum is b and whose product is c.

◉ If the product of two numbers is 0, then at least one of these numbers must be 0. For example, if $cd = 0$ then either $c = 0$ or $d = 0$ or they are both 0.

◉ A quadratic equation has two solutions (or roots). Sometimes these solutions may be equal.

Example 5   Solve

a $2x^2 = 6x$ b $y^2 - y - 20 = 0$

ResultsPlus
Watch Out!

When there is a power of x on both sides of an equation, do not simply divide both sides by one of the powers of x because the solution $x = 0$ may be lost.

a $2x^2 = 6x$

$2x^2 - 6x = 0$ ← Rearrange into the form $ax^2 + bx + c = 0$.

$2x(x - 6) = 0$ ← Factorise.

So either $2x = 0$ or $(x - 6) = 0$ ← Solve the linear equations.
giving the two solutions $x = 0$ and $x = 6$.

b $y^2 - y - 20 = 0$ ← Factorise into two bracketed terms.
$(y - 5)(y + 4) = 0$
So either $(y - 5) = 0$ or $(y + 4) = 0$. ← Remember: You are looking for two numbers whose
The two solutions are $y = 5$ and $y = -4$.   product is -20 and whose sum is -1 (i.e. -5 and $+4$).

Example 6   Solve $q(q + 4) + 4 = 6q + 3$.

$q(q + 4) + 4 = 6q + 3$
$q^2 + 4q + 4 = 6q + 3$ ← Expand the brackets and rearrange into the form $ax^2 + bx + c = 0$.
$q^2 + 4q + 4 - 6q - 3 = 0$
$q^2 - 2q + 1 = 0$
$(q - 1)(q - 1) = 0$ ← Factorise.

So either $q - 1 = 0$ or $q - 1 = 0$, giving
the two equal solutions $q = 1$ and $q = 1$. ← The solutions are both the same.
We say the solution is $q = 1$.

Example 7   Solve $4x^2 - 25 = 0$.

Method 1
$4x^2 - 25 = 0$
$4x^2 = 25$
$x^2 = 25 \div 4 = 6.25$ ← Take the square root of both sides.
$x = \pm\sqrt{6.25}$
So the two solutions are $x = 2.5$ or $x = -2.5$.

Method 2
$4x^2 - 25 = 0$
$(2x - 5)(2x + 5) = 0$ ← Factorise by the difference of two squares
So either $(2x - 5) = 0$ or $(2x + 5) = 0$.   method (see Section 9.4).
So the two solutions are $x = 2.5$ or $x = -2.5$.

Exercise 22D

B

1 Solve

a $x(x - 4) = 0$
b $(a + 5)(a - 3) = 0$
c $(2m - 1)(4m - 9) = 0$
d $y^2 + 2y = 0$
e $t^2 - t = 0$
f $4p^2 - 7p = 0$

2 Solve

a $x^2 - 6x + 8 = 0$
b $x^2 + 7x + 6 = 0$
c $x^2 + x - 12 = 0$
d $x^2 - 6x + 9 = 0$
e $x^2 - 5x - 36 = 0$
f $x^2 - 16 = 0$
g $x^2 + 10x + 25 = 0$
h $x^2 - 100 = 0$

A

3 Solve

a $5x^2 + 26x + 5 = 0$
b $3x^2 - 11x + 6 = 0$
c $2x^2 + 7x - 4 = 0$
d $5x^2 + 14x - 3 = 0$

A☆

4 Solve

a $x^2 - x = 6$
b $x^2 - 10 = 3x$
c $x(x - 3) = x + 21$
d $x^2 - 36 = 2x - 1$
e $x(3x - 1) = x^2 + 15$
f $6(x^2 + 3) = 31x$
g $(x + 4)(x - 3) = 4(2x - 1)$
h $(4x - 1)^2 = 10 - x$

22.5 Completing the square

⊚ Objective

● You can complete the square for a quadratic expression.

⑦ Why do this?

Completing the square can help you to find the coordinates of the minimum (or maximum) point on a quadratic curve, for example, the maximum height of a bouncing ball.

⬥ Get Ready

Expand and simplify

1. $(x + 3)^2$
2. $(x - 5)^2$
3. $(x + a)^2$

◗ Key Points

⊚ Expressions such as $(x + 1)^2$, $(x + 4)^2$ and $(x + \frac{1}{2})^2$ are all called **perfect squares**.

⊚ Expressions like $x^2 + bx + c$ can be written in the form $\left(x + \frac{b}{2}\right)^2 - \left(\frac{b}{2}\right)^2 + c$.

This process is called **completing the square**.

⊚ Expressions like $ax^2 + bx + c$ are rewritten as $a\left(x^2 + \frac{b}{a}x\right) + c$ before completing the square for the expression inside the brackets.

perfect square   completing the square

Example 8   Write $x^2 + 4x + 5$ in the form $(x + p)^2 + q$, stating the values of p and q.

$x^2 + 4x = (x + 2)^2 - 4$ ←   Ignore the **constant term**. Find the perfect square which will give the correct terms in x^2 and x, then subtract 4 to make the identity true.

So
$x^2 + 4x + 5 = (x + 2)^2 - 4 + 5$ ←   Add 5 to obtain $x^2 + 4x + 5$.

$= (x + 2)^2 + 1$ ←   Simplify the expression.

$p = 2, q = 1$ ←   Compare $(x + 2)^2 + 1$ with $(x + p)^2 + q$ and write down the values of p and q.

Example 9
a   Write the expression $2y^2 - 12y - 5$ in the form $p(y + q)^2 + r$.
b   Hence write down the minimum possible value of $2y^2 - 12y - 5$.
c   Find the value of y for which $2y^2 - 12y - 5$ has its minimum value.

a $2y^2 - 12y - 5 = 2(y^2 - 6y) - 5$ ←   Take out the coefficient of y^2 for the y^2 and y terms. Leave the constant term separate.

$= 2[(y - 3)^2 - 9] - 5$ ←   Complete the square for $y^2 - 6y$.

$= 2(y - 3)^2 - 18 - 5$ ←   Multiply out the square brackets.

$= 2(y - 3)^2 - 23$ ←   Simplify the expression so that it is in the required form.

b   The minimum possible value of any square number is zero so $(y - 3)^2 \geq 0$ for any value of y.

The minimum possible value of
$2(y - 3)^2 - 23 = 2 \times 0 - 23$.   ←   Substitute $(y - 3)^2 = 0$ into the answer to part a.
The minimum value of
$2y^2 - 12y - 5$ is therefore -23.

c   The minimum value of $2y^2 - 12y - 5$ occurs   ←   Find the value of y which makes $(y - 3)^2 = 0$.
when $y = 3$.

Exercise 22E

1   Write the following in the form $(x + p)^2 + q$.

   A

a $x^2 + 4x$ b $x^2 + 10x$ c $x^2 + 12x$ d $x^2 - 2x$

e $x^2 - 14x$ f $x^2 - 24x$ g $x^2 + x$ h $x^2 - 3x$

i $x^2 + 4x + 7$ j $x^2 + 8x + 17$ k $x^2 + 10x - 20$ l $x^2 - 6x + 11$

m $x^2 - 20x + 80$ n $x^2 - 26x - 1$ o $x^2 - x + 1$ p $x^2 + 5x - 5$

A

2 Write the following in the form $a(x + p)^2 + q$.

a $2x^2 + 12x$

b $2x^2 - 4x + 5$

c $3x^2 - 12x + 10$

d $5x^2 + 50x + 100$

A02
A03

3 For all values of x, $x^2 + 8x + 24 = (x + p)^2 + q$.

a Find the value of the constants p and q.

b Write down the minimum value of $x^2 + 8x + 24$.

A03

4 The diagram shows a sketch of the curve with equation $y = x^2 + 6x + 10$.

a Write down the coordinates of the point A, at which the curve crosses the y-axis.

b By completing the square for $x^2 + 6x + 10$, find the coordinates of the minimum point B.

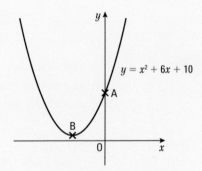

A☆
A03

5 a By writing $1 + 4x - x^2$ as $-(x^2 - 4x - 1)$ find the value of r and the value of s for which $1 + 4x - x^2 = r - (x - s)^2$.

b Use your answer to part **a** to write down the maximum value of $1 + 4x - x^2$.

c For what value of x does this occur?

22.6 Solving quadratic equations by completing the square

Objective

○ You can solve quadratic equations by completing the square.

Why do this?

The path of a cricket ball can be modelled using a quadratic equation.

Get Ready

Solve

1. $x^2 - 4 = 0$

2. $x^2 - x - 6 = 0$

3. $x^2 - 3x - 28 = 0$

Key Points

◉ By completing the square, any quadratic expression can be written in the form $p(x + q)^2 + r$.

◉ Similarly, any quadratic equation can be written in the form $p(x + q)^2 + r = 0$.

Example 10 Solve $x^2 - 12x + 9 = 0$.

Give your solutions: **a** in surd form

b correct to 3 significant figures.

> $x^2 - 12x + 9$ will not factorise into two brackets since no two integers have a product of 9 and a sum of -12.

$x^2 - 12x + 9 = 0$

$(x - 6)^2 - 36 + 9 = 0$

$(x - 6)^2 - 27 = 0$

> Complete the square for $x^2 - 12x$.
> Comparing this with $p(x + q)^2 + r = 0$ gives, $p = 1$, $q = -6$ and $r = -27$.

$(x - 6)^2 = 27$

$x - 6 = \pm\sqrt{27}$

> Take the square root of both sides.

$x - 6 = \pm 3\sqrt{3}$

> Add 6 to both sides.

$x = 6 \pm 3\sqrt{3}$

a The two solutions are $x = 6 + 3\sqrt{3}$ and $x = 6 - 3\sqrt{3}$.

b The two solutions are $x = 11.2$ and $x = 0.804$.

Example 11 Solve $2x^2 - 5x + 1 = 0$.

Give your solutions correct to 2 decimal places.

$2x^2 - 5x + 1 = 0$

> Divide both sides by 2, the coefficient of x^2.

$x^2 - 2.5x + 0.5 = 0$

$(x - 1.25)^2 - 1.25^2 + 0.5 = 0$

> Complete the square for $x^2 - 2.5x$.
> Evaluate $-1.25^2 + 0.5$

$(x - 1.25)^2 - 1.0625 = 0$

$(x - 1.25)^2 = 1.0625$

$x - 1.25 = \pm\sqrt{1.0625}$

> Take the square root of both sides.

$x - 1.25 = \pm 1.030\,776\,406$

The two solutions are $x = 2.28$ and $x = 0.22$.

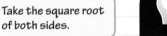

ResultsPlus

Exam Tip

Write down more digits than are required from your calculator display before you do any rounding.

Exercise 22F

1 Solve these quadratic equations, giving your solutions in surd form.

  a $x^2 - 6x - 2 = 0$ **b** $x^2 + 4x + 1 = 0$ **c** $x^2 + 10x - 12 = 0$

  d $x^2 - 2x - 7 = 0$ **e** $2x^2 - 6x - 3 = 0$ **f** $5x^2 + 12x + 3 = 0$

2 Solve these quadratic equations, giving your solutions correct to 2 decimal places.

  a $x^2 + 8x + 5 = 0$ **b** $x^2 - 9x + 6 = 0$ **c** $x^2 + x - 8 = 0$

  d $2x^2 + 4x - 5 = 0$ **e** $6x^2 - 3x - 2 = 0$ **f** $10x^2 - 5x - 4 = 0$

A

22.7 Solving quadratic equations using the formula

◎ Objective

○ You can use the formula $x = \dfrac{-b \pm \sqrt{b^2 - 4ac}}{2a}$ to solve quadratic equations.

◈ Why do this?

Many physical situations can be modelled by quadratic equations. For example, the time it takes a high diver to dive into a pool from a particular springboard is the solution of $7t^2 - t - 4 = 0$.

◈ Get Ready

Solve these equations by completing the square.

1. $x^2 + 4x - 10 = 0$ **2.** $x^2 + 6x - 5 = 0$ **3.** $x^2 - 2x - 7 = 0$

◈ Key Points

◉ You can use the method of solving the general quadratic equation $ax^2 + bx + c = 0$ by completing the square (see section 22.5) to develop a formula which can be used to solve all quadratic equations. This is called a **quadratic formula**.

$ax^2 + bx + c = 0$

$x^2 + \dfrac{b}{a}x + \dfrac{c}{a} = 0$ ← Divide both sides by a.

$\left(x + \dfrac{b}{2a}\right)^2 - \left(\dfrac{b}{2a}\right)^2 + \dfrac{c}{a} = 0$ ← Complete the square on $x^2 + \dfrac{b}{a}x$.

$\left(x + \dfrac{b}{2a}\right)^2 = \left(\dfrac{b}{2a}\right)^2 - \dfrac{c}{a}$ ← Rearrange.

$\left(x + \dfrac{b}{2a}\right)^2 = \dfrac{b^2}{4a^2} - \dfrac{c}{a} = \dfrac{b^2 - 4ac}{4a^2}$

$x + \dfrac{b}{2a} = \pm\sqrt{\dfrac{b^2 - 4ac}{4a^2}} = \dfrac{\sqrt{b^2 - 4ac}}{2a}$ ← Take the square root of both sides.

$x = -\dfrac{b}{2a} \pm \dfrac{\sqrt{b^2 - 4ac}}{2a}$ ← Make x the subject of the formula and simplify.

$x = \dfrac{-b \pm \sqrt{b^2 - 4ac}}{2a}$

◉ If the value of $b^2 - 4ac$ is negative, the quadratic equation does not have any real solutions.

Example 12 Solve $x^2 - 5x + 3 = 0$.

Give your solutions correct to 2 decimal places.

$x^2 - 5x + 3 = 0$ ← Compare with $ax^2 + bx + c = 0$ and write down the values of a, b and c.
$a = 1, b = -5, c = 3$

ResultsPlus
Exam Tip

In equations like $x^2 + bx + c = 0$ it is helpful to write it as $1x^2 + bx + c = 0$ so that the value of a is clearly 1.

$x = \dfrac{-(-5) \pm \sqrt{(-5)^2 - 4 \times 1 \times 3}}{2 \times 1}$ ← Substitute a, b and c into the quadratic formula.

$x = \dfrac{5 \pm \sqrt{25 - 12}}{2}$

$x = \dfrac{5 + \sqrt{13}}{2}$

or $x = \dfrac{5 - \sqrt{13}}{2}$

The solutions are $x = 4.30$ or $x = 0.70$

Example 13 Solve $5x^2 + x - 3 = 0$.

Give your solutions correct to 2 decimal places.

$5x^2 + x - 3 = 0$
$a = 5, b = 1, c = -3$ ← Compare with $ax^2 + bx + c = 0$ and write down the values of a, b and c.

$x = \dfrac{-1 \pm \sqrt{1^2 - 4 \times 5 \times -3}}{2 \times 5}$ ← Substitute a, b and c into the quadratic formula.

$x = \dfrac{-1 \pm \sqrt{1 + 60}}{10}$

$x = \dfrac{-1 + \sqrt{61}}{10}$ or $x = \dfrac{-1 - \sqrt{61}}{10}$

The solutions are $x = 0.68$ or $x = -0.88$.

Exercise 22G

Solve these quadratic equations. Give your solutions correct to 3 significant figures.

1 $x^2 + 4x + 2 = 0$
2 $x^2 + 7x + 5 = 0$
3 $x^2 + 6x - 4 = 0$

4 $x^2 + x - 10 = 0$
5 $x^2 - 4x - 7 = 0$
6 $x^2 - 5x + 3 = 0$

7 $2x^2 + 4x + 1 = 0$
8 $5x^2 - 9x + 2 = 0$
9 $6x^2 - 5x - 8 = 0$

10 $10x^2 + 3x - 2 = 0$
11 $4x^2 - 7x + 2 = 0$
12 $x(x - 1) = x + 5$

13 $x(2x + 3) = 4 - 7x$
14 $(x + 2)(x - 3) = 15$
15 $3x + 5 = 5x^2 + 2(x - 4)$

A

22.8 Solving algebraic fraction equations leading to quadratic equations

◎ Objective

○ You can solve equations involving both algebraic fractions and quadratic equations.

◈ Get Ready

Multiply out:

1. $(x + 6)(2x - 3)$ **2.** $(2x - 2)(3x - 6)$ **3.** $(x + 6)(x + 1) \times \dfrac{2}{(x + 1)}$

▸ Key Points

◉ Equations with algebraic fractions often occur in mathematics.

◉ These sometimes lead to quadratic equations which you can solve using one of the methods already described.

Example 14   Solve $\dfrac{5}{2x + 1} + \dfrac{6}{x + 1} = 3$.

$(2x + 1)(x + 1) \times \dfrac{5}{2x + 1} +$ ← *Multiply both sides by $(2x + 1)(x + 1)$ and cancel.*

$(2x + 1)(x + 1) \times \dfrac{6}{x + 1} =$

$(2x + 1)(x + 1) \times 3$

$5(x + 1) + 6(2x + 1) = 3(2x + 1)(x + 1)$

$5x + 5 + 12x + 6 = 3(2x^2 + 3x + 1)$ ← *Expand the brackets. Simplify both sides.*

$17x + 11 = 6x^2 + 9x + 3$

$6x^2 - 8x - 8 = 0$ ← *Rearrange into the form $ax^2 + bx + c = 0$.*

$3x^2 - 4x - 4 = 0$

$(3x + 2)(x - 2) = 0$ ← *Solve by factorisation.*

So either $3x + 2 = 0$ or $x - 2 = 0$.

The solutions are $x = -\dfrac{2}{3}$ and $x = 2$.

Results Plus

Exam Tip

When the denominators have no common factor, multiply by the product of the denominators.

⚙ Exercise 22H

Solve these quadratic equations.

A☆

1 $\dfrac{2}{x - 1} + \dfrac{3}{x + 1} = 1$ **2** $\dfrac{6}{x} = \dfrac{5x - 1}{3}$ **3** $\dfrac{6}{x} - \dfrac{5}{x + 1} = 2$

4 $\dfrac{2}{x + 1} + \dfrac{4}{3x - 1} = 3$ **5** $\dfrac{3}{x} - \dfrac{2}{2x + 1} = 5$ **6** $\dfrac{3}{2x - 1} + \dfrac{4}{x + 2} = 2$

In questions 7–12, give your solutions correct to 3 significant figures.

7 $\dfrac{3}{x} - \dfrac{1}{1+x} = 1$

8 $\dfrac{1}{x+4} - \dfrac{2}{x-5} = 4$

9 $\dfrac{2}{x-1} + \dfrac{5}{x+4} = 1$

10 $\dfrac{3}{x-1} - \dfrac{2}{x+3} = 1$

11 $\dfrac{2x+1}{x} = \dfrac{1-x}{6}$

12 $\dfrac{2}{x-1} - \dfrac{1}{1-2x} = 1$

22.9 Setting up and solving quadratic equations

◎ Objective

○ You can solve practical problems which involve quadratic equations.

❓ Why do this?

You can find the best positioning for a satellite dish by setting up and solving a quadratic equation involving the diameter and depth of the dish.

◈ Get Ready

Multiply out:

1. $2x(2-x)$ **2.** $(x+1)(1+x)$ **3.** $(3x+2)(4-x)$

🔍 Key Points

◎ To find the equation to represent a problem:
- ◉ where relevant, draw a diagram and put all of the information you are given on it
- ◉ use x to represent the unknown which you have been asked to find
- ◉ use other letters to identify any other relevant unknowns
- ◉ look for information in the question which links these letters to x and write them down
- ◉ try simple numbers for the unknowns and see if this helps you to find a method
- ◉ make sure that the units on both sides of your equation are the same.

📌 Example 15

The diagram shows a rectangular lawn surrounded on three sides by flower beds.
Each flower bed is 2 m wide.
The area of the lawn is 14 m².
Find the length of the lawn.

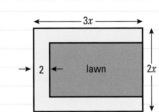

A02
A03

Length of lawn = $3x - 2$ ← Write down expressions for the length and width of the lawn.
Width of lawn = $2x - 4$

Area of lawn = $(3x - 2)(2x - 4)$ ← Using area of rectangle = length × width.
            $= 6x^2 - 16x + 8$.

But the area of the lawn is given as 14 m².

So $6x^2 - 16x + 8 = 14$ ← Set up a quadratic equation.
$6x^2 - 16x - 6 = 0$
$3x^2 - 8x - 3 = 0$
$(3x + 1)(x - 3) = 0$

So either $x = -\dfrac{1}{3}$ or $x = 3$. ← Solve using the method of factorisation.

The only acceptable solution is $x = 3$. ← The solution cannot be negative since x is a measurement of length.

Length of lawn $= 3x - 2 = 3 \times 3 - 2$

$\qquad\qquad\qquad = 7$ m.

Substitute $x = 3$ into the expression for the length of the lawn.

Example 16 Angela drove 300 km to the seaside.

Her average speed was 10 km/h less than she expected.

The journey therefore took 1 hour longer than she had planned.

Find Angela's actual average speed.

Let Angela's actual average speed be x km/h. ← Define the letters to be used.

Her expected average speed was $(x + 10)$ km/h.

The time taken for the journey $= \dfrac{300}{x}$ hours. ← Time $= \dfrac{\text{distance}}{\text{speed}}$

Her expected time for the journey $= \dfrac{300}{x + 10}$ hours.

$\dfrac{300}{x} - \dfrac{300}{x + 10} = 1$ ← Set up an equation using the information that the difference in the times is 1 hour.

$x(x + 10) \times \dfrac{300}{x} - x(x + 10) \times \dfrac{300}{x + 10} = x(x + 10) \times 1$ ← Multiply by $x(x + 10)$ and cancel.

$300(x + 10) - 300x = x(x + 10)$

$300x + 3000 - 300x = x^2 + 10x$ ← Expand brackets and rearrange.

$x^2 + 10x - 3000 = 0$

$(x + 60)(x - 50) = 0$ ← Solve by factorising.

Since x cannot be negative, $x = 50$.

Angela's actual average speed is 50 km/h.

Exercise 22I

1 The sum of the square of an integer and 2 times itself is 24.
   Find the two possible values of the integer.

2 The product of three numbers 5, $2x$ and $x - 8$ is -160.
   Find the value of the integer x.

3 A man is four times as old as his son, and 8 years ago the product of their ages was 160.
   Find their present ages as integers.

4 The length of a rectangular wall is 5 m greater than the height of the wall.
   The area of the wall is 16 m². Work out the length of the wall. Give your answer correct to 3 significant figures.

5 The sum of the squares of two consecutive integers is 41.
   a If x is one of the integers, show that $x^2 + x - 20 = 0$.
   b Solve $x^2 + x - 20 = 0$ to find the two consecutive integers.

6 Find the length of each side of this right-angled triangle.
The measurements are given in cm.
Give your answers correct to 2 d.p.

A02
A03

A☆

7 192 square tiles are needed to tile a kitchen wall. If the tiles had measured 2 cm less each way, 300 tiles would have been needed. Find the size of the larger tiles.

A02
A03

8 A farmer uses 60 m of fencing to make three sides of a rectangular sheep pen. The fourth side of the pen is a wall. Work out the length of the shorter sides of the pen if the area enclosed is 448 m².

A02
A03

9 The diameters of two circles are $4x$ cm and $(x + 3)$ cm.

The area of the shaded region is 84π cm².
Work out the value of x.

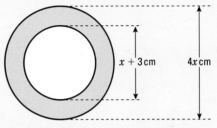

A03

10 On a journey of 420 km, a train driver calculates that the journey would take 40 minutes less if he increased his average speed by 5 km/h.
Work out his present average speed. Give your answer correct to the nearest whole number.

A02
A03

22.10 Constucting graphs of simple loci

◎ Objectives

● You can construct the graphs of simple loci including the circle $x^2 + y^2 = r^2$ for a circle of radius r centred at the origin of the coordinate plane.
● You can select and apply construction techniques and use your understanding of loci to draw graphs based on circles and perpendiculars of lines.

⑦ Why do this?

Planets and stars move in paths that can be described by simple loci. Scientists model these paths using these techniques.

⬥ Get Ready

1. Which of the following are Pythagorean triples?
   a 3, 4, 5   **b** 4, 5, 6   **c** 6, 8, 12   **d** 5, 12, 13   **e** 8, 15, 17
2. Construct a circle of radius 5 cm.
3. On your circle in question 2, draw **a** a tangent to the circle, **b** a perpendicular from the centre of the circle to the tangent in **a**.

🔑 Key Points

◉ The locus of a circle is the path of all points equidistant from a given fixed point.

◉ To construct the graph of $x^2 + y^2 = r^2$ consider any point P with coordinates (x, y) at a distance r from the origin O $(0, 0)$.
Draw the lines PQ and OQ to form the right-angled triangle PQO.

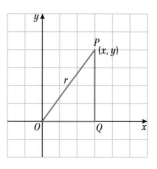

◉ By Pythagoras, $x^2 + y^2 = r^2$.
This is true for any point P, at a distance r from the fixed point O.
This is the definition of a circle.

◉ Joining all of the points A, B, C, D, E, etc gives a circle of radius r.

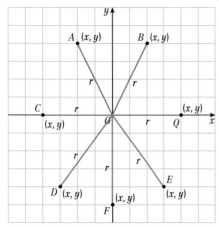

So $x^2 + y^2 = r^2$ is the equation of a circle of radius r, centre $(0, 0)$.

Example 17

a Construct the graph of the locus of all points distance 3 units from the line $y = x$.
b Find the equation of this locus.

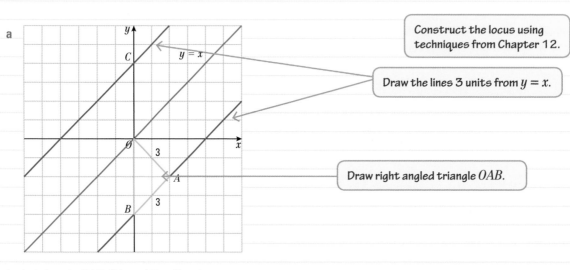

a

> Construct the locus using techniques from Chapter 12.

> Draw the lines 3 units from $y = x$.

> Draw right angled triangle OAB.

b In triangle OAB, $OA = AB = 3$ units.
 By Pythagoras, $OB\,2 = 3^2 + 3^2 = 9 + 9 = 18$
 $OB = \sqrt{18} = 4.2426\ldots$
Equation of locus through B is $y = x - 4.24\ldots$
Equation of locus through C is $y = x + 4.24\ldots$

> Apply Pythagoras to triangle OAB to find the length of OB.

Example 18
a Construct the graph of $x^2 + y^2 = 25$.
b Find the equation of the tangent to the curve at the point $(3, -4)$.

$r = \sqrt{25} = 5$

> Compare $x^2 + y^2 = 25$ with $x^2 + y^2 = r^2$.

a

> Draw a circle, centre O, with compasses set to 5 units.

$(3, -4)$

> Draw tangent through $(3, -4)$ and read off y-intercept (-6.5).

4.1

> Estimate the gradient: $4.1 \div 5$.

-6.5

5

b Gradient of tangent $= 4.1 \div 5$
$= 0.82$

> Using $y = mx + c$

y-intercept $= -6.5$
The equation is $y = 0.82x - 6.5$

Exercise 22J

1 On graph paper, draw the graphs of the following equations.
   a $x^2 + y^2 = 4$ b $x^2 + y^2 = 16$ c $x^2 + y^2 = 36$ d $x^2 + y^2 = 64$ e $x^2 + y^2 = 100$

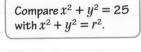

2 Using your graph of $x^2 + y^2 = 16$, construct a tangent parallel to the line $y = x$.
   Write down the coordinates of the point where this tangent touches the graph.

A03

3 Find the equation of the locus of points 5 units from the following lines.
   a $y = 6$ b $y = -4$ c $x = 3$ d $x = -5$

A03

4 Find the equation of the locus of points 6 units from the line with equation $y = x + 4$.

A03

22.11 Solving simultaneous equations when one is linear and the other is quadratic

⊙ Objective

○ You can solve a pair of simultaneous equations in two unknowns when one equation is linear and the other is quadratic.

⟡ Get Ready

Use a graphical method to find the solution of these pairs of simultaneous equations.

1. $x + y = 10$
$y = 2x + 1$

2. $y = -3x + 3$
$y = 2x - 7$

◉ Key Points

◉ You can solve a pair of simultaneous equations where one equation is linear and the other quadratic by an algebraic method or a graphical approach.

◉ The solution of a pair of simultaneous equations where one is linear and one is quadratic is represented by the points of intersection of a straight line and a quadratic curve.

Example 19 Solve the simultaneous equations **a** graphically **b** algebraically.

$$x^2 + 2y = 1$$
$$y = x - 1$$

a $y = \dfrac{1}{2} - \dfrac{x^2}{2}$ ← Make y the subject of $x^2 + 2y = 1$.

x	-3	-2	-1	0	1	2	3
y	-4	-1.5	0	0.5	0	-1.5	-4

For $y = x - 1$,
when $x = 3, y = 2$
when $x = 0, y = -1$
when $x = -4, y = -5$

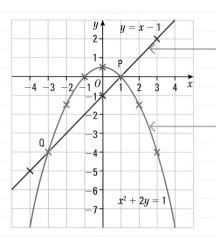

Find and plot any three points then draw the graph.

Construct a table of values, plot the points and draw the graph.

Results**Plus**
Exam Tip

The graph of the linear equation crosses the graph of the quadratic at **two** points. So the pair of simultaneous equations has **two** solutions.

The solutions are $x = 1, y = 0$ and $x = -3, y = -4$. ← The points of intersection of the two graphs are $P(1, 0)$ and $Q(-3, -4)$.

b $x^2 + 2y = 1$ (1)

          $y = x - 1$ (2)

> Label the equations (1) and (2).

$x^2 + 2(x - 1) = 1$

$x^2 + 2x - 2 = 1$

> Substitute (2) into (1) and rearrange.

$x^2 + 2x - 3 = 0$

$(x + 3)(x - 1) = 0$

> Solve the quadratic equation using the method of factorisation. See Section 22.4.

So either $(x + 3) = 0$ or $(x - 1) = 0$

$x = -3$ or $x = 1$

When $x = -3$,

> Substitute values of x into (2).

$y = -3 - 1$

$y = -4$.

When $x = 1$,

$y = 1 - 1$

$y = 0$.

So the solutions are $x = -3, y = -4$ and $x = 1, y = 0$.

Exercise 22K

For each of these pairs of simultaneous equations:

a draw a quadratic graph and a linear graph on the same grid and use them to solve the simultaneous equations (use a scale of -10 to $+10$ on each axis)

b solve them using an algebraic method. You must show all of your working.

1 $x^2 + y = 6$

     $y = x$

2 $x^2 - 2y = 2$

     $y = x + 3$

3 $x^2 + 4y = 7$

     $2y + x = 2$

4 $y = 3x^2 - 2$

     $y = 3 - 2x$

5 $y = 3 - x^2$

     $y = 5 - 3x$

6 $2y = 4x^2 - 7$

     $y = 6x$

A

22.12 Solving simultaneous equations when one is linear and one is a circle

Objective

- You can solve a pair of simultaneous equations in two unknowns when one equation is linear and the other is of the form $x^2 + y^2 = r^2$ (i.e. a circle).

Get Ready

On graph paper, draw graphs of the following equations:

1. $x^2 + y^2 = 16$
2. $x^2 + y^2 = 64$
3. $x^2 + y^2 = 49$

Key Points

- The **equation of a circle** with centre $(0, 0)$ and radius r can be written as $x^2 + y^2 = r^2$.
- You can solve simultaneous equations where one is linear and one is the equation of a circle graphically and algebraically.
- If the solutions are not integer values, they can only be estimated using the graphical method and can be calculated using the quadratic formula (see Section 22.7).

Example 20   Solve the simultaneous equations   a   graphically   b   algebraically.

$$x^2 + y^2 = 25$$
$$y = x + 1$$

a   The graph of $x^2 + y^2 = 25$ is a circle,   ⟵   Compare with $x^2 + y^2 = r^2$.
    centre $(0, 0)$ of radius 5 units.

    For $y = x + 1$,   ⟵   Plot the points and draw the line $y = x + 1$.
    when $x = 2, y = 3$
    when $x = 0, y = 1$
    when $x = -2, y = -1$.

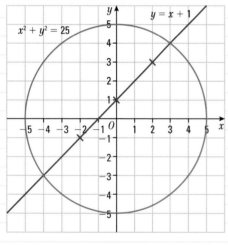

The solutions are the coordinates of the points of intersection.

The solutions are
$$x = 3, y = 4 \text{ and } x = -4, y = -3.$$

b $x^2 + y^2 = 25$ (1)
            $y = x + 1$ (2)
      $x^2 + (x + 1)^2 = 25$ ⟵   Substitute $y = x + 1$ into (1).

    $x^2 + x^2 + 2x + 1 = 25$ ⟵   Expand and simplify.
      $2x^2 + 2x - 24 = 0$
        $x^2 + x - 12 = 0$
    $(x - 3)(x + 4) = 0$ ⟵   Solve the quadratic equation by the factorisation method.

    So $x = 3$ or $x = -4$
    $y = 3 + 1 = 4$ or $y = -4 + 1 = -3$ ⟵   Substitute values of x to find the corresponding values of y.

    The solutions are $x = 3, y = 4$ and $x = -4, y = -3$.

Example 21 Draw suitable graphs to find estimates of the solutions of:
$$x^2 + y^2 = 16$$
$$y = x - 1.$$

The graph of $x^2 + y^2 = 16$ is a circle, centre $(0, 0)$ of radius 4 units. ← Compare with $x^2 + y^2 = r^2$.

For $y = x - 1$, ← Plot the points and draw the line $y = x - 1$.
when $x = 2, y = 1$
when $x = 0, y = -1$
when $x = -2, y = -3$

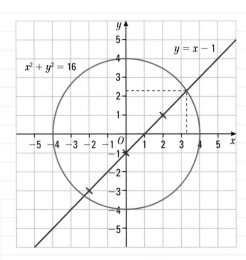

The solutions can only be estimated.

The estimated solutions are
$x = 3.3, y = 2.3$ and $x = -2.3, y = -3.3$.

Example 22 Solve these simultaneous equations
$$x^2 + y^2 = 16$$
$$y = x - 1$$
Give your answers correct to 2 decimal places.

$$x^2 + y^2 = 16 \quad (1)$$
$$y = x - 1 \quad (2)$$
$$x^2 + (x - 1)^2 = 16 \quad ← \text{Substitute (2) in (1) and rearrange to give a quadratic equation.}$$
$$x^2 + x^2 - 2x + 1 = 16$$
$$2x^2 - 2x - 15 = 0$$

$$x = \frac{2 \pm \sqrt{2^2 - 4 \times 2 \times (-15)}}{2 \times 2} \quad ← \text{Solve using the quadratic formula.}$$

$$x = \frac{2 \pm \sqrt{4 + 120}}{2 \times 2} = \frac{2 \pm \sqrt{124}}{4}$$

$$x = 3.28 \text{ or } -2.28$$

The solutions are $x = 3.28, y = 2.28$ ← Write the solutions correct to 2 decimal places.
and $x = -2.28, y = -3.28$ to 2 d.p.

Exercise 22L

1 On graph paper, draw the graph of the circle with equation $x^2 + y^2 = 36$. On the same axes, draw the straight line with equation $y = 2x$.

Hence find estimates of the solutions of the simultaneous equations $x^2 + y^2 = 36$ and $y = 2x$.

2 Draw suitable graphs to find estimates of the solutions of the simultaneous equations $x^2 + y^2 = 19$ and $y = x + 3$.

3 Draw suitable graphs to find estimates of the solutions of the simultaneous equations $x^2 + y^2 = 49$ and $x + y = 5$.

A02

4 Solve these simultaneous equations.

a $x^2 + y^2 = 13$ b $x^2 + y^2 = 20$ c $x^2 + y^2 = 34$
$\ \ y = x + 1$ $\ \ y = 2 - x$ $\ \ y = 1 + 2x$

A02

5 Solve these simultaneous equations. Give your answers correct to 3 significant figures.

a $x^2 + y^2 = 20$ b $x^2 + y^2 = 32$ c $x^2 + y^2 = 100$
$\ \ y = x + 4$ $\ \ y = 1 + 3x$ $\ \ y = 2x - 3$

Chapter review

- When there are two unknowns you need two equations. These are called **simultaneous equations**.
- Simultaneous equations can be solved using elimination or substitution.
- To **eliminate** an unknown, multiply the equations so that the coefficients of that unknown are the same. Add or subtract the equations to eliminate the chosen unknown.
  Sometimes the equations have to be multiplied by numbers before an unknown can be eliminated.
- To substitute an unknown, rearrange one of the equations to make the unknown the subject, then substitute its value (in terms of the second unknown) into the other equation.
- Once you know one unknown, you can use substitution to find the other.
- When setting up your simultaneous equations, clearly define the unknowns used.
- Simultaneous equations can be solved graphically, by drawing the graphs of the two equations and finding the coordinates of their point of **intersection**.
- A **quadratic equation** can always be written in the form $ax^2 + bx + c = 0$ where $a\ (\neq 0)$, b and c represent numbers.
- To factorise a quadratic equation you need to find two numbers whose sum is b and whose product is c.
- If the product of two numbers is 0, then at least one of these numbers must be 0. For example, if $cd = 0$ then either $c = 0$ or $d = 0$ or they are both 0.
- A quadratic equation always has two solutions (or roots). Sometimes these solutions may be equal.
- Expressions such as $(x + 1)^2$, $(x + 4)^2$ and $(x + \frac{1}{2})^2$ are all called **perfect squares**.
- Expressions like $x^2 + bx + c$ can be written in the form $\left(x + \frac{b}{2}\right)^2 - \left(\frac{b}{2}\right)^2 + c$.
  This process is called **completing the square**.
- Expressions like $ax^2 + bx + c$ are rewritten as $a\left(x^2 + \frac{b}{a}x\right) + c$ before completing the square for the expression inside the brackets.

- By completing the square, any quadratic expression can be written in the form $p(x + q)^2 + r$.
- Similarly, any quadratic equation can be written in the form $p(x + q)^2 + r = 0$.
- All quadratic equations can be solved by the formula

$$x = \frac{-b \pm \sqrt{b^2 - 4ac}}{2a}$$

- If the value of $b^2 - 4ac$ is negative, the quadratic equation does not have any real solutions.
- Equations with algebraic fractions sometimes lead to quadratic equations.
- To find the equation to represent a problem:
  - where relevant, draw a diagram and put all of the information on it
  - use x to represent the unknown which you have been asked to find
  - use other letters to identify any other relevant unknowns
  - look for information in the question which links these letters to x and write them down
  - try simple numbers for the unknowns and see if this helps you to find a method
  - make sure that the units on both sides of your equation are the same.
- The locus of a circle is the path of all points equidistant from a given fixed point.
- The **equation of a circle** with centre (0, 0) and radius r can be written as $x^2 + y^2 = r^2$.
- You can solve a pair of simultaneous equations where one equation is linear and the other quadratic by an algebraic method or a graphical approach.
- The solution of a pair of simultaneous equations where one is linear and one is quadratic is represented by the points of intersection of a straight line and a quadratic curve.
- You can solve simultaneous equations where one is linear and one is the equation of a circle graphically and algebraically.
- If the solutions are not integer values they can only be estimated using the graphical method and can be calculated using the quadratic formula.

Review exercise

1. Solve the following equations   **a** $x^2 = 9$ **b** $2x^2 = 72$ **c** $2x^2 - 108 = 0$

2. Solve the following equations   **a** $4 - y^2 = 0$ **b** $\dfrac{t^2}{4} = 1$ **c** $\dfrac{p^2}{3} - 3 = 0$

3. The diagram shows a trapezium.
   The lengths of three sides of the trapezium are
   $x - 5$, $x + 2$ and $x + 6$.
   All measurements are given in centimetres.
   The area of the trapezium is 36 cm².
   Find the length of the shortest side of the trapezium.

   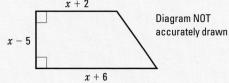

   Diagram NOT accurately drawn

4. Solve the following equations
   a $y^2 - 5y - 6 = 0$ **b** $4t^2 - 16 = 0$ **c** $4 - 3p - p^2 = 0$

5. Solve these simultaneous equations.
   a $2x + y = 11$ **b** $3x - y = -4$ **c** $4x - 3y = 7$
        $y = x + 7$ $y = 3 + 2x$ $y = 2x - 1$

6. For each of these pairs of simultaneous equations, draw two linear graphs on the same grid and use them to solve the simultaneous equations. Use a scale of -10 to $+10$ on each axis.
   a $y = 8 - 3x$ **b** $2x + y = 4$
        $x + y = 4$ $3x + 4y = 12$

B A03

7 A rectangular room is 2 metres longer than it is wide.
If its area is 52 m², what is its perimeter? Give your answer to 2 decimal places.

A03

8 The height of a ball above the ground in metres can be calculated from the formula:

$$h = 30t - 5t^2$$

where t = time in seconds after being thrown.
Find:
a the total time that the ball was in the air
b the maximum height of the ball above the ground
c the time at which the ball was 25 cm above the ground.

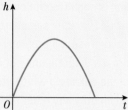

A

9 Solve the following equations
a $2k^2 - 11k + 5 = 0$ **b** $4m^2 - 4m = 3$ **c** $(2n - 1)(3n + 2) = 24$

10 $x^2 + 8x + 5$ can be written in the form $(x + p)^2 + q$.
a Find the value of p and the value of q.
b Use your answer to part **a** to solve the equation $x^2 + 8x + 5 = 0$.
Give your solutions to 3 significant figures.

11 Solve this quadratic equation $x^2 - 5x - 8 = 0$.
Give your answers correct to 3 significant figures.

June 2006

12 a Solve the equation $x^2 - 2x - 1 = 0$.
Give your answer correct to 3 significant figures.
Hence, or otherwise
b solve the equation $3x^2 - 6x - 3 = 0$.

June 2009

A02
A03

13 a $x^2 - 2x + 3$ is to be written in the form $(x + a)^2 + b$.
Find the values of a and b.
b Use your answer to part **a** to find the minimum value of $x^2 - 2x + 3$.
c Write down the value of x for which $x^2 - 2x + 3$ has a minimum value.
d Sketch the graph of $y = x^2 - 2x + 3$.

A02

14 Write $4x^2 + 24x$ in the form $a(x + p)^2 + q$. State the values of a, p and q.

A02

15 Solve these simultaneous equations.
a $2x + 3y = 10$ **b** $5x + 4y = 8$ **c** $2x + 3y = 1$
 $3x + 5y = 16$ $2x - 3y = -6$ $7x + 8y = -4$

A02
A03

16 A gas bill consists of a fixed charge (£F) and a charge (g pence) for each unit used.
Mrs Anwar used 350 units and paid £30. Mr White used 450 units and paid £35. Find the fixed charge and the charge per unit.

A

17 Write down the pair of simultaneous equations that are solved by the coordinates of the point of intersection of the two lines shown in the diagram.

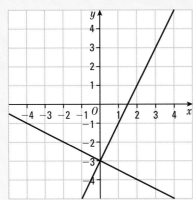

ResultsPlus

Exam Question Report

58% of students answered this sort of question well. They used all the information in the question.

18 Solve this pair of simultaneous equations

$4x + 3y = 4$

$2y = 1 - 3x$

a by drawing two linear graphs on the same grid
(use a scale of -10 to $+10$ on each axis)

b by using an algebraic method (you must show all of your working).

19 $x^2 - 8x + 23 = (x - p)^2 + q$ for all values of x.

a Find the value of p and the value of q.

Here is a sketch of the curve with equation

$\quad y = x^2 - 8x + 23$

B is the minimum point on the curve.

b Find the coordinates of B.

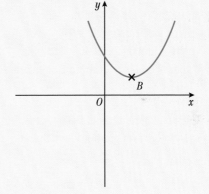

A02

June 2006

20 Kate buys 2 lollies and 5 choc ices for £6.50.

Pete buys 2 lollies and 3 choc ices for £4.30.

Work out the cost of 1 lolly.

Give your answer in pence.

A03

June 2007

21 The diagram shows a sketch of the graph of $y = 3(x^2 - x)$.

The line $y = 4 - 4x$ intersects the curve $y = 3(x^2 - x)$ at the points A and B.

Use an algebraic method to find the coordinates of A and B.

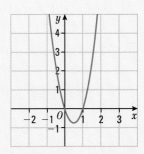

A02

Nov 2005

445

A☆

22 **a** Show that the equation $\dfrac{5}{x+2} = \dfrac{4-3x}{x-1}$ can be rearranged to give $3x^2 + 7x - 13 = 0$.

b Solve $3x^2 + 7x - 13 = 0$.

Give your solutions correct to 2 decimal places.

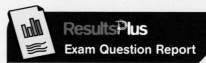

ResultsPlus
Exam Question Report

80% of students answered this question poorly because they did not take care with the brackets in the denominators.

June 2008

A02

23 The diagram shows a 6-sided shape.

All the corners are right angles.

All the measurements are given in centimetres.

Diagram **NOT** accurately drawn

The area of the shape is 95 cm².

a Show that $2x^2 + 6x - 95 = 0$.

b Solve the equation $2x^2 + 6x - 95 = 0$.

Give your answers correct to 3 significant figures.

Nov 2008

A02
A03

24 Sean runs in a 20 km fun run. He runs the first 10 km at a speed of x km per hour. He runs the second 10 km at a speed 1 km per hour less than the first 10 km. His total time for the fun run is 4 hours.

a Show that $\dfrac{10}{x} + \dfrac{10}{x-1} = 4$.

b Show x satisfies the quadratic equation $4x^2 - 24x + 10 = 0$.

c Find the two solutions of the equation $4x^2 - 24x + 10 = 0$.

Give your answers correct to 3 significant figures.

d What was Sean's speed for the second 10 km?

25 Solve these simultaneous equations. Give your answers correct to 3 significant figures.

a $x^2 + y^2 = 19$
   $y = x + 5$

b $x^2 + y^2 = 45$
   $y = 6 + 2x$

A02

26 Solve the equation $x^2 - 4x + 2 = 0$.

A03

27 Show that any straight line that passes through the point $(1, 2)$ must intersect the curve with equation $x^2 + y^2 = 16$ at two points.

June 2006

28 The diagram shows a circle, radius 5 units, centre the origin.

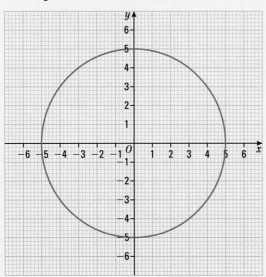

Use the diagram to find estimates of the solutions to the equations:

$x^2 + y^2 = 25$

$y = 2x + 1.$

Nov 2005, adapted

The photo shows a work of art by the artists Christo and Jeanne-Claude in which they wrapped the Pont Neuf Bridge in Paris in 40,876 m² (454,178 sq ft) of silky golden fabric. To wrap these buildings, they needed to work out the surface area and calculate the amount of fabric required.

◉ Objectives

In this chapter you will:
- find the length of an arc, the area of a sector of a circle, and answer problems involving circles in terms of π
- convert between units of area and volume
- work out the volume and surface area of 3D shapes
- specify and find points in three dimensions.

◈ Before you start

You need to:
- be able to solve problems involving perimeters and areas
- know and be able to use the formulae for the circumference and area of a circle
- be able to work out the volume of cuboids, prisms and cylinders.

23.1 Sectors of circles

Objectives

- You can work out the length of an arc of a circle.
- You can work out the area of a sector of a circle.
- You can solve problems involving arc lengths and areas of sectors of circles, including finding the area of a segment of a circle.

Why do this?

Astronomers often use the arc length of a circle when working out distances in space.

Get Ready

1. Here is a circle and a sector with an angle of 60°.
   a How many of these sectors will fill the circle without overlapping?
   b What fraction of the circle is the sector?
   c What fraction of the circle is a sector with an angle of 40°?

Key Points

- For a sector with angle $x°$ of a circle with radius r:

  sector $= \dfrac{x}{360}$ of the circle so

  area of sector $= \dfrac{x}{360} \times \pi r^2$

  and arc length $= \dfrac{x}{360} \times 2\pi r$

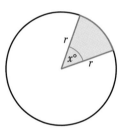

Example 1

For this sector of a circle, work out:
a the arc length
b the perimeter.
Give your answers correct to 3 significant figures.

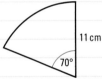

a arc length $= \dfrac{70}{360} \times 2 \times \pi \times 11$ ⟵ Use arc length $= \dfrac{x}{360} \times 2\pi r$ with $x = 70$ and $r = 11$ cm.

arc length $= 13.4390\ldots$ ⟵ Write down at least 4 figures of the calculator display.

arc length $= 13.4$ cm ⟵ Give the answer correct to 3 significant figures. The units are the same as the radius (cm).

b perimeter $= 13.4390 + 2 \times 11$ ⟵ perimeter = arc length + two radii.

perimeter $= 35.4390\ldots$ ⟵ Use the unrounded value for the arc length.

perimeter $= 35.4$ cm ⟵ Give the answer correct to 3 significant figures.

Example 2 Calculate the area of this sector.
Give your answer correct to 3 significant figures.

$$\text{area} = \frac{130}{360} \times \pi \times 9^2$$

area of sector $= \dfrac{x}{360} \times \pi r^2$
with $x = 130$ and $r = 9$ m.

$$\text{area} = 91.8915\ldots$$

$$\text{area of sector} = 91.9\,\text{m}^2$$

r is in metres so area is in m^2.

Exam Tip

Always write down the angle at the centre of the circle as a fraction of 360.

Exercise 23A

Questions in this chapter are targeted at the grades indicated.

In this exercise, if your calculator does not have a π button, take the value of π to be 3.142.
Give answers correct to 3 significant figures unless the question says differently.

A

1 Calculate the arc length of each of these sectors.

a 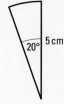 **b** 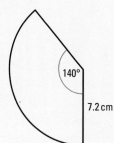 **c**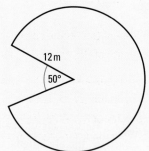

2 Calculate the area of each of these sectors.

a 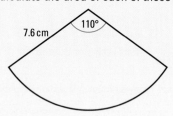 **b** 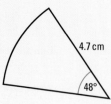 **c**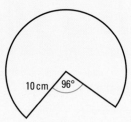

3 Calculate the perimeter of each of these sectors.

a 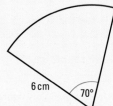 **b**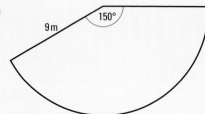

4 The diagram shows a sector of a circle of radius 5.5 cm.
The length of the arc of the sector is 6.72 cm.
Work out the size of the angle, x, of the sector.
Give your answer to the nearest degree.

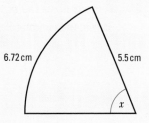

A02
A03
A☆

5 The diagram shows a sector of a circle.
The area of the sector is 17.453 cm².
Work out the radius of the circle.
Give your answer correct to the nearest cm.

A02
A03

6 The diagram shows information about the throwing circle and the landing area for a discus competition.
The radius of the throwing circle is 1.25 m.
Distances are measured from the front of the circle as shown.
The discus must land in the sector shown green in the diagram. The angle of the sector is 40°.
The winning throw in the men's discus in the 2008 Olympics was 68.82 m.
Calculate:

a the area of the region shown green

b the length of the arc of the landing area sector.

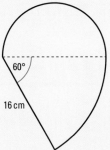

A02
A03

7 Here is a shape made from a sector of a circle of radius 16 cm and a semicircle.
The angle of the sector is 60°.
The diameter of the semicircle is a radius of the sector.

a Work out the perimeter of the shape.

b Work out the area of the shape.

A02
A03

23.2 **Problems involving circles in terms of π**

◎ Objectives

- ◯ You can give answers in terms of π.
- ◯ You can solve problems when the information is given in terms of π.

⟨?⟩ Why do this?

It is impossible to calculate the exact value of π so when scientists or architects need an exact answer it has to be given in terms of π.

⟨⟩ Get Ready

1. Simplify **a** $2 \times a \times 5$ **b** $\dfrac{12 \times b}{4}$
2. Find the value of x when $4ax = 18a$

Example 3 A circle has a radius of 6 cm. Find: **a** the circumference of the circle **b** the area of the circle. Give your answers in terms of π.

a $C = 2 \times \pi \times 6 = 12\pi$ ⟵ Use $C = 2\pi r$ with $r = 6$ cm.

Circumference $= 12\pi$ cm ⟵ $2 \times 6 = 12$
Do not forget the units.

b $A = \pi \times 6^2 = \pi \times 36 = 36\pi$ ⟵ Use $A = \pi r^2$ with $r = 6$ cm.

Area $= 36\pi$ cm^2 ⟵ $6^2 = 36$

A02 A03

Example 4 The perimeter of this sector is $(2r + 12\pi)$ m.
Find the radius, r m, of the sector.

r m

arc length $= \frac{90}{360} \times 2 \times \pi \times r = \frac{1}{4} \times 2\pi \times r$ ⟵ Find the arc length of the sector.

arc length $= \frac{1}{2}\pi r$

perimeter $= 2r + \frac{1}{2}\pi r$ ⟵ Add $2r$ to the arc length to find the perimeter of the sector.

$2r + \frac{1}{2}\pi r = 2r + 12\pi$ ⟵ Use the given expression $2r + 12\pi$.

$\frac{1}{2}\pi r = 12\pi$ ⟵ Solve for r.

$\frac{1}{2}r = 12$

$r = 2 \times 12 = 24$
Radius of sector $= 24$ m

Exercise 23B

D

1 Giving your answer in terms of π, find the circumference of a circle:
   a with diameter 9 cm **b** with diameter 1 m **c** with radius 26 mm

2 Giving your answer in terms of π, find the area of a circle:
   a with radius 2 cm **b** with radius 10 m **c** with diameter 40 m

C **A02 A03**

3 The diagram shows two circles with the same centre.
   a Show that the area of the coloured region between the two circles is 64π cm^2.
   b Find, in terms of π, the circumference of a circle whose area is the same as the area of the coloured region.

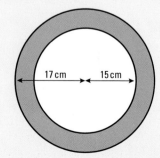

17 cm ⟷ 15 cm

4 The blue arc and the red arc make a complete circle.
The length of the blue arc is 6π cm and the length of the red arc is 10π cm.
   a Find the radius of the circle.
   b Find, in terms of π, the area of the circle.

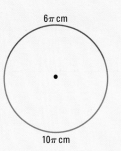

6π cm

10π cm

B

5 Find, in terms of π:
   a the area of this sector
   b the arc length of the sector
   c the perimeter of the sector.

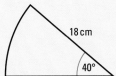

18 cm

40°

A03

A

6 A shape is made by drawing three semicircles
on the line AB as shown. Two of the semicircles
have diameter X cm and Y cm as shown.
Show that the perimeter of the shape is the same
as the circumference of a circle with diameter
$(X + Y)$ cm.

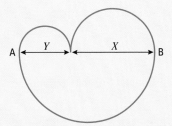

A   Y   X   B

A02

23.3 Units of area

◉ Objective

◉ You can change between m² and cm², between
cm² and mm² and between km² and m².

❓ Why do this?

It is sometimes necessary to change from one unit
to another, for example, a garden design may have
measurements in both metres and centimetres.

⬆ Get Ready

1. Work out   **a** 10×10 **b** 100×100 **c** 1000×1000
2. Work out   **a** 5×100 **b** $13 \times 10\,000$ **c** $7.6 \times 1\,000\,000$
3. Work out   **a** $7000 \div 100$ **b** $3500 \div 10\,000$ **c** $49\,000 \div 1\,000\,000$

◉ Key Points

◉ The diagram shows two identical squares.
The sides of square A are measured in metres and the sides of square B are measured in centimetres.

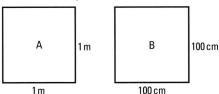

A   1 m

B   100 cm

1 m   100 cm

Square A is 1 m by 1 m so the area of square A is 1×1 m² = 1 m².
Square B is 100 cm by 100 cm so the area of square B is 100×100 cm² = 10 000 cm².
The squares have the same area so 1 m² = 100×100 cm² = 10 000 cm².

There are similar results for other units.

Length	Area
1 cm = 10 mm	1 cm² = 10 × 10 = 100 mm²
1 m = 100 cm	1 m² = 100 × 100 = 10 000 cm²
1 km = 1000 m	1 km² = 1000 × 1000 = 1 000 000 m²

divide by 100　　　divide by 10 000　　　divide by 1 000 000

mm² 　　　 cm² 　　　 m² 　　　 km²

multiply by 100　　multiply by 10 000　　multiply by 1 000 000

Example 5 Convert 4.6 m² to cm².

$$4.6\,\text{m}^2 = 4.6 \times 10\,000\,\text{cm}^2$$
$$= 46\,000\,\text{cm}^2$$

1 m² = 10 000 cm²
Multiply the number of m² by 10 000.

ResultsPlus
Watch Out!

Remember that to change from a larger unit to a smaller unit, you multiply.

Example 6 Convert 870 mm² to cm².

$$870\,\text{mm}^2 = 870 \div 100\,\text{cm}^2 = 8.7\,\text{cm}^2$$

1 cm² = 100 mm²
So 1 mm² = $\frac{1}{100}$ cm²
Divide the number of mm² by 100.

Exercise 23C

D

1 Work out the area of this circle in:
   a cm²　　　　b m².

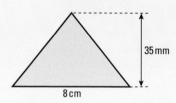

120 cm

2 Work out the area of this triangle in:
   a cm²　　　　b mm².

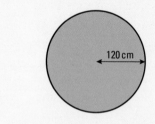

35 mm

8 cm

C

3 Convert to cm².
   a 4 m²　　　　b 6.9 m²　　　　c 600 mm²　　　　d 47 mm²

4 Convert to m².
   a 5 km²　　　　b 0.3 km²　　　　c 40 000 cm²　　　　d 560 cm²

5 a How many mm are there in 1 m?　　　b How many mm² are there in 1 m²?
   c Convert 8.3 m² to mm².

6 Find, in cm², the area of a rectangle:
   a 3.2 m by 1.4 m　　　　　　　　　　b 45 mm by 8 mm.

23.4 Volume of a pyramid and a cone

◎ You can work out the volume of a pyramid and a cone.

? **Why do this?**

A manufacturer of ice-cream cones would use the volume of a cone formula to work out the volume of ice-cream that each different-sized cone requires.

◈ **Get Ready**

1. Work out the area of:
   a a rectangle with sides 1.1×2.01 cm
   b a triangle of height 3.2 cm and base 9.1 mm
   c a circle with diameter 9.1 mm.

🔍 **Key Points**

◎ Volume of **pyramid** $= \frac{1}{3} \times$ area of base $\times$ vertical height
◎ Volume of **cone** $= \frac{1}{3} \times$ area of base $\times$ vertical height
$= \frac{1}{3}\pi r^2 h$

where r is the radius and h is the height.

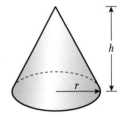

🔍 **Example 7**

A pyramid has a square base of side 3.6 m and a vertical height of 5 m.
Work out the volume of the pyramid.

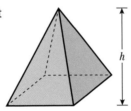

Volume of pyramid $= \frac{1}{3} \times 3.6 \times 3.6 \times 5$ ⟵ Use volume of pyramid $= \frac{1}{3} \times$ area of base $\times$ vertical height.

$= \frac{1}{3} \times 64.8 = 21.6 \text{ m}^3$ ⟵ Here the base is a square of side 3.6 m, so the area of the base $= 3.6 \times 3.6 \text{ m}^2$. The vertical height of the pyramid is 5 m.

🔍 **Example 8**

A cone has a circular base of radius 7 cm and a vertical height of 16.2 cm.
Work out the volume of the cone.
Give your answer correct to 1 decimal place.

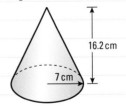

Volume of cone $= \frac{1}{3} \times \pi \times 7^2 \times 16.2$

$= \frac{1}{3} \times 3.142 \times 7 \times 7 \times 16.2$

$= \frac{1}{3} \times 2493.796248\ldots$

$= 831.2654161\ldots$

$= 831.3 \text{ cm}^3 \text{ (1 d.p.)}$

A cone is a pyramid with a circular base.
Use volume of pyramid $= \frac{1}{3} \times$ area of base $\times$ vertical height.
Here area of base $= \pi \times 7^2 \text{ cm}^2$ and vertical height $= 16.2$ cm.

Exercise 23D

A

1 Work out the volumes of these pyramids.

a

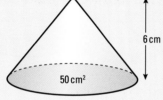

b

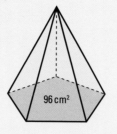

c

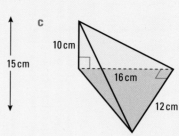

2 The Great Pyramid at Giza has height of 262 m and a square base of side 434 m.
Work out the volume of the Great Pyramid.

3 Work out the volumes of these cones. Use $\pi = 3.142$.
Give your answers correct to 3 significant figures.

a

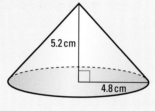

b

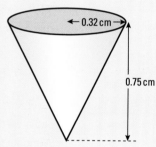

c

A03

4 The diagram shows a shape made from a cone with
base radius 4 cm and height 4 cm joined to a cone
with base radius 4 cm and height 6 cm.
Work out the total volume of the shape.
Give your answer in terms of π.

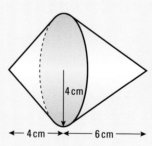

A★ **A02**
A03

5 The diagram shows a cone cut into two parts,
part A and part B. Work out the volume of part B.

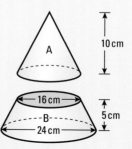

23.5 Volume of a sphere

◎ Objective

● You can work out the volume of a sphere.

⊘ Why do this?

A factory making footballs would have to work out the volumes of their different types of footballs to know how much to inflate them.

◈ Get Ready

1. Work out the area of a circle with diameter 6.4cm:

   a   in terms of π b   with $\pi = 3.142$

2. Calculate

   a 4^3 b 6^3 c 11^3

Key Point

● Volume of **sphere** $= \frac{4}{3}\pi r^3$, where r is the radius.

Example 9   The radius of a spherical raindrop is 1.8 mm. Work out the volume of the raindrop. Give your answer correct to 3 significant figures.

Volume of raindrop $= \frac{4}{3} \times \pi \times 1.8^3$

⟵ Use volume of sphere $= \frac{4}{3}\pi r^3$.
Here $r = 1.8$ mm.
Remember 1.8^3 means $1.8 \times 1.8 \times 1.8$.

$= \frac{4}{3} \times 3.142 \times 1.8 \times 1.8 \times 1.8$

$= \frac{4}{3} \times 18.324144$

$= 24.432192$

$= 24.4 \text{ mm}^3$ (3 s.f.)

Example 10   The volume of a spherical ball is 0.86 m³. Work out the radius of the ball. Give your answer correct to 3 significant figures. Use $\pi = 3.142$.

Let the radius of the ball $= r$ metres. ⟵ Write down an equation in terms of r, the radius of the spherical ball.

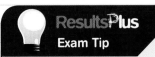
ResultsPlus
Exam Tip

Show all stages in your working.

So, $\frac{4}{3}\pi r^3 = 0.86$ ⟵ The volume of the ball is 0.86 m³, so $\frac{4}{3}\pi r^3 = 0.86$.

$3 \times \frac{4}{3}\pi r^3 = 2.58$

$\pi r^3 = 0.645$

$r^3 = 0.2052832591$

$r = \sqrt[3]{0.2052832591}$

$= 0.5899083062$

radius $= 0.590$ m (3 s.f.)

⟵ To make r the subject of the equation, multiply both sides by 3, divide both sides by 4, divide both sides by 3.142 and then take the cube root of both sides.

Exercise 23E

In the following questions, use the value of π on your calculator and give your answers correct to 3 significant figures.

1 A spherical soap bubble has a radius of 4 cm. Work out the volume of the bubble.

2 A hemispherical dome has a diameter of 25 m. Work out the volume of the dome.

3 The volume of a sphere is 2400 cm³. Work out the radius of the sphere.

4 The dimensions of a cuboid are 30 cm × 20 cm × 40 cm. The volume of a sphere has the same volume as the cuboid. Work out the radius of the sphere.

5 A spherical ball is made from plastic. The external and internal diameters of the ball are 50 cm and 49.5 cm, respectively. Work out the volume of plastic used to make the ball.

6 The volume of a sphere of radius r metres is twice the volume of a sphere of radius 4 metres. Work out the value of r.

23.6 Further volumes of shapes

◉ Objective

● You can work out the volumes of harder shapes made from cuboids, cylinders, cones, pyramids and spheres.

⍰ Why do this?

Many toys, such as a wooden toy trains, are made up of combinations of simple shapes, such as cuboids and cylinders.

⬥ Get Ready

1. Calculate the volume of:

a a cylinder of radius 2 cm and length 5 mm

b a triangular pyramid where the length of the base of the triangle is 12 cm, the height of the base triangle is 6 cm and the height of the pyramid is 24 cm.

◉ Key Point

◉ To work out the volume of a composite shape, work out the volumes of the shapes it is made from and add the volumes together.

🔍 Example 11

Work out the volume of this shape.

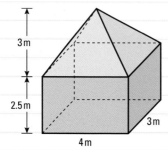

3 m

2.5 m

4 m

3 m

> The shape is made from a cuboid and a pyramid.

Volume of cuboid $= 4 \times 3 \times 2.5 = 30 \, m^3$

> Work out the volume of the cuboid.
> Use volume of cuboid $= a \times b \times c$.
> Here $a = 4$ m, $b = 3$ m and $c = 2.5$ m.

Volume of pyramid $= \frac{1}{3} \times 4 \times 3 \times 3 = 12 \, m^3$

> Work out the volume of the pyramid.
> Use volume of pyramid $= \frac{1}{3} \times$ area of base $\times$ vertical height.
> Here area of the base $= 4 \times 3 \, m^2$ and vertical height is 3 m.

Total volume $= 30 + 12 = 42 \, m^3$ ←

> Work out the total volume. Add the volume of the
> cuboid and the volume of the pyramid.

Example 12 ▶ Work out the volume of this shape. Leave your answer in terms of π.

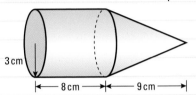

> The shape is made from a cylinder
> and a cone.

3 cm

|← 8 cm →|← 9 cm →|

Volume of cylinder $= \pi \times 3^2 \times 8 = 72\pi \, cm^3$

> Work out the volume of the cylinder.
> Use volume of cylinder $= \pi r^2 h$. Here $r = 3$ cm and $h = 8$ cm.

Volume of cone $= \frac{1}{3} \times \pi \times 3^2 \times 9 = 27\pi \, cm^3$

> Work out the volume of the cone.
> Use volume of cone $= \frac{1}{3} \times$ area of base $\times$ vertical height
> $= \frac{1}{3}\pi r^2 h$.
> Here $r = 3$ cm and $h = 9$ cm.

Total volume $= 72\pi + 27\pi = 99\pi \, cm^3$ ←

> Work out the total volume in terms of π. Add the
> volume of the cylinder and the volume of the cone.

Exercise 23F

1 Work out the volumes of these shapes. Give your answers correct to 3 significant figures.

A02
A03
A

a

5 cm
5 cm
5 cm
5 cm
5 cm

b

6 m
7 m
5 m

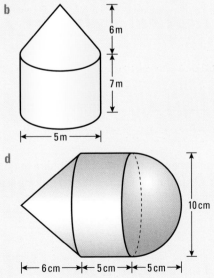

c

7.5 m
1.25 m
1.75 m
8 m

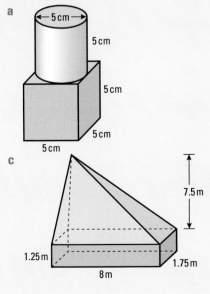

d

10 cm
|← 6 cm →|← 5 cm →|← 5 cm →|

A AO2 AO3

2 A cone is joined to a cylinder, as shown in the diagram.
Work out the total volume of the shape. Give your answer in terms of π.

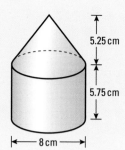

5.25 cm

5.75 cm

8 cm

AO2 AO3

3 A hemisphere is joined to a cylinder, as shown in the diagram.
Work out the volume of the shape.
Give your answer in terms of π.

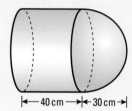

40 cm ── 30 cm

A★ AO2

4 A wooden cube of side 15 cm is used to make a spherical ball.
Work out the volume of wood that must be cut from the cube
to make a ball with the largest possible radius.

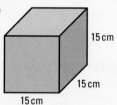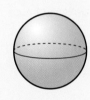

15 cm

15 cm

15 cm

AO2 AO3

5 A conical hole is cut into a cylinder to make the shape shown in the diagram.
Work out the volume of the shape. Give your answer in terms of π.

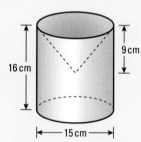

16 cm

9 cm

15 cm

AO2 AO3

6 A shape is made by joining a hemisphere of radius r cm to a cone of radius r cm.
The height of the cone is $2r$ cm. Find an expression, in terms of r and π, for
the volume of the shape.

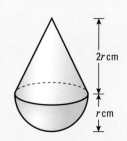

2r cm

r cm

23.7 Units of volume

◎ Objectives

○ You can convert between volume measures in metric units
○ You can convert between units of volume and units of capacity.

? Why do this?

Factories need to convert between units of volume when they pack small boxes, like DVDs, into larger boxes for transporting.

◈ Get Ready

1. Calculate:

a $1000 \times 1000 \times 1000$ **b** 4.62×1000 **c** $35\,\text{m} \times 125\,\text{cm} \times 2.1\,\text{m}$

Key Points

◉ $1\,\text{m}^3 = 1\,000\,000\,\text{cm}^3$ $1\,\text{cm}^3 = \frac{1}{1\,000\,000}\,\text{m}^3$

◉ $1\,\text{cm}^3 = 1000\,\text{mm}^3$ $1\,\text{mm}^3 = \frac{1}{1000}\,\text{cm}^3$

◉ Litres are often used to measure the capacity or amount a container can hold.

◉ 1 litre $= 1000\,\text{cm}^3$

◉ $1\,\text{cm}^3 = 1\,\text{m}l$

Example 13   Convert $4\,\text{m}^3$ to cm^3.

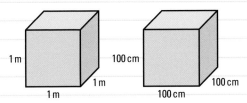

Convert m^3 to cm^3.
Draw a cube of side 1 m.
Now 1 m = 100 cm,
so 1 m × 1 m × 1 m =
100 cm × 100 cm × 100 cm.

$1\,\text{m}^3 = 100\,\text{cm} \times 100\,\text{cm} \times 100\,\text{cm} = 1\,000\,000\,\text{cm}^3$ ← Replace $1\,\text{m}^3$ with $1\,000\,000\,\text{cm}^3$.

So $4\,\text{m}^3 = 4 \times 1\,000\,000\,\text{cm}^3 = 4\,000\,000\,\text{cm}^3$

Example 14   Convert $358\,\text{mm}^3$ to cm^3.

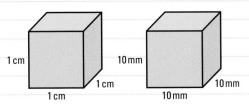

Convert cm^3 to mm^3.
Draw a cube of side 1 cm.
Now 1 cm = 10 mm,
so 1 cm × 1 cm × 1 cm =
10 mm × 10 mm × 10 mm.

$1\,\text{cm}^3 = 10\,\text{mm} \times 10\,\text{mm} \times 10\,\text{mm} = 1000\,\text{mm}^3$ ←

So $1\,\text{mm}^3 = \frac{1}{1000}\,\text{cm}^3$

Convert mm^3 to cm^3.
$1\,\text{cm}^3 = 1000\,\text{mm}^3$, so $1\,\text{mm}^3 = \frac{1}{1000}\,\text{cm}^3$.
Replace $1\,\text{mm}^3$ with $\frac{1}{1000}\,\text{cm}^3$,
so $358\,\text{mm}^3 = 358 \times \frac{1}{1000}\,\text{cm}^3$.

$358\,\text{mm}^3 = 358 \times \frac{1}{1000}\,\text{cm}^3$

$= 0.358\,\text{cm}^3$

Example 15
  a Convert 3.5 litres to cm³.

  b Convert 17 000 cm³ to litres.

a $3.5 \times 1000 = 3500\,cm^3$

b $17\,000 \div 1000 = 17$ litres

Exercise 23G

C

1 Convert these to cm³.
    a 2 m³   b 6.75 m³   c 450 mm³   d 6.8 mm³

2 Convert these to mm³.
    a 7 cm³   b 3.75 cm³   c 0.025 cm³

3 Convert these to m³.
    a 75 000 cm³   b 800 cm³   c 125 000 mm³

4 Convert to litres.
    a 830 m*l*   b 5600 cm³   c 1 m³   d 3540 mm³

A02

5 A swimming pool has length 50 m, width 9 m and depth 1.6 m.
How much water does it hold?
Give your answer in litres.

A02
A03

6 A cylinder hold 34.5 litres of molten metal.
The metal is to be made into cubes of side 3 cm.
How many cubes can be made?

B
A03

7 The table gives the volumes of three shapes.
Which shape has the greatest volume?
Give a reason for your answer.

Shape	Volume
A	1.25×10^7 cm³
B	2.45 m³
C	3.75×10^8 mm³

23.8 Surface area of a prism

◎ Objective

● You can work out the surface area of a prism.

② Why do this?

Tents are an example of triangular prisms. Larger tents have a greater surface area of fabric and can accommodate more people.

⬆ Get Ready

1. Work out the area of a triangle with base length 5 cm and height 6 cm.
2. Work out the area of a circle with radius 3.4 mm.

🔍 **Key Point**

⦿ To work out the surface area of a shape, work out the surface area of each of the sides and add them together.

🔖 **Example 16** Work out the surface area of this cuboid.

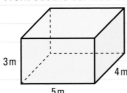

A cuboid has 6 faces. Each face is a rectangle. Work out the areas of the rectangles. Opposite faces of a cuboid have the same area so there are 3 repeated areas.

Surface area = (3 × 5) + (3 × 5) + (3 × 4)
           + (3 × 4) + (5 × 4) + (5 × 4)
         = 15 + 15 + 12 + 12 + 20 + 20
         = 94 m²

Give the unit with your answer. The lengths of the sides are in m, so the unit of area is m².

🔖 **Example 17** Work out the surface area of this triangular prism.

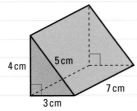

A triangular prism has 5 sides: 2 triangles (with equal areas) and 3 rectangles. Work out the area of the triangular side. Use area of triangle = $\frac{1}{2}$ × base × height. Here base = 3 cm and height = 4 cm.

Surface area = ($\frac{1}{2}$ × 3 × 4) + ($\frac{1}{2}$ × 3 × 4)
           + (5 × 7) + (4 × 7) + (3 × 7)
         = 6 + 6 + 35 + 28 + 21
         = 96 cm²

Results**Plus**
Watch Out!

Make sure you are adding the areas of the right number of sides.

⚙ **Exercise 23H**

1 Work out the surface areas of these shapes. Give the units with your answers.

D

a
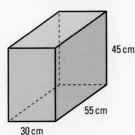

b
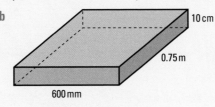

D

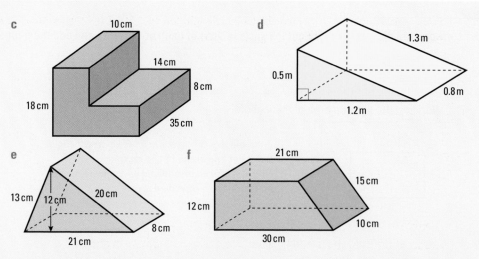

c

10 cm

14 cm

8 cm

18 cm

35 cm

d

1.3 m

0.5 m

0.8 m

1.2 m

e

13 cm 12 cm 20 cm

8 cm

21 cm

f

21 cm

15 cm

12 cm

10 cm

30 cm

C

A03

2 A room has dimensions 6.8 metres × 9.2 metres × 2.5 metres. Stephanie wants to paint the walls of the room. A large tin of paint covers 20 m² and costs £11.75. Ignoring windows and doors, and allowing for two coats of paint, how much will it cost Stephanie to paint the room?

23.9 **Further surface area of shapes**

◎ Objective

● You can work out the surface area of harder shapes made from cylinders, cones and spheres.

⍰ Why do this?

If you have an oddly shaped present that you need to wrap, you can work out the surface area of the present so that you know how much wrapping paper you will need.

⬆ Get Ready

1. What is the radius of a circle if the circumference is $6.4\,\pi$ cm?
2. What is the diameter of a circle if the area is $12.96\,\pi$ cm²?
3. What is the radius of a cylinder which has a volume of 44.064 mm³ and length 3.4 mm.

🔑 Key Points

● Total surface area of cylinder $= 2\pi rh + 2\pi r^2$, where r is the radius and h is the height.

● Total surface area of cone $= \pi r^2 + \pi rl$, where r is the radius and l is the slant height.

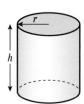

● Surface area of sphere $= 4\pi r^2$, where r is the radius.

Example 18

A cylindrical can has a radius of 4 cm and a height of 12 cm. Work out the surface area of the can. Use $\pi = 3.142$. Give your answer correct to 3 significant figures.

Surface area $= 2 \times \pi \times 4 \times 12 + 2 \times \pi \times 4^2$

$= 2 \times 3.142 \times 4 \times 12 +$
$\quad 2 \times 3.142 \times 4 \times 4$

$= 301.632 + 100.544$

$= 402.176$

$= 402 \text{ cm}^2 \text{ (3 s.f.)}$

> Use total surface area of cylinder $= 2\pi rh + 2\pi r^2$.
> Here $r = 4$ cm and $h = 12$ cm.
> Replace π with 3.142.
> Remember $4^2 = 4 \times 4$.
> Write down all the figures on your calculator display.
> Give your final answer correct to 3 significant figures.
> Remember to give the units with your answer.

Example 19

A cone has a radius of 3 cm and a vertical height of 4 cm.

Work out the total surface area of the cone. Give your answer in terms of π.

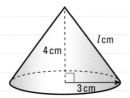

> To work out the total surface area of a cone you need to find the slant height of the cone. Label the slant height, l, in the diagram.

Let the slant height $= l$ cm.

Using Pythagoras' Theorem,

$l^2 = 4^2 + 3^2$

$\quad = 16 + 9 = 25$

$l = \sqrt{25} = 5 \text{ cm}$

> Remember Pythagoras' theorem: $c^2 = a^2 + b^2$.
> Here $c = l$, $a = 4$ cm and $b = 3$ cm.

Total surface area of cone

$= \pi \times 3^2 + \pi \times 3 \times 5$

$= 9\pi + 15\pi$

$= 24\pi \text{ cm}^2$

> Use total surface area of cone $= \pi r^2 + \pi l$.
> Here $r = 3$ cm and $l = 5$ cm.

Example 20

A hemisphere has a radius r cm.

Show that the total surface area of the hemisphere is $3\pi r^2 \text{ cm}^2$.

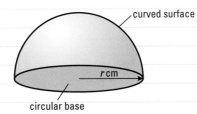

circular base

curved surface

> A hemisphere is half a sphere, so the area of the curved surface $= \frac{1}{2} \times 4\pi r^2$.
> The area of the circular base $= \pi r^2$, so the total surface area of the hemisphere $= \frac{1}{2} \times 4\pi r^2 + \pi r^2$.

Total surface area $= \frac{1}{2} \times 4\pi r^2 + \pi r^2$

$= 2\pi r^2 + \pi r^2$

$= 3\pi r^2 \text{ cm}^2$

Exercise 23I

1 Work out the total surface areas of these cylinders. Give your answers correct to 1 decimal place.

a
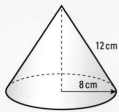
4.5 cm
4 cm

b
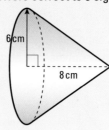
0.78 m
0.95 m

2 Work out the total surface areas of these cones. Give your answers correct to 3 significant figures.

a
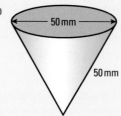
12 cm
8 cm

b
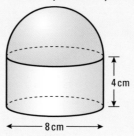
50 mm
50 mm

c
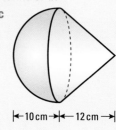
6 cm
8 cm

3 Work out the total surface areas of these shapes. Give your answers in terms of π.

a
5 cm

b
4 cm
8 cm

c
10 cm 12 cm

4 Work out the total surface areas of these prisms.

a
3 cm
5 cm
3 cm
5 cm
5 cm

b
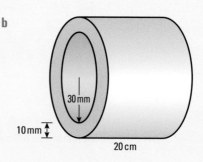
30 mm
10 mm
20 cm

23.10 Coordinates in three dimensions

◎ Objective

● You can use axes and coordinates to specify and find points in three dimensions.

⦿ Why do this?

GPS systems make their calculations in three dimensions.

◈ Get Ready

1. AB is a section of a line between A (6, 1) and B (−3, 5). Find the midpoint.

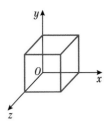

Key Points

- To locate a point in two dimensions, two perpendicular axes are used, the x-axis and the y-axis, and two coordinates are given, the x-coordinate and the y-coordinate.
- In three dimensions, an extra axis is needed, the z-axis.
- The three axes are perpendicular to each other.
- The position of a point is given by three coordinates: the x-coordinate, the y-coordinate and the z-coordinate.
- The coordinates of a point are written (x, y, z).

Example 21

The diagram represents a cuboid on a 3D grid.
$OR = 2$ units, $OP = 4$ units and $OS = 3$ units.
Find the coordinates of

a S　　b P　　c R
d V　　e U　　f O

g Find the midpoint of PV.

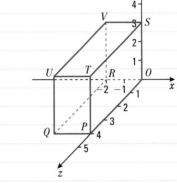

a $S = (0, 3, 0)$ ← To get to S from O you go 0 along the x-axis, 3 units up the y-axis and O units parallel to the z-axis.

b $P = (0, 0, 4)$ ← Remember to give the coordinates in the order (x, y, z).

c $R = (-2, 0, 0)$

d $V = (-2, 3, 0)$

e $U = (-2, 3, 4)$ ← To get to U from O you go -2 units along the x-axis, 3 units parallel to the y-axis and 4 units parallel to the z-axis.

f $O = (0, 0, 0)$

g Midpoint of $PV = \left(\dfrac{0 + -2}{2}, \dfrac{0 + 3}{2}, \dfrac{4 + 0}{2} \right) = (-1, 1\tfrac{1}{2}, 2)$

Exercise 23J

1 Write down the coordinates of each vertex of this cuboid.

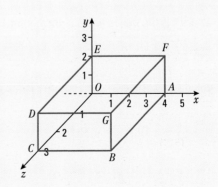

C

C

2 Draw a diagram to show the points $A(1, 0, 0)$, $B(1, 0, 3)$ and $C(1, 2, 3)$.

3 **a** Write down the coordinates of each vertex of this cuboid.
   b Write down the midpoints of
      i DG
      ii EB

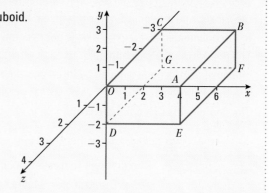

B **AO3**

4 The coordinates of five of the corners of a cuboid are
$(1, 0, 0)$, $(1, -3, 0)$, $(1, -3, -1)$, $(1, 0, -1)$ and $(-2, 0, 0)$.
Find the coordinates of the other three corners.

Chapter review

- For a sector with angle $x°$ of a circle with radius r:

   area of sector $= \dfrac{x}{360} \times \pi r^2$

   and arc length $= \dfrac{x}{360} \times 2\pi r$

Length	Area
1 cm = 10 mm	1 cm² = 10 × 10 = 100 mm²
1 m = 100 cm	1 m² = 100 × 100 = 10 000 cm²
1 km = 1000 m	1 km² = 1000 × 1000 = 1 000 000 m²

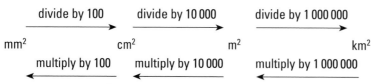

- Volume of **pyramid** $= \frac{1}{3} \times$ area of base $\times$ vertical height
- Volume of **cone** $= \frac{1}{3} \times$ area of base $\times$ vertical height
   $= \frac{1}{3}\pi r^2 h$

where r is the radius and h is the height.

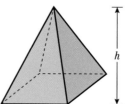
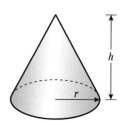

- Volume of **sphere** $= \frac{4}{3}\pi r^3$, where r is the radius.

- To work out the volume of a composite shape, work out the volumes of the shapes it is made from and add the volumes together.

- $1\,m^3 = 1\,000\,000\,cm^3$ $1\,cm^3 = \frac{1}{1\,000\,000}\,m^3$

- $1\,cm^3 = 1000\,mm^3$ $1\,mm^3 = \frac{1}{1000}\,cm^3$

- Litres are often used to measure the capacity or amount a container can hold.

- 1 litre $= 1000\,cm^3$

- $1\,cm^3 = 1\,ml$

- To work out the surface area of a shape, work out the surface areas of each side of the shape and add them together.

- Total surface area of cylinder $= 2\pi rh + 2\pi r^2$, where r is the radius and h is the height.

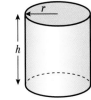

- Total surface area of cone $= \pi r^2 + \pi rl$, where r is the radius and l is the slant height.

- Surface area of sphere $= 4\pi r^2$, where r is the radius.

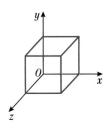

- To locate a point in three dimensions, three perpendicular axes are used, the x-axis, the y-axis and the z-axis.

- The position of a point is given by three coordinates: the x-coordinate, the y-coordinate and the z-coordinate.

- The coordinates of a point are written (x, y, z).

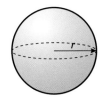

Review exercise

1   Work out the total surface area of the triangular prism.

D

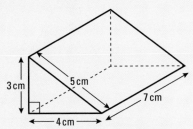

Give the units with your answer.

June 2008

D

2   Work out the total surface area of the L-shaped prism.
State the units with your answer.

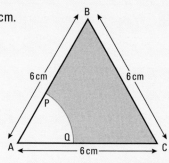

Diagram **NOT**
accurately drawn

June 2007

C

3   The volume of this cube is 8 m³
Convert 8 m³ to cm³.

June 2007

A02

4   Convert to cm².
   a   450 mm²   b   6 m²

5   The area of a large farm is 6 540 000 m².
Convert 6 540 000 m² to km².

B

6   The diagram shows a storage tank.
The storage tank consists of a hemisphere on top of a cylinder.
The height of the cylinder is 30 metres.
The radius of the cylinder is 3 metres.
The radius of the hemisphere is 3 metres.
   a   Calculate the total volume of the storage tank.
      Give your answer correct to 3 significant figures.
A sphere has a volume of 500 m³.
   b   Calculate the radius of the sphere.
      Give your answer correct to 3 significant figures.

Nov 2008

A02
A03

7   Rainfall on a flat rectangular roof 10 m by 5.5 m flows into a cylindrical tub of diameter 3 m.
Find, in cm, the increase in depth of the water in the tub caused by a rainfall of 1.2 cm.
Give your answer correct to 2 significant figures.

A
A03

8   The volume of a cone with base radius $2x$ cm and height $5x$ cm is equal to the total surface area of a
cylinder with radius $3x$ cm and height h cm. Find an expression for h in terms of x.

9   The diagram shows an equilateral triangle ABC with sides of length 6 cm.
P is the midpoint of AB.
Q is the midpoint of AC.
APQ is a sector of a circle, centre A.

Calculate the area of the shaded region.
Give your answer correct to 3 significant figures.

June 2009

10 The diagram shows a sector of a circle, centre O.
The radius of the circle is 6 cm.
Angle $AOB = 120°$.

Work out the perimeter of the sector.
Give your answer in terms of π in its simplest form.

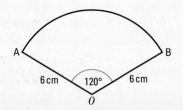

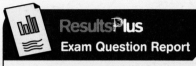

 11 OAD is a sector of a circle centre O radius 6 cm.
OBC is a sector of a circle centre O radius 8 cm.
OAB and ODC are straight lines.
Angle $COB = 45°$.
Find, in terms of π:

a the area of $ABCD$

b the perimeter of $ABCD$.

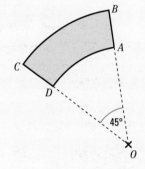

A02
A03

 12 A cylindrical bowl has a radius of 15 cm.
It is filled with water to a depth of 12 cm.
Work out the volume of water in the bowl.
Give your answer in litres as a multiple of π.

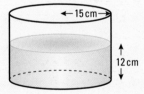

13 The diagram shows a cuboid drawn on a 3D grid.
Vertex A has coordinates (5, 2, 3).

a Write down the coordinates of vertex E.

B and D are vertices of the cuboid.

b Work out the coordinates of the midpoint of BD.

Diagram **NOT** accurately drawn

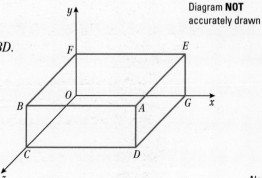

Nov 2008

A03

14 The diagram shows a cylinder and a sphere.

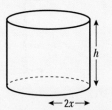 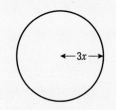

Diagram **NOT**
accurately drawn

The radius of the base of the cylinder is $2x$ cm and the height of the cylinder is h cm.
The radius of the sphere is $3x$ cm.
The volume of the cylinder is equal to the
volume of the sphere.
Express h in terms of x.
Give your answer in its simplest form.

ResultsPlus
Exam Question Report

91% of students answered this question poorly
because they did not apply the power to all the
parts of the expression.

June 2007

A02
A03

15 The diagram represents a large cone of height 30 cm
and base diameter 15 cm. The large cone is made by
placing a small cone A of height 10 cm and
base diameter 5 cm on top of a frustum B.
Calculate the volume of the frustum B.
Give your answer correct to 3 significant figures.

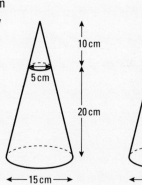 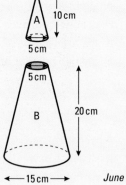

June 2003

A02
A03

*16 The diagram shows a pyramid and a cone. The volumes are
equal. Hamud says that the heights of the solids are equal.
Is he right? Give a reason for your answer.

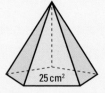 

A02
A03

17 Calculate the area of the shaded segment of this quarter circle.

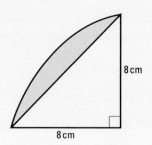

Life expectancy over time is one variable often represented using a line graph. The line for life expectancy in the UK shows a continual increase from 1980 to the present day. In 1980, a man could expect to live to an age of about 71 years whilst the average life expectancy for a woman was 77. By 2009, the life expectancy for both sexes had gone up considerably with average life expectancy for a baby girl at 81.5 years and for a baby boy at 77.2 years.

◎ Objectives

In this chapter you will:
- draw and interpret line graphs and scatter graphs
- distinguish between positive, negative and zero correlation
- draw and use lines of best fit.

◈ Before you start

You need to:
- understand how to draw, label and scale axes
- substitute numbers in simple algebraic expressions.

24.1 Drawing and using line graphs

⊕ **Get Ready**

1. Look at this set of axes.
   What does one small division represent on
   a the x-axis
   b the y-axis?

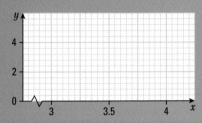

◐ **Key Points**

◉ Sometimes when sampling you take two observations from each selected member of the population. We call these bivariate data.
◉ Bivariate data consists of pairs of related variables.
◉ Pairs of observations can be plotted on a **line graph**.
◉ Time is often the variable along the horizontal axis in line graphs.

🔍 **Example 1** ▶ The table gives information about the close-of-day price for buying a share in a particular company during one week in June.

Day (June)	Wed 13th	Thur 14th	Fri 15th	Sat 16th	Sun 17th	Mon 18th	Tues 19th
Price per share (pence)	138	144	141	Closed		137	143

    a Draw a line graph for these data.
    b On which day was the share price at its highest?
    c On which day was the share price at its lowest?

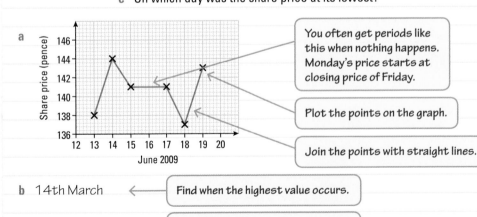

a

You often get periods like this when nothing happens. Monday's price starts at closing price of Friday.

Plot the points on the graph.

Join the points with straight lines.

b 14th March ← Find when the highest value occurs.

c 18th March ← Find when the lowest value occurs.

Example 2   The line graph shows the temperature, in °C, at different times of the day during a day in April.

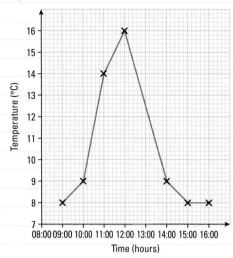

The temperature for 13:00 hours was missed.

a Estimate the temperature at 13:00 hours.

b Estimate the times when the temperature was 8.5°C.

c What was the highest temperature reached and at what time did it occur?

d In which hour did the temperature rise most quickly?

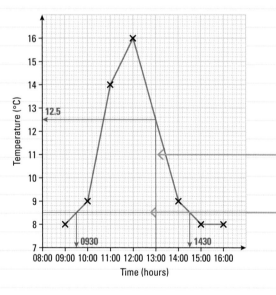

To estimate the temperature at 13:00 hours, draw a line up from 13:00 until it meets the line graph. Draw a horizontal line from here to meet the vertical axis. Read off the answer.

To estimate the times at which the temperature was 8.5°C draw a line horizontally from 8.5. Draw lines down from where this crosses the line graph. Read off the answers.

a 12.5°C

b 09:30 and 14:30 hours. ←   It is 8.5°C twice during the day.

c 16°C at 12:00 hours. ←   This will be the highest point.

d The temperature rose most quickly between 10:00 and 11:00 hours. ←   The steeper the gradient of the line, the more quickly the temperature changed.

Exercise 24A

Questions in this chapter are targeted at the grades indicated.

A03

1 The line graph shows the time it took a parent to take her child to school.

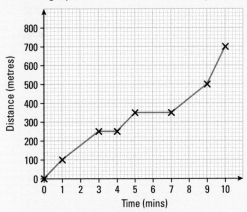

a How far was the journey to school?

b How many times did they stop on the journey and for how long did they stop?

c Use the line graph to estimate how long it took them to walk 600 metres.

d Use the line graph to estimate how far they had walked in 8 minutes.

2 The line graph shows the depth, in metres, of water in a reservoir each month for one year.

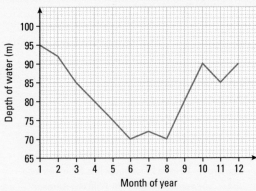

a In which month was the water at its deepest? What was the depth?

b Use the line graph to estimate the depth of water in the fourth month.

A03

c In which months was the water at its lowest? Suggest a reason for this.

3 The line graph shows the power supplied, for domestic power consumption, by Yorkoft power station over a 12-hour period in September.

A03

a At what time was domestic consumption of power at its highest?
Suggest a reason why demand is high at this time.

b What was the consumption at 16:00 hours?

c Use the line graph to find an estimate of the times when the consumption was 150 000 kilowatts.

d Write down the time when consumption was at its lowest. Suggest a reason for this.

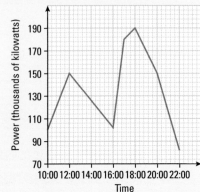

24.2 Drawing and using scatter graphs

◉ Objective

◉ You can use a scatter graph to see if there is any relationship between pairs of variables.

⊙ Why do this?

You could use a scatter graph to plot the data on local speed limits and the number of children involved in traffic accidents, and see if there is a relationship between the two.

◈ Get Ready

1. What are the values of points A to E?

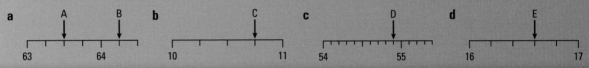

◉ Key Points

◉ When taking pairs of observations from members of a sample, we plot points on a graph to see if there is any relationship between the two variables being observed.
  The resulting graph is called a scatter diagram or **scatter graph**.

◉ A scatter graph enables you to see how scattered pairs of values are when plotted.

Example 3

In a certain city council area the proportion of open space was 2% and the percentage of accidents involving children was 40%.
The table shows the figures for seven other council areas.

Open space (%)	5	1.4	2.5	5.2	12.2	15	6.3
% accidents involving children	43	40	36	33	30	25	32

a Draw a scatter diagram for these data.
b Describe how the variables are related.

A03

a

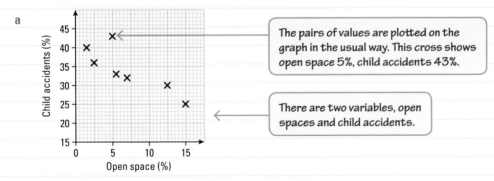

The pairs of values are plotted on the graph in the usual way. This cross shows open space 5%, child accidents 43%.

There are two variables, open spaces and child accidents.

b We can see that the crosses are in a roughly downward sloping line.
  So there seems to be a relationship between the amount of open space and the percentage of accidents involving children.
  The relationship is not perfect but a general trend can be seen.
  The greater the percentage of open space, the fewer children involved in accidents.

Example 4

The table below shows females' years of birth and the age they could expect to live to.

Year of birth	1986	1991	1996	2001	2006
Life expectation (years)	77.7	78.7	79.4	80.4	81.5

Data source: Stats.gov.uk

A03

a Draw a scatter graph for these data.

b Comment on the relationship between birth year and life expectancy.

a

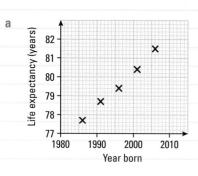

b There seems to be a close relationship.
The later a person was born the greater their life expectancy at birth.

24.3 Recognising correlation

⊙ **Objectives**

○ You can distinguish between positive, negative and zero correlation.

○ You appreciate that correlation is a measure of the strength of the association.

⊘ **Why do this?**

If a scatter diagram shows points that seem to have a relationship then they are said to be correlated. The year you were born and your life expectancy are correlated.

Key Points

◎ If every time one variable changes the other variable changes as well, we say the variables are correlated. If the points lie almost in a straight line they are said to be linearly correlated. Linear means 'in a straight line'. A relationship between pairs of variables is called a **correlation**. We will only consider linearly correlated variables here.

◎ If one variable increases as the other one increases the correlation is said to be positive.

◎ If one variable decreases as the other increases the correlation is said to be negative.

◎ If there is no relationship between the variables then there is no correlation and the correlation is said to be zero.

These three possibilities are shown in these graphs.

Positive correlation

As one value increases the
other one increases.

Negative correlation

As one value increases the
other decreases.

No correlation

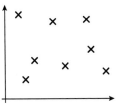

The points are random and
widely spaced.

- If there is perfect **positive correlation** between two variables the correlation is given a value of $+1$.
- If there is perfect **negative correlation** between two variables the correlation is given a value of -1.
- If there is **no correlation** then the correlation is zero and is given the value 0.

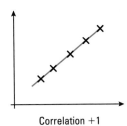

Correlation $+1$

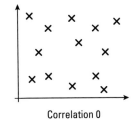

Correlation 0

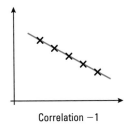

Correlation -1

Example 5

A bar is supported at each end. A weight is hung in the middle and the amount that the middle of the bar sags is measured.

The scatter diagram shows the resulting sag for different weights.

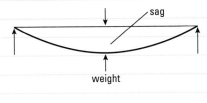

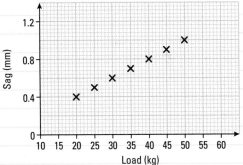

a Describe the correlation.
b Describe the relationship between the load and the amount of sag in the middle of the bar.

a Positive correlation. ← | As one variable increases so does the other one.

b The greater the weight the greater the sag.

Example 6

Which of the following pairs of variables are related?

a A child's height and weight
b A car's engine size and the number of seats
c The number of MP3 players and the number of flat screen televisions sold by a store
d The number of speed cameras and the number of speeding fines

The variables of **a** and **d** are related.

| The number of seats in a car doesn't affect the size of the engine. Sales of MP3 players don't affect sales of flat screen televisions.

Exercise 24B

D

1 The table gives information about the engine size, in litres, and the petrol consumption, in miles per litre, of eight cars.

Engine size (litres)	1.6	2.6	1.4	1.0	2.2	1.2	3.0	1.7
Petrol consumption (mp*l*)	10	7	11	12	8	13	6	9

   a Draw a scatter graph for these data.

   b Describe the correlation.

A03   **c** Describe the relationship between engine size and petrol efficiency.

2 The table gives information about the selling price of computers and their screen size.

Screen size (inches)	10	17	17	15	12	14	16	11	16
Selling price (£)	700	700	550	400	450	480	300	350	880

   a Draw a scatter graph for these data.

   b Describe the correlation.

A03   **c** What can you say about the relationship between screen size and price?

3 In the year 2000 a river was restocked with fish.
The estate with the fishing rights kept a record of the number of fish caught on a particular stretch of the river for the next 10 years. The data collected are shown in the table.

Year after restocking	1	2	3	4	5	6	7	8	9	10
Number of fish caught	180	165	168	155	158	150	145	148	140	135

   a Draw a scatter graph for these data.

   b Describe the correlation.

A03   **c** Describe the relationship between years after restocking and number of fish caught.

A03 **4** Which of the following pairs of variables are related? Give a reason for your answer.

   a A car's maximum speed and its weight

   b The length of a motorway and the number of petrol stations on the motorway

   c The number of washing machines and the number of computers sold by an electrical store

   d The number of bicycles and the number of cycle helmets sold by a shop

24.4 Drawing lines of best fit

◉ You can draw lines of best fit by eye.
◉ You can explain an isolated point on a scatter diagram.

A diver could plot a graph of pressure at certain depths, with a line of best fit and use the line to estimate a formula relating depth to pressure.

🔍 Key Points

◉ If the points on a scatter graph lie approximately in a straight line the correlation is said to be linear.

◉ If the points are roughly in a straight line you can draw a **line of best fit** through them.

◉ A line of best fit is a straight line that passes as near as possible to the various points so as to best represent the **trend** of the graph.

◉ A line of best fit does not have to pass through any of the points, but it may pass through some of them.

◉ When drawing a line of best fit, draw it so that roughly the same numbers of points are either side of the line and so that the line drawn best represents the trend of the points.

If lines of best fit are added to the scatter graphs in Example 3 and Example 4 they will look like this.

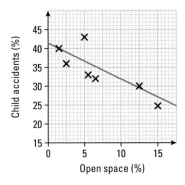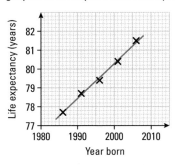

◉ An isolated point is an extreme point that lies outside the normal range of values.

◉ When drawing lines of best fit, or reaching conclusions, isolated points should be omitted from your data set.

🔍 Example 7

The table shows information about the percentage of people unemployed in a country and the percentage rise in wages over a number of years.

Unemployed (%)	2.0	2.2	3.0	1.8	1.6	4.0	1.6	1.8	2.0	1.4
Rise in wages (%)	2.7	3.0	2.7	3.5	1.4	1.6	4.2	3.5	3.8	3.9

a Draw a scatter graph of these data.
b Describe the correlation between the percentage of unemployed and the rise in wages.
c Draw a line of best fit on your scatter graph.
An economist thinks that when there is a lot of unemployment wage rises will be lower.
d Is the economist right? Give reasons.

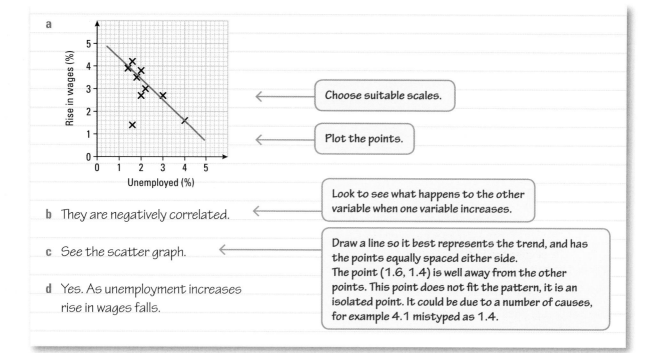

a

b They are negatively correlated.

c See the scatter graph.

d Yes. As unemployment increases rise in wages falls.

Choose suitable scales.

Plot the points.

Look to see what happens to the other variable when one variable increases.

Draw a line so it best represents the trend, and has the points equally spaced either side.
The point (1.6, 1.4) is well away from the other points. This point does not fit the pattern, it is an isolated point. It could be due to a number of causes, for example 4.1 mistyped as 1.4.

Exercise 24C

D

1 The scatter diagram shows the energy consumption and the GNP (Gross National Product, a measure of economic prosperity) for nine countries.

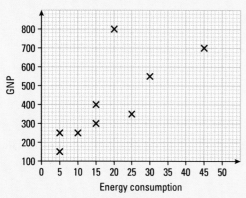

a Copy the scatter diagram.

b One point seems to be an isolated point. Circle the isolated point and write down its coordinates.

c Ignoring the isolated point, draw a line of best fit on your diagram.

2 The table shows the ages and prices of 10 second-hand cars.

Age (years)	1	2	3	3	5	2	7	8	9	12
Price (£1000)	10	7.5	7	6.5	4.5	8	2	1.5	1	0.5

a Draw a scatter graph for these data.

b One point seems to be an isolated point. Circle it and suggest a reason why this might have occurred.

c Ignoring the isolated point, draw a line of best fit on your scatter graph.

3 The table shows information that has been recorded by a researcher about the mean high and the mean low temperatures for some cities. The first six cities have been plotted on the scatter graph.

	Amsterdam	Berlin	Mumbai	Dublin	Hong Kong	London
High temp.	54	55	87	56	77	58
Low temp.	46	40	74	42	68	44

	Madrid	Oslo	Ottawa	Paris	Rangoon	Rome
High temp.	65	50	32	59	89	71
Low temp.	54	35	51	43	73	51

a Copy the scatter diagram and complete it by plotting the last six points.

b There is one point that seems to be an isolated point. Circle this point.

c Write down the name of the city that is the isolated point. Suggest a reason for this isolated point.

d Ignoring the isolated point, draw a line of best fit on your scatter graph.

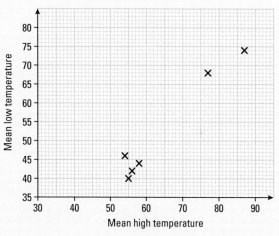

24.5 Using lines of best fit to make predictions

◎ Objective

You can use a line of best fit to predict a value of one of a pair of variables given a value for the other variable.

❓ Why do this?

A graph with a line of best fit for the number of drinks that a shop sold in the spring, at cooler temperatures, might help the shop predict how many drinks they might sell in the summer, as temperatures increase.

🖐 Key Points

⦿ If a value of one of the variables is known, you can estimate the corresponding value of the other variable by using the line of best fit.

For example, to estimate the likely rise in wages given that 3.6% were unemployed, you draw a vertical line at 3.6% until it hits the line of best fit. You then draw a horizontal line from there and read off where it comes on the vertical scale. In this case, you read off 2%.

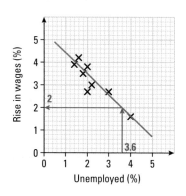

⦿ Using the line of best fit when the value you are finding is within the range of values on the scatter diagram is called **interpolation** which is usually reasonably accurate.

⦿ Using the line of best fit to find values outside the range of values on the scatter diagram is called **extrapolation** which may not be very accurate.

Example 8 The scatter graph gives information about the population density, in people per hectare, and the distance, in kilometres, from a city centre.

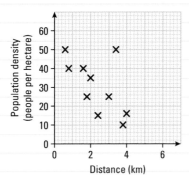

a   Identify any possible outliers.

b   Draw a line of best fit.

c   Estimate the population density at 1.2 km from the centre.

d   Estimate how far an area with a density of 30 people per hectare is from the centre.

a   Point (3.4, 50) is an isolated point.

b

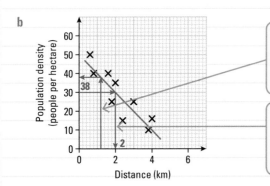

For **c** draw a line from 1.2 on the horizontal axis up to the line of best fit. From where it hits the line draw a horizontal line across to the vertical axis and read off the required value.

For **d** draw a horizontal line from 30 on the vertical axis across to the line of best fit. From where it hits the line draw a vertical line down to the horizontal axis and read off the required value.

c   38 people per hectare.

d   2 km.

ResultsPlus

Exam Tip

Always draw the lines on your diagram. Even if you get the wrong answer you might get marks for the correct method.

✿ **Exercise 24D**

1　The scatter diagram shows the marks in statistics and mathematics of a group of students.

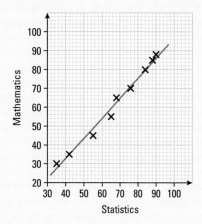

　a　A student gets a mark of 50 in statistics.
　　Use the line of best fit to find the mark he is likely to get in mathematics.

　b　A student gets a mark of 60 in mathematics.
　　Use the line of best fit to find the mark he is likely to get in statistics.

2　The scatter graph shows the latitudes and the mean highest temperatures of 10 countries of the world.

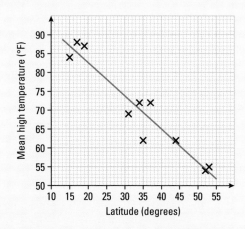

　a　Use the line of best fit to work out an estimate of the mean high temperature at a latitude of:
　　i　40 degrees　　ii　25 degrees.

　b　Use the line of best fit to work out an estimate of the latitude where you are likely to get a mean high temperature of:
　　i　60°F　　　　ii　85°F.

3　An engineer measured the length of a copper rod at various temperatures.
The table represents the data collected.

Temperature (°C)	20.5	27.5	40	43	55	60	65	70
Length in metres	2.4611	2.4614	2.4619	2.462	2.4623	2.4629	2.463	2.4636

　a　Draw a scatter graph for this data.

　b　Work out an estimate of the temperature when the length is
　　i　2.462 metres
　　ii　2.463 metres.

　c　Work out an estimate of the length when the temperature is 30°C.

Chapter review

◉ Bivariate data consists of pairs of related variables.

◉ Pairs of observations can be plotted on a **line graph**.

◉ A **scatter graph** enables you to see how scattered pairs of points are when plotted.

◉ A relationship between pairs of variables is called a **correlation**.

◉ If one variable increases as the other one increases, the correlation is said to be positive.

◉ If one variable decreases as the other increases, the correlation is said to be negative.

◉ If there is no relationship between the variables then there is **no correlation** and the correlation is said to be zero.

◉ A **line of best fit** is a straight line that passes as near as possible to the various points so as to best represent the **trend** of the graph.

◉ An isolated point is an extreme point that lies outside the normal range of values.

◉ If a value of one of the variables is known, you can estimate the corresponding value of the other variable by using the line of best fit.

◉ Using the line of best fit when the value you are finding is within the range of values on the scatter diagram is called **interpolation** which is usually reasonably accurate.

◉ Using the line of best fit to find values outside the range of values on the scatter diagram is called **extrapolation** which may not be very accurate.

Review exercise

D

1 The table gives information about the number of people who were unemployed in a seaside town over the course of a year.

Month	Jan	Feb	Mar	Apr	May	Jun	Jul	Aug	Sep	Oct	Nov	Dec
No. of unemployed	110	98	56	50	48	34	40	30	45	–	85	105

a Plot a line graph for these data.

b Estimate the number of unemployed people in October.

A03

c For which month of the year was the number of unemployed lowest? Give a reason for this.

2 The table gives some information about the science and art marks of some students.

Student	A	B	C	D	E	F	G	H	I	J
Science mark	78	65	48	68	89	95	46	38	56	70
Art mark	45	58	60	50	43	70	52	58	50	50

a Draw a scatter graph for these data.

b One point appears to be an isolated point. Circle that point.

A03

c Ignoring the isolated point, describe the correlation.

d Describe the relationship between the science marks and the art marks.

3 The table shows the height, in metres, above sea level and the temperature, in °C, at 06:00 hours at 10 places in Austria on one day in July.

Height (100s metres)	11	15	10	5	4	5	9	12	18	16
Temperature (°C)	11	6	10	16	14	13	8	9	5	6

a Draw a scatter graph for these data.

b Describe the correlation.

c Describe the relationship between the height and the temperature.

d Draw a line of best fit on your scatter graph.

4 The table gives some information about the length of a metal rod at different temperatures.

Temperature (x °C)	60	65	70	75	80	85
Length (y mm)	90.2	90.8	91.7	92	93	94.2

a Draw a scatter diagram for these data. Use values from 90 to 94 for the y axis and 55 to 85 for the x axis.

b Describe the correlation.

c Draw a line of best fit on your graph.

d Work out an estimate for the gradient of the line of best fit.

e Interpret the gradient in terms of length and temperature.

5 A superstore sells the Clicapic digital camera. The price of the camera changes each week. Each week the manager records the price of the camera and the number of cameras sold that week.
The scatter graph shows this information.

a Describe how the price of the camera and the number of cameras sold are related.

b Draw a line of best fit on a copy of the scatter graph.

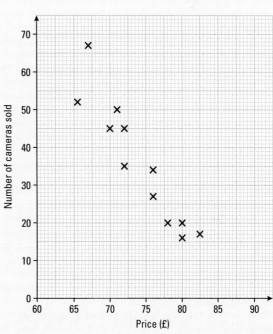

Nov 2008

D
A03

6 The scatter graph shows some information about the ages and values of fourteen cars. The cars are the same make and type.

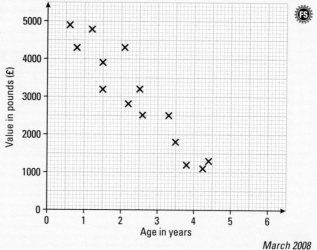

March 2008

a Describe the relationship between the age of a car and its value in pounds.

b Draw a line of best fit on a copy of the scatter graph.

A car is 3 years old.

c Find an estimate of its value.

A car has a value of £3500.

d Find an estimate of its age.

A03

7 Jake recorded the weight, in kg, and the height, in cm, of each of ten children.
The scatter graph shows information about his results.

a Describe the relationship between the weight and the height of these children.

b Draw a line of best fit on a copy of the scatter graph.

c Estimate the height of a child whose weight is 47 kg.

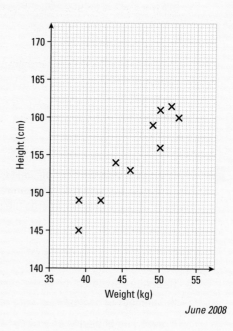

June 2008

C

8 The table gives some information about the heights and weights of 10 athletes.

| Height (cm) | 180 | 165 | 185 | 190 | 178 | 184 | 168 | 188 | 192 | 200 |
| Weight (kg) | 73 | 70 | 80 | 86 | 75 | 75 | 72 | 83 | 85 | 86 |

a Using a horizontal scale from 160 to 210 cm and a vertical scale from 65 to 90 kg, draw a scatter graph for these data.

b Describe the correlation.

A02
A03

c Describe the relationship between height and weight and estimate the weight of an athlete who is 175 cm tall.

9 In a study to see how effective a weight-reducing drug was, data regarding the weight loss, in lbs, and the length of treatment, in months, was collected.

Some of the results are shown in the scatter graph.

a Copy the scatter graph.

Three more people's records are taken.

Jackie lost 70 lbs in 22 months. Joan lost 60 lbs in 18 months and Tim lost 55 lbs in 14 months.

b Add these pieces of data to your scatter graph.

c One piece of data seems to be an isolated point. Circle it. Do you think this is a genuine piece of data? Give a reason for your answer.

d Ignoring the isolated point, describe the correlation.

e Write down whether or not you think the drug is effective. Give a reason for your answer.

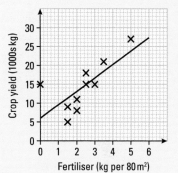

10 The scatter diagram shows the amount of fertiliser used and the crop yields on 10 equal-size plots at a crop regulatory centre.

a Describe the correlation.

b Describe the relationship between crop yield and amount of fertiliser used.

c Estimate the crop yield when 4 kg per 80 m² of fertiliser is used.

d Estimate the amount of fertiliser used to give a crop yield of 15 000 kg.

e Nassim says he will use the line of best fit to find out what the crop would be if 20 kg of fertiliser per 80 m² was put on a plot. Will Nassim get a sensible result? Explain your answer.

11 The numbers of fleas kept in an enclosed environment were counted every 6 days. The results are shown in the table.

Day	0	1	2	3	4	5	6
Number of fleas	50	100	196	390	780	1550	3000

a Draw a scatter diagram of these data.

b Draw in a curve of best fit.

c Suggest a suitable general equation for the relationship between these data.

12 A scatter diagram is drawn to show the height above sea level (x) and air temperature (y). The equation of the line of best fit is $y = -0.01x + 19$.

a Interpret the gradient in context.

b What does the number 19 in the equation tell you about the height above sea level and air temperature?

25 INDICES, STANDARD FORM AND SURDS

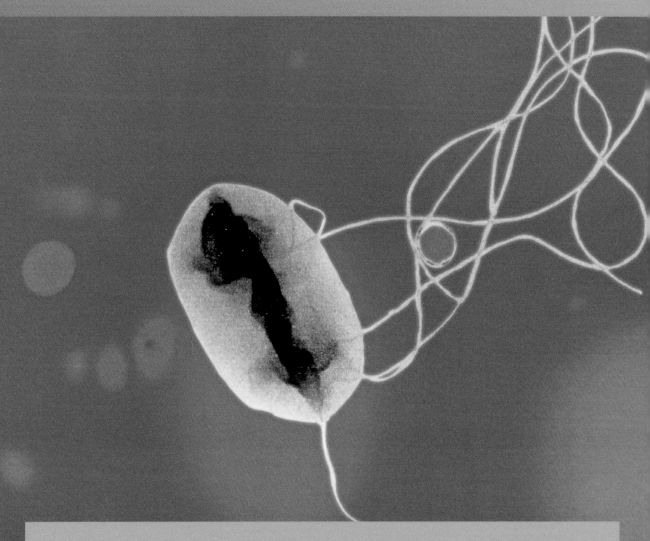

The photo shows a male *Escheria coli* bacteria. You may have heard of e-coli. These bacteria are commonly known in relation to food poisoning as they can cause serious illness. Each bacterium is about a millionth of a metre, or 0.000001 m long. Standard form allows us to write both very large and very small numbers in a more useful form.

◎ Objectives

In this chapter you will:
- work out the value of an expression with zero, negative or fractional indices
- convert between standard form and ordinary numbers
- calculate with numbers in standard form
- make estimates to calculations using standard form
- manipulate surds.

◈ Before you start

You need to be able to:
- use the index laws
- round numbers to one significant figure.

25.1 Using zero and negative powers

Objectives

- You know that $n^0 = 1$ when $n \neq 0$.
- You know the meaning of negative indices.

Why do this?

If you are x metres from a live band, the volume of sound they are producing is directly proportional to x^{-2}. This means that if you halve your distance from the band, the music will get four times as loud.

Get Ready

Work out

1. $\frac{1}{4} \times \frac{1}{4}$

2. $\frac{3}{5} \times \frac{3}{5}$

3. -2^3

Key Points

- For non-zero values of a
  $$a^0 = 1$$
- For any number n
  $$a^{-n} = \frac{1}{a^n}$$

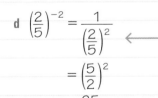

Example 1   Work out the value of   **a** 3^0 **b** 5^{-1} **c** 6^{-2} **d** $\left(\frac{2}{5}\right)^{-2}$

a $3^0 = 1$ ← Any number to the power of zero is 1.

b $5^{-1} = \frac{1}{5}$ ← Use the rule $a^{-n} = \frac{1}{a^n}$

c $6^{-2} = \frac{1}{6^2}$
      $= \frac{1}{36}$ ← $6^2 = 6 \times 6 = 36$

d $\left(\frac{2}{5}\right)^{-2} = \frac{1}{\left(\frac{2}{5}\right)^2}$ ← To work out the reciprocal of a fraction, turn the fraction upside down. Square the number on the top and the number on the bottom of the fraction.

      $= \left(\frac{5}{2}\right)^2$

      $= \frac{25}{4}$

ResultsPlus
Exam Tip

Do not convert the fraction to a decimal. It is much easier to square the numbers in a fraction than it is to square a decimal.

Exercise 25A

Questions in this chapter are targeted at the grades indicated.

1 Write down the value of these expressions.

   a 7^0 **b** 8^{-1} **c** 5^{-1} **d** 4^0

   e $(-2)^{-3}$ **f** 9^{-2} **g** 10^{-4} **h** 145^0

   i $(-3)^{-2}$ **j** $(-8)^0$ **k** 16^0 **l** 10^{-6}

B

B

2 Work out the value of these expressions.

a $\left(\frac{1}{3}\right)^{-1}$ b $\left(\frac{2}{7}\right)^{-1}$ c $\left(\frac{1}{7}\right)^{-2}$ d $\left(\frac{1}{4}\right)^{-3}$

e $(0.25)^{-2}$ f $\left(\frac{2}{5}\right)^{-3}$ g $\left(\frac{5}{3}\right)^{0}$ h $\left(\frac{9}{5}\right)^{-1}$

i $\left(1\frac{2}{5}\right)^{-2}$ j $\left(1\frac{1}{3}\right)^{-3}$ k $(0.1)^{-4}$ l $(0.2)^{-3}$

25.2 Using standard form

⊙ Objectives

● You can convert between ordinary numbers and standard form.
● You can calculate with numbers in standard form.
● You can convert to standard form to make sensible estimates for calculations.

⊘ Why do this?

Astronomers use standard form to record large measurements. The Sun's diameter is about 1.392×10^6 km. Biologists working with micro-organisms sometimes use standard form to record their very small sizes, like 2.1×10^{-4} cm.

⊕ Get Ready

1. Work out   a 10^3 b 10^{-2}
2. Write 10 000 as a power of 10.
3. Work out $2.35 \times 10\,000$.

Key Points

● **Standard form** is used to represent very large (or very small) numbers.
  A number is in standard form when it is in the form $a \times 10^n$ where $1 \le a < 10$ and n is an integer.

● A number in standard form looks like this.

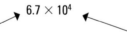

This part is written as a number between 1 and 10.

This part is written as a power of 10.

● These numbers are all in standard form: 4.5×10^2, 9×10^{-8}, 1.2657×10^6.

● These numbers are not in standard form because the first number is not between 1 and 10: 67×10^9, 0.087×10^3.

● It is often easier to multiply and divide very large or very small numbers, or estimate a calculation, if the numbers are written in standard form.

● To input numbers in standard form into your calculator, use the $\boxed{10^x}$ or $\boxed{\text{EXP}}$ key.
  To enter 4.5×10^7 press the keys $\boxed{4}$ $\boxed{\cdot}$ $\boxed{5}$ $\boxed{\times}$ $\boxed{10^x}$ $\boxed{7}$.

🔍 Example 2   Write these numbers in standard form.   a 50 000   b 34 600 000   c 682.5

a $50\,000 = 5 \times 10\,000$
        $= 5 \times 10^4$

b $34\,600\,000 = 3.46 \times 10\,000\,000$ ←   Use 3.46 not 34.6 or 346 as 3.46 is between 1 and 10.
             $= 3.46 \times 10^7$

c $682.5 = 6.825 \times 100$
       $= 6.825 \times 10^2$

Example 3 Write as an ordinary number   a 8.1×10^5 b 6×10^8

a $8.1 \times 10^5 = 8.1 \times 100\,000$
   $= 810\,000$
b $6 \times 10^8 = 6 \times 100\,000\,000$
   $= 600\,000\,000$

Exercise 25B

B

1 Write these numbers in standard form.
   a 700 000   b 600   c 2000   d 900 000 000   e 80 000

2 Write these as ordinary numbers.
   a 6×10^5 b 1×10^4 c 8×10^5 d 3×10^8 e 7×10^1

3 Write these numbers in standard form.
   a 43 000   b 561 000   c 56   d 34.7   e 60

4 Write these as ordinary numbers.
   a 3.96×10^4 b 6.8×10^7 c 8.02×10^3 d 5.7×10^1 e 9.23×10^0

5 In 2008 there were approximately 7 000 000 000 people in the world. Write this number in standard form.

6 The circumference of Earth is approximately 40 000 km. Write this number in standard form.

Example 4 Write these in standard form
   a 0.000 000 006   b 0.000 56

a $0.000\,000\,006 = 6 \times 0.000\,000\,001$ ← 0.000 000 001 is equivalent to $\frac{1}{1\,000\,000\,000}$.
   $= 6 \times \frac{1}{1\,000\,000\,000}$
   $= 6 \times \frac{1}{10^9}$ ← Using $a^{-n} = \frac{1}{a^n}$
   $= 6 \times 10^{-9}$
b $0.000\,56 = 5.6 \times 0.0001$
   $= 5.6 \times \frac{1}{10\,000}$
   $= 5.6 \times \frac{1}{10^4}$ ← Use 5.6 rather than 56 as 5.6 is between 1 and 10.
   $= 5.6 \times 10^{-4}$

Example 5 Write these as ordinary numbers
   a 3×10^{-6} b 1.5×10^{-3}

a $3 \times 10^{-6} = \frac{3}{10^6}$ b $1.5 \times 10^{-3} = \frac{1.5}{10^3}$

   $= \frac{3}{1\,000\,000}$ $= \frac{15}{10\,000}$

   $= 0.000\,003$ $= 0.0015$

Exercise 25C

1 Write these numbers in standard form.
    a 0.005   **b** 0.04   **c** 0.000 007   **d** 0.9   **e** 0.0008

2 Write these as ordinary numbers.
    a 6×10^{-5} **b** 8×10^{-2} **c** 5×10^{-7} **d** 3×10^{-1} **e** 1×10^{-8}

3 Write these numbers in standard form.
    a 0.0047   **b** 0.987   **c** 0.000 803 4   **d** 0.000 15   **e** 0.601

4 Write these as ordinary numbers.
    a 8.43×10^{-5} **b** 2.01×10^{-2} **c** 4.2×10^{-7} **d** 7.854×10^{-1} **e** 9.4×10^{-4}

5 Write these numbers in standard form.
    a 457 000   **b** 0.0023   **c** 0.0003   **d** 2 356 000   **e** 0.782
    f 89 000   **g** 200   **h** 0.005 26   **i** 6034   **j** 0.000 008 73

6 Write these as ordinary numbers.
    a 4.12×10^{-4} **b** 3×10^{3} **c** 2.065×10^{7} **d** 4×10^{-6} **e** 3.27×10^{8}
    f 7.5×10^{-1} **g** 1.5623×10^{2} **h** 5.12×10^{-7} **i** 2.7×10^{5} **j** 6.12×10^{-1}

7 1 micron is 0.000 001 of a metre. Write down the size of a micron, in metres, in standard form.

8 A particle of sand has a diameter of 0.0625 mm. Write this number in standard form.

Example 6   Write in standard form
    a 40×10^{2} **b** 0.008×10^{-2}

Results Plus
Exam Tip

The power of 10 tells you how many 0s there are.
$10^{2} = 100$ 2 zeros
$10^{-2} = 0.01$ 2 zeros

Method 1
a $40 \times 10^{2} = 4 \times 10^{1} \times 10^{2}$ ← Write 40 in standard form. Use the rule $a^{m} \times a^{n} = a^{m+n}$.
             $= 4 \times 10^{1+2}$
             $= 4 \times 10^{3}$

b $0.008 \times 10^{-2} = 8 \times 10^{-3} \times 10^{-2}$ ← Write 0.008 in standard form. Use $a^{m} \times a^{n} = a^{m+n}$.
                  $= 8 \times 10^{-3 + -2}$
                  $= 8 \times 10^{-5}$

Method 2
a $40 \times 10^{2} = 40 \times 100$
            $= 4000$
            $= 4 \times 10^{3}$ ← Work out the calculation. Change the answer into standard form.

b $0.008 \times 10^{-2} = 0.008 \times \frac{1}{100}$
                $= 0.008 \div 100$
                $= 0.000 08$ ← Use the rule $a^{-n} = \frac{1}{a^{n}}$. Multiplying by $\frac{1}{100}$ is the same as dividing by 100.
                $= 8 \times 10^{-5}$

B

Exercise 25D

1 Write these in standard form.
   a 45×10^3
   b 980×10^{-3}
   c 3400×10^{-2}
   d 186×10^{10}

2 Write these in standard form.
   a 0.009×10^5
   b 0.045×10^6
   c 0.3708×10^{-12}
   d 0.006×10^{-7}

3 Some of these numbers are not in standard form. State if a number is in standard form.
   If a number is not in standard form then rewrite it so that it is in standard form.
   a 7.8×10^4
   b 890×10^6
   c 13.2×10^{-5}
   d 0.56×10^9
   e $60\,000 \times 10^{-8}$
   f 8.901×10^{-7}
   g $0.040\,05 \times 10^{-10}$
   h 9080×10^{15}
   i 6.002×10^5
   j 0.0046×10^8
   k $67\,000 \times 10^{-3}$
   l 0.004×10^3

4 Write these numbers in order of size. Start with the smallest number.
   $6.3 \times 10^6, 0.637 \times 10^7, 6\,290\,000, 63.4 \times 10^5$

5 Write these numbers in order of size. Start with the smallest number.
   $0.034 \times 10^{-2}, 3.35 \times 10^{-5}, 0.000\,033, 37 \times 10^{-4}$

Example 7 Work out $(3 \times 10^6) \times (4 \times 10^3)$ giving your answer in standard form.

$$(3 \times 10^6) \times (4 \times 10^3) = 3 \times 4 \times 10^6 \times 10^3$$
$$= 12 \times 10^9$$
$$= 1.2 \times 10^1 \times 10^9$$
$$= 1.2 \times 10^{10}$$

Rearrange the expression so the powers of 10 are together.
Multiply the numbers.
Use $a^m \times a^n = a^{m+n}$ to multiply the powers of 10.
12×10^9 is not in standard form.
Write your final answer in standard form.

Example 8 By writing $760\,000\,000$ and $0.000\,19$ in standard form correct to one significant figure, work out an approximation for $760\,000\,000 \div 0.000\,19$.

$760\,000\,000 = 8 \times 10^8$ correct to one significant figure.
$0.000\,19 = 2 \times 10^{-4}$ correct to one significant figure.

$$\frac{760\,000\,000}{0.000\,19} \approx \frac{8 \times 10^8}{2 \times 10^{-4}}$$
$$= \frac{8}{2} \times \frac{10^8}{10^{-4}}$$
$$= 4 \times 10^{8--4}$$
$$= 4 \times 10^{12}$$

Rearrange the expression so the powers of 10 are together.
Divide the numbers.
Use $a^m \div a^n = a^{m-n}$ to divide the powers of 10.

Exercise 25E

A

1 Work out and give your answer in standard form.
   a $(4 \times 10^8) \times (2 \times 10^3)$
   b $(6 \times 10^5) \times (1.5 \times 10^3)$
   c $(4 \times 10^{-7}) \times (3 \times 10^5)$
   d $(8.6 \times 10^8) \div (2 \times 10^{13})$
   e $(1 \times 10^{12}) \div (4 \times 10^3)$
   f $(7 \times 10^{-9}) \div (7 \times 10^{-5})$

2 Write these in standard form.
   a $(2 \times 10^5)^2$
   b $(5 \times 10^{-5})^2$
   c $(4 \times 10^6)^2$
   d $(7 \times 10^{-8})^2$

A

3 By writing these numbers in standard form correct to one significant figure, work out an estimate of the value of these expressions. Give your answer in standard form.
    a $600\,008 \times 598$ **b** $78\,018 \times 4180$ **c** $699\,008 \div 198$ **d** $8\,104\,660\,000 \div 0.000\,078$

4 The base of a microchip is in the shape of a rectangle. Its length is 2×10^3 mm and its width is 1.6×10^{-3} mm. Find the area of the base. Give your answer in mm² in standard form.

A03

5 The distance of the Earth from the Sun is approximately 149 000 000 kilometres.
Light travels at a speed of approximately 300 000 kilometres per second.
Work out an estimate of the time it takes light to travel from the Sun to the Earth.

A02
A03

6 An atomic particle has a lifetime of 3.86×10^{-5} seconds. It travels at a speed of 4.2×10^6 metres per second. Calculate an approximation for the distance it travels in its lifetime.

Example 9   Use a calculator to work out
         a $(3.4 \times 10^6) \times (7.1 \times 10^4)$
         b $(4.56 \times 10^8) \div (3.2 \times 10^{-3})$

a $(3.4 \times 10^6) \times (7.1 \times 10^4)$ $= 2.414 \times 10^{11}$ ⟵   Use the $\boxed{\text{EXP}}$ or $\boxed{10^x}$ button on your calculator.
b $(4.56 \times 10^8) \div (3.2 \times 10^{-3})$ $= 1.425 \times 10^{11}$

Example 10 $x = 3.1 \times 10^{12}, y = 4.7 \times 10^{11}$
         Use a calculator to work out the value of $\dfrac{x + y}{xy}$.
         Give your answer in standard form correct to 3 significant figures.

$\dfrac{(3.1 \times 10^{12} + 4.7 \times 10^{11})}{(3.1 \times 10^{12} \times 4.7 \times 10^{11})}$ ⟵   Substitute the values into the expression.

$= \dfrac{3.57 \times 10^{12}}{1.457 \times 10^{24}}$ ⟵   Write the number from your calculator correctly in standard form showing more than 3 significant figures.

$= 2.4502\ldots \times 10^{-12}$ ⟵

$= 2.45 \times 10^{-12}$ ⟵   Give your answer correct to 3 significant figures.

Results Plus

Exam Tip

Include brackets here to ensure that the answer from the calculation on the top of the fraction is divided by the answer to the calculation on the bottom of the fraction.

Exercise 25F

A

1 Find the value of these expressions, giving your answers in standard form.
Give your answers to 4 significant figures where necessary.
    a $500 \times 600 \times 700$ **b** 0.006×0.004 **c** $\dfrac{65 \times 120}{1500}$
    d $\dfrac{8.82 \times 5.007}{10\,000}$ **e** $(12.8)^4$ **f** $(2.46 \times 10^{10}) \div (2.5 \times 10^6)$
    g $(3.6 \times 10^{20}) \div (3.75 \times 10^6)$ **h** $(2.46 \times 10^{-10}) \div (2.5 \times 10^6)$
    i $(3.6 \times 10^{-20}) \div (3.75 \times 10^{-6})$

2 Evaluate these expressions. Give your answers in standard form correct to 3 significant figures.

    a $(3.5 \times 10^{11}) \div (6.5 \times 10^{6})$

    b $(1.33 \times 10^{10}) \times (4.66 \times 10^{4})$

    c $(3.5 \times 10^{11}) \div (6.5 \times 10^{-6})$

    d $(1.33 \times 10^{-10}) \times (4.66 \times 10^{4})$

3 $x = 3.5 \times 10^{9}, y = 4.7 \times 10^{5}$

Work out the following. Give your answer in standard form correct to 3 significant figures.

    a $\dfrac{x}{y}$

    b $x(x + 800y)$

    c $\dfrac{xy}{x + 800y}$

    d $\left(\dfrac{x}{2000}\right)^{2} + y^{2}$

4 $x = 2.4 \times 10^{-5}, y = 9.6 \times 10^{-6}$

Evaluate these expressions.

Give your answer in standard form correct to 3 significant figures where necessary.

    a $\dfrac{x^{2}}{y}$

    b $\dfrac{x^{2} + y^{2}}{x + y}$

    c $\dfrac{xy}{x - y}$

5 The distance of the Earth from the Sun is 1.5×10^{8} km.

The distance of the planet Neptune from the Sun is 4510 million km.

Write in the form $1 : n$ the ratio

distance of the Earth from the sun : distance of the planet Neptune from the Sun

6 The mass of a uranium atom is 3.98×10^{-22} grams.

Work out the number of uranium atoms in 2.5 kilograms of uranium.

25.3 Working with fractional indices

◎ Objective

◦ You know the meaning of fractional indices.

❖ Why do this?

Fractional indices are used when you model the rates at which things vibrate, such as your voice box.

◈ Get Ready

Work out

1. $\sqrt[3]{1000}$ **2.** $\sqrt[3]{8}$ **3.** $\sqrt[3]{-27}$

◯ Key Points

◉ Indices can be fractions. In general,

$$a^{\frac{1}{n}} = \sqrt[n]{a}$$

◉ In particular, this means that

$$a^{\frac{1}{2}} = \sqrt{a} \text{ and } a^{\frac{1}{3}} = \sqrt[3]{a}$$

Example 11 | Find the value of the following

a $25^{\frac{1}{2}}$ b $(-1000)^{\frac{1}{3}}$ c $16^{-0.25}$

a $25^{\frac{1}{2}} = \sqrt{25}$ ← | The square root of 25 is 5 because $5 \times 5 = 25$.
   $= 5$

b $(-1000)^{\frac{1}{3}} = \sqrt[3]{-1000}$ ← | The cube root of -1000 is -10 because $-10 \times -10 \times -10 = -1000$.
   $= -10$

c $16^{-0.25} = 16^{-\frac{1}{4}}$ ← | Change the decimal into a fraction $0.25 = \frac{1}{4}$. Use the rule $a^{-n} = \frac{1}{a^n}$.

$= \dfrac{1}{16^{\frac{1}{4}}}$

$= \dfrac{1}{\sqrt[4]{16}}$ ← | $16^{\frac{1}{4}} = \sqrt[4]{16} = 2$ because $2^4 = 16$

$= \dfrac{1}{2}$

Example 12 | Work out the value of   a $8^{\frac{2}{3}}$ b $16^{-\frac{3}{4}}$

a $8^{\frac{2}{3}} = (8^{\frac{1}{3}})^2$ | Use the rule $(a^m)^n = a^{mn}$.
   $= 2^2$ ← | Work out the cube root of 8 first.
   $= 4$ | Then square your answer.

b $16^{-\frac{3}{4}} = \dfrac{1}{16^{\frac{3}{4}}}$

$= \dfrac{1}{(16^{\frac{1}{4}})^3}$ ← | Use $a^{-n} = \frac{1}{a^n}$.

$= \dfrac{1}{2^3}$

$= \dfrac{1}{8}$

ResultsPlus
Exam Tip

It is easier to work out the root first as this makes the numbers smaller and easier to manage.

Exercise 25G

B

1 Work out the value of the following.
   a $9^{\frac{1}{2}}$ b $49^{\frac{1}{2}}$ c $100^{\frac{1}{2}}$ d $4^{\frac{1}{2}}$ e $\left(\frac{1}{4}\right)^{\frac{1}{2}}$

2 Work out the value of
   a $27^{\frac{1}{3}}$ b $1000^{\frac{1}{3}}$ c $(-64)^{\frac{1}{3}}$ d $125^{\frac{1}{3}}$ e $\left(\frac{1}{8}\right)^{\frac{1}{3}}$

3 Work out the value of
   a $16^{-\frac{1}{4}}$ b $4^{-\frac{1}{2}}$ c $125^{-\frac{1}{3}}$ d $\left(\frac{1}{32}\right)^{-\frac{1}{5}}$ e $\left(\frac{4}{9}\right)^{-\frac{1}{2}}$

A

4 Work out the value of
   a $27^{\frac{2}{3}}$ b $1000^{\frac{2}{3}}$ c $64^{\frac{2}{3}}$ d $16^{\frac{3}{4}}$ e $25^{\frac{3}{2}}$

5 Work out, as a single fraction, the value of

a $125^{-\frac{2}{3}}$ **b** $10\,000^{-\frac{3}{4}}$ **c** $27^{-\frac{1}{3}}$ **d** $8^{-\frac{2}{3}}$ **e** $64^{-\frac{3}{2}}$

f $125^{-\frac{2}{3}} \times \left(\frac{1}{5}\right)^2$ **g** $8^{-\frac{1}{3}} \times \left(\frac{2}{5}\right)^2$

6 Find the value of n.

a $\frac{1}{8} = 8^n$ **b** $64 = 2^n$ **c** $\frac{1}{\sqrt{5}} = 5^n$ **d** $(\sqrt{7})^5 = 7^n$ **e** $(\sqrt[3]{2})^{11} = 2^n$

A

A☆

25.4 Using surds

◉ Objectives

- ○ You can simplify surds.
- ○ You can expand expressions involving surds.
- ○ You can rationalise the denominator of a fraction.

❓ Why do this?

Surds occur in nature. The golden ratio $\frac{1+\sqrt{5}}{2}$ occurs in the arrangement of branches along the stems of plants, as well as veins and nerves in animal skeletons.

◈ Get Ready

1. Write down the first 10 square numbers.
2. Write down the value of   **a** $\sqrt{36}$ **b** $\sqrt{100}$
3. Which of these have an exact answer: $\sqrt{5}, \sqrt{9}, \sqrt{37}, \sqrt{64}$?

Key Points

- ◉ A number written exactly using square roots is called a **surd**.
  $\sqrt{2}$ and $\sqrt{3}$ are both surds.

- ◉ $2 - \sqrt{3}$ and $5 + \sqrt{2}$ are examples of numbers written in surd form.
  $\sqrt{4}$ is not a surd as $\sqrt{4} = 2$.

- ◉ These two rules can be used to simplify surds.
  $\sqrt{m} \times \sqrt{n} = \sqrt{mn}$ $\frac{\sqrt{m}}{\sqrt{n}} = \sqrt{\frac{m}{n}}$

- ◉ Simplified surds should never have a surd in the denominator.

- ◉ To **rationalise the denominator** of a fraction means to get rid of any surds in the denominator.

- ◉ To rationalise the denominator of $\frac{a}{\sqrt{b}}$ you multiply the fraction by $\frac{\sqrt{b}}{\sqrt{b}}$. This ensures that the final fraction has an integer as the denominator.
  $$\frac{a}{\sqrt{b}} = \frac{a}{\sqrt{b}} \times \frac{\sqrt{b}}{\sqrt{b}} = \frac{a \times \sqrt{b}}{\sqrt{b} \times \sqrt{b}} = \frac{a\sqrt{b}}{b}$$

Example 13   Simplify $\sqrt{12}$.

$$\sqrt{12} = \sqrt{4 \times 3}$$
$$= \sqrt{4} \times \sqrt{3}$$
$$= 2\sqrt{3}$$

⟵ Use $\sqrt{m} \times \sqrt{n} = \sqrt{mn}$.
$\sqrt{4} = 2$.

Example 14 Expand and simplify $(2 + \sqrt{3})(4 + \sqrt{3})$.

$(2 + \sqrt{3})(4 + \sqrt{3}) = 8 + 2\sqrt{3} + 4\sqrt{3} + \sqrt{3} \times \sqrt{3}$ ← Multiply out the brackets.

$\qquad\qquad\qquad = 8 + 6\sqrt{3} + 3$ ← Simplify the expression.

$\qquad\qquad\qquad = 11 + 6\sqrt{3}$

Exercise 25H

A

1 Find the value of the integer k.

   a $\sqrt{8} = k\sqrt{2}$ b $\sqrt{18} = k\sqrt{2}$ c $\sqrt{50} = k\sqrt{2}$ d $\sqrt{80} = k\sqrt{5}$

2 Simplify

   a $\sqrt{200}$ b $\sqrt{32}$ c $\sqrt{20}$ d $\sqrt{28}$

3 Solve the equation $x^2 = 30$, leaving your answer in surd form.

A*

4 Expand these expressions. Write your answers in the form $a + b\sqrt{c}$ where a, b and c are integers.

   a $\sqrt{3}(2 + \sqrt{3})$ b $(\sqrt{3} + 1)(2 + \sqrt{3})$ c $(\sqrt{5} - 1)(2 + \sqrt{5})$

   d $(\sqrt{7} + 1)(2 - \sqrt{7})$ e $(2 - \sqrt{3})^2$ f $(\sqrt{2} + 5)^2$

5 The area of a square is 40 cm². Find the length of one side of the square.

   Give your answer as a surd in its simplest form.

6 The lengths of the sides of a rectangle are $(3 + \sqrt{5})$ cm and $(3 - \sqrt{5})$ cm.

   Work out, in their simplified forms:

   a   the perimeter of the rectangle   b   the area of the rectangle.

7 The length of the side of a square is $(1 + \sqrt{2})$ cm. Work out the area of the square.

   Give your answer in the form $(a + b\sqrt{2})$ cm² where a and b are integers.

Example 15 Rationalise the denominator of $\frac{2}{\sqrt{3}}$.

$\dfrac{2}{\sqrt{3}} = \dfrac{2}{\sqrt{3}} \times \dfrac{\sqrt{3}}{\sqrt{3}}$ ← Multiply the fraction by $\frac{\sqrt{3}}{\sqrt{3}}$.

$\qquad = \dfrac{2 \times \sqrt{3}}{\sqrt{3} \times \sqrt{3}}$ ← Simplify the denominator by using the fact that $\sqrt{3} \times \sqrt{3} = 3$.

$\qquad = \dfrac{2\sqrt{3}}{3}$

Example 16 Rationalise the denominator of $\dfrac{15 - \sqrt{5}}{\sqrt{5}}$ and give your answer in the form $a + b\sqrt{5}$.

$$\dfrac{15 - \sqrt{5}}{\sqrt{5}} = \dfrac{15 - \sqrt{5}}{\sqrt{5}} \times \dfrac{\sqrt{5}}{\sqrt{5}}$$

$$= \dfrac{15\sqrt{5} - \sqrt{5} \times \sqrt{5}}{\sqrt{5} \times \sqrt{5}}$$

$$= \dfrac{15\sqrt{5} - 5}{5}$$

$$= -1 + 3\sqrt{5}$$

ResultsPlus

Watch Out!

Remember to multiply both parts of the expression on the top of the fraction.

Simplify the fraction by dividing both parts of the expression on the top of the fraction by 5.

Exercise 25I

1. Rationalise the denominators and simplify your answers, if possible.

   a $\dfrac{1}{\sqrt{2}}$ b $\dfrac{1}{\sqrt{5}}$ c $\dfrac{5}{\sqrt{10}}$ d $\dfrac{2}{\sqrt{2}}$ e $\dfrac{4}{\sqrt{12}}$

2. Rationalise the denominators and give your answers in the form $a + b\sqrt{c}$ where a, b and c are integers.

   a $\dfrac{2 + \sqrt{2}}{\sqrt{2}}$ b $\dfrac{6 - \sqrt{2}}{\sqrt{2}}$ c $\dfrac{10 + \sqrt{5}}{\sqrt{5}}$ d $\dfrac{12 - \sqrt{3}}{\sqrt{3}}$ e $\dfrac{14 + \sqrt{7}}{\sqrt{7}}$

3. The diagram shows a right-angled triangle.
   The lengths are given in centimetres.
   Work out the area of the triangle.
   Give your answer in the form $a + b\sqrt{c}$ where a, b and c are integers.

   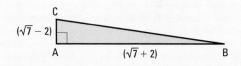

4. Solve these equations leaving your answers in surd form.

   a $x^2 - 6x + 2 = 0$ b $x^2 + 10x + 14 = 0$

5. The diagram represents a right-angled triangle ABC.
   AB = $(\sqrt{7} + 2)$ cm   AC = $(\sqrt{7} - 2)$ cm.
   Work out, leaving any appropriate answers in surd form:

   a the area of triangle ABC

   b the length of BC.

Chapter review

- For non-zero values of a
  $a^0 = 1$
- For any number n
  $a^{-n} = \dfrac{1}{a^n}$
- **Standard form** is used to represent very large or very small numbers.
- A number is in standard form when it is in the form $a \times 10^n$ where $1 \leqslant a < 10$ and n is an integer.
- It is often easier to multiply and divide very large or very small numbers, or estimate a calculation, if the numbers are written in standard form.
- To input numbers in standard form into your calculator, use the $\boxed{10^x}$ or $\boxed{\text{EXP}}$ key.

- Indices can be fractions. In general,

$$a^{\frac{1}{n}} = \sqrt[n]{a}$$

- A number written exactly using square roots is called a **surd**.

- These two laws can be used to simplify surds.

$$\sqrt{m} \times \sqrt{n} = \sqrt{mn} \qquad \frac{\sqrt{m}}{\sqrt{n}} = \sqrt{\frac{m}{n}}$$

- Simplified surds should never have a surd in the denominator.

- To **rationalise the denominator** of a fraction means to get rid of any surds in the denominator.

- To rationalise the denominator of $\frac{a}{\sqrt{b}}$ you multiply the fraction by $\frac{\sqrt{b}}{\sqrt{b}}$, this ensures that the final fraction has an integer as the denominator.

Review exercise

1 Work out the values of

   a 4^0 b 4^{-1} c 2^0 d 2^{-3}

2 Work out the values of

   a 3^0 b $(-3)^0$ c 3^{-1} d $\left(\frac{1}{3}\right)^0$

3 Work out the values of

   a $\frac{1}{3^{-1}}$ b $\left(\frac{1}{3}\right)^{-1}$ c 2×4^{-1} d $\frac{2}{4^{-1}}$

4 Work out

   a $9^{\frac{1}{2}}$ b $100^{\frac{1}{2}}$ c $8^{\frac{1}{3}}$ d $64^{\frac{1}{3}}$

5 Work out

   a $9^{-0.5}$ b $49^{-\frac{1}{2}}$ c $125^{-\frac{1}{3}}$ d $8^{-\frac{1}{3}}$

6 Work out

   a $4^{\frac{1}{2}}$ b $8^{-\frac{1}{3}}$

June 2009

7

Planet	Average distance from the Sun in km
Mercury	5.8×10^7
Venus	1.1×10^8
Earth	1.5×10^8
Mars	2.3×10^8
Jupiter	7.8×10^9
Saturn	1.4×10^9
Uranus	2.9×10^9
Neptune	4.5×10^9
Pluto	5.9×10^9

The table above gives the average distance in kilometres of the nine major planets from the Sun.

   a Which planet is approximately 4 times further away from the Sun than Mercury?

   b How far apart are the orbits of Neptune and Pluto?

   c Which planet is about half the distance from the Sun as Uranus?

   d Which planet is 40 times further away from the Sun than Venus?

   e A probe was sent from Earth to Mars. If it took one year to reach Mars, what average speed would it have to travel? Give your answer in km/h.

8   **a**   **i**   Write 7900 in standard form   **ii**   Write 0.000 35 in standard form.

    b   Work out $\dfrac{4 \times 10^3}{8 \times 10^{-5}}$ Give your answer in standard form.

9   In 2003 the population of Great Britain was 6.0×10^7.
In 2003 the population of India was 9.9×10^8.
Work out the difference between the population
of India and the population of Great Britain in
2003.
Give your answer in standard form.

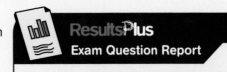

80% of students answered this question well.
They knew how to use their calculators properly.

June 2007

10 $3 \times \sqrt{27} = 3^n$ Find the value of n.   *June 2006*

11 $8\sqrt{8}$ can be written in the form 8^k.

    a   Find the value of k.

    $8\sqrt{8}$ can also be expressed in the form $m\sqrt{2}$ where m is a positive integer.

    b   Find the value of m.

    c   Rationalise the denominator of $\dfrac{1}{8\sqrt{8}}$.

      Give your answer in the form $\dfrac{\sqrt{2}}{p}$ where p is a positive integer.   *June 2006*

12   Work out

$$\frac{2 \times 2.2 \times 10^{12} \times 1.5 \times 10^{12}}{2.2 \times 10^{12} - 1.5 \times 10^{12}}$$

    Give your answer in standard form correct to 3 significant figures.   *Nov 2007*

13 $x = \sqrt{\dfrac{p+q}{pq}}$

    $p = 4 \times 10^8$
    $q = 3 \times 10^6$

    Find the value of x.
    Give your answer in standard form correct to 2 significant figures.   *Mar 2005*

 14   A nanosecond is 0.000 000 001 seconds.

    a   Write the number 0.000 000 001 in standard form.

    A computer does a calculation in 5 nanoseconds.

    b   How many of these calculations can the computer do in 1 second?
      Give your answer in standard form.   *June 2004*

15   Solve

    a $4^x = \dfrac{1}{16}$ **b** $2^x = \dfrac{1}{16}$ **c** $2 \times 2^{-x} = \dfrac{1}{4}$ **d** $2^{2x} = \dfrac{1}{2}$

16   Calculate $\dfrac{1}{\sqrt{2}+1} + \dfrac{1}{\sqrt{3}+\sqrt{2}} + \dfrac{1}{\sqrt{4}+\sqrt{3}} + \ldots\ldots \dfrac{1}{10+\sqrt{99}}$

26 SIMILAR SHAPES

The Angel of the North, a sculpture by Antony Gormley, is located in Gateshead. Before work is begun on a sculpture of this size, several maquettes are made – these are small-scale models of the sculpture with the same proportions. They are similar to the finished sculpture. Several maquettes were made of the Angel of the North, including one human-sized one which sold at auction for £2m in July 2008.

◎ Objective

In this chapter you will:
- use similar shapes to solve problems involving lengths, areas and volumes
- understand and use the relationship between length, area and volume scale.

◇ Before you start

You should already know how to:
- recognise congruent shapes
- use the angle properties associated with parallel lines
- use ratios to compare lengths.

26.1 Areas of similar shapes

Objective

○ You can solve problems involving the areas of similar shapes.

Why do this?

Designers, architects and surveyors use scale drawings and need to be able to calculate accurate areas from maps, plans and diagrams.

Get Ready

1. Find the square root of **a** 81 **b** 256 **c** 8100
2. Find the squares of **a** 14 **b** 25 **c** 19.6

Key Points

◉ The diagram shows squares of side 1 cm, 2 cm, 3 cm and 4 cm.

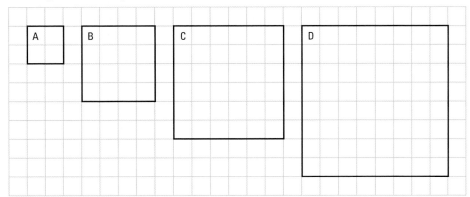

◉ The squares are all similar.
  ◉ It takes 4 squares of side 1 cm to fill the square with side 2 cm.
  ◉ It takes 9 squares of side 1 cm to fill the square with side 3 cm.
  ◉ It takes 16 squares of side 1 cm to fill the square with side 4 cm.
  ◉ If the ratio of the corresponding sides is k then the ratio of the areas is k^2.
◉ k is sometimes called the linear scale factor.
◉ k^2 is called the area scale factor.

Example 1

The diagram shows a rectangle with side 2 cm and area 2 cm². A second similar rectangle is drawn with side 6 cm. Calculate the area of the new rectangle.

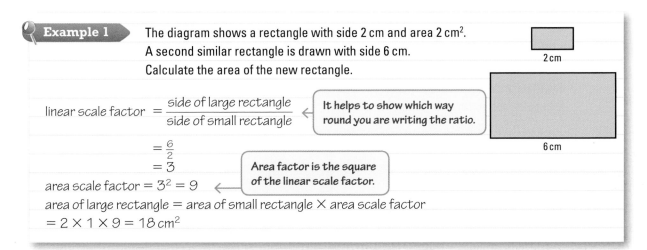

$$\text{linear scale factor} = \frac{\text{side of large rectangle}}{\text{side of small rectangle}}$$

It helps to show which way round you are writing the ratio.

$$= \frac{6}{2}$$
$$= 3$$

Area factor is the square of the linear scale factor.

area scale factor $= 3^2 = 9$

area of large rectangle $=$ area of small rectangle $\times$ area scale factor
$= 2 \times 1 \times 9 = 18\,\text{cm}^2$

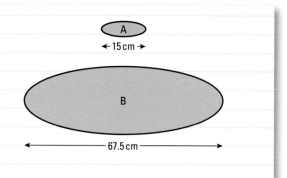

Example 2

Shape A is similar to shape B.
The area of A is 150 cm².
Find the area of shape B.

linear scale factor = $\dfrac{67.5}{15}$ = 4.5

area scale factor = 4.5² = 20.25

area of B = 20.25 × 150 = 3037.5 cm²

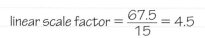

Exercise 26A

> Questions in this chapter are targeted at the grades indicated.

A

1 Shapes A and B are similar.
Shape A has area 20 cm² and length 4.5 cm.
Shape B has length 27 cm.
Calculate the area of shape B.

A02

2 Shapes A and B are similar.
Shape A has area 30 cm² and height 5 cm.
Shape B has height 15 cm.
Calculate the area of shape B.

A02

3 Cuboids A and B are similar.
The surface area of cuboid B is 108 cm².
Calculate the surface area of cuboid A.

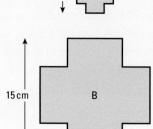

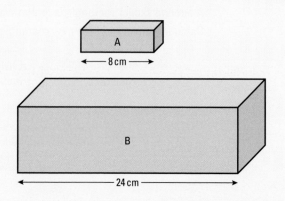

4   The diagram shows a small box of chocolates with surface area 500 cm².
    The box has a piece of ribbon 50 cm long wrapped around it.
    A similar box has a similar piece of ribbon wrapped around it of length 75 cm.
    Calculate the surface area of the larger box.

A02   A

Example 3   Shape C is similar to shape D.
                Calculate the length of shape D.

C
Area 7 cm²
←— 3 cm —→

D
Area 252 cm²

area scale factor $= \frac{252}{7} = 36$ ←——   Linear scale factor is the square root of the area scale factor.

linear scale factor $= \sqrt{36} = 6$

shape D has length $6 \times 3 = 18$ cm

Exercise 26B

1   Triangle A is similar to triangle B.
    The area of triangle A is 90 cm².
    The area of triangle B is 202.5 cm².
    Calculate the value of
    a x
    b y.

A03   A

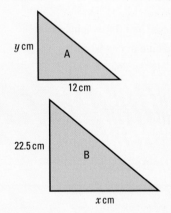

y cm

A

12 cm

22.5 cm

B

x cm

2   Two similar triangles have areas 36 cm² and 64 cm² respectively.
    The base of the smaller triangle is 6 cm.
    Find the base of the larger triangle.

A03   A☆

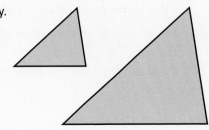

26.2 Volumes of similar shapes

⊙ **Objective**

● You can solve problems involving the volumes of similar shapes.

⊘ **Why do this?**

When designing packaging the manufacturer needs to be able to design similar shapes for large, medium and small sizes which will hold specific quantities.

⬆ **Get Ready**

1. Find the cube roots of **a** 343 **b** 1000
2. Find the cubes of **a** 2.5 **b** 11

🔍 **Key Points**

⊙ The diagram shows three similar shapes, a cube of side 1 cm, a cube of side 2 cm and a cube of side 3 cm.
The volume of the cube side 1 cm = $1 \times 1 \times 1 = 1$ cm³
The volume of the cube side 2 cm = $2 \times 2 \times 2 = 8$ cm³
The volume of the cube side 3 cm = $3 \times 3 \times 3 = 27$ cm³

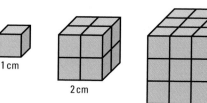

⊙ If k is the ratio of the lengths, the ratio of the volumes is k^3.

When the length is multiplied by k, the volume is multiplied by k^3.

⊙ k^3 is called the volume scale factor.

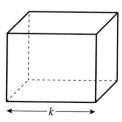

📌 **Example 4**

Cuboids P and Q are similar.
The volume of P is 60 cm³.
Calculate the volume of Q.

linear scale factor = $\dfrac{\text{large}}{\text{small}} = \dfrac{15}{5} = 3$

volume scale factor = $3^3 = 27$
volume of Q = $27 \times 60 = 1620$ cm³

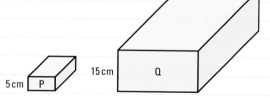

⚙ **Exercise 26C**

A
A02

1 Prisms A and B are similar.
The volume of A is 10 cm³.
Calculate the volume of prism B.

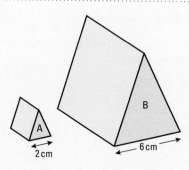

2 Cylinders C and D are similar.
The volume of C is 5 cm³.
Calculate the volume of cylinder D.

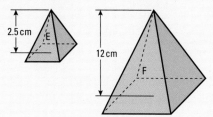

3 Square-based pyramids E and F are similar.
The volume of E is 12 cm³.
Calculate the volume of square-based pyramid F.

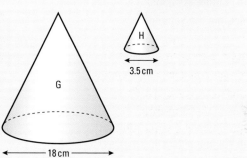

4 Cones G and H are similar.
The volume of G is 306 cm³.
Calculate the volume of cone H.

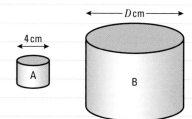

Example 5

Cylinders A and B are similar.
The diameter of A is 4 cm.
The volume of A is 120 cm³.
The volume of B is 405 cm³.
Work out the diameter D of cylinder B.

$volume\ scale\ factor = \dfrac{large}{small} = \dfrac{405}{120} = \dfrac{27}{8}$

> Volume scale factor is the cube of the linear scale factor.

$linear\ scale\ factor = \sqrt[3]{\dfrac{27}{8}} = \dfrac{\sqrt[3]{27}}{\sqrt[3]{8}} = \dfrac{3}{2} = 1.5$

$D = 1.5 \times 4 = 6\ cm$

Exercise 26D

1 Sphere K is similar to sphere J.
(All spheres are similar to each other.)
The volume of K is **64** times the volume of J.
Calculate the diameter of J.

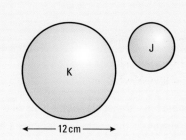

A A02 A03

2 Prisms K and L are similar.
The volume of L is 216 times the volume of K.
Calculate the value of
a x
b y.

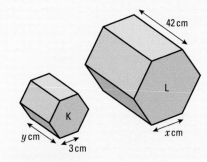
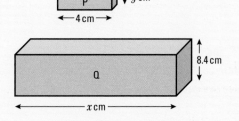

A02 A03

3 Cuboids P and Q are similar.
The volume of cuboid P is 48 cm³.
The volume of cuboid Q is 16 464 cm³.
Calculate the value of
a x
b y.

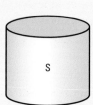

A02 A03

4 Cylinders R and S are similar.
The volume of R is 12π cm².
The volume of S is 40.5π cm².
Calculate the length of the radius of cylinder S.

A★ A02 A03

5 A bakery sells two sizes of doughnuts.
The small size has a mass of 50 g.
The large size has a mass of 168.75 g.
If the larger doughnut has a diameter of 15 cm,
find the diameter of the small size.

26.3 Lengths, areas and volumes of similar shapes

◎ Objective

● You understand and use the
relationship between length, area
and volume scale.

⦾ Why do this?

A manufacturer designing a new carton for a drink would
experiment with different heights and widths of the packaging to
find the best shape to hold a certain amount of their product.

◈ Get Ready

Express these ratios in their simplest form **1.** $14:63$ **2.** $10^2:5^3$ **3.** $2.5:15$ **4.** $4^3:8^2$

◖ Key Points

● A length has 1 dimension – the scale factor is used once.
● An area has 2 dimensions – the scale factor is used twice.
● A volume has 3 dimensions – the scale factor is used three times.

Example 6

Two similar cylinders have masses of 32 kg and 108 kg.
The area of the label on the small cylinder is 10 cm².
Calculate the area of the label on the large cylinder.

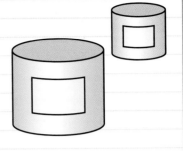

volume scale factor = $\dfrac{\text{large}}{\text{small}} = \dfrac{108}{32} = 3.375$

linear scale factor = $\sqrt[3]{3.375} = 1.5$
area scale factor = $1.5^2 = 2.25$
area of the label = $10 \times 2.25 = 22.5$ cm²

Exercise 26E

Give your answers to the following questions correct to 3 significant figures.

1 Cones P and Q are similar.
The volume of cone P is 125 times the volume of cone Q.
If the surface area of P is 40 cm²,
calculate the surface area of Q.

A02
A03 **A**

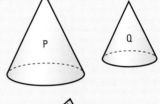

2 Prisms G and H are similar.
The surface area of G is 64 times the surface area of H.
If the volume of G is 2000 cm³,
calculate the volume of H.

A02
A03

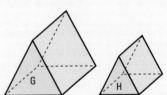

3 A container has a surface area of 5000 cm² and a capacity of 10.6 litres.
Find the surface area of a similar container which has a capacity of 4.8 litres.

A02
A03

4 The volume of a toy is 900 cm³.
A similar larger toy has volume 13 500 cm³.
The surface area of the larger toy is 2700 cm².
Find the surface area of the smaller toy.

A02
A03

5 A detergent manufacturer makes bottles of detergent in three sizes.
The bottles are mathematically similar.
The large bottle holds 5 l, the medium bottle holds 3 l and
the small bottle holds 1 l.
The 1 l bottle has a label with an area of 100 cm².
Calculate the area of the labels on

a the large bottle

b the medium bottle.

A02 **A☆**

6 A recycling bin holds 50 l of rubbish.
A smaller bin will hold 30 l of rubbish.
The surface area of the larger bin is 0.142 m².
Calculate the surface area of the smaller bin.

A02

Chapter review

- k is called the linear scale factor.
- k^2 is called the area scale factor.
- k^3 is called the volume scale factor.
- A length has 1 dimension – the scale factor is used once.
- An area has 2 dimensions – the scale factor is used twice.
- A volume has 3 dimensions – the scale factor is used three times.

Review exercise

C
AO2

1 A car is 4 m long and 1.8 m wide.
A model of the car, similar in all respects, is 5 cm long. How wide is it?

AO2

2 A model of a car is 12 cm long and 5.2 cm high.
If the real car is 3.36 m long, how high is it?

A
AO2

3 The volumes of two mathematically similar solids are in the ratio 27 : 125.
The surface area of the smaller solid is 36 cm².
Work out the surface area of the larger solid.

Nov 2007

AO2

4 The diagram shows two quadrilaterals that are mathematically similar.

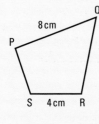 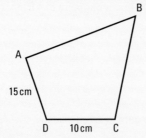

In quadrilateral PQRS, PQ = 8 cm, SR = 4 cm.
In quadrilateral ABCD, AD = 15 cm, DC = 10 cm.
Angle PSR = angle ADC.
Angle SPQ = angle DAB.
a Calculate the length of AB.
b Calculate the length of PS.

ResultsPlus
Exam Question Report

91% of students answered this sort of question
well. They showed all of their working.

June 2007

5 Two solid shapes, **A** and **B**, are mathematically similar.
The base of shape **A** is a circle with radius 4 cm.
The base of shape **B** is a circle with radius 8 cm.
The surface area of shape **A** is 80 cm².

a Work out the surface area of shape **B**.

The volume of shape **B** is 600 cm³.

b Work out the volume of shape **A**.

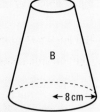

A02

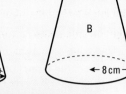

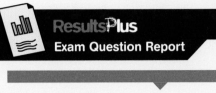

82% of students answered this question poorly
because they used the wrong scale factor.

June 2008

6 Two cones, **P** and **Q**, are mathematically similar.
The total surface area of cone **P** is 24 cm².
The total surface area of cone **Q** is 96 cm².
The height of cone **P** is 4 cm.

a Work out the height of cone **Q**.

The volume of cone **P** is 12 cm³.

b Work out the volume of cone **Q**.

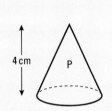

June 2007

A02

7 Two prisms, **A** and **B**, are mathematically similar.
The volume of prism **A** is 12 000 cm³.
The volume of prism **B** is 49 152 cm³.
The total surface area of prism **B** is 9728 cm².
Calculate the total surface area of prism **A**.

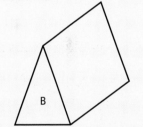

Nov 2006

A03

8 A cone is divided by a cut parallel to the base halfway between the top and the base.
What is the ratio of

a the area of the base of the small cone to the area of the base of the large cone

b the volume of the small cone to the volume of the large cone?

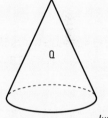

A02

9 a Prove that all cubes are similar.

b Two cubes have edges 2 cm and 5 cm.
What is the ratio of the total surface areas of the two cubes?

c 'Two cuboids are similar' – This statement is not always true.
Explain why.

A02

A☆
A03

10 A manufacturer makes pots of cream in two sizes.

The small size contains 300 g and has a diameter of 10 cm.

The large size contains 500 g.

The pots are similar.

a Find the diameter of the larger pot.

The front of the large pot has a rectangular label

of area 36 cm².

b Find the area of the label on the smaller pot.

27 PROPORTION 2

On 6 August 1945, a nuclear bomb known as 'Little Boy' was dropped on Hiroshima, Japan. The energy of the explosion caused was equivalent to around 18 kilotons of TNT, caused by 600 mg of uranium within the weapon converting into energy. Three days later a second bomb, 'Fat Man', was detonated over Nagasaki, Japan. This bomb was 15% larger, and the energy released in the explosion was equivalent to about 21 kilotons of TNT.

Objectives

In this chapter you will:

- learn to solve problems involving direct proportion
- discover how to write down the statement of proportionality and the formula for a variety of problems
- find out how to solve problems involving inverse, square and cubic proportion.

Before you start

You need to be able to:

- understand, use and calculate proportion
- derive simple formulae and simple rules for number sequences
- plot points on a graph
- derive quadratic and cubic formulae.

27.1 Direct proportion

Objective

○ You can use graphs to solve problems involving direct proportion.

Why do this?

Scientists use proportion to work out the amount of energy produced by a nuclear explosion.

Get Ready

1. A pen costs 25p. What is the cost of:
   a 2   **b** 3   **c** 4   **d** 5 pens?
2. A photocopier takes 10 seconds to copy 2 letters. How long does it take to copy:
   a 4   **b** 8   **c** 16   **d** 32 letters?

Key Point

◉ When a graph of two quantities is a straight line through the origin, one quantity is directly proportional to the other.

Example 1

The cost of buying 5 litres of fuel is £11.
   a Show that the cost, £C, of buying the fuel is directly proportional to the amount, x litres, of fuel bought.
   b Find a formula for C in terms of x.

a The cost of buying 5 litres of fuel is £11. ←
   So, the cost of buying 10 litres of fuel is £22,
   the cost of buying 15 litres of fuel is £33,
   the cost of buying 20 litres of fuel is £44,
   and so on.

> 10 litres cost twice as much as 5 litres. 5 litres cost £11, so 10 litres cost 2 × £11 = £22.
>
> 15 litres cost three times as much as 5 litres. 5 litres cost £11, so 15 litres cost 3 × £11 = £33 (and so on). Summarise the information in a table.

x	5	10	15	20
C	11	22	33	44

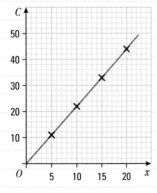

> Plot a graph of C against x.
> Draw a line through the points.
> The graph is a straight line which passes through the point O. So C is directly proportional to x.

The graph shows that the cost, £C, of buying the fuel is directly proportional to the amount, x litres, bought.

b $C = kx$

The point $(11, 5)$ lies on the line, so

$11 = k \times 5$

$k = \frac{11}{5} = 2.2$

The formula is $C = 2.2x$.

> The graph passes through the origin O. The equation of a straight line which passes through the origin has the form $y = mx$.
> Here $y = C$ and $m = k$ (the constant of proportionality).
> Find the constant k. Substitute $C = 11$ and $x = 5$ into the formula.
> Write down the formula, substituting $k = 2.2$ into $C = kx$.

Exercise 27A

Questions in this chapter are targeted at the grades indicated.

1 The table gives information about the variables x and Y.

x	2	4	6	8	10
Y	3	6	9	12	15

a Plot the graph of Y against x.

b Is Y directly proportional to x? Give a reason for your answer.

2 Here is a graph of $Y = kd$. Use the information in the graph to work out the value of k.

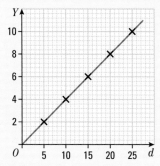

3 The table gives information about the variables x and M.

x	4	10	16	20
M	9	22.5	36	45

a Show that M is directly proportional to x.

b Given that $M = kx$, work out the value of k.

c Use your formula to work out the value of M when $x = 32$.

4 Y is directly proportional to t. $Y = 10$ when $t = 20$.

a Sketch a graph of Y against t.

b Work out a formula for Y in terms of t.

c Use your formula to work out the value of Y when $t = 100$.

5 The cost (£C) of a bottle of white correcting fluid is directly proportional to the volume (v cm³) of fluid in the bottle. A bottle containing 4 cm³ of fluid costs £1.80.

a Work out the cost of **i** 8, **ii** 12, **iii** 16 cm³ of the liquid.

b Work out a formula for C in terms of v.

c The cost of a bottle of correcting fluid is £5.85. Work out the volume of fluid in the bottle.

C

B

A

27.2 **Further direct proportion**

◎ Objective

○ You can use formulae to solve problems involving direct proportion.

⊘ Why do this?

If you know the currency exchange rate for the country you're on holiday in then you can convert local prices into pounds, to see roughly how expensive goods are.

◈ Get Ready

1. Solve for x

   a $x = \frac{6}{7}a$ when $a = 21$

   b $\frac{x}{3} = \frac{6}{7}a$ when $a = 28$

   c $\frac{2}{5}x = \frac{6}{7}a$ when $a = 35$

◷ Key Points

◉ The symbol $\propto$ means 'is proportional to'.

◉ When y is directly proportional to x:

  ◉ $y \propto x$ is the statement of proportionality.

  ◉ $y = kx$ is the formula for direct proportion, where k is the **constant of proportionality**.

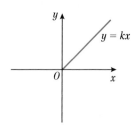

⟁ Example 2

W is directly proportional to x. $W = 18$ when $x = 1.5$.
Work out the value of W when $x = 7$.

$W \propto x$ ← Write down the statement of proportionality. W is 'directly proportional' to x.

$W = kx$ ← Replace $\propto$ with '$= k$'.

$18 = 1.5k$ ← Work out the value of k. Substitute $W = 18$ and $x = 1.5$.

$k = \frac{18}{1.5} = 12$ ← Divide both sides by 1.5.

So, $W = 12x$ ← Write down the formula, putting in the value of k.

When $x = 7$
$W = 12 \times 7 = 84$ ← Put $x = 7$ into $W = 12x$.

⚙ Exercise 27B

B

1 y is directly proportional to x so that $y = kx$. $y = 12$ when $x = 8$. Work out the value of k.

2 B is directly proportional to t so that $B = kt$. $B = 1.75$ when $t = 2.5$. Work out the value of k.

A

3   P is directly proportional to h. $P = 40.5$ when $h = 18$.
  a   Show that $P = \frac{9}{4}h$.
  b   Work out the value of P when $h = 32$.
  c   Work out the value of h when $P = 27$.

4   The voltage V across a resistor (in volts) is directly proportional to the current c flowing through it (in amps).
  a   Show that $V = 250c$.
  b   Work out the value of c when $V = 9$ volts.

A02

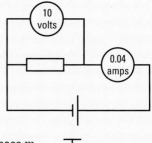

5   The extension E of an elastic string (in mm) is directly proportional to the mass m on the string (in grams).
  a   Find a formula for E in terms of m.
  b   Find the extension of the string when the mass is 450 g.
  c   Find the mass that will extend the string by 52.5 mm.

A02

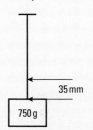

6   The volume, V cm^3, of mercury in a tube is directly proportional to the height, h cm, of the tube. When the height of the tube is 12 cm, the volume of mercury is 40 cm^3. Work out the volume of mercury in the tube when the height of the tube is 20 cm.

A02

27.3 Writing statements of proportionality and formulae

Objective

○ You can write down the statement of proportionality and the formula for a variety of problems.

Why do this?

When scientists are trying to establish a formula, they will often write a statement of proportionality before they perform an experiment and work out the exact results.

Get Ready

1. a   The volume, V, of a gas is directly proportional to its temperature, T.
    Find a formula for V in terms of T.
   b   The pressure, P, is directly proportional to its temperature, T. Find a formula for P in terms of T.

Key Points

◉ Sometimes quantities are proportional to the square, cube or other power of another quantity.
  ◉ 'y is proportional to the square of x' so $y \propto x^2$ means $y = k \times x^2$ or $y = kx^2$, where k is the constant of proportionality.
  ◉ 'y is proportional to the cube of x' so $y \propto x^3$ means $y = k \times x^3$ or $y = kx^3$, where k is the constant of proportionality.
  ◉ 'y is proportional to the square root of x' so $y \propto \sqrt{x}$ means $y = k \times \sqrt{x}$ or $y = k\sqrt{x}$, where k is the constant of proportionality.

◉ Quantities can also be inversely proportional to each other.

  ◉ 'y is inversely proportional to x' so $y \propto \dfrac{1}{x}$ means $y = k \times \dfrac{1}{x}$ or $y = \dfrac{k}{x}$, where k is the constant of proportionality.

◉ Some common proportional graphs are shown below.

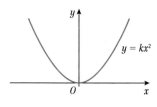
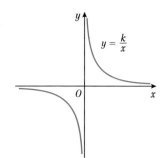
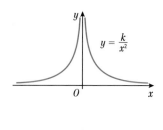

🔍 **Example 3**   Write down **i** the statement of proportionality, **ii** the formula, for each of the following.

    a H is proportional to the square of d.   **b** T is proportional to the cube of f.
    c G is proportional to the square root of p.   **d** R is inversely proportional to x.
    e L is inversely proportional to the square of n.

a i $H \propto d^2$ **ii** $H = kd^2$ ← The symbol $\propto$ means 'is proportional to', so replace the words with the symbol $\propto$. The 'square of d' is the same as 'd squared', i.e. d^2.
Replace the symbol $\propto$ with '$= k$'. So $H \propto d^2$ becomes $H = kd^2$.

b i $T \propto f^3$ **ii** $T = kf^3$
c i $G \propto \sqrt{p}$ **ii** $G = k\sqrt{p}$

d i $R \propto \dfrac{1}{x}$ **ii** $R = \dfrac{k}{x}$ ← You write an inverse proportion as a reciprocal (i.e. 'one over...'). The reciprocal of x is $\dfrac{1}{x}$.

e i $L \propto \dfrac{1}{n^2}$ **ii** $L = \dfrac{k}{n^2}$ ← The reciprocal of n^2 is $\dfrac{1}{n^2}$.

🔍 **Example 4**   Write these statements of proportionality in words.

    a $P \propto w^2$ **b** $A \propto g^3$ **c** $F \propto \sqrt[3]{t}$ **d** $Y \propto \dfrac{1}{d}$ **e** $B \propto \dfrac{1}{r^3}$

a P is proportional to the square of w.   ← This could also be written as 'P is proportional to w squared'.
b A is proportional to the cube of g.
c F is proportional to the cube root of t.
d Y is inversely proportional to d.   ← $\dfrac{1}{d}$ is the reciprocal of d. So Y is inversely proportional to d.

e B is inversely proportional to the cube of r.   ← This could also be written as B is inversely proportional to r cubed.

Exercise 27C

1. Write down **i** the statement of proportionality, **ii** the formula, for each of the following.
   Use the symbol ∝.
   a M is directly proportional to n.
   b L is proportional to the square of h.
   c P is proportional to the cube of t.
   d Q is proportional to the square root of y.
   e W is proportional to the cube root of x.
   f A is inversely proportional to b.
   g H is inversely proportional to the square of g.
   h U is inversely proportional to the square of f.
   i E is inversely proportional to the cube of w.
   j V is inversely proportional to the square root of r

2. The number of ants in an ants' nest, N, is proportional to the square of the diameter, d cm, of the ants' nest.
   a Write down the statement of proportionality.
   b Write down a formula for N in terms of d and k (the constant of proportionality).

3. The height, in metres, of a tree is proportional to the square of the circumference, of the tree in metres. Write down a formula for H in terms of c.

4. The quantity of fuel, Q tonnes, used by a rocket is proportional to the cube of the time, t seconds, that the fuel burns. Write down a formula for Q in terms of t.

5. A formula for the force of attraction (F newtons) between two objects is given by $F = \dfrac{k}{r^2}$ where r is the distance between the objects (in metres) and k is a constant. What is the relationship between F and r?

6. A formula for the period (T seconds) of a pendulum is given by $T = k\sqrt{l}$, where l is the length of the pendulum (in metres) and k is a constant. Tom says that T is inversely proportional to the square of l. He is wrong. Explain why.

27.4 Problems involving square and cubic proportionality

◉ Objective

○ You can solve problems involving square and cubic proportions.

◈ Why do this?

The power output of a wind turbine is proportional to the cube of the wind speed.

◈ Get Ready

1. Solve
   a 15^2
   b 8^3
   c 11^3
   d $\sqrt{81}$
   e $\sqrt[3]{512}$
   f $\sqrt{4096}$

◉ Key Points

◉ To solve problems involving square or cubic proportion, first write the statement of proportionality, then the formula.

◉ Substitute given values into the formula to find the solution.

Example 5

The resistance (R newtons) to the motion of a racing car is directly proportional to the square of the speed (v m/s) of the racing car. Given that $R = 5000$ when $v = 10$, work out the value of R when $v = 30$.

$R \propto v^2$ ← Write down the statement of proportionality. R is proportional to the square of v.

$R = kv^2$ ← Write down the formula. Replace the symbol $\propto$ with '$= k$'.

$5000 = k \times 10^2$
$5000 = k \times 100$ ← Work out the value of k. Substitute $R = 5000$ and $v = 10$ into $R = kv^2$.

$k = \dfrac{5000}{100} = 50$ ← Divide both sides of the equation by 100.

So, $R = 50v^2$
When $v = 30$ ← Work out the value of R when $v = 30$. Substitute $v = 30$ into $R = 50v^2$.
$R = 50 \times 30^2$
$R = 50 \times 900 = 45\,000\,\text{N}$

Example 6

$T \propto m^3$. $T = 10$ when $m = 2$. Find the value of m when $T = 80$.

$T \propto m^3$
$T = km^3$ ← Replace the symbol $\propto$ with '$= k$'.

$T = 10$ when $m = 2$, so ← Work out the value of k. Substitute $T = 10$ and $m = 2$ into $T = km^3$.
$10 = k \times 2^3$
$10 = 8k$
$k = 1.25$
So, $T = 1.25m^3$ ← Write down the formula. Substitute $k = 1.25$ into $T = km^3$.

When $T = 80$, ← You need to find m when $T = 80$, so substitute $T = 80$ into $T = 1.25m^3$.
$1.25m^3 = 80$
$m^3 = \dfrac{80}{1.25} = 64$ ← Divide both sides by 1.25.

$m = \sqrt[3]{64} = 4$ ← Take the cube root of both sides.

Exercise 27D

A

1. A is proportional to the square of x so that $A = kx^2$. $A = 20$ when $x = 5$. Work out the value of k.

2. G is proportional to the cube of f so that $G = kf^3$. $G = 54$ when $f = 1.5$. Work out the value of k.

3. $Z \propto b^2$. $Z = 32$ when $b = 5$. Find a formula for Z in terms of b.

4. $L \propto v^3$. $L = 120$ when $v = 4$. Find a formula for L in terms of v.

5. W is proportional to the square of p. $W = 8$ when $p = 4$. Work out the value of W when $p = 3$.

6 A stone is thrown vertically upwards with a speed of v m/s. The height, H metres, reached by the stone is proportional to the square of v. When $v = 20$ m/s, $H = 20$ m. Work out the value of:

a H when $v = 15$ m/s

b v when $H = 10$ m.

7 The volume, V mm^3, of a raindrop is proportional to the cube of its radius, r mm. When the radius of a raindrop is 2 mm its volume is 33 mm^3. Work out the radius of a raindrop which has a volume of 65 mm^3. Give your answer to 2 decimal places.

8 The mass of an earthworm (in grams) is proportional to the cube of its length (in cm). Copy and complete this table.

Mass (grams)	Length (cm)
20	5
	6
43.94	
50	

A

A02

A02

27.5 Problems involving inverse proportion

Objective

○ You can solve problems involving inverse proportion.

Why do this?

You can use inverse proportion to work out how much each person that holds a winning ticket in a lottery draw actually wins.

Get Ready

1. If $V \propto T$ and $P \propto \dfrac{1}{V}$, write P in terms of T.

Key Points

◉ To solve problems involving inverse proportion, first write the statement of proportionality, then the formula.
◉ Substitute given values into the formula to find the solution.

Example 7 The number of hours (H) needed to dig a certain hole is inversely proportional to the number of men (x) available to dig the hole. If it takes 5 men 8 hours to dig the hole, how long will it take 6 men to dig the hole?

$H \propto \dfrac{1}{x}$ ← Write down the statement of proportionality. H is inversely proportional to x.

$H = \dfrac{k}{x}$ ← Write down the formula. Replace the symbol $\propto$ with '$= k$'.

$8 = \dfrac{k}{5}$ ← Work out the value of k. Substitute $H = 8$ and $x = 5$ into $H = \dfrac{k}{x}$.

$k = 40$ ← Multiply both sides of the equation by 5.

So, $H = \dfrac{40}{x}$ ← Substitute the value of k into the formula.

When $x = 6$, $H = \dfrac{40}{6} = 6\frac{2}{3}$ hours. ← Work out the value of H when $x = 6$. Substitute $x = 6$ into $H = \dfrac{40}{x}$.

Example 8

The force of attraction (F newtons) between two magnets is inversely proportional to the square of the distance (d cm) between them. When the magnets are 1.5 cm apart, the force of attraction is 32 newtons.

   a Find a formula for F in terms of d.

   b Work out the distance between the magnets when the force of attraction is 1.125 newtons.

a $F \propto \dfrac{1}{d^2}$

   $F = \dfrac{k}{d^2}$

When $d = 1.5$, $F = 32$

So, $32 = \dfrac{k}{1.5^2}$ ← Multiply both sides by 1.5^2.

$k = 1.5^2 \times 32$

   $= 2.25 \times 32$

   $= 72$

The formula is $F = \dfrac{72}{d^2}$.

b $F = 1.125$

So $1.125 = \dfrac{72}{d^2}$ ← Multiply both sides by d^2.

$d^2 \times 1.125 = 72$ ← Divide both sides by 1.125.

$d^2 = \dfrac{72}{1.125}$

$d^2 = 64$ ← Take the square root of both sides.

$d = \sqrt{64} = 8$ cm

Exercise 27E

A

1 L is inversely proportional to d so that $L = \dfrac{k}{d}$.

   $L = 2.25$ when $d = 20$. Work out the value of k.

2 $A \propto \dfrac{1}{x}$

   $A = 3.75$ when $x = 8$.

   a Work out the value of A when $x = 12$.

   b Work out the value of x when $A = 5$.

3 $G \propto \dfrac{1}{d}$

   $G = 4.8$ when $d = 6$.

   a Work out the value of G when $d = 8$.

   b Work out the value of d when $G = 10.8$.

4   The volume V (m³) of a gas is inversely proportional to the pressure P (N/m²). $V = 4$ m³ when $P = 500$ N/m². Work out the volume of the gas when the pressure is 750 N/m².

A02

5   The frequency F of sound (in hertz) is inversely proportional to the wavelength w (in metres). The musical note of middle C has a frequency of 256 hertz and a wavelength of 1.29 m.

   a Work out the frequency of a note with a wavelength of 0.86 m.

   b Work out the wavelength of a note with frequency 344 hertz.

6 $M \propto \dfrac{1}{f^2}$

   $M = 0.625$ when $f = 8$.

   a Work out the value of M when $f = 3.5$.

   b Work out the value of f when $M = 1$.

7 The shutter speed (S) of a camera is inversely proportional to the square of the aperture setting (f).
When $f = 8$, $S = 125$.

    a Find a formula for S in terms of f.

    b Work out the value of S when $f = 4$.

Chapter review

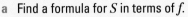

- When a graph of two quantities is a straight line through the origin, one quantity is directly proportional to the other.
- The symbol $\propto$ means 'is proportional to'.
- $y \propto x$ means 'y is directly proportional to x'.
- When y is directly proportional to x:
  - $y \propto x$ is the **statement of proportionality**.
  - $y = kx$ is the formula, where k is the constant of proportionality.
- Where k is the constant of proportionality:
  - $y = kx^2 \equiv y \propto x^2$ means y is proportional to the square of x.
  - $y = kx^3 \equiv y \propto x^3$ means y is proportional to the cube of x.
  - $y = k\sqrt{x} \equiv y \propto \sqrt{x}$ means y is proportional to the square root of x.
- When y is inversely proportional to x:
  - $y \propto \dfrac{1}{x}$ is the statement of proportionality.
  - $y = k \times \dfrac{1}{x}$ or $y = \dfrac{k}{x}$ are ways of writing the formula, where k is the constant of inverse proportionality.
- To solve problems involving proportion:
  - write the statement of proportionality
  - write the formula
  - substitute given values into the formula to find the solution.

Review exercise

1 In an experiment, the value of V was measured for different values of d. The results are given in this table. Show that V is directly proportional to d.

d	0.5	1.5	3.5	4.5	6.0
V	0.75	2.25	5.25	6.75	9.0

2 The time, T seconds, it takes a water heater to boil some water is directly proportional to the mass of water, m kg, in the water heater.
When $m = 250$, $T = 600$.

    a Find T when $m = 400$.

The time, T seconds, it takes a water heater to boil a costant mass of water is inversely proportional to the power, P watts, of the water heater.
When $P = 1400$, $T = 360$.

    b Find the value of T when $P = 900$.

June 2006

A

3 Copy and complete this table.

a	T is directly proportional to b	
b		$R \propto a$
c	P is proportional to the square of m	
d	Z is proportional to the cube of g	
e		$H \propto \dfrac{1}{y}$

4 $S \propto p$. $S = 12$ when $p = 8$.

a Find a formula for S in terms of p.

b Sketch a graph of S against p.

ResultsPlus

Exam Question Report

79% of students answered this sort of question well. They knew how to move from a statement of proportionality to a formula.

A03

5 For an oil company, the cost (£C) of drilling a hole is directly proportional to the depth (d metres) of the hole. The cost of drilling a hole to a depth of 100 metres is £8500. Work out the cost of drilling a hole to a depth of 825 metres.

A03

*__**6** The pressure P of water on a diver (in bars) is directly proportional to his depth d (in metres). When the diver is at a depth of 5 metres the pressure on the diver is 0.5 bars. For safety reasons the pressure on a particular diver must not exceed 6.5 bars. The diver wants to dive to a depth of 60 m. Can the diver do this safely? Give a reason for your answer.

A03

7 For batteries having the same length, the energy stored in a battery is proportional to the square of the circumference of the battery. When the circumference is 3.5 cm the energy stored in the battery is 5 units. The radius of a battery is 5 cm. Work out the energy stored in the battery.

8 When a stone is dropped from a cliff it travels a distance D (in metres) after a time t (in seconds). This table gives information about the stone.

t	0	1	2	3	4	5
D	0	5	20			

a Show that $D \propto t^2$, and complete the table.

b Work out the time taken for the stone to travel 15 metres. Give your answer correct to 2 decimal places.

9 The resistance, R ohms, of a particular cable is inversely proportional to the square of its radius, r mm. Copy and complete this table. Give your answers correct to 2 decimal places.

Radius (r mm)	Resistance (R ohms)
10	500
15	
17.5	
	250

10 X is proportional to the square root of h. $X = 3$ when $h = 16$.

a Show that $X = 0.75\sqrt{h}$.

b Work out the value of X when $h = 25$.

c Work out the value of h when $X = 6$.

11 In a particular industrial process, Germalex is added to increase the speed of a chemical reaction The time taken, T seconds, for the reaction is inversely proportional to the square root of the mass, m grams, of Germalex added. When 50 grams of Germalex is added the time taken for the reaction is 20 seconds.

  a Show that $T = 100\sqrt{\dfrac{2}{m}}$.

  b It is required that the time taken for a particular reaction should be 15 seconds. Work out the mass of Germalex that needs to be added to achieve this reaction time. Give your answer to 3 significant figures.

12 Here are 4 sketch graphs.

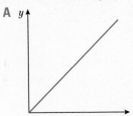 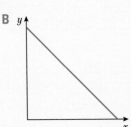 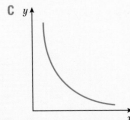 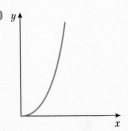

  a Write down the letter of the graph which shows y is directly proportional to x.

  b Write down the letter of the graph which shows y is inversely proportional to x.

13 M is directly proportional to L^3.
When $L = 2$, $M = 160$.

  a Find the value of M when $L = 3$.

  b Find the value of L when $M = 120$.

June 2009, adapted

14 q is inversely proportional to the square of t.
When $t = 4$, $q = 8.5$.

  a Find a formula for q in terms of t.

  b Calculate the value of q when $t = 5$.

ResultsPlus
Exam Question Report

78% of students answered this question poorly because they did not read the details of the question properly.

June 2008

15 The time, T seconds, for a hot sphere to cool is proportional to the square root of the surface area, $A\ m^2$, of the sphere.
When $A = 100$, $T = 40$.
Find the value of T when $A = 60$.
Give your answer correct to 3 significant figures.

June 2009

16 y is directly proportional to the square of x.
x is directly proportional to the square root of z.

  a Find a formula for y in terms of z and a constant of proportionality.

  u is directly proportional to the square of v.
v is inversely proportional to the square root of w.

  b Show that the product of u and w is constant.

28 PROBABILITY

Philippe Petit is a French high-wire artist famous for walking along cables suspended from well-known buildings. His most celebrated and perilous exploit was negotiating the 43 m gap between the World Trade Center towers in New York, despite the fact that the probability of him surviving a fall was zero.

⊙ Objectives

In this chapter you will:
- learn how to use a number to represent a probability
- learn how to find and estimate probabilities
- learn how to add and multiply probabilities.

◇ Before you start

You need to be able to:
- add, subtract, multiply and divide fractions and decimals.

28.1 Writing probabilities as numbers

⦿ Objectives

- You can use a number to represent a probability.
- You can use a sample space diagram to record all possible outcomes.

⦵ Why do this?

Banks use probability to assess risk when they plan takeovers of other companies.

◈ Get Ready

Represent how likely each of these events is on a probability scale:

1. A man will jump over the Eiffel tower.
2. When you add 2 and 2 together you get the answer 4.
3. When you spin a coin you will get a head.

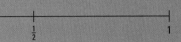

Key Points

- The **probability** P that an **event** will happen is a number in the range $0 \leqslant P \leqslant 1$.
- For an event which is **certain** $P = 1$.
- For an event which is **impossible** $P = 0$.
- A probability can be written as a fraction, a decimal or a percentage.
- For **equally likely** outcomes, the probability that an event will happen is

$$\text{Probability} = \frac{\text{number of successful outcomes}}{\text{total number of possible outcomes}}$$

- A sample space is all the possible **outcomes** of one or more events.
- These outcomes can be presented in a **sample space diagram**.

Example 1

This 5-sided spinner is spun.

The spinner is fair.

> A fair spinner is one that has an equal chance of landing on any of its sides, so each side is equally likely.

Work out the probability that the spinner will land on red.

> The spinner has 2 red sectors, so the number of successful outcomes = 2.

> $\frac{2}{5}$ can also be expressed as 40% or 0.4.

$$p(red) = \frac{2}{5}$$

> The spinner has 5 sides altogether, so the total number of possible outcomes = 5.

⚙ Exercise 28A

Questions in this chapter are targeted at the grades indicated.

1   Rashid spins this 7-sided spinner. The spinner is fair.
Work out the probability that the spinner will land on
a   yellow
b   red
c   white.

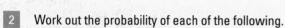

2   Work out the probability of each of the following.
a   rolling the number 4 with an ordinary dice
b   rolling an even number with an ordinary dice
c   taking an ace from an ordinary pack of cards
d   taking a diamond from an ordinary pack of cards
e   taking a black king from an ordinary pack of cards

3   A bag contains 5 red balls and 4 green balls. A ball is taken at random from the bag.
Work out the probability that the ball will be:
a   red   b   green   c   yellow.

4   The faces of an 8-sided dice are numbered from 1 to 8.
Work out the probability of rolling each of the following.
a   an odd number   b   an even number   c   a 4 or a 5
d   a prime number   e   a factor of 10

5   2500 tickets are sold in a school raffle. Chelsy buys 5 tickets in the raffle.
Work out the probability that she will win the raffle.

6   A box contains 3 bags of salt and vinegar crisps, 4 bags of cheese and onion crisps and 2 bags of beef crisps. One of these bags of crisps is taken from the box at random.
Work out the probability that the bag of crisps will be:
a   salt and vinegar   b   cheese and onion
c   beef   d   cheese and onion or beef.

7   A letter is chosen at random from the word PROBABILITY. Write down the probability that it will be:
a   B   b   Y   c   R or I   d   a vowel   e   G.

8   The table gives the numbers of boys and the numbers of girls in a primary school and whether they are left-handed or right-handed.

	Left-handed	Right-handed	Total
Boys	47	135	
Girls	61	119	
Total			362

a   Copy and complete the table.
b   One of these children is chosen at random.
Use the information in your table to work out the probability that the child will be:
   i   a boy
   ii left-handed
   iii a right-handed girl.

9   The pie chart gives information about how
    some students travelled to school one day.
    One of these students is chosen at random.
    Use the information in the pie chart to
    work out the probability that the student:
    a   travelled to school by bus
    b   walked to school.

10 In a group of students
    55% are boys
    65% prefer to watch film A
    10% are girls who prefer to watch film B.

    One of these students is picked at random.
    Work out the probability that the student is a boy who prefers to watch film A.

Example 2   Two ordinary dice are rolled. Work out the probability
that the total score on the two dice will be 8.

	6	(1,6)	(2,6)	(3,6)	(4,6)	(5,6)	(6,6)
	5	(1,5)	(2,5)	(3,5)	(4,5)	(5,5)	(6,5)
Red	4	(1,4)	(2,4)	(3,4)	(4,4)	(5,4)	(6,4)
dice	3	(1,3)	(2,3)	(3,3)	(4,3)	(5,3)	(6,3)
	2	(1,2)	(2,2)	(3,2)	(4,2)	(5,2)	(6,2)
	1	(1,1)	(2,1)	(3,1)	(4,1)	(5,1)	(6,1)
		1	2	3	4	5	6

Blue dice

Draw a sample space diagram.
A sample space diagram shows
all the possible outcomes, for
example (6, 4) represents the
outcome of throwing a 6 on the
blue dice and a 4 on the red dice.

There are a total of 36 possible
outcomes.

Identify all the outcomes that
give a total score of 8.
There are 5 outcomes that give
a total score of 8: (2, 6), (3, 5),
(4, 4), (5, 3) and (6, 2).

$$P(8) = \frac{\text{number of successful outcomes}}{\text{total number of possible outcomes}}$$

$$= \frac{5}{36}$$

Exercise 28B

1   Two ordinary dice are rolled.
    Use the sample space diagram in Example 2 to work out the probability of getting each of the following
    outcomes.
    a   a total score of
        i 2   ii 5   iii 10 or more
    b   the same number on each dice
    c   a number on the blue dice exactly 2 more than the number on the red dice

C

2 An ordinary dice is rolled and a fair coin is spun.

a Copy and complete the following sample space diagram to show all the possible outcomes.

	H	(1,H)					
Coin	T	(1,T)	(2,T)				
		1	2	3	4	5	6

Dice

b Work out the probability of getting:

   i a 1 on the dice and a head on the coin

   ii a number greater than 3 on the dice

   iii a number less than 3 on the dice and a tail.

3 Two fair 4-sided spinners are spun and the difference between the numbers is calculated.

a Copy and complete this sample space diagram to show all the possible outcomes.

Spinner A

		1	2	3	4
	1	0	1		
Spinner B	2	1			
	3				
	4				

b Work out the probability of getting a difference of:

   i 0   **ii** 3   **iii** 4.

B

4 An ordinary dice is rolled and a fair 3-sided spinner is spun.

a Draw a sample space diagram to show all the possible outcomes.

b Use your sample space diagram to work out the probability of getting a total score of:

   i 7   **ii** 3   **iii** less than 5.

5 Sunti has two boxes of crayons, A and B.

In box A he has a red crayon, a blue crayon, a yellow crayon and a black crayon.

In box B he has a red crayon, a yellow crayon and a blue crayon.

Sunti takes a crayon at random from each box.

a Draw a sample space diagram to show all the possible outcomes.

b Work out the probability that the crayons will be:

   i both red   **ii** the same colour   **iii** different colours.

A

6 Andy, Brigitta, Carrie, Dean and Eli are playing in a tennis competition. Each player in the competition plays every other player. There are ten matches altogether.

Two players are picked at random to play the first game.

Work out the probability that the first game will be played by a male player and a female player.

A*

A03

7 A fair 10-sided dice and an ordinary 6-sided dice are each rolled. The numbers rolled by the 10-sided dice are used for the x-coordinates, and the numbers rolled by the 6-sided dice are used for y-coordinates.

Find the probability that the point generated by the numbers on the two dice lies on each of the following lines.

   a $y = 1$ **b** $x + y = 7$ **c** $y = x + 5$ **d** $y = 2x - 5$ **e** $y = \frac{1}{2}x + 1$

28.2 Mutually exclusive outcomes

◎ Objectives

○ You can add probabilities for mutually exclusive events.
○ You can work out the probability that something will not happen given that you know the probability that it will happen.

⑦ Why do this?

If you are planning a barbeque and you know that there is a 1 in 3 chance of it raining on that day, you can work out the probability that it won't rain.

⬆ Get Ready

1. Work out the missing numbers.

a $0.3 + ? = 0.7$ **b** $1 - ? = 0.35$ **c** $\frac{2}{5} + ? = 1$ **d** $? + \frac{1}{3} = \frac{1}{2}$

◉ Key Points

⊙ Two events are **mutually exclusive** when they cannot occur at the same time.
⊙ For mutually exclusive events A and B,
$P(A \text{ or } B) = P(A) + P(B)$.
⊙ For 3 or more events,
$P(A \text{ or } B \text{ or } C \text{ or } \ldots) = P(A) + P(B) + P(C) + \ldots$.
⊙ For mutually exclusive events A and not A,
$P(\text{not } A) = 1 - P(A)$.
⊙ If a set of mutually exclusive events contains all possible outcomes, then the sum of their probabilities must come to 1. $\Sigma p = 1$.
⊙ For three mutually exclusive events that cover all possible outcomes, $P(A) + P(B) + P(C) = 1$.

Example 3

A card is taken at **random** from an ordinary pack of cards. Work out the probability that the card will be an ace or the 10 of clubs.

> Use $P(A) = \dfrac{\text{number of successful outcomes}}{\text{total number of possible outcomes}}$

$P(\text{ace or } 10\clubsuit) = P(\text{ace}) + P(10\clubsuit)$
$= \dfrac{4}{52} + \dfrac{1}{52}$
$= \dfrac{5}{52}$

> There are 52 cards in an ordinary pack of cards, so the total number of possible outcomes = 52.
> There are four ways to pick an ace, so there are four successful outcomes A♣, A♦, A♥, A♠.
> So $P(\text{ace}) = \dfrac{4}{52}$.

> There is only one 10 of clubs, so $P(10\clubsuit) = \dfrac{1}{52}$.

⚙ Exercise 28C

1   Here is a number of shapes.

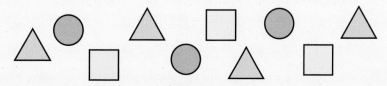

One of these shapes is chosen at random. Work out the probability that the shape will be:

a   a square   **b**   a triangle   **c**   a square or a triangle.

D

2 A bag contains some counters. The colour of each counter is either red, or black, or white. A counter is taken at random from the bag. The probability that the counter will be red is 0.3. The probability that the counter will be white is 0.6. Work out the probability that the counter will be red or white.

3 The table gives the probability of getting each of 1, 2, 3 and 4 on a biased 4-sided spinner.

Number	1	2	3	4
Probability	0.2	0.35	0.15	0.3

Work out the probability of getting:

a 1 or 4   b 2 or 3   c 2 or 4   d 1 or 2 or 3

4 Below is a number of lettered tiles.

One of these tiles is selected at random.
Work out the probability of getting:

a an A   b an L   c an O

d an A or an L   e an A or an O   f an L or an O.

5 Below are some cards with coloured letters.

One of these cards is picked at random. Work out the probability that the letter on the card will be:

a red   b an X   c a red X   d red or an X.

C

6 A card is taken at random from an ordinary pack of cards.
Work out the probability of getting:

a a 3 of hearts or a 5 of spades

b a heart or a spade

c a king of clubs or a queen of any suit

d a diamond or the ace of hearts

e a picture card (jack, queen or king) or a red 10.

B

7 A and B are two mutually exclusive events. P(A) = 0.45 and P(A or B) = 0.8.
Work out the value of P(B).

A

* **8** Paul rolls an ordinary dice. He says that the probability of getting a 6 on the dice is $\frac{1}{6}$, and the probability of getting another 6 when the dice is rolled again is $\frac{1}{6}$, so the probability of getting two sixes is $\frac{1}{6} + \frac{1}{6} = \frac{2}{6} = \frac{1}{3}$. Is he correct? Explain why.

Example 4

A bag contains 10 balls. Three of the balls are green. A ball is taken at random from the bag. Work out the probability that the ball will be:

  a green

  b not green.

> Use $P(A) = \dfrac{\text{number of successful outcomes}}{\text{total number of possible outcomes}}$

> There are 10 balls altogether, so the total number of possible outcomes = 10.

a $P(\text{green}) = \dfrac{3}{10}$ ← 3 of these outcomes result in successfully taking a green ball, so $P(\text{green}) = \dfrac{3}{10}$

b Using $P(\text{not } A) = 1 - P(A)$,

$$P(\text{not green}) = 1 - P(\text{green})$$
$$= 1 - \dfrac{3}{10}$$
$$= \dfrac{7}{10}$$

> Subtract $\dfrac{3}{10}$ from 1.
> $1 - \dfrac{3}{10} = \dfrac{10}{10} - \dfrac{3}{10}$
> $\qquad = \dfrac{10 - 3}{10} = \dfrac{7}{10}$

Exercise 28D

1 The pie chart shows the proportions of people voting Labour (L), Conservative (C) and Liberal Democrat (LD) in a town.

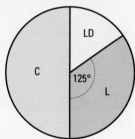

One of these voters is chosen at random for an opinion poll.
Work out the probability that the person voted Liberal Democrat.

2 The probability of rolling a 5 on a biased dice is $\frac{1}{5}$.
Work out the probability of not rolling a 5.

3 The probability that it will rain tomorrow is 0.65.
Work out the probability that it will not rain tomorrow.

4 Harry has a 75% chance of hitting a treble 20, on a dartboard, with a dart.
Work out the probability that Harry will not hit the treble 20 with a dart.

5 A card is taken at random from an ordinary pack of cards.
Work out the probability that the card will be:

  a an ace

  b not an ace.

D

C

C

6   The sectors of a 3-sided spinner are coloured brown, blue and black.
The table gives information about the probability of getting brown, and blue, on the spinner.

Colour	Brown	Blue	Black
Probability	0.35	0.15	

The spinner is spun.
  a Work out the probability of getting a colour that is:
      i not brown   ii not blue.
  b Jamie says the probability of getting black with this spinner is 0.5. He is right. Explain why.

B

7   A bag contains red balls, blue balls and green balls in the ratio $2:3:4$.
A ball is taken at random from the bag. Work out the probability that the ball will be:
  a red   b not red   c not green.

A
A03

8   For two mutually exclusive events A and B, P(B) = 0.3 and P(A or B) = 0.7.
Work out P(not A).

28.3 Estimating probability from relative frequency

◎ Objective

● You can find an estimate for a probability from the results of an experiment.

⟨?⟩ Why do this?

If you tossed a coin a large number of times you could calculate the probability of getting a head. If this isn't 0.5 then it might show that the coin is biased.

⟨↔⟩ Get Ready

1. Write these fractions in their simplest form.
  a $\frac{10}{20}$ b $\frac{12}{16}$ c $\frac{72}{81}$ d $\frac{39}{169}$

⟨◉⟩ Key Points

● You can use **relative frequency** to find an estimate of a probability.

Estimated probability $= \dfrac{\text{number of successful trials}}{\text{total number of trials}}$

● The estimated probability may be different to the theoretical probability.
● The greater the number of **trials**, the more accurate the estimated probability.

🔍 Example 5

Suki rolls a dice 120 times. Here are the results of her experiment.

Number	1	2	3	4	5	6
Frequency	19	18	23	22	21	17

Work out an estimate for the probability of rolling a 6 on Suki's dice.

Estimated probability $= \dfrac{17}{120}$ ⟵ From the table, the number of trials which successfully result in rolling a 6 = 17, and the total number of trials = 120.

Exercise 28E

1   Tania spins a coin 100 times and gets 45 heads.
    Work out an estimate for the probability of getting a head on Tania's coin.

2   A gardener plants 60 seeds. 52 of these seeds germinate.
    Work out an estimate for the probability that this type of seed will germinate.

3   The sectors of a 3-sided spinner are each coloured red or orange or green.
    The table gives the results when the spinner is spun 300 times.

Colour	Red	Orange	Green
Frequency	154	56	90

    a   Use the information in the table to find an estimate for getting red.

    b   Is this a fair spinner? Give a reason for your answer.

4   Drop a drawing pin 50 times and record whether it lands on its head or on its tail.

    a   Use your results to find an estimate for the probability of the drawing pin landing on its head.

    b   How could you improve on your answer to part a?

    Head   Tail

5   Domenique records the numbers of 6s she gets when she rolls a dice 10, 100 and 1000 times.
    The table below shows her results.

Number of rolls	10	100	1000
Number of 6s	1	15	165

    Use this information to work out the best estimate for getting a 6 on Domenique's dice.
    Give a reason for your answer.

*6   Malik says that when he drops a piece of toast it always lands butter-side down.
    Carry out an experiment to find an estimate for the probability that a piece of toast will land
    butter-side down.
    Explain all stages of your work.

28.4 Finding the expected number of outcomes

◎ Objective

○ You can find the expected number of outcomes in an experiment.

◈ Why do this?

Knowing the probability that a small sample of people, such as your class, have a pet means that you can estimate the number of people in your school who have a pet.

⬥ Get Ready

1. Work out the following.

a $\frac{1}{2} \times 100$ b $\frac{1}{3} \times 54$ c $\frac{2}{5} \times 75$ d $\frac{3}{8} \times 108$

Example 6 The probability of winning a prize in a raffle is $\frac{1}{25}$. Jaqui buys 100 tickets in the raffle. How many prizes can she expect to win?

The probability that Jaqui will win a prize is $\frac{1}{25}$.
So she can expect to win 1 prize in every 25 tickets she buys.

> Expected number of outcomes
> = Number of trials × Probability.

Jaqui buys 100 tickets so she can expect
to win $\frac{1}{25} \times 100 = 4$ prizes.

⚙ Exercise 28F

1 Lauren spins an ordinary coin 100 times. How many heads can she expect to get?

2 Yousif rolls an ordinary dice 60 times. How many 6s can he expect to get?

D

3 The table gives the probability of spinning the numbers 1, 2, 3 and 4 on a 4-sided spinner.

Number	1	2	3	4
Probability	0.2	0.35	0.15	0.3

Vicky spins the spinner 200 times. Work out an estimate for the number of 3s that Vicky will get.

B

4 A bag contains 1 red peg, 5 white pegs and 4 yellow pegs.
A peg is taken at random from the bag and then replaced. This is done 250 times.
Copy and complete the table to show the expected numbers of red, white and yellow pegs that will be taken from the bag.

Colour of peg	Red	White	Yellow
Expected number			

A02

5 Fatima spins two coins and records the result. She does this 120 times. One possible outcome is (head, head). Find an estimate for the number of times she will get two heads.

*6 The probability of winning a prize in a lottery is $\frac{1}{50}$. Austin says that if he buys 50 tickets in the lottery he will win a prize. Is he right? Give a reason for your answer.

A

7 A card is taken from an ordinary pack of cards. It is then replaced. If this is done 260 times, work out the expected number of the following that will be taken from the pack.
  a jacks
  b spades
  c aces or clubs

* 8 A doctor estimates that the probability a patient will come to see her about a bad back is 0.125.
Of the next 240 patients who come to see her, 20 have a bad back.
How good is the doctor's estimate of this probability? Explain your answer.

28.5 Independent events

Objectives

○ You can find the probability of independent events.
○ You can multiply probabilities.

Why do this?

If you pick one sweet from a bag and replace it because you don't like the flavour, then you pick again, you have exactly the same probability of choosing the same flavour again.

Get Ready

1. Work out the following. Give your answers in their simplest form.
  a $\frac{1}{2} \times \frac{1}{3}$ **b** $\frac{1}{3} \times \frac{6}{7}$ **c** $\frac{2}{3} \times \frac{3}{5}$ **d** $\frac{5}{12} \times \frac{9}{20}$

Key Points

◉ Two events are independent if one event does not affect the other event.
◉ For two **independent events** A and B,
P(A and B) = P(A) × P(B)
◉ For 3 or more events,
P(A and B and C and …) = P(A) × P(B) × P(C) × ….

Example 7

An ordinary coin is spun and an ordinary dice is rolled.
Work out the probability of getting a head on the coin and a 6 on the dice.

P(head and 6) = P(head) × P(6)

The event of getting a head on the coin cannot affect the event of getting a 6 on the dice. They are independent events. Use P(A and B) = P(A) × P(B). Here A = head and B = 6.

$= \frac{1}{2} \times \frac{1}{6}$ ← *Multiply the fractions.*

$= \frac{1}{12}$ ← *So $\frac{1}{2} \times \frac{1}{6} = \frac{1 \times 1}{2 \times 6} = \frac{1}{12}$*

Exercise 28G

C

1   Dan and Nicole play a game of chess and a game of draughts.
    The probability that Dan will win the game of chess is 0.4.
    The probability that Dan will win the game of draughts is 0.8.
    Work out the probability that Dan will win both games.

2   The probability that it will rain tomorrow is $\frac{2}{3}$.
    The probability that Jaleel will forget his umbrella tomorrow is $\frac{3}{4}$.
    Work out the probability that it will rain tomorrow and Jaleel will forget his umbrella.

3   The probability that a postman will deliver mail to Yasmin's house tomorrow is 0.8.
    The probability that Yasmin's dog will bark when the postman delivers the mail is 0.75.
    Work out the probability that a postman will deliver mail to Yasmin's house tomorrow and
    her dog will bark.

4   The probability that Joshua will forget his protractor for an examination is 0.35.
    The probability that he will forget his calculator for the examination is 0.15.
    Work out the probability that he will:
    a   not forget his protractor
    b   not forget his calculator
    c   not forget his protractor and not forget his calculator.

B

5   A card is taken at random from each of two ordinary packs of cards, pack A and pack B.
    Work out the probability of getting:
    a   a red card from pack A and a red card from pack B
    b   a diamond from pack A and a club from pack B
    c   a king from pack A and a picture card (king, queen, jack) from pack B
    d   a 10 from pack A and a 10 of clubs from pack B
    e   an ace of hearts from each pack.

A

6   Tony and Hannah each have some coins. The table below gives information about these coins.

	Number of coins			
	1p	2p	5p	20p
Tony	5	2	3	1
Hannah	3	1	2	2

Tony and Hannah each pick one of their own coins at random.
Work out the probability that:
a   Tony picks a 5p coin and Hannah picks a 5p coin
b   they both pick a silver coin
c   Tony does not pick a 20p coin and Hannah does not pick a 20p coin.

28.6 Probability tree diagrams

◉ Objective

○ You can use a tree diagram to work out probabilities.

⟡ Why do this?

Probability trees are used differently to family trees but they both show all the outcomes from a single starting point.

⟡ Get Ready

1. Work out

a $\frac{1}{2} + \frac{1}{4}$ b $\frac{1}{3} + \frac{1}{5}$ c $\frac{2}{5} + \frac{1}{4}$ d $\frac{3}{7} + \frac{4}{9}$

🔍 Key Point

◉ A **probability tree diagram** shows all possible outcomes of an experiment.

Example 8	Box A contains 3 red balls and 4 blue balls.

Box A contains 3 red balls and 4 blue balls.
Box B contains 2 red balls and 3 blue balls.
One ball is taken at random from each box.

a Draw a tree diagram to show all the outcomes.

b Work out the probability that the balls will have the same colour.

A   B

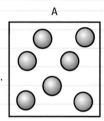 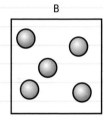

a

Work out the probability of getting each event and show these on the tree diagram. For example, the probability of getting a ball from box A = $\dfrac{\text{number of successful outcomes}}{\text{total number of possible outcomes}} = \frac{3}{7}$.

Box A	Box B	Outcome	Probability
	$\frac{2}{5}$ red	red, red	$\frac{3}{7} \times \frac{2}{5} = \frac{6}{35}$
$\frac{3}{7}$ red	$\frac{3}{5}$ blue	red, blue	$\frac{3}{7} \times \frac{3}{5} = \frac{9}{35}$
$\frac{4}{7}$ blue	$\frac{2}{5}$ red	blue, red	$\frac{4}{7} \times \frac{2}{5} = \frac{8}{35}$
	$\frac{3}{5}$ blue	blue, blue	$\frac{4}{7} \times \frac{3}{5} = \frac{12}{35}$

A tree diagram shows all the possible outcomes of an experiment. For example, this branch represents taking a red ball from box A followed by a red ball from box B.

Taking a red ball from box A and taking a red ball from box B are independent events.
Use P(A and B) = P(A) × P(B). Here A = a red ball from box A, and B = a red ball from box B.
So P(red and red) = P(red) × P(red) = $\frac{3}{7} \times \frac{2}{5}$.
Multiply the fractions:
$\frac{3}{7} \times \frac{2}{5} = \frac{3 \times 2}{7 \times 5} = \frac{6}{35}$

b P(same colour) = P(both red or both blue)

The events are mutually exclusive; you cannot take two red balls and two blue balls from the box at the same time.

    = P(red, red) + P(blue, blue)

Use P(A or B) = P(A) + P(B). Here A = both red (i.e. red, red) and B = both blue (i.e. blue, blue).

    = $\dfrac{6}{35} + \dfrac{12}{35}$

From the tree diagram, P(red, red) = $\dfrac{6}{35}$ and P(blue, blue) = $\dfrac{12}{35}$

    = $\dfrac{18}{35}$

Add the fractions.

✱ **Exercise 28H**

A

1   Bag A contains 2 blue counters and 3 white counters.
    Bag B contains 3 blue counters and 4 white counters.
    A counter is taken at random from each bag.

    a   Copy and complete the tree diagram to show all the
        possible outcomes.

    b   Work out the probability that the counters will both be:

        i   white

        ii blue

        iii the same colour.

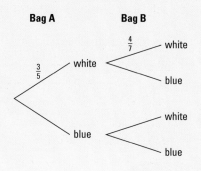

2   There are 10 pencils in a pencil case.
    3 of the pencils are HB pencils.
    A pencil is taken at random from the pencil case and
    then returned.
    A second pencil is now taken at random from the pencil case
    and then returned.

    a   Copy and complete the tree diagram to show all the possible outcomes.

    b   Work out the probability that only one of the pencils will be an HB pencil.

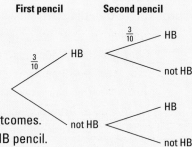

3   Ryan and Ibrahim each have a bag of sweets. In Ryan's bag there are 3 orange sweets and 5 red
    sweets.
    In Ibrahim's bag there are 2 orange sweets and 3 red sweets.
    The boys each take a sweet at random from their own bag.

    a   Draw a tree diagram to show all the possible outcomes.

    b   Use your tree diagram to work out the probability that the sweets will:

        i   both be orange

        ii each have a different colour.

4   On her way home from work Taylor must drive through two sets of traffic lights.
    The probability that she will be stopped at the first set of traffic lights is 0.4.
    The probability that she will be stopped at the second set of traffic lights is 0.7.

    a   Work out the probability that she will not be stopped at:

        i   the first set of traffic lights

        ii the second set of traffic lights.

    b   Draw a tree diagram to show all the possible outcomes.

    c   Work out the probability that she will be stopped by:

        i   both sets of traffic lights

        ii only one set of traffic lights

        iii at least one set of traffic lights.

5   Will spins two spinners, A and B. The probability of getting a 6 on spinner A is 0.3.
    The probability of getting a 6 on spinner B is 0.45.

    a   Draw a tree diagram to show all the possible outcomes.

    b   Work out the probability of getting a 6 on:

        i   neither spinner

        ii only one spinner

        iii spinner B only.

6   The probability that Steph will be late for school on Monday is $\frac{1}{4}$. The probability that Steph will be late for school on Tuesday is $\frac{2}{9}$. Work out the probability that she will be late on at least one of these days.

A02
A03
A

7   A card is taken at random from an ordinary pack of cards. It is then replaced.
Another card is now taken at random from the pack of cards.
Work out the probability of the following.
   a   Both cards are kings.
   b   Neither of the cards is a heart.
   c   One of the cards is a spade.
   d   One of the cards is the ace of clubs.
   e   At least one of the cards is a diamond.

8   Three ordinary coins are spun.
   a   Show all the possible outcomes.
   b   Work out the probability of getting:
      i   3 heads
      ii 2 heads and 1 tail (in any order).

A02

9   The probability that an egg has a double yolk is 0.1.
Nick has three eggs. Work out the probability that exactly one of his eggs has a double yolk.

A02

10   There is a 95% chance that a Gleemo light bulb is faulty. A shop sells Gleemo light bulbs in packets of three. The shop has 400 packets of Gleemo light bulbs in stock.
Find an estimate for the number of packets that will have exactly 2 faulty light bulbs.

A03
A*

28.7 Conditional probability

◎ Objective

● You can find the probability of events that are not independent.

❓ Why do this?

If you pick two socks from a drawer, you can work out the probability that they would match using conditional probability.

◈ Get Ready

1. There are 10 buttons in a box. Four of these buttons are red. Tony takes a red button from the box and sews it on his shirt. He now takes at random another button from the box. What is the probability that this button will be red?

🌐 Key Point

◉ A **conditional probability** is when one outcome affects another outcome, so that the probability of the second outcome depends on what has already happened in the first outcome.

Example 9

A bag contains 5 counters. 3 counters are white and 2 counters are black.

Two counters are taken at random from the bag.

a Draw a tree diagram to show all the possible outcomes.

b Find the probability that:

   i both counters will be black

   ii only one counter will be white.

a

First counter	Second counter	Outcome	Probability
	$\frac{2}{4}$ W	W, W	$\frac{3}{5} \times \frac{2}{4} = \frac{6}{20}$
$\frac{3}{5}$ W	$\frac{2}{4}$ B	W, B	$\frac{3}{5} \times \frac{2}{4} = \frac{6}{20}$
$\frac{2}{5}$ B	$\frac{3}{4}$ W	B, W	$\frac{2}{5} \times \frac{3}{4} = \frac{6}{20}$
	$\frac{1}{4}$ B	B, B	$\frac{2}{5} \times \frac{1}{4} = \frac{2}{20}$

Taking two counters from the bag is the same as taking one counter from the bag followed by taking another counter from the bag (the first counter is not put back before the second counter is taken).

If the first counter is white, then the second counter will be taken from a bag containing 2 white counters and 2 black counters.
If the first counter is black, then the second counter will be taken from a bag containing 3 white counters and 1 black counter.

ResultsPlus
Exam Tip

It is not necessary to simplify a probability fraction in the examination.

From the tree diagram:

b i $P(B, B) = \frac{2}{5} \times \frac{1}{4} = \frac{2}{20}$

  ii $P(W, B \text{ or } B, W) = P(W, B) + P(B, W)$

Only one counter is white, i.e. either the first counter is white and the second counter is black or the first counter is black and the second counter is white.

$= \frac{3}{5} \times \frac{2}{4} + \frac{2}{5} \times \frac{3}{4}$

$= \frac{6}{20} + \frac{6}{20}$

$= \frac{12}{20} = \frac{3}{5}$

Exercise 28I

1 A box contains 7 black balls and 3 white balls.
A ball is taken at random from the box and it is not replaced.
A second ball is now taken at random from the box.

a Copy and complete the tree diagram to show all the possible outcomes.

b Work out the probability that the balls will be:

   i black

   ii the same colour

   iii one of each colour.

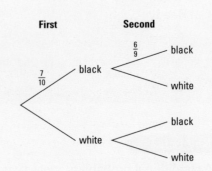

2 5 boys and 7 girls want to be chosen for a school council.
  Two of these students are picked at random.

  a Copy and complete the tree diagram to show all the possible outcomes.

  b Work out the probability of getting:
    i 2 boys
    ii 1 boy
    iii 1 or more boys.

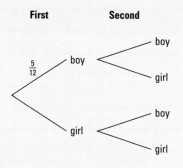

3 Eight cards numbered 1 to 8 are shuffled thoroughly. The top two cards are turned face up on a table. Draw a probability tree diagram and use it to work out the probability that the numbers will:

  a both be even

  b add up to an odd number.

4 On any school day, the probability that Josh oversleeps is $\frac{1}{5}$. If he oversleeps, the probability that he will remember all his books is $\frac{2}{9}$. If he does not oversleep, the probability that he will remember all his books is $\frac{5}{7}$. Use a tree diagram to work out the probability that Josh will not remember all of his books tomorrow.

5 Alexis travels to school by bus or by train. The probability that she travels by bus is 0.45.
  If she travels to school by bus, the probability that she will be late is 0.15. If she travels to school by train, the probability that she will be late is 0.35. Work out the probability that she will not be late.

6 Widgets Electronic Company makes two types of microprocessors, X and Y, in equal numbers. The probability that type X will be faulty is 0.25. The probability that type Y will be faulty is 0.15. If a type X microprocessor is faulty, the probability that it will be recycled is 0.85. If a type Y microprocessor is faulty, the probability that it will be recycled is 0.35. The company only recycles faulty microprocessors. A microprocessor is picked at random. Work out the probability that it will be recycled.

7 Lizzie has seven coins in her purse. Three of the coins are 1 euro coins and four of the coins are £1 coins.
  Lizzie drops her purse and two coins fall out. Work out the probability that the coins will be:

  a both £1 coins

  b not 1 euro coins.

8 Two cards are taken at random from an ordinary pack of cards.
  Work out the probability that the cards will be:

  a both aces

  b both hearts

  c from different suits.

9 The probability that it will rain today is $\frac{3}{7}$. If it does not rain today the probability that it will rain tomorrow is $\frac{4}{7}$. Work out the probability that it will rain today or tomorrow.

10 A doctor diagnoses that a patient has a virus. She does not know which type of virus, X, Y or Z, the patient has. The probability that the patient will have a virus of type X, or Y, or Z is 0.56, or 0.28, or 0.16, respectively. The probability that the patient will not recover from each type of virus is 0.18, 0.22 or 0.35, respectively. Work out the probability that the patient will recover from the virus.

Chapter review

- The **probability** P that an **event** will happen is a number in the range $0 \leqslant P \leqslant 1$.
- For an event which is **certain** P = 1.
- For an event which is **impossible** P = 0.
- A probability can be written as a fraction, a decimal or a percentage.
- For **equally likely** outcomes, the probability that an event will happen is

  $$\text{Probability} = \frac{\text{number of successful outcomes}}{\text{total number of possible outcomes}}$$
- A sample space is all the possible **outcomes** of one or more events.
- These outcomes can be presented in a **sample space diagram**.
- Two events are **mutually exclusive** when they cannot occur at the same time.
- For mutually exclusive events A and B,

  P(A or B) = P(A) + P(B)
- For mutually exclusive events A and not A,

  P(not A) = 1 − P(A)
- If a set of mutually exclusive events contains all possible outcomes, then the sum of their probabilities must come to 1. $\Sigma p = 1$.
- You can use **relative frequency** to find an estimate of a probability.

  $$\text{Estimated probability} = \frac{\text{number of successful trials}}{\text{total number of trials}}$$
- The estimated probability may be different to the theoretical probability.
- Two events are independent if one event does not affect the other event.
- For two **independent events** A and B,

  P(A and B) = P(A) × P(B)
- A **probability tree diagram** shows all possible outcomes of an experiment.
- A **conditional probability** is when one outcome affects another outcome, so that the probability of the second outcome depends on what has already happened in the first outcome.

 Review exercise

1   A letter is picked at random from the word MISSISSIPPI.
    Work out the probability that the letter will be:

    a an S
    b an I
    c not an S
    d not an I
    e an S or an I
    f neither an S nor an I.

2   Some students are asked which topic from algebra, geometry and statistics they like best.
    The results are given in the table below.

	Algebra	Geometry	Statistics
Boys	14	14	16
Girls	9	13	24

    One of these students is picked at random. Work out the probability that the student:

    a is a girl
    b likes geometry best
    c is a boy who likes statistics best
    d is a girl who does not like algebra best
    e is a boy who does not like algebra or geometry best.

3 A jar of sweets contains toffees, truffles and creams in the ration 3 : 7 : 8.
A sweet is chosen at random. Write down the probability that it is a truffle.

4 Mike rolls an ordinary dice and spins a fair 4-sided spinner.
By drawing a sample space diagram or otherwise, work out the
probability that the total score will be:

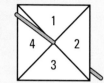

a 7
b less than 5
c a prime number.

5 A card is taken at random from an ordinary pack of cards.
It is then replaced. This is done 390 times.
How many times would you expect to see:
a a heart
b the ace of spades
c a jack?

ResultsPlus
Exam Question Report

71% of students answered this sort of question
poorly because they chose the wrong approach
to calculating the probability.

* 6 Amy has a dice. She thinks it is biased. Describe what Amy could do to see if her dice is biased.
Give as much detail as you can.

7 Chris spins three coins and records the results. He does this 240 times. One possible outcome is (head,
head, head). Find an estimate for the number of times he will get 2 heads and 1 tail (in any order).

8 Megan buys 10 tickets in a raffle. Three of these tickets win a prize.
She says that the probability of winning the raffle is 0.3.
a Give a reason why Megan may be right.
b Give a reason why Megan may be wrong.

9 A ball is taken at random from a bag containing 12 balls, of which b are black.
a Write down, in terms of b, the probability that the ball will be:
i black **ii** not black.
b When a further 6 black balls are added to the bag, the probability of getting a black ball is doubled.
Work out the value of b.

10 A fair tetrahedral dice (4-sided, numbered 1 to 4) and an ordinary dice are each rolled. A win occurs
when the number on the ordinary dice is greater than or equal to the number on the tetrahedral dice.
Find the probability of a win.

11 There are two parts to a driving test: the theory test and the practical test.
You must pass the theory test before you pass the practical test.
Salma plans to take her driving test.
The probability that Salma will pass her theory
test is 0.85.
The probability that she will pass her practical
test is 0.65.
Work out the probability that Salma will pass
her driving test.

ResultsPlus
Exam Question Report

90% of students answered this sort of question
poorly.

B **A02** **A03**

12 A naturalist wants to find out an estimate of the number of adult rabbits there are in a rabbit warren.

He catches 24 of the adult rabbits and marks them with some dye.

He returns the rabbits to the warren.

A few days later he catches 24 adult rabbits from the warren and finds that 6 are marked with the dye.

a Work out an estimate of the number of rabbits in the warren.

b Suggest the sort of assumptions you needed to make when you worked out this estimate.

A **A03**

13 A fruit machine has three independent reels and pays out a jackpot of £1000 when three raspberries are obtained. Each reel has 12 pictures of fruit. The first reel has four pictures of raspberries; the second reel has three pictures of raspberries and the third reel has five pictures of raspberries.

Find the probability of winning the jackpot.

14 Sarah and Jim each have a number of flower bulbs. Sarah has 4 daffodil bulbs and 5 hyacinth bulbs, and Jim has 3 daffodil bulbs and 7 hyacinth bulbs. They each pick one of their bulbs at random.

a Draw a tree diagram to show all the possible outcomes.

b Use your tree diagram to work out the probability that neither of the bulbs is a hyacinth.

A03

15 A fair game is one in which everyone has the same probability of winning.

For example, if you toss a fair coin, you have an equal chance of getting heads or tails.

Consider the following games. Are they fair? You should perhaps play them to understand how the game works. You must explain your answer.

a Three horses run a race over ten lengths. You toss two coins to see which horse moves.

- Horse A moves 1 length if you toss 2 heads.
- Horse B moves 1 length if you toss 2 tails.
- Horse C moves 1 length if you toss a head and a tail.

The horse that completes ten lengths first is the winner.

b Twelve horses run a race over ten lengths. You roll two dice to see which horse moves. For example:

- if you throw a 4 and a 3, Horse 7 moves 1 length
- if you throw a 6 and a 4, Horse 10 moves 1 length.

The horse that reaches ten lengths first is the winner.

A03

16 To play a lottery game, you choose six different numbers between 1 and 40.

Show that the probability of choosing all six numbers correctly is about 1 in 4 million.

A02 **A03**

17 A local garage donated a prize to Probability Junior School for its summer fair, thinking that it would be impossible to win.

To win the prize, someone had to roll six 6s from six dice. The entry fee was 10p, with money going to the school fund. If no one won then the garage kept the car.

Calculate the chances of the car being won.

A03

18 The diagram shows a circle drawn inside a square.

A point is chosen at random from inside the square.

Work out the probability that the point lies inside the circle.

19 The names Justin, Kayla, Hasan, Jessica, Amanda and Dave are each written on a piece of paper and placed in a hat. Two names are taken at random from the hat.
Work out the probability that the names are both boys' names.

20 A bag contains 3 red sweets, 2 green sweets and 4 yellow sweets. Two sweets are taken at random from the bag. Work out the probability that:
a both sweets will be red
b both sweets will be the same colour
c the sweets will be different colours.

ResultsPlus
Exam Question Report

72% of students answered this sort of question poorly because they assumed the sweets were replaced, or they did not consider every scenario.

21 A fruit machine has three reels. Each reel has 10 pictures of fruit.
The table below gives information about the numbers of pictures of apples, pears, cherries and lemons on each of the reels.

	Apple	Pear	Cherry	Lemon
Reel 1	2	3	4	1
Reel 2	2	2	3	3
Reel 3	1	3	3	3

The fruit machine can be programmed to give a prize when particular fruit show on the reels, e.g. when three cherries show on the reels the fruit machine pays 10 tokens.

Programme the fruit machine to give prizes. Decide on the different combinations of fruit that will get a prize, the amount of the prize, and how many tokens a player will need to pay to play the fruit machine.

29 PYTHAGORAS' THEOREM AND TRIGONOMETRY 2

GPS works by sending and receiving signals to up to four satellites in orbit around the Earth. The paths of the signals create triangles, and the GPS uses trigonometry to work out your distance from each point. The satellites are 150 miles away from Earth but can still work out your position to within 15 metres.

◉ Objectives

In this chapter you will:

- use Pythagoras' Theorem and trigonometry in three dimensions
- find the size of an angle between a line and a plane
- draw, sketch and recognise graphs of the trigonometric functions $y = \sin x$ and $y = \cos x$
- work out the area of a triangle using $\frac{1}{2}ab\sin C$
- use the sine rule and cosine rule to solve problems.

◈ Before you start

- You should know the trigonometric ratios and how to apply Pythagoras' Theorem to right-angled triangles.

29.1 Pythagoras' Theorem and trigonometry in three dimensions

Objective

○ You can use Pythagoras' Theorem and trigonometry to solve problems in three dimensions.

Why do this?

You could calculate the diagonal distance across a room. For example, in adventure training centres, zip lines are often attached between diagonally opposite corners of a room.

Get Ready

1. Here is a cuboid.
   Name as many right-angled triangles in this cuboid as you can.
   You should try to find at least six triangles.

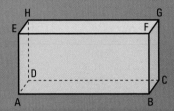

Key Points

○ Problems on cuboids and other three-dimensional shapes involve identifying right-angled triangles and using Pythagoras' Theorem and trigonometry. It is important to draw the relevant triangles separately.

○ The length of the longest diagonal of a cuboid with dimensions a, b, c is
$d = \sqrt{a^2 + b^2 + c^2}$

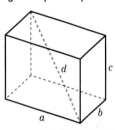

Example 1

ABCDEFGH is a cuboid, with length 8 cm, width 6 cm and height 9 cm.

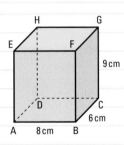

ResultsPlus
Exam Tip

Remember that you need to know Pythagoras' Theorem and the trigonometric results.

a Calculate the length of
  i AC **ii** AG. Give your answers correct to 3 significant figures.

b Calculate the size of angle FAB.
  Give your answer correct to the nearest degree.

a i

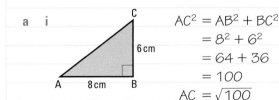

$AC^2 = AB^2 + BC^2$
$= 8^2 + 6^2$
$= 64 + 36$
$= 100$
$AC = \sqrt{100}$
$= 10$ cm

Triangle ABC is a right-angled triangle with AC as one side and the lengths of the other two sides known. Draw a sketch of triangle ABC. Use Pythagoras' Theorem for triangle ABC.

ii

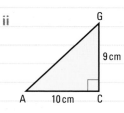

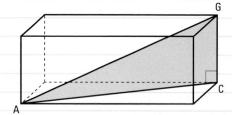

Triangle ACG is a right-angled triangle with AG as one side and the lengths of the other two sides known. Use Pythagoras' Theorem for triangle ACG.

$$AG^2 = AC^2 + CG^2$$
$$= 10^2 + 9^2$$
$$= 100 + 81$$
$$= 181$$
$$AG = \sqrt{181} = 13.4536$$
$$AG = 13.5 \text{ cm (to 3 s.f.)}$$

ResultsPlus
Exam Tip

Write down at least four figures of the calculator display.

b

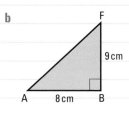

$$\tan \text{ angle FAB} = \frac{9}{8} = 1.125$$

For angle FAB
9 cm is the opposite side
8 cm is the adjacent side
$$\tan = \frac{\text{opp}}{\text{adj}}$$

$$\text{angle FAB} = 48.366\ldots°$$

$$\text{angle FAB} = 48° \text{ (to the nearest degree)}$$

Exercise 29A

Questions in this chapter are targeted at the grades indicated.

Where necessary give lengths correct to 3 significant figures and angles correct to one decimal place.

A

1 ABCDEFGH is a cuboid of length 8 cm, width 4 cm and height 13 cm.
  a Calculate the length of:
    i AC **ii** GB **iii** FA **iv** GA.
  b Calculate the size of:
    i angle FAB **ii** angle GBC.

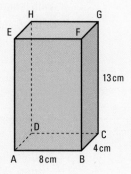

A03

2 For the cuboid ABCDEFG, show that
$$AG^2 = AB^2 + BC^2 + CG^2.$$

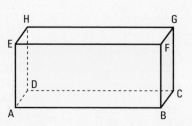

3 A box is in the shape of a cuboid. The length of the box is 12 cm, the width of the box is 6 cm and the height of the box is 4 cm. The length of a needle is 15 cm. The needle cannot be broken. Can the needle fit inside the box?

4 The diagram shows a cylinder and a stick.
The cylinder has a base but no top.
The cylinder is standing on a horizontal table.
The radius of the cylinder is 8 cm.
The height of the cylinder is 12 cm.
The length of the stick is 25 cm.
The stick rests in the cylinder as shown so that as much of the stick is inside the cylinder as possible.
Work out the length of the stick that is not inside the cylinder.

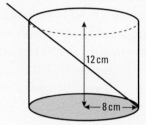

5 The diagram shows a square-based pyramid.
The lengths of sides of the square base, ABCD, are 10 cm and the base is on a horizontal plane.
The centre of the base is the point M and the vertex of the pyramid is O, so that OM is vertical.
The point E is the midpoint of the side AB.
OA = OB = OC = OD = 15 cm

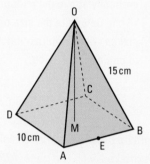

a Calculate the length of **i** AC **ii** AM.
b Calculate the length of OM.
c Calculate the size of angle OAM.　　d Hence find the size of angle AOC.
e Calculate the length of OE.　　f Calculate the size of angle OAB.

29.2 **Angle between a line and a plane**

◎ Objective

- You can find the size of the angle between a line and a plane.

◈ Why do this?

You can use knowledge of angles between lines and planes to set up correctly the guy-lines for a tent.

◈ Get Ready

1. Look at these three diagrams. Each diagram shows a thin pole stuck into the ground.

It looks as if these are three diagrams of three different poles. Explain how these diagrams could be of the same pole.

Key Points

◉ The angle between the pole and the ground seems to depend
on how you look at the pole. So it is necessary to define what
is meant by the angle between a line and a plane.
Imagine a light shining directly above AB onto the **plane**.
AN is the shadow of AB on the plane.
AN is called the **projection** of AB on the plane.
A line drawn from the point B perpendicular to the
plane will meet the line AN and form a right angle with this line.

◉ Angle BAN is the angle between the line AB and the plane.

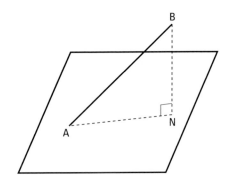

Example 2 The diagram shows a pyramid.
The base, ABCD, is a horizontal rectangle in which
AB = 12 cm and AD = 9 cm. The vertex, O, is
vertically above the midpoint of the base and OB =
18 cm. Calculate the size of the angle that OB
makes with the horizontal plane. Give your answer
to one decimal place.

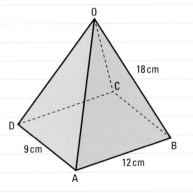

The base ABCD of the pyramid is horizontal so the angle that OB makes with the horizontal plane is
the angle that OB makes with the base, ABCD.
Let M, directly below O, be the midpoint of the base and join O to M
and M to B.
As OM is perpendicular to the base of the pyramid, the angle OBM
is the angle between OB and the base and is the required angle.

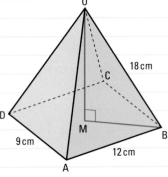

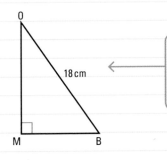

Draw the right-angled triangle OBM.
To find the size of angle OBM, find the
length of either MB or OM.
The length of MB is $\frac{1}{2}$DB.

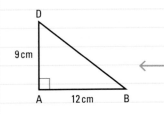

Draw the right-angled triangle ABD and work out the length of DB.

$DB^2 = 9^2 + 12^2 = 81 + 144 = 225$ ← Use Pythagoras' Theorem to calculate the length of DB.
$DB = \sqrt{225} = 15$

O

18 cm

M   7.5 cm   B

$MB = \frac{1}{2}DB = 7.5$ cm

$\cos(\text{angle OBM}) = \frac{7.5}{18}$ ← For angle OBM, 18 cm is the hypotenuse and 7.5 cm is the adjacent side.

angle OBM = 65.37568...° ← $\cos = \dfrac{\text{adj}}{\text{hyp}}$

The angle between OB and the horizontal plane is 65.4° (to 1 d.p.).

Exercise 29B

Where necessary give lengths correct to 3 significant figures and angles correct to one decimal place.

1   The diagram shows a pyramid.
The base, ABCD, is a horizontal rectangle in which AB = 15 cm
and AD = 8 cm. The vertex, O, is vertically above the centre of the
base and OA = 24 cm. Calculate the size of the angle that OA
makes with the horizontal plane.

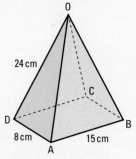

A

2   ABCDEFGH is a cuboid with a rectangular base in which AB = 12 cm and BC = 5 cm.
The height, AE, of the cuboid is 15 cm.
a   Calculate the size of the angle:
  i   between FA and ABCD
  ii   between GA and ABCD
  iii   between BE and ADHE.
b   Write down the size of the angle between HE and ABFE.

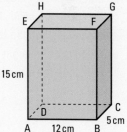

3   The diagram shows a learners' ski slope, ABCD, of length, AB, 500 m.
Triangles BAF and CDE are congruent right-angled
triangles and ABCD, AFED and BCEF are rectangles.
The rectangle BCEF is horizontal and the rectangle AFED
is vertical.
The angle between AB and the horizontal is 20° and the
angle between AC and the horizontal is 10°. Calculate:
a   the length FB
b   the height of A above F
c   the distance AC
d   the width, BC, of the ski slope.

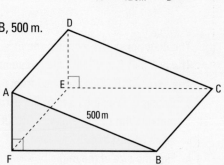

A02
A03

4 ABCD is a horizontal rectangular lawn in a garden and TC is a
vertical pole. Ropes run from the top of the pole, T, to the corners
A, B and D of the lawn.

a Calculate the length of the rope TA.

b Calculate the size of the angle made with the lawn by:
  i the rope TB ii the rope TD iii the rope TA.

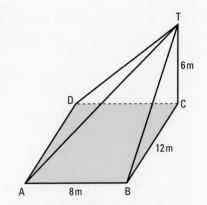

5 Diagram 1 shows a square-based pyramid, OABCD.
Each side of the square is of length 60 cm and
OA = OB = OC = OD = 50 cm.

Diagram 1

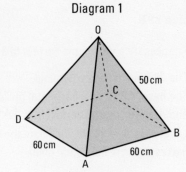

Diagram 2 shows a cube, ABCDEFGH, in which
each edge is of length 60 cm.

A solid is made by placing the pyramid on top of the cube
so that the base, ABCD, of the pyramid is on the top, ABCD,
of the cube. The solid is placed on a horizontal table with
the face EFGH on the table.

a Calculate the height of the vertex O above the table.

b Calculate the size of the angle between OE and the
horizontal.

Diagram 2

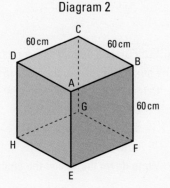

6 The diagram shows a solid cube of side 12 cm.
The points P, Q and R are the midpoints of the edges on which they lie.
The pyramid OPQR is removed from the cube.

a Taking OPQ as the base of the pyramid, draw a sketch of the pyramid,
  marking the size of angles POQ, QOR and ROP and the lengths of
  sides OP, OQ and OR.

b Find the size of the angle between:
  i RP and the plane OPQ
  ii RQ and the plane OPQ.

c Work out the volume of the solid remaining when the pyramid is removed from the cube.

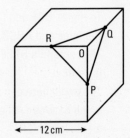

29.3 Trigonometric ratios for any angle

Objective

- You can draw, sketch and recognise graphs of the trigonometric functions $y = \sin x$ and $y = \cos x$.

Why do this?

The sine and cosine wave patterns can be seen in light waves, sound waves and ocean waves.

Get Ready

1. P, Q, R and S are points on a circle.
   The coordinates of P are (u, v).
   What are the coordinates of:
   a Q   b R   c S?

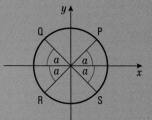

2. Using your calculator, find:
   a sin 30° and sin 150°.
   b cos 50° and cos 130°.
   c What do you notice?

Key Points

- Values of sine and cosine can be found for any angle.
- The diagram shows a circle, centre the origin O and radius 1 unit. A line, OP, of length 1 unit fixed at O, rotates in an anticlockwise direction about O, starting from the x-axis. The diagram shows OP when it has rotated through 40°.

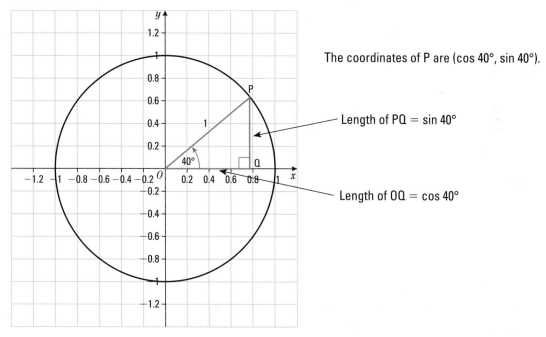

The coordinates of P are (cos 40°, sin 40°).

Length of PQ = sin 40°

Length of OQ = cos 40°

- Trigonometry can be used to determine the lengths of side PQ and side OQ in the right-angled triangle OPQ, and the coordinates of point P.
- In general, when OP rotates through the angle $\theta°$, the position of P on the circle, radius = 1 is given by $x = \cos \theta°$, $y = \sin \theta°$. The coordinates of P are (cos $\theta°$, sin $\theta°$).
- A rotation of 400° is one complete revolution of 360° plus a further rotation of 40°. The position of P is the same in the previous diagram so (cos 400°, sin 400°) is the same point as (cos 40°, sin 40°), therefore cos 400° = cos 40° and sin 400° = sin 40°.
- A rotation through $-40°$ means the line OP rotates through 40° in a clockwise direction.

◉ For $\theta° = 136$, $\theta° = 225$, $\theta° = 304$ and $\theta° = -40$ the position of P is shown on the diagram.

◉ In this **quadrant**:
  ◉ $\sin x$ is positive
  ◉ $\cos x$ is negative

The coordinates of **P** are ($\cos 136°$, $\sin 136°$).

The coordinates of **P** are ($\cos 225°$, $\sin 225°$).

◉ In this quadrant:
  ◉ $\sin x$ is negative
  ◉ $\cos x$ is negative

◉ In this quadrant:
  ◉ $\sin x$ is positive
  ◉ $\cos x$ is positive

The coordinates of **P** are ($\cos -40°$, $\sin -40°$).

The coordinates of **P** are ($\cos 304°$, $\sin 304°$).

◉ In this quadrant:
  ◉ $\sin x$ is negative
  ◉ $\cos x$ is positive

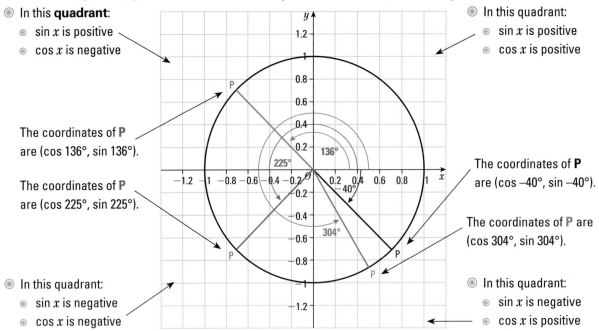

◉ The sine and cosine of any angle can be found using your calculator. Using these values the graphs of $y = \sin \theta°$ and $y = \cos \theta°$ can be drawn.

◉ **Graph of $y = \sin \theta°$**

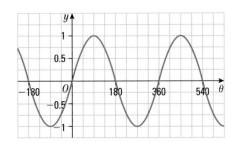

◉ The graph of $y = \sin \theta°$:
  ◉ cuts the θ-axis at ..., $-180, 0, 180, 360, 540, ...$
  ◉ repeats itself every 360°, that is, it has a **period** of 360°
  ◉ has a maximum value of 1 at $\theta° = ..., 90, 450, ...$
  ◉ has a minimum value of -1 at $\theta° = ..., -90, 270, ...$

◉ **Graph of $y = \cos \theta°$**

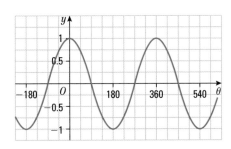

◉ The graph of $y = \cos \theta°$:
  ◉ cuts the θ-axis at ..., $-90, 90, 270, 450, ...$
  ◉ repeats itself every 360°, that is, it has a period of 360°
  ◉ has a maximum value of 1 at $\theta° = ..., 0, 360, ...$
  ◉ has a minimum value of -1 at $\theta° = ...,$ $-180, 180, 540, ...$

◉ The graph of $y = \sin \theta°$ and the graph of $y = \cos \theta°$ are horizontal translations of each other.

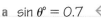 **Example 3** For values of θ in the interval -180 to 360 solve the equation

 a $\sin \theta° = 0.7$

 b $5\cos \theta° = 2$.

Give each answer correct to one decimal place.

 a $\sin \theta° = 0.7$ ←———— [Use a calculator to find one value of θ.]

 $\theta = 44.4$

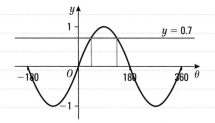

[To find the other solutions draw a sketch of $y = \sin \theta$ for $\theta°$ from -180 to 360.]

[The sketch shows that there are two values of θ in the interval -180 to 360 for which $\sin \theta° = 0.7$.]

 $\theta = 44.4,\ 180 - 44.4$ ←———— [One solution is $\theta = 44.4$ and by symmetry the other solution is $\theta = 180 - 44.4$.]

 $\theta = 44.4,\ 135.6$

 b $5\cos \theta° = 2$ ←———— [Divide each side of the equation by 5.]

 $\cos \theta° = \dfrac{2}{5} = 0.4$

 $\theta = 66.4$ ←———— [Use a calculator to find one value of θ.]

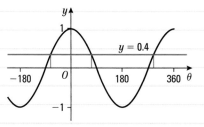

[To find the other solutions draw a sketch of $y = \cos \theta°$ for θ from -180 to 360.]

[The sketch shows that there are three values of θ in the interval -180 to 360 for which $\cos \theta° = 0.4$.]

 $\theta = 66.4,\ -66.4,\ 360 + -66.4$ ←———— [One solution is $\theta = 66.4$ and by symmetry another solution is $\theta = -66.4$. Using the period of the graph the other solution is $\theta = 360 + -66.4$.]

 $\theta = 66.4,\ -66.4,\ 293.6$

Exercise 29C

1. For $-360 \leqslant \theta \leqslant 360$, sketch the graph of
   a $y = \sin \theta°$
   b $y = \cos \theta°$.

2. Find all values of θ in the interval 0 to 360 for which
   a $\sin \theta° = 0.5$
   b $\cos \theta° = 0.1$.

3. a Show that one solution of the equation $3\sin \theta° = 1$ is 19.5, correct to 1 decimal place.
   b Hence solve the equation $3\sin \theta° = 1$ for values of θ in the interval 0 to 720.

4. a Show that one solution of the equation $10\cos \theta° = -3$ is 107.5 correct to 1 decimal place.
   b Hence find all values of θ in the interval -360 to 360 for which $10\cos \theta° = -3$.

A

A☆

29.4 Finding the area of a triangle using $\frac{1}{2}ab\sin C$

⊚ Objectives

- You can work out the area of a triangle using $\frac{1}{2}ab\sin C$.
- You can work out the area of a segment of a circle.

⦾ Why do this?

You could use the formula $\frac{1}{2}ab\sin C$ to work out the area of a triangular lake if you could measure the lengths of the edges but not the distance across.

⬆ Get Ready

1. Work out the area of these triangles.

a

10 cm
6 cm

b
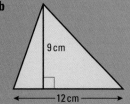
9 cm
12 cm

c
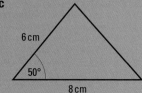
6 cm
50°
8 cm

🌐 Key Points

- The vertices of a triangle are labelled with capital letters.
  The triangle shown is triangle ABC.
  The sides opposite the angles are labelled so that a is the length of the side opposite angle A, b is the length of the side opposite angle B and c is the length of the side opposite angle C.

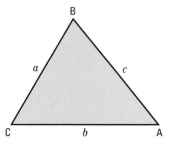

- Area of a triangle $= \frac{1}{2}$ base $\times$ height

  Area of triangle ABC $= \frac{1}{2}bh$

  In the right-angled triangle BCN, $h = a\sin C$

  So area of triangle ABC $= \frac{1}{2}b \times a\sin C$

  that is

  area of triangle ABC $= \frac{1}{2}ab\sin C$

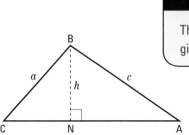

ResultsPlus
Exam Tip

The formula $\frac{1}{2}ab\sin C$ will be given on the formula sheet.

- The angle C is the angle between the sides of length a and b and is called the included angle.
  The formula for the area of a triangle means that

  area of a triangle $= \frac{1}{2}$ product of two sides $\times$ sine of the included angle

  For triangle ABC, there are two more formulae for the area.

  Area of triangle ABC $= \frac{1}{2}ab\sin C = \frac{1}{2}bc\sin A = \frac{1}{2}ac\sin B$

  These formulae give the area of a triangle, whether the included angle is acute or obtuse.

Example 4   Calculate the area of each of the triangles correct to 3 significant figures.

a

B

6.8 cm

66°

C   7.6 cm   A

b

5.1 m

108°

6.2 m

a   Area $= \frac{1}{2} \times 6.8 \times 7.6 \times \sin 66$ ← | Substitute $a = 6.8$, $b = 7.6$, $C = 66°$ into area $= \frac{1}{2}ab\sin C$.
Area $= 23.606...$ Give the area correct to 3 significant figures and give the units.
Area $= 23.6\,\text{cm}^2$ (3 s.f.)

b   Area $= \frac{1}{2} \times 5.1 \times 6.2 \times \sin 108$ ← | Substitute into area of a triangle
Area $= 15.036...$ $= \frac{1}{2}$ product of two sides $\times$ sine of the included angle.
Area $= 15.0\,\text{m}^2$ (3 s.f.)

Example 5   The diagram shows a circle of radius 6 cm
and centre O.
AB is a chord of the circle and angle AOB $= 56°$.
Work out the area of the shaded segment.
Give your answer correct to 3 significant figures.

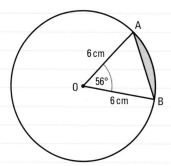

Area of segment $=$ area of sector OAB $-$ area of triangle OAB ← | See section 23.1 for the area of a sector.

Area of sector OAB $= \frac{56}{360} \times \pi \times 6^2 = 17.5929...\,\text{cm}^2$ ← | Write down at least four figures from your calculator display.

Area of triangle OAB $= \frac{1}{2} \times 6 \times 6 \times \sin 56 = 14.9226...\,\text{cm}^2$ ← | Use area of triangle $= \frac{1}{2}ab\sin C$.
Area of segment $= 17.5929... - 14.9226... = 2.6703$

Area of segment $= 2.67\,\text{cm}^2$ (3 s.f.) ← | Give your answer correct to 3 significant figures.

Exercise 29D

Give lengths and areas correct to three significant figures and angles correct to one decimal place.

1 Work out the area of each of these triangles.

a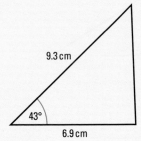

9.3 cm

43°

6.9 cm

b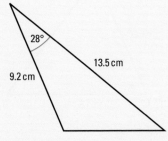

28°

13.5 cm

9.2 cm

c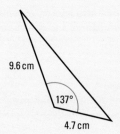

9.6 cm

137°

4.7 cm

d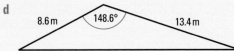

8.6 m   148.6°   13.4 m

2 The area of triangle ABC is 60.7 m².
Work out the length of BC.

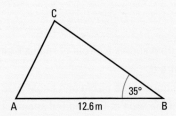

C

35°

A   12.6 m   B

3 The area of triangle ABC is 15 cm².
Angle A is acute.
Work out the size of angle A.

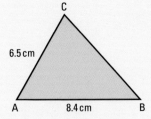

C

6.5 cm

A   8.4 cm   B

4 a Triangle ABC is such that $a = 6$ cm, $b = 9$ cm and angle C = 25°.
Work out the area of triangle ABC.

b Triangle PQR is such that $p = 6$ cm, $q = 9$ cm and angle R = 155°.
Work out the area of triangle PQR.

c What do you notice about your answers? Why do you think this is true?

5 The diagram shows a regular octagon, with centre O.

a Work out the size of angle AOB.
OA = OB = 6 cm.

b Work out the area of triangle AOB.

c Hence work out the area of the octagon.

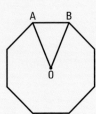

A   B

O

6   The diagram shows a sector, OAB, of a circle, centre O.
The radius of the circle is 8 cm and the size of angle AOB is 50°.
Work out the area of the segment of the circle shown shaded
in the diagram.

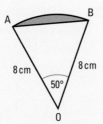

A03   A★

29.5 The sine rule

Objective

You can use the sine rule to work out the length of
a side in a triangle.

Why do this?

You can use the sine rule to find a length that isn't
easily accessible to measure, for example, the
height of a mountain, using the angle of elevation
to the top of the mountain from two points a
certain distance apart.

Get Ready

1. Work out the value of   **a** $\dfrac{7.9 \times \sin 67°}{8.4}$ **b** $\dfrac{14.8 \times \sin 58°}{\sin 67}$

Key Points

The last section showed that

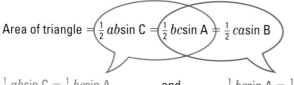

Area of triangle $= \frac{1}{2} ab\sin C = \frac{1}{2} bc\sin A = \frac{1}{2} ca\sin B$

$\frac{1}{2} ab\sin C = \frac{1}{2} bc\sin A$ and $\frac{1}{2} bc\sin A = \frac{1}{2} ca\sin B$

cancelling $\frac{1}{2}$ and b from both sides   cancelling $\frac{1}{2}$ and c from both sides

$a\sin C = c\sin A$ and $b\sin A = a\sin B$

or

$\dfrac{a}{\sin A} = \dfrac{c}{\sin C}$ and $\dfrac{b}{\sin B} = \dfrac{a}{\sin A}$

These results are combined to get the **sine rule**
which can be used in any triangle.

$\dfrac{a}{\sin A} = \dfrac{b}{\sin B} = \dfrac{c}{\sin C}$

ResultsPlus

Exam Tip

The sine rule will be given on
the formula sheet.

To use the sine rule to find a length in a triangle, it is necessary to know any two angles and a side (ASA).

Example 6

Find the length of the side marked a in the triangle. Give your answer correct to three significant figures.

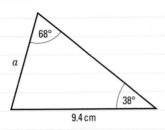

Results Plus
Watch Out!

You can only use Pythagoras' Theorem in a right-angled triangle.

$$\frac{a}{\sin 38°} = \frac{9.4 \text{ cm}}{\sin 68°}$$

Substitute A = 38°, b = 9.4 cm, B = 68° into $\frac{a}{\sin A} = \frac{b}{\sin B}$

Results Plus
Exam Tip

$$a = \frac{9.4 \times \sin 38°}{\sin 68°}$$

Multiply both sides by sin 38°.

Check that your answer is sensible: the greater length is always opposite the greater angle.

$a = 6.2417\ldots$

$a = 6.24 \text{ cm (3 s.f.)}$

Example 7

Find the length of the side marked x in the triangle. Give your answer correct to three significant figures.

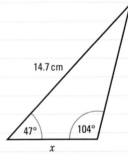

Missing angle = $180° - (47° + 104°) = 29°$

The angle opposite x must be known before the sine rule can be used. Use the angle sum of a triangle.

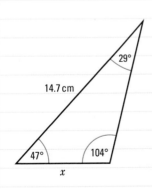

$$\frac{x}{\sin 29°} = \frac{14.7}{\sin 104°}$$

Write down the sine rule with x opposite 29° and 14.7 cm opposite 104°.

$$x = \frac{14.7 \times \sin 29°}{\sin 104°}$$

Multiply both sides by sin 29°.

$x = 7.3448\ldots$

$x = 7.34 \text{ cm (3 s.f.)}$

Exercise 29E

Give lengths correct to three significant figures.

1 Find the lengths of the sides marked with letters in these triangles.

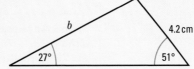

a

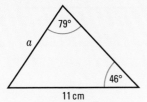

b

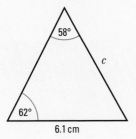

c

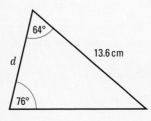

d

e

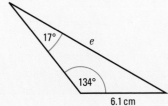

f

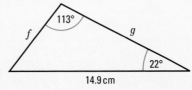

29.6 Using the sine rule to calculate an angle

⊙ Objective

○ You can use the sine rule to work out the size of an angle in a triangle.

⟐ Why do this?

Astronomers sometimes use the sine rule to calculate unknown angles between stars in space.

⟐ Get Ready

1. Find the size of the acute angle x when $\sin x = \dfrac{14 \times 0.7}{17}$.

Key Points

◉ To use the sine rule to find an angle in a triangle, it is necessary to know two sides and the non-included angle (SSA).

◉ When the sine rule is used to calculate an angle, it is a good idea to turn each fraction upside down (the reciprocal).

This gives $\dfrac{\sin A}{a} = \dfrac{\sin B}{b} = \dfrac{\sin C}{c}$

Example 8 Find the size of the acute angle x in the triangle.

Give your answer correct to one decimal place.

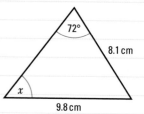

$$\frac{\sin x}{8.1 \text{ cm}} = \frac{\sin 72°}{9.8 \text{ cm}}$$ ← Write down the sine rule with x opposite 8.1 cm and 72° opposite 9.8 cm.

$$\sin x = \frac{8.1 \times \sin 72°}{9.8}$$ ← Multiply both sides by 8.1.

$$\sin x = 0.7860\ldots$$ ← Work out the value of $\sin x$.

$$x = 51.820\ldots°$$

$$x = 51.8° \text{ (1 d.p.)}$$

Exercise 29F

Give lengths and areas correct to three significant figures and angles correct to one decimal place.

1 Calculate the size of each of the acute angles marked with a letter.

a

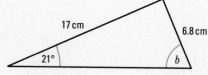

b

c

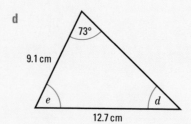

d

2 The diagram shows quadrilateral ABCD and its diagonal AC.
 a In triangle ABC, work out the length of AC.
 b In triangle ACD, work out the size of angle DAC.
 c Work out the size of angle BCD.

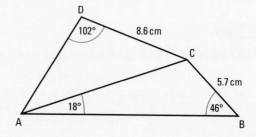

3 In triangle ABC, BC = 8.6 cm, angle BAC = 52° and angle ABC = 63°.

  a Calculate the length of AC.

  b Calculate the length of AB.

  c Calculate the area of triangle ABC.

4 In triangle PQR all the angles are acute. PR = 7.8 cm and PQ = 8.4 cm. Angle PQR = 58°.

  a Work out the size of angle PRQ.

  b Work out the length of QR.

5 The diagram shows the position of a port (P), a lighthouse (L) and a buoy (B). The lighthouse is due east of the buoy.

The lighthouse is on a bearing of 035° from the port and the buoy is on a bearing of 312° from the port.

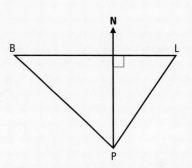

  a Work out the size of: **i** angle PBL **ii** angle PLB.

The lighthouse is 8 km from the port.

  b Work out the distance PB.

  c Work out the distance BL.

  d Work out the shortest distance from the port (P) to the line BL.

29.7 The cosine rule

⊙ Objective

○ You can use the cosine rule to work out the length of a side in a triangle.

? Why do this?

You can use the cosine rule to find a length that isn't easily accessible to measure. For example, mapmakers can calculate the distance between two towns separated by a lake using the distance to each town from a fixed point and the angle between these two lengths.

◈ Get Ready

1. Work out the value of **a** $5 + 3 \times 2$ **b** $9 + 8 - 2 \times 7$ **c** $5^2 + 6^2 - 2 \times 5 \times 6 \times \cos 50°$

2. Work out the positive value of x when $x^2 = 7^2 + 4^2 - 2 \times 7 \times 4 \times 0.9$

🌑 Key Points

◉ The diagram shows triangle ABC.

The line BN is perpendicular to AC and meets the line AC at N so that AN = x and NC = $(b - x)$. The length of BN is h.

In triangle ANB,
Pythagoras' Theorem gives
$c^2 = x^2 + h^2$ (1)

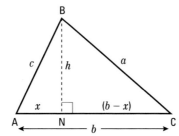

In triangle BNC,
Pythagoras' Theorem gives
$a^2 = (b - x)^2 + h^2$
$a^2 = b^2 - 2bx + x^2 + h^2$
Using (1), substitute c^2 for $x^2 + h^2$
$a^2 = b^2 - 2bx + c^2$
$\quad = b^2 + c^2 - 2bx$ (2)

In the right-angled triangle ANB, $x = c \cos A$

Substituting this into (2)

$a^2 = b^2 + c^2 - 2bc\cos A$

This result is known as the **cosine rule** and can be used in any triangle.

⊙ Similarly $b^2 = a^2 + c^2 - 2ac\cos B$ and $c^2 = a^2 + b^2 - 2ab\cos C$

⊙ To use the cosine rule to find a length in a triangle, it is necessary to know the other two sides and the angle between these sides (SAS).

ResultsPlus
Exam Tip

The cosine rule will be given on the formula sheet.

Example 9   Find the length of the side marked with a letter in each triangle.
Give your answers correct to three significant figures.

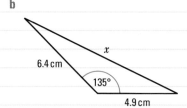

a

B

a 9 cm

36°

C   11 cm   A

b

6.4 cm

x

135°

4.9 cm

a $a^2 = 11^2 + 9^2 - 2 \times 11 \times 9 \times \cos 36°$

$a^2 = 121 + 81 - 160.1853\ldots$

$a^2 = 41.8146$

$a = \sqrt{41.8146}$ ← Take the square root.

$a = 6.4664$

$a = 6.47\ cm$

> Substitute $b = 11$ cm, $c = 9$ cm, A = 36° into $a^2 = b^2 + c^2 - 2bc\cos A$.
> Evaluate each term separately.

b $x^2 = 6.4^2 + 4.9^2 - 2 \times 6.4 \times 4.9 \times \cos 135°$

$x^2 = 40.96 + 24.01 - 62.72 \times (-0.70710\ldots)$

$x^2 = 64.97 + 44.3497\ldots$

$x^2 = 109.3197\ldots$

$x = \sqrt{109.3197\ldots}$ ← Take the square root.

$x = 10.4556\ldots$

$x = 10.5\ cm$

> Substitute the two given lengths and the included angle into the cosine rule.

> The cosine of an obtuse angle is negative so $\cos 135° < 0$.

Exercise 29G

Give lengths correct to three significant figures.

1   Calculate the length of the sides marked with letters in these triangles.

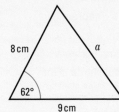

a

8 cm a

62°

9 cm

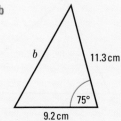

b

b 11.3 cm

75°

9.2 cm

c

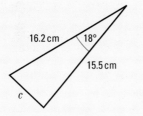

16.2 cm   18°

15.5 cm

c

d

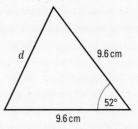

d   9.6 cm

52°

9.6 cm

e

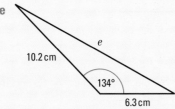

e

10.2 cm

134°

6.3 cm

f

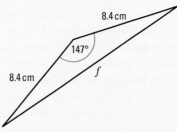

8.4 cm

147°

8.4 cm   *f*

2 In triangle XYZ, XY = 20.3 cm, XZ = 14.5 cm and angle YXZ = 38°.
Calculate the length of YZ.

A02

29.8 Using the cosine rule to calculate an angle

◎ Objective

○ You can use the cosine rule to work out the size of an angle in a triangle.

⍰ Why do this?

The cosine rule can be used in sailing to calculate bearings to a destination, given different wind speeds and the angle at which it strikes the boat.

◈ Get Ready

1. Find the size of angle A when:

   a $\cos A = \dfrac{12}{19}$ **b** $\cos A = \dfrac{7}{23}$ **c** $\cos A = \dfrac{11^2 + 6^2 - 9^2}{2 \times 11 \times 6}$

⬤ Key Points

◉ To use the cosine rule to find the size of an angle in a triangle, it is necessary to know all three sides (SSS).

◉ To find an angle using the cosine rule, rearrange $a^2 = b^2 + c^2 - 2bc\cos A$

$2bc\cos A = b^2 + c^2 - a^2$

$\cos A = \dfrac{b^2 + c^2 - a^2}{2bc}$

Similarly $\cos B = \dfrac{a^2 + c^2 - b^2}{2ac}$ and $\cos C = \dfrac{a^2 + b^2 - c^2}{2ab}$

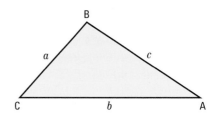

Example 10

Find the size of:　**a** angle BAC　　**b** angle x.

Give your answers correct to one decimal place.

a

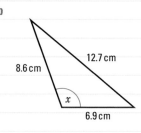

b

a $\cos A = \dfrac{11^2 + 16^2 - 13^2}{2 \times 11 \times 16} = \dfrac{208}{352}$

Substitute $b = 11$ cm, $c = 16$ cm, $a = 13$ cm into $\cos A = \dfrac{b^2 + c^2 - a^2}{2bc}$.

$\cos A = 0.590909\ldots$

$A = 53.77\ldots$

$A = 53.8°$ (1 d.p.)

b $\cos x = \dfrac{8.6^2 + 6.9^2 - 12.7^2}{2 \times 8.6 \times 6.9}$

Substitute the three lengths into the cosine rule, noting that 12.7 cm is opposite the angle to be found.

$\cos x = -\dfrac{39.72}{118.68}$

$\cos x = -0.33468\ldots$

The value of $\cos x$ is negative so x is an obtuse angle.

$x = 109.553\ldots$

$x = 109.6°$ (1 d.p.)

Exercise 29H

Give lengths and areas correct to three significant figures and angles correct to one decimal place.

1 Calculate the size of each of the angles marked with a letter in these triangles.

a

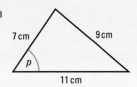

b

c

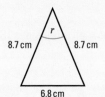

d

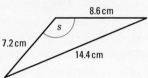

2 AB is a chord of a circle with centre O.
The radius of the circle is 7 cm and the
length of the chord is 11 cm.
Calculate the size of angle AOB.

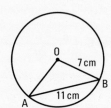

3 The region ABC is marked on a school field.
The point B is 70 m from A on a bearing of 064°.
The point C is 90 m from A on a bearing of 132°.
  a Work out the size of angle BAC.
  b Work out the length of BC.

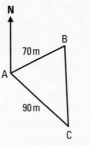

A02
A03

A

4 The diagram shows the quadrilateral ABCD.
  a Work out the length of DB.
  b Work out the size of angle DAB.
  c Work out the area of quadrilateral ABCD.

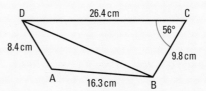

A★

A03

5 Chris ran 4 km on a bearing of 036° from P to Q. He then ran in a straight line from Q to R, where R is
7 km due east of P. Chris then ran in a straight line from R to P.
Calculate the total distance that Chris ran.

A02
A03

6 The diagram shows a parallelogram.
Work out the length of each diagonal of the parallelogram.

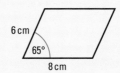

A03

29.9 Using trigonometry to solve problems

◎ Objectives

- You can identify whether to use the sine rule or the cosine rule when solving 2D and 3D problems.
- You can solve problems involving non right-angled triangles.

⑦ Why do this?

Some engineers use the sine and cosine rules to find unknown lengths and angles when they create maps of large features on the Earth's surface, such as a mountain range or an ocean floor.

◈ Get Ready

1. Look at these triangles. To find the length of the lettered side or the size of the lettered angle, which one of the sine rule or the cosine rule should be used?

a

b

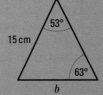

c

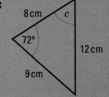

🔍 Key Points

- Use the sine rule when a problem involves two sides and two angles.
- Use the cosine rule when a problem involves three sides and one angle.
- The formulae for the sine rule and the cosine rule are given on the formula sheet.

Example 11

The area of triangle ABC is 12 cm².

AB = 3.8 cm and angle ABC = 70°.

a Find the length of: i BC ii AC.

b Find the size of angle BAC.

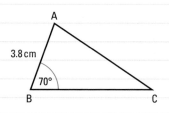

a   i $\frac{1}{2} \times BC \times 3.8 \times \sin 70° = 12$ ← Substitute $c = 3.8$ cm, B = 70° into area $= \frac{1}{2}ac\sin B$.

$BC = \dfrac{2 \times 12}{3.8 \times \sin 70°}$

$BC = 6.721...$
$BC = 6.72$ cm (3 s.f.)

   ii $b^2 = 6.721...^2 + 3.8^2 - 2 \times 6.721... \times 3.8 \times \cos 70°$

   $= 59.613... - 17.470...$

   $= 42.142...$

   $b = 6.491...$

   $b = 6.49$ cm (3 s.f.)

Substitute $a = 6.721...$ cm, $c = 3.8$ cm and B = 70° into $b^2 = a^2 + c^2 - 2ac\cos B$.

Results Plus
Exam Tip

Remember to use uncorrected values of your answers for subsequent calculations.

b $\dfrac{\sin A}{6.721...} = \dfrac{\sin 70°}{6.491...}$ ← Substitute $a = 6.721...$ cm, $b = 6.491$ cm... and B = 70° into $\dfrac{\sin A}{a} = \dfrac{\sin B}{b}$.

   $\sin A = \dfrac{6.721... \times \sin 70°}{6.491...}$

   $\sin A = 0.9728...$

   $A = 76.62...°$

   $A = 76.7°$ (1 d.p.)

Exercise 29I

Where necessary, give lengths and areas correct to three significant figures and angles correct to one decimal place, unless the question states otherwise.

A02
A03

1   A triangle has sides of lengths 9 cm, 10 cm and 11 cm.

   a Calculate the size of each angle of the triangle.

   b Calculate the area of the triangle.

A*
A03

2   In the diagram, ABC is a straight line.

   a Calculate the length of BD.

   b Calculate the size of angle DAB.

   c Calculate the length of AC.

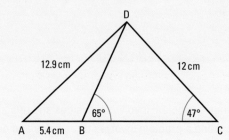

3 The area of triangle ABC is 15 cm².
AB = 4.6 cm and angle BAC = 63°.

a Work out the length of AC.

b Work out the length of BC.

c Work out the size of angle ABC.

4 ABCD is a kite, with diagonal DB.

a Calculate the length of DB.

b Calculate the size of angle BDC.

c Calculate the value of x.

d Calculate the length of AC.

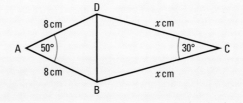

5 James walked 9 km due south from point A to point B.
He then changed direction and walked 5 km to point C.
James was then 6 km from his starting point A.

a Work out the bearing of point C from point B.
Give your answer correct to the nearest degree.

b Work out the bearing of point C from point A.
Give your answer correct to the nearest degree.

6 The diagram shows a pyramid. The base of the pyramid, ABCD,
is a rectangle in which AB = 15 cm and AD = 8 cm.
The vertex of the pyramid is O where
OA = OB = OC = OD = 20 cm.
Work out the size of angle DOB,
correct to the nearest degree.

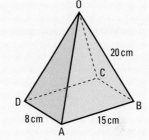

7 The diagram shows a vertical pole, PQ, standing on a hill.
The hill is at an angle of 8° to the horizontal.
The point R is 20 m downhill from Q and the line PR is
at 12° to the hill.

a Calculate the size of angle RPQ.

b Calculate the length, PQ, of the pole.

8 A, B and C are points on horizontal ground so that AB = 30 m,
BC = 24 m and angle CAB = 50°.
AP and BQ are vertical posts, where AP = BP = 10 m.

a Work out the size of angle ACB.

b Work out the length of AC.

c Work out the size of angle PCQ.

d Work out the size of the angle between QC and the ground.

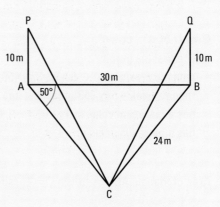
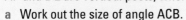

9 The diagram shows a port P and two buoys A and B.
The buoy A is at a distance of 15 km and on a bearing of 020° from P.
The buoy B is at a distance of 20 km and on a bearing of 310° from P.
Calculate the distance between A and B.
Give your answer in km correct to 3 significant figures.

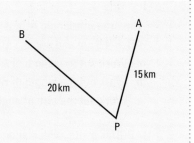

Chapter review

⚬ Problems on cuboids and other 3D shapes involve identifying right-angled triangles and using Pythagoras' Theorem and trigonometry. It is important to draw these triangles separately.

⚬ AN is called the **projection** of AB on the **plane**.

⚬ Angle BAN is the angle between the line AB and the plane.

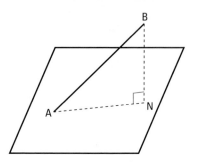

⚬ The diagram shows for each **quadrant** whether the sine and cosine of angles in that quadrant are positive or negative.

	sin+ cos−	sin+ cos+
	2nd	1st
	3rd	4th
	sin− cos−	sin− cos+

Graph of $y = \sin \theta°$

Notice that the graph:

⚬ cuts the θ-axis at …, −180, 0, 180, 360, 540, …
⚬ repeats itself every 360°, that is, it has a **period** of 360°
⚬ has a maximum value of 1 at $\theta° = …, 90, 450, …$
⚬ has a minimum value of −1 at $\theta° = …, -90, 270, ….$

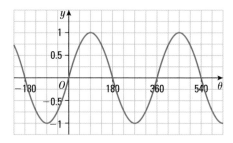

Graph of $y = \cos \theta°$

Notice that the graph:

⚬ cuts the θ-axis at …, −90, 90, 270, 450, …
⚬ repeats itself every 360°, that is, it has a period of 360°
⚬ has a maximum value of 1 at $\theta° = …, 0, 360, …$
⚬ has a minimum value of −1 at $\theta° = …, -180, 180, 540, ….$

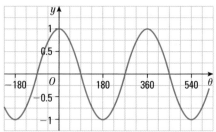

⚬ Area of triangle ABC $= \frac{1}{2}ab\sin C = \frac{1}{2}bc\sin A = \frac{1}{2}ac\sin B.$

⊙ The **sine rule** can be used in any triangle: $\dfrac{a}{\sin A} = \dfrac{b}{\sin B} = \dfrac{c}{\sin C}$

which can also be written as $\dfrac{\sin A}{a} = \dfrac{\sin B}{b} = \dfrac{\sin C}{c}$

⊙ The **cosine rule** can be used in any triangle: $a^2 = b^2 + c^2 - 2bc\cos A$

Similarly $b^2 = a^2 + c^2 - 2ac\cos B$ and $c^2 = a^2 + b^2 - 2ab\cos C$

which can also be written as

$$\cos A = \dfrac{b^2 + c^2 - a^2}{2bc} \qquad \cos B = \dfrac{a^2 + c^2 - b^2}{2ac} \qquad \cos C = \dfrac{a^2 + b^2 - c^2}{2ab}$$

⊙ Use the sine rule when a problem involves two sides and two angles.

⊙ Use the cosine rule when a problem involves three sides and one angle.

Review exercise

1 A cuboid has length 3 cm, width 4 cm and height 12 cm.
Work out the length of PQ.

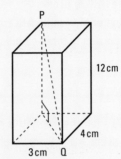

Nov 2007

2 ABC is a triangle.

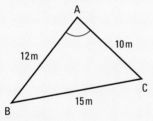

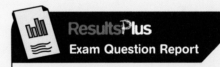

ResultsPlus
Exam Question Report

80% of students answered this question poorly because they did not take care substituting values into the equation.

AB = 12 m.

AC = 10 m.

BC = 15 m.

Calculate the size of angle BAC.

Give your answer correct to one decimal place.

June 2008

3 ABC is a triangle.

AC = 8 cm.

BC = 9 cm.

Angle ACB = 40°.

Calculate the length of AB.

Give your answer correct to 3 significant figures.

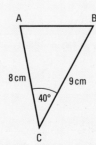

June 2007

A
A02

4 The diagram represents a cuboid ABCDEFGH.

AB = 5 cm.

BC = 7 cm.

AE = 3 cm.

a Calculate the length of AG.
  Give your answer to 3 significant figures.

b Calculate the size of the angle between AG
  and the face ABCD.
  Give your answer correct to 1 decimal place.

Nov 2004

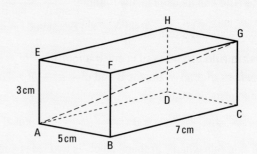

5 AB = 3.2 cm.

BC = 8.4 cm.

The area of triangle ABC is 10 cm².

Calculate the perimeter of triangle ABC.

Give your answer correct to three significant figures.

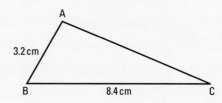

June 2004

A☆
A02
A03

6 The diagram shows a tetrahedron. AD is perpendicular to both AB and AC.

AB = 10 cm. AC = 8 cm. AD = 5 cm. Angle BAC = 90°.

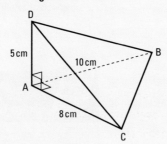

Calculate the size of angle BDC.

Give your answer correct to 1 decimal place.

Nov 2007

7 The diagram shows a sketch of the curve $y = \sin x°$ for $0 \leqslant x \leqslant 360$.

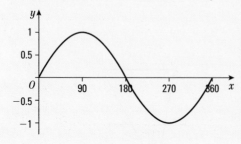

Results Plus

Exam Question Report

92% of students answered this question poorly
because they did not use the information given in
the question.

The exact value of $\sin 60° = \dfrac{\sqrt{3}}{2}$.

Write down the exact value of: a $\sin 120°$ b $\sin 240°$.

Nov 2008, adapted

8 Here is a graph of the curve $y = \cos x°$
for $0 \leqslant x \leqslant 360$.
Use the graph to solve $\cos x° = 0.75$
for $0 \leqslant x \leqslant 360$.

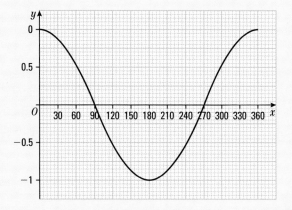

June 2007

9 The diagram shows an equilateral triangle ABC with sides of length 6 cm.
P is the midpoint of AB.
Q is the midpoint of AC.
APQ is a sector of a circle, centre A.
Calculate the area of the shaded region.
Give your answer correct to 3 significant figures.

June 2009

10 The lengths of the sides of a triangle are 4.2 cm, 5.3 cm and 7.6 cm.
a Calculate the size of the largest angle of the triangle.
Give your answer correct to 1 decimal place.
b Calculate the area of the triangle.
Give your answer correct to 3 significant figures.

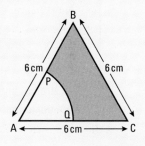

Nov 2006

11 The diagram represents a prism.
AEFD is a rectangle.
ABCD is a square.
EB and FC are perpendicular to plane ABCD.
AB = 60 cm.
AD = 60 cm.
Angle ABE = 90°.
Angle BAE = 30°.
Calculate the size of the angle that the line DE makes with the plane ABCD.
Give your answer correct to 1 decimal place.

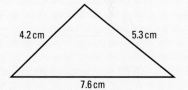

June 2004

577

12 The diagram shows a sector OABC of a circle with centre O.

OA = OC = 10.4 cm.

Angle AOC = 120°.

  a Calculate the length of the arc ABC of the sector.
Give your answer correct to 3 significant figures.

  b Calculate the area of the shaded segment ABC.
Give your answer correct to 3 significant figures.

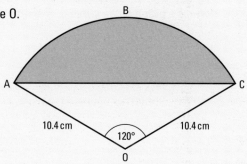

June 2006

13

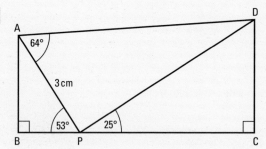

Diagram **NOT** accurately drawn

BPC is a straight line. Angles ABP = angle DCP = 90°.

Calculate the length of PD. Give your answer correct to 3 significant figures.

14

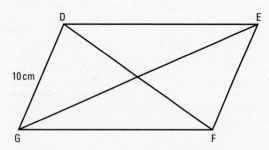

Diagram **NOT** accurately drawn

DEFG is a parallelogram with DG = 10 cm.

The diagonals DF and EG are of length 16 cm and 24 cm respectively.

  a Calculate the size of angle DGE.

  b Calculate the length of DE.

15 The diagram shows a pyramid HABCD standing on horizontal ground. The points A, B, C and D are the corners of its square base. The length of a side of the square is 12 m and its diagonals intersect at O. Each sloping edge makes an angle of 28° with the ground. Calculate:

  a the height, OH, in metres to 3 significant figures

  b the size, to the nearest degree, of the angle which the plane HCB makes with the ground.

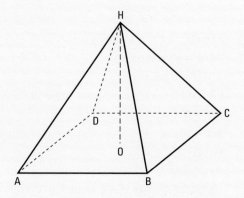

30 TRANSFORMATIONS OF FUNCTIONS

It can take just a transformation to make a new song successful. Music producers use sound mixers to change the pitch and volume of different sounds, as well as to add different layers of sounds and other effects.

⊙ Objectives

In this chapter you will:
- ⊙ use function notation
- ⊙ learn the relationship between simple transformations of curves and their effect on the equations of curves.

◇ Before you start

You should be able to:
- ⊙ identify transformations
- ⊙ solve an equation in x for a given value of x.

30.1 Using function notation

Objective

- You can use function notation.

Why do this?

Computer programmers use function notation as a means of shorthand for long lines of code.

Get Ready

1. Work out the value of $2x^2$ when $x = 3$.
2. Find the value of x if $4x - 3 = 8$.
3. If $y = x^2$ and $x = t + 1$, express y in terms of t.

Key Point

- A **function** $y = f(x)$ is a rule for working out values of y when given values of x.

A03

Example 1 $y = f(x) = 2x + 3$

Find the values of

    a $f(5)$ b $f(-4)$ c a where $f(a) = 5$.

a $f(5) = 2 \times 5 + 3 = 13$ ← | Substitute $x = 5$ in $2x + 3$. |

b $f(-4) = 2 \times (-4) + 3 = -5$

c $f(a) = 2a + 3 = 5$ ← | Solve the equation to find the value of a. |

    so $a = 1$

A03

Example 2 $y = g(x) = 2x^2 + 1$

    a Find the value of $g(-3)$.

    b Find the value of $2g(1)$.

    c What is the algebraic expression for $3g(x - 1)$?

a $g(-3) = 2 \times (-3)^2 + 1 = 19$

b $2g(1) = 2 \times (2 \times 1^2 + 1) = 6$ ← | Work out $g(1)$ first and then multiply by 2. |

c $g(x - 1) = 2(x - 1)^2 + 1 = 2x^2 - 4x + 3$ ← | Replace x by $(x - 1)$ in the expression $2x^2 + 1$. Then multiply by 3. |

    $3g(x - 1) = 3(2x^2 - 4x + 3)$

              $= 6x^2 - 12x + 9$

Exercise 30A

Questions in this chapter are targeted at the grades indicated.

B

1 $f(x) = 3x^2$, $g(x) = \frac{4}{x}$

Find the values of

    a $f(2)$ b $f(0)$ c $f(-4)$ d $g(4)$

    e $g(-1)$ f $g\left(\frac{1}{2}\right)$

2 $f(x) = 2x^3$, $g(x) = x^2 + x$

Find the values of

   a $f(1) + g(1)$ **b** $f(2) + g(3)$ **c** $f(2) \times g(2)$ **d** $\dfrac{f(4)}{g(4)}$

B

3 $f(x) = 2x + 2$

   a   Find the value of f(3).

   b $f(a) = 6$ Find the value of a.

A

4 $g(x) = x^2 - 4$

   a   Find the values of   **i**   g(0)   **ii**   g(1)   **iii**   g(−2).

   b $g(k) = 12$ Find the values of k.

A02   A☆

5 $g(x) = (x - 3)(x + 4)$

   a   Find the values of   **i**   g(5)   **ii**   g(0)   **iii**   3g(−2).

   b $g(a) = 0$ Find the values of a.

A02

6 $f(x) = x(x - 4)$

   a   Find the values of   **i**   f(1)   **ii**   f(2)   **iii**   2f(−1).

   b $f(k) = 0$ Find the values of k.

   c $f(m) = 5$ Find the values of m.

A02 A03

7 $f(x) = x^2$

   a   Find f(4).   **b**   Write out in full f(x) − 4.   **c**   Write out in full f(x − 4).

A02

8 $g(x) = 4(x + 1)$

   a   Find g(−3).   **b**   Write out in full g(2x).   **c**   Write out in full 3g(x).

A02

30.2   **Translation of a curve parallel to the axes**

◎ Objective

○ You understand the relationship between the translation of a curve parallel to an axis and the change in its function form.

❓ Why do this?

You could draw the graphs of the paths of two balls being juggled on the same axes. The graph of the second ball is a translation along the x-axis of the graph of the first ball.

◈ Get Ready

1.   The point P (2, 3) is translated by 2 units parallel to the y-axis. Find the new coordinates.

2.   The point P (2, 3) is translated by 2 units parallel to the x-axis. Find the new coordinates.

3.   The point P (2, 3) is translated by $\begin{pmatrix} -3 \\ 0 \end{pmatrix}$. Find the new coordinates.

◈ Key Points

◎ The relationship between the curves given by $y = f(x)$ and $y = f(x) + a$ is a translation of a units parallel to the y-axis or a translation by $\begin{pmatrix} 0 \\ a \end{pmatrix}$.

◎ The relationship between the curves given by $y = f(x)$ and $y = f(x + a)$ is a translation of $-a$ units parallel to the x-axis or a translation of $-\begin{pmatrix} a \\ 0 \end{pmatrix}$.

Example 3

Here is the graph of $y = f(x)$ where
$$f(x) = x^2$$

a Draw the graph with equation $y = f(x) + 3$.

b Describe the transformation that maps $y = f(x)$ to $y = f(x) + 3$.

c Write the algebraic form for $y = f(x) + 3$.

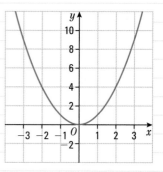

a

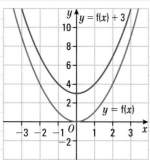

Each point on the curve $y = f(x)$ is moved up by 3 units.

b The transformation is a translation by $\begin{pmatrix} 0 \\ 3 \end{pmatrix}$.

c $f(x) = x^2$ so $f(x) + 3 = x^2 + 3$

The equation of the transformed curve is $y = x^2 + 3$.

Example 4

Here is a sketch of the graph of $y = f(x)$ where
$f(x) = x^2$.

a Describe the transformation which maps
$y = f(x)$ to $y = f(x - 2)$.

b Sketch the curve with equation $y = f(x - 2)$.
The coordinates of the minimum point of
$y = f(x)$ are $(0, 0)$.

c Write down the coordinates of the minimum
point of $y = f(x - 2)$.

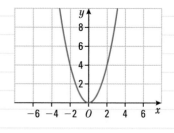

a Translation by $+2$ units parallel to the x-axis or translation by $\begin{pmatrix} 2 \\ 0 \end{pmatrix}$.

b

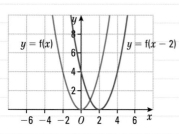

c $(0, 0)$ is mapped to $(0, 2)$.

Exercise 30B

1 Here is the graph of $y = f(x) = x^2$.

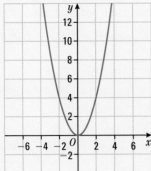

Draw the graphs of:
a $y = f(x) + 3$ b $y = f(x) - 4$
c $y = f(x + 2)$ d $y = f(x - 1)$.

A03

A*

2 Here is a sketch of the graph of $y = f(x) = x^3$.
a Draw sketches of the graphs of: i $y = f(x) + 3$ ii $y = f(x - 1)$.
b Write down the coordinates of the point to which the point
(0, 0) is mapped in each case.

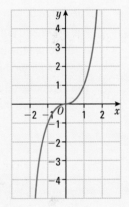

A03

3 Here is a sketch of the graph of $y = f(x) = \dfrac{1}{x}$.
The curve $y = f(x)$ is translated by $\begin{pmatrix} 0 \\ 2 \end{pmatrix}$.

a Sketch the graph of the new curve.
b Write down the coordinates to which the point (2, 0.5)
is mapped.
c Write down the equation of the translated curve:
  i in function form
  ii in algebraic form.
The curve $y = f(x)$ is now translated by $\begin{pmatrix} -2 \\ 0 \end{pmatrix}$
d Sketch the transformed curve.
e The point (1, 4) is mapped to the point (p, q).
  Write down the values of p and q.
f Write down the equation of the translated curve:
  i in function form
  ii in algebraic form.

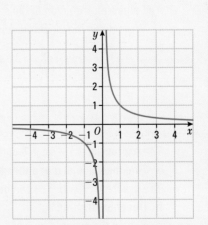

A03

4 A03

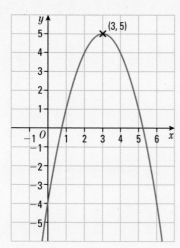

Here is a curve with equation $y = f(x)$.
The maximum point of the curve is (3, 5).

a Write down the coordinates of the maximum point of $y = f(x) - 3$.

b Write down the coordinates of the maximum point of $y = f(x + 2)$.

5 A03

Here are two curves, C_1 and C_2. The equation of the curve C_1 is $y = f(x)$.
The curve C_1 can be mapped to the curve C_2 by a translation.
The maximum point of C_1 is (2, 4) and the maximum point of C_2 is (2, 7).

a Describe the translation.

b Write down the equation of the curve C_2 in function form.

The algebraic equation of the curve C_1 is $y = 4x - x^2$.

c Write down the algebraic equation of the curve C_2.

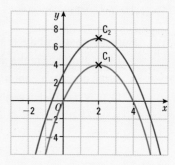

6 A03

Here are two curves, C_1 and C_2. The equation of the curve C_1 is $y = f(x)$.
The curve C_1 can be mapped to the curve C_2 by a translation.

a Describe the translation.

b Write down the equation of the curve C_2 in function form.

The algebraic equation of the curve C_1 is $y = x^2$.

c Write down the algebraic equation of the curve C_2.

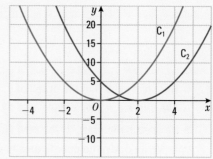

7 The expression $x^2 + 4x + 9$ can be written in the form $(x + a)^2 + b$ for all values of x.

a Find the value of a and the value of b.

The graph of $x^2 + 4x + 9$ can be obtained from the graph of $y = x^2$ by a translation.

b Describe this translation.

A03

c Sketch the graph of $y = x^2$.

d Sketch the graph of $x^2 + 4x + 9$ on the same axes.

8 A02 A03 Describe fully the transformation that will map the curve with equation $y = x^2$ to the curve with equation $y = x^2 - 6$.

30.3 Stretching a curve parallel to the axes

◎ Objective

○ You understand the effect that stretching a curve parallel to one of the axes has on its function form.

⊘ Why do this?

If you were designing a bridge in the style of the Golden Gate Bridge, you could experiment with the length of the support struts. Taller struts would stretch the shape of the curved cable.

◈ Get Ready

1. Draw the triangles with coordinates (1, 1), (3, 1), (3, 2) and (2, 1), (6, 1), (6, 2). What is the relationship between the two triangles?
2. Draw the triangles with coordinates (1, 1), (3, 1), (3, 2) and (1, 3), (3, 3), (3, 6). What is the relationship between the two triangles?

◔ Key Points

◉ The relationship between the curves $y = f(x)$ and $y = af(x)$ (where a is a constant) is that of a **stretch** of magnitude a parallel to the y-axis.

◉ The relationship between the curves $y = f(x)$ and $y = f(ax)$ (where a is a constant) is that of a stretch of magnitude $\frac{1}{a}$ parallel to the x-axis.

◔ Example 5

Here is the graph of $y = f(x)$.

  a Sketch the graph of $y = 2f(x)$.

  b To what point is the point (2, 1) mapped?

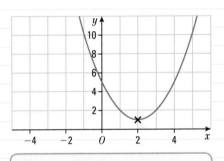

a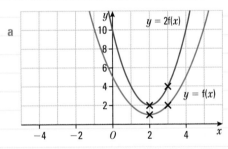

b The minimum point (2, 1) of $y = f(x)$, is mapped to the minimum point (2, 2) of $y = f(2x)$.

> The transformation is a stretch parallel to the y-axis of magnitude 2.

> The y-coordinates of all points on the curve $y = f(x)$ are doubled.

◔ Example 6

Here is the graph of $y = f(x)$.

  a Sketch the graph of $y = f(2x)$.

  b To what point is the point (2, 1) mapped?

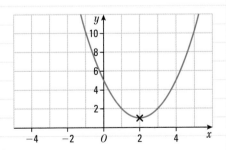

a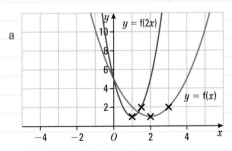

> The x-coordinates of all points on the curve $y = f(x)$ are halved.

b The point (2, 1) of $y = f(x)$ is mapped to the point (0.5, 1) of $y = f(2x)$.

Example 7 Here is the graph of $y = f(x) = \sin x°$.

    a Draw the graph of $y = 2f(x)$.

    b Write down the algebraic equation of $y = 2f(x)$.

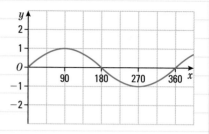

a

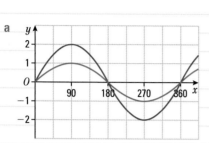

The graph is stretched parallel to the y-axis.

b $y = 2\sin x°$

Exercise 30C

1 Here is the graph of $y = f(x)$. It has a minimum point at $(2, 2)$.

Draw the graph of $y = f(2x)$.

To which point is the minimum point of $y = f(x)$ mapped?

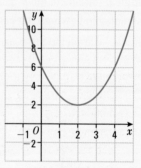

2 Here is a sketch of the curve C_1 $y = f(x) = x^2$.

    a Sketch the curve C_2 with equation $y = 4f(x)$.

    b Write down the equation of the curve C_2 in algebraic form.

    c Give two different transformations that will each map the curve C_1 to the curve C_2.

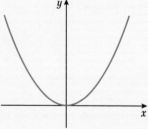

3 Here is the graph of $y = f(x) = \cos x°$.

    a Draw the graph of $y = 2f(x)$.

       Write the equation of $y = 2f(x)$ in algebraic form.

    b Draw the graph of $y = f(2x)$.

       Write the equation of $y = f(2x)$ in algebraic form.

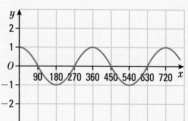

4 Here is the graph of $y = f(x)$.

The graph crosses the y-axis at $(0, 4)$ and the x-axis at $\left(-\frac{2}{3}, 0\right)$.

    a Sketch the graph with equation $y = 3f(x)$.

    b Write down the coordinates of the points to which $(0, 4)$ and $\left(-\frac{2}{3}, 0\right)$ are mapped.

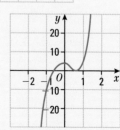

5   Here is the graph of $y = f(x) = 1 + \sin x°$.
    On separate graphs, sketch the curves with equations: a $y = 2f(x)$
                                                             b $y = f(2x)$
                                                             c $y = f(x) + 2$.

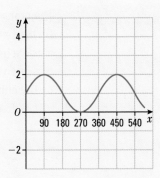

6   Here is the graph of $y = f(x) = \cos x°$.
    a   Sketch the graph with equation $y = f\left(\dfrac{x}{2}\right)$.
    b   How many solutions does the equation
        $f\left(\dfrac{x}{2}\right) = 0.5$ have in the range $0 < x < 360$?

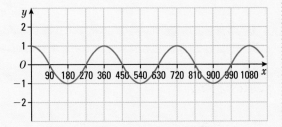

A03

7   Here is a sketch of the curve C_1, $y = f(x) = x(x - 4)$.
    The curve C_1 has a minimum point at $(2, -4)$.
    The curve C_1 is mapped to the curve C_2 by a stretch.
    The minimum point on C_2 is $(4, -4)$.
    The minimum point on C_1 is mapped to the minimum point of C_2.
    a   Describe the stretch fully.
    b   Draw a sketch of C_2.
    c   Write the equation of C_2:   i   using functional form
                                       ii in algebraic form.

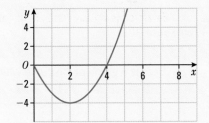

A03

8   The expression $x^2 - 8x + 5$ can be written in the form $(x - p)^2 + q$.
    a   Find the values of p and q.
    b   Write down the coordinates of the point P where the curve $y = f(x) = x^2 - 8x + 5$ crosses the
        y-axis.
    The curve $y = f(x)$ is mapped by a stretch parallel to the y-axis, so that the point P is mapped to the
    point $(0, 3)$.
    c   Describe the stretch and write down the equation of the new curve.

A02
A03

⬤ **Key Points**

◉ The relationship between the curves $y = f(x)$ and $y = af\left(\dfrac{x}{a}\right)$ (where a is a constant) is that of an
   enlargement, centre the origin and scale factor a.

Example 8

The graph of the curve C_1 $y = f(x) = x^2 + 1$ is shown on the grid.

On the same grid,

    a sketch the curve C_2 with equation $y = 2f(x)$

    b sketch the curve C_3 with equation $y = 2f\left(\dfrac{x}{2}\right)$

    c Describe fully the transformation that maps

        i C_1 to C_2 **ii** C_2 to C_3 **iii** C_1 to C_3

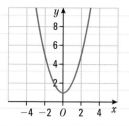

a C_2 is in black.

b C_3 is in blue.

c **i** Stretch parallel to the y-axis of magnitude 2.

    ii Stretch parallel to the x-axis of magnitude 2.

   iii Enlargement with scale factor 2 and centre the origin.

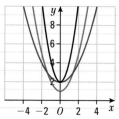

Exercise 30D

1 Here is a graph of the curve with equation $y = \sin x°$.

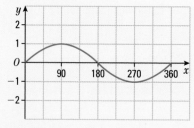

    a Copy the graph and sketch the curve C_2,
the enlargement of C_1 with scale factor 2.

    b Write down the equation of the curve C_2.

2 **a** Draw a sketch of the U shaped curve C_1, with equation $y = x(x + 2)$.
The curve C_1 is enlarged with a scale factor 2, centre O, to give the curve C_2.

    b Find the equation of the curve C_2 in function form.

    c Sketch the curve C_2.

3 **a** Sketch the curve with equation $y = f(x) = \cos x°$. for values of x from 0 to 1080.

    b On the same axes, sketch the graph of the curve with equation $y = 3f\left(\dfrac{x}{3}\right)$.

4 Write down the algebriac equation of the given curve after it has been enlarged by scale factor 4 and centre O.

    a $y = x^2 + 3$ **b** $y = \dfrac{1}{x} + 1$ **c** $y = 2^x$ **d** $y = 2\sin(2x)$

30.4 Rotation about the origin and reflection in the axes

Objectives

- You understand the effect that a reflection of a curve in one of its axes has on its function form.
- You understand the effect that a rotation of a curve by 180° about the origin has on its function form.

Why do this?

If two cars race between two towns, but one starts in one town and the other car starts in another, the graph of one car's displacement against time will be a reflection of the other car's graph.

Get Ready

1. Draw the line joining the origin to the point (4, 5). Reflect this line in the x-axis.
2. Draw the line joining (1, 0) to the point (4, 3). Reflect this line in the y-axis.
3. Draw the line joining (1, 0) to the point (4, 3). Rotate this line by 180° about the origin.

Key Points

- The curve $y = f(-x)$ is a reflection in the y-axis of the curve $y = f(x)$.
- The curve $y = -f(x)$ is a reflection in the x-axis of the curve $y = f(x)$.
- The curve $y = -f(-x)$ is a rotation by 180° about the origin of the curve $y = f(x)$.

Example 9

Here is the graph of $y = f(x)$.

Sketch the curves with equations:

a $y = f(-x)$ **b** $y = -f(x)$ **c** $y = -f(-x)$.

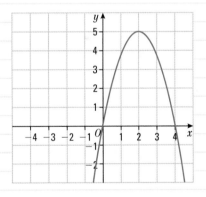

a
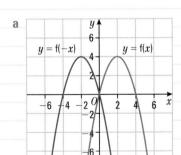

b
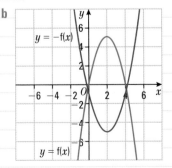

c
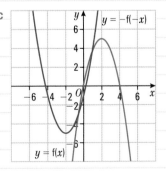

Exercise 30E

1 The graph of $y = f(x) = x^2 + 3$ has been drawn.

  a Sketch the graph of $y = -f(x)$.

  b Write down the equation of the new graph in algebraic form.

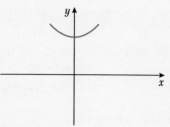

2 The graph of $y = f(x) = x^3$ has been drawn.

  a Sketch the graph of $y = f(-x)$.

  The point $(2, 8)$ on $y = f(x) = x^3$ has been mapped to the point (p, q).

  b Write down the values of p and q.

  c Write down the equation of the new curve.
  Give your answer in algebraic form.

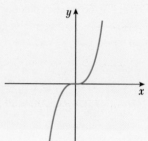

3 The graph of $y = f(x) = 2^x$ has been drawn.

  a Sketch the graph of $y = -f(-x)$.

  The point $(1, 2)$ has been mapped to the point (r, t).

  b Write down the values of r and t.

  c Write down the equation of the new curve.
  Give your answer in algebraic form.

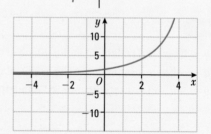

Chapter review

- A **function** $y = f(x)$ is a rule for working out values of y when given values of x.
- You should know the following transformations:
  - $y = f(x) + a$ is a translation by $+a$ units, parallel to the y-axis of $y = f(x)$
  - $y = f(x + a)$ is a translation by $-a$ units, parallel to the x-axis of $y = f(x)$
  - $y = af(x)$ is a **stretch** of magnitude a units parallel to the y-axis of $y = f(x)$
  - $y = f(ax)$ is a stretch of magnitude $\frac{1}{a}$ units parallel to the x-axis of $y = f(x)$
  - $y = af\left(\frac{x}{a}\right)$ is an enlargement, centre the origin and scale factor a, of $y = f(x)$
  - $y = f(-x)$ is a reflection in the y-axis of $y = f(x)$
  - $y = -f(x)$ is a reflection in the x-axis of $y = f(x)$
  - $y = -f(-x)$ is a rotation by $180°$ about the origin of $y = f(x)$.

Review exercise

1 $f(x) = x^2 + 2$

  Work out   **a** $f(2)$ **b** $f(-3)$ **c** a where $f(a) = 2$.

2 $f(x) = x^2 + 3x$ Show that $f(x-1) = (x + 2)(x - 1)$.

3 Sketch the graph of $y = x^2$. Hence sketch the graph of $y = -x^2$.

4 The graph of $y = f(x)$ is shown on the grid.
  a Copy the graph and then draw the graph of $y = f(x) + 2$ on the same axes.
  b On another copy of the graph, draw the graph of $y = -f(x)$.

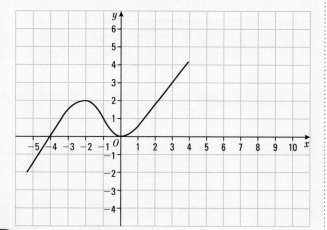

June 2007, adapted

5 The curve with equation $y = f(x)$ is translated so that the point at (0, 0) is mapped onto the point (4, 0).

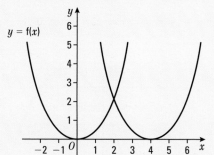

Find an equation of the translated curve.

Nov 2007, adapted

6 The diagram shows a sketch of the graph of $y = x^2 - x$.

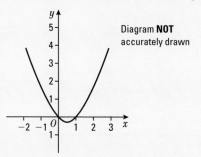

Diagram **NOT** accurately drawn

  a On the same diagram, sketch and label the graph of
  $y = (x - 1)^2 - (x - 1)$.
  Show clearly where this graph crosses the x-axis and where it crosses the y-axis.
  b On the same diagram sketch and label the graph of $y = 3(x^2 - x)$.
  c Write down the solutions of the equation $(x - 1)^2 - (x - 1) = 0$.

Nov 2005

7 The diagram shows a sketch of part of the curve $y = \sin x°$.
  a Write down the coordinates of the point A.
  b On the same diagram, sketch the graph of $y = \sin 2x°$.

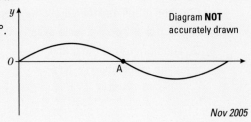

Diagram **NOT** accurately drawn

Nov 2005

8 The diagram shows a sketch of part of the curve $y = \sin x°$.

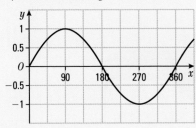

On a suitable grid, draw the graph of $y = 4\sin 2x°$

9 Describe the transformation that will map the curve $y = x^2$ to the curve $y = x^2 - 8x + 11$.

10 The curve $y = x^2$ can be mapped to the curve $y = 4x^2$ using a single transformation.
  a Describe one possible transformation.
  b Describe another possible transformation.

11 The equation of the curve C_1 is $y = f(x) = 8 + 4x - x^2$.
  a Write $8 + 4x - x^2$ in the form $q - (x - p)^2$
     where p and q are numbers to be found.
  Here is a sketch of the curve $y = 8 + 4x - x^2$.
  b Write down the coordinates of the maximum point of the curve.
  The curve C_1 is stretched to the curve C_2 so that the maximum point of C_1 is mapped to (2, 24).
  c Describe the stretch.
  d Write down the equation of C_2 in function form.

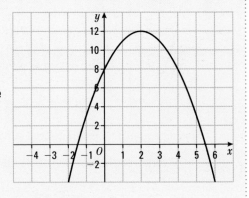

31 CIRCLE GEOMETRY

The O$_2$ dome is a circle when looked at from above. It has a diameter of 365 m – a metre for every day of the year. The roof structure is incredibly light, weighing less than the air contained within the building. However, it is not strictly a dome as it is not self-supporting.

◎ **Objective**

In this chapter you will:
- learn about the geometric properties of circles and tangents.

◈ **Before you start**

You need to know that:
- the angles in a triangle add up to 180°
- the base angles in an isosceles triangle are equal
- angles on a straight line add up to 180°.

31.1 Isosceles triangle in a circle

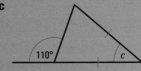

⊙ Objective

○ You can use the properties of angles in a circle.

⑦ Why do this?

Ferris wheels are constructed using the properties of an isosceles triangle in a circle.

⬆ Get Ready

1. Calculate the size of the angles marked a, b and c.

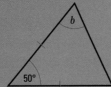

Key Point

⊙ The radii of a circle are all the same length.
This means that a triangle in a circle where two of its sides are radii is an isosceles triangle.

🔍 Example 1

P and Q are points on the circumference of a circle, centre O.
PQR is a straight line.
Angle OQR = 150°
Calculate the size of angle POQ.
Give reasons for your answer.

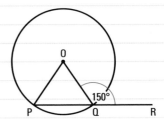

Angle OQP = 180° − 150° ← **Adjacent angles on a straight line.**
$\qquad$ = 30°
Reason: The angles on a straight line add up to 180°.

🔍 ResultsPlus
Exam Tip

Write down the reason; this is the rule that you have used and should be in words.

OP = OQ ← **Radii equal, thus isosceles triangle.**

Angle QPO = Angle OQP = 30° ← **Mark the angles on the diagram.**

Reason: In an isosceles triangle the angles opposite the equal sides (radii equal) are the same size.

Angle POQ = 180° − 30° − 30° ← **Subtract the two angles you know from 180°.**
$\qquad$ = 120°
Reason: The angles in a triangle add up to 180°.

Exercise 31A

| Questions in this chapter are targeted at the grades indicated. |

The diagrams all show circles, centre O.
Work out the size of each angle marked with a letter.

1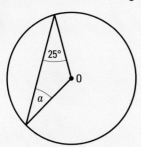

2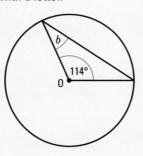

3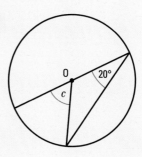

D

In questions 4–6 give reasons for your answers.

* **4**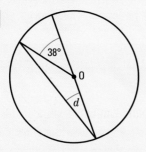

* **5**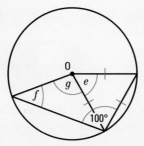

* **6**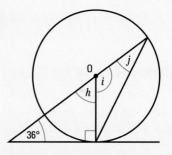

A03 **C**

31.2 Tangents to a circle

 **Objective**

 You can use the properties of tangents
to a circle.

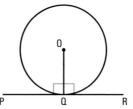 **Why do this?**

You would need to understand tangents to a circle
to design the gear system on a bike.

 Get Ready

1. What are the values of angles a, b and c in this rectangle?

 Key Points

 The angle between a tangent and a radius of a circle is 90°.
Angle OQP = angle OQR = 90°

 Tangents to a circle from a point outside the circle are equal in length.
AB = BC

A03

Example 2

QR is a tangent to the circle, centre O.

PQ is a chord of the circle.

Angle POQ = 108°

Work out the value of x.

Give reasons for your answer.

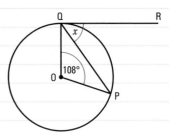

Angle OQP = (180° − 108°) ÷ 2 ← Isosceles triangle, radii equal.

        = 72° ÷ 2

        = 36° ← Put 36° on the diagram.

OQP is an isosceles triangle as the radii are equal. ← Write all the different reasons in words.

In an isosceles triangle the angles opposite the

equal sides are the same size and the angles in a

triangle add up to 180°.

ResultsPlus

Exam Tip

When you have calculated an angle, you can mark

it on the diagram to help you answer the question.

$x = 90° −$ angle OQP

  = 90° − 36°

  = 54°

The angle between the tangent and the radius is 90°. ← Write this reason too.

Example 3

RP and RQ are tangents to the circle, centre O.

PQ is a chord.

Work out the value of y.

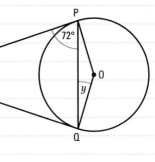

RP = RQ ← Tangents are the same length.

Tangents to a circle from a point outside

the circle are equal in length.

Angle RQP = angle RPQ = 72° ← Isosceles triangle.

In an isosceles triangle, the angles opposite

the equal sides are the same size.

Angle OQR = 90° ← Tangent to radius = 90°.

The angle between the tangent and the radius is 90°. ← Mark the 90° angle on the diagram using a square.

$y = 90° − 72°$

  = 18°

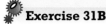

Exercise 31B

The diagrams all show circles, centre O.
Work out the size of each angle marked with a letter.

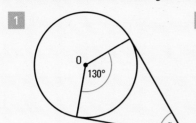

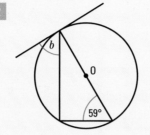

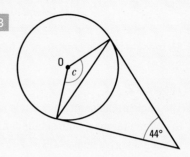

In questions 4–6 give reasons for your answers.

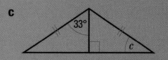

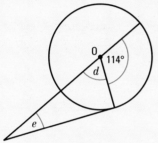

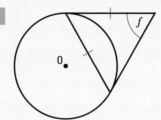

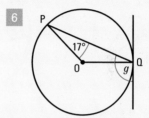

31.3 Circle theorems

◎ Objectives

○ You know three theorems about circles and
how to prove and apply them.

❓ Why do this?

You can use circle theorems in engineering as
many machines contain circular parts like cogs.

◈ Get Ready

1. Calculate the size of the angles marked a, b and c.

a

b

c

◉ Key Points

◎ **Theorem 1**

The perpendicular from the centre of a circle to a chord bisects the chord (and vice versa:
the line drawn from the centre of a circle to the midpoint of a chord is perpendicular
to the chord).

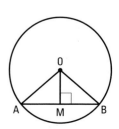

Proof

In triangles OAM and OBM

OA = OB $\longleftarrow$ Radii equal.

Angle OMA = angle OMB $\longleftarrow$ Both given as 90°.

OM = OM $\longleftarrow$ Common side.

So triangle OAM is congruent to triangle OBM (RHS).

So AM = MB.

⊚ **Theorem 2**

The angle at the centre of a circle is twice the angle at the circumference, both subtended by the same arc.

In each diagram angle AOB = 2 × angle ACB.

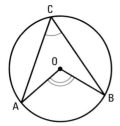 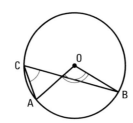 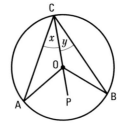

Proof

Draw in the line CO and extend it to P as shown in the diagram above on the right.

Let angle ACO = x and angle BCO = y.

Since triangle CAO and triangle CBO are both isosceles (radii equal), the angles opposite the equal sides are equal. $\longleftarrow$ Give the reason in words as well as giving the sizes of the angles.

So angle CAO = x and angle CBO = y.

The exterior angle of a triangle is equal to the sum of the two interior opposite angles. $\longleftarrow$ This rule was learnt in Section 5.3.

So angle AOP = $2x$ and angle BOP = $2y$.

Angle ACB = $x + y$ $\longleftarrow$ We can see this from the diagram.

Angle AOB = $2x + 2y = 2(x + y)$
= 2(angle ACB)

⊚ **Theorem 3**

The angle in a semicircle is a right angle.

Angle ACB = 90°

Proof

The angle at the centre of the circle is twice the angle at the circumference. $\longleftarrow$ We can use the rule proved in theorem 2.

So angle AOB = 2 × angle ACB.

But angle AOB = 180° as it is a straight line. $\longleftarrow$ AOB is the diameter.

So angle ACB = $\frac{1}{2}$(180°)
= 90°

Example 4

PQ is a chord of the circle, centre O.
N is the midpoint of PQ.
Angle QPO = 27°
Work out the size of angle POQ.

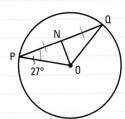

Angle PNO = 90° ← Line from centre to midpoint of chord is perpendicular to chord (Theorem 1).

Angle PON = 180° − 90° − 27° ← The angles in a triangle add up to 180°.
       = 63°

Similarly angle QON = 63° ← Angle QNO = 90° and triangle PQO is isosceles.

Angle POQ = angle PON + angle QON
So angle POQ = 126°

Example 5

P, Q and R are points on a circle, centre O.
Angle POQ = 142°.
Work out the size of angle PRQ.
Give a reason for your answer.

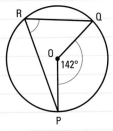

Angle PRQ = 142° ÷ 2 ← As angle POQ = 2 × angle PRQ (Theorem 2) we can say that angle PRQ = $\frac{1}{2}$ angle POQ.
       = 71°

Reason: The angle at the centre of the circle is twice the angle at ← Give the reason in words.
the circumference, so angle POQ = 2 × angle PRQ.

Example 6

P, Q and R are points on a circle, centre O.
Angle RQO = 20°

Work out the size of angle PRO.

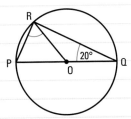

Angle QRO = 20° ← OQ = OR radii, thus triangle ORQ is isosceles.

Angle PRO = angle PRQ − angle QRO ← Angle PRO = 90° as it is in a semicircle (Theorem 3).
       = 90° − 20°
       = 70°

A03

⚙ **Exercise 31C**

Find the size of each of the angles marked with a letter.

O is the centre of the circle in each case.

A

1
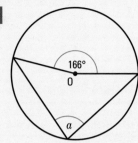

2
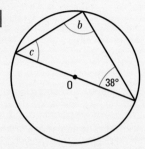

3
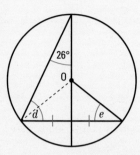

4
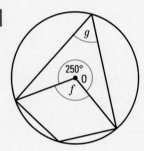

5
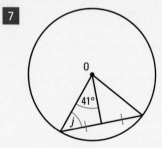

6
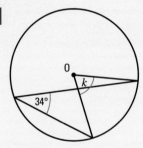

In questions 7–9 give reasons for your answers.

A03

7
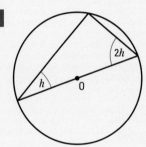

8
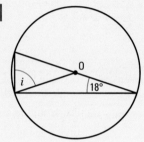

9
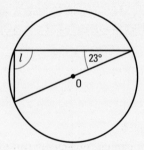

A03

10 A, B, C and D are points on the circle, centre O.

Angle DOB = 130°

a Work out the size of angle DAB.

Give a reason for your answer.

b Work out the size of reflex angle DOB.

Give a reason for your answer.

c Work out the size of angle BCD.

Give a reason for your answer.

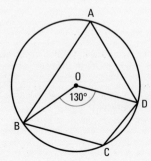

31.4 More circle theorems

⊙ Objectives

○ You know three more theorems about circles and how to prove and apply them.
○ You can use all you have learnt about circles to work out angles in more complex problems.

? Why do this?

There are many applications for circle geometry. Some people believe that the mysterious crop circles that sometimes appear in our fields are created using circle geometry.

◈ Get Ready

1. Calculate the size of the angles marked a, b and c.

a
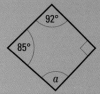

b

c

◉ Key Points

○ **Theorem 4**

Angles in the same segment are equal.

Angle ACB = angle ADB

Proof

Angle AOB = 2 × angle ACB ⟵ | Draw in the angle at the centre and use Theorem 2.

The angle at the centre is twice the angle at the circumference.

Angle AOB = 2 × angle ADB ⟵ | Now do the same for the other angle.

The angle at the centre is twice the angle at the circumference.

So angle ACB = angle ADB. ⟵ | Both are half of angle AOB.

○ **Theorem 5**

Opposite angles of a **cyclic quadrilateral** add up to 180°.
(A cyclic quadrilateral is a quadrilateral that has all four vertices on the circumference of a circle.)

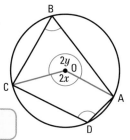

Angle ABC + angle ADC = 180°

Proof

Angle ABC = x ⟵ | First draw in angle AOC.

The angle at the centre of the circle is twice the angle at the circumference.

Angle ADC = y ⟵ | The reflex angle is twice angle ADC.

The angle at the centre of the circle is twice the angle at the circumference.

But $2x + 2y = 2(x + y) = 360°$. ⟵ | The angles at a point add up to 360°.

So $x + y = 180°$ ⟵ | Divide each side of the equation by 2.

cyclic quadrilateral

◉ **Theorem 6**

The angle between a tangent and a chord is equal to the angle in the alternate segment.

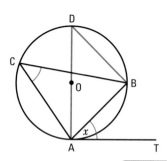

Angle BAT = angle ACB

Proof

Angle DAT = 90° (angle between the tangent and the radius is 90°) ← Draw the diameter AD.

So angle DAB = 90° − x

Angle DBA = 90° (angle in a semicircle = 90°) ← Theorem 3

So angle ADB = 180° − 90° − (90° − x)

= x (angles in a triangle add up to 180°)

Angle ACB = angle ADB (angles in the same segment) ← Theorem 4

So angle ACB = x

So angle BAT = angle ACB. ← Both equal to x.

A03

Example 7 P, Q, R and S are points on the circle, centre O.

Angle RSQ = 43°

Work out the size of angle RPQ.

Give a reason for your answer.

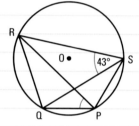

Angle RPQ = 43°

Reason: The angles in the same segment are equal. ← Write down the reason in words (Theorem 4).

Example 8 P, Q, R and S are points on the circle, centre O.

Angle ROP = 102°

Work out the size of angle RQP.

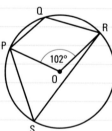

Angle RSP = $\frac{1}{2}$(102°)

= 51° ← The angle at the centre is twice the angle at the circumference.

Angle RQP = 180° − 51°

= 129° ← Opposite angles of cyclic quadrilateral add to 180° (Theorem 5).

Example 9

PQR is a tangent to the circle, centre O.
S and T are points on the circle.
Angle TQS = 47°
Angle QTS = 76°
Work out the size of angle PQT.
Give reasons for your answer.

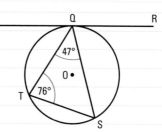

Angle QST = 180° − 47° − 76°
        = 57°
The angles in a triangle add up to 180°. ← Always give the reason.
Angle PQT = angle QST = 57°

Give the second reason too (Theorem 6).

The angle between a tangent and a chord is equal to the angle in the alternate segment.

Exercise 31D

Find the size of each of the angles marked with a letter.
O is the centre of the circle in each case.

1

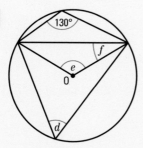

2

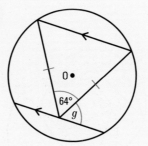

3

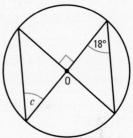

4

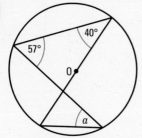

5

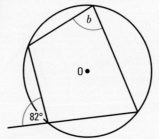

6

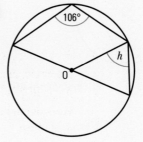

In questions 7–9 give reasons for your answers.

***7**

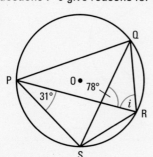

***8**

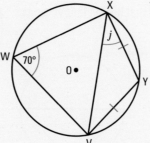

***9**

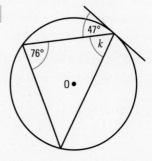

603

Chapter review

◉ The radii of a circle are all the same length. This means that a triangle in a circle where two of its sides are radii is an isosceles triangle.

◉ The angle between a tangent and a radius of a circle is 90°.

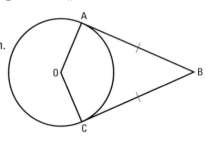

◉ Tangents to a circle from a point outside the circle are equal in length.

Circle theorems

1. The perpendicular from the centre of a circle to a chord bisects the chord (and vice versa: the line drawn from the centre of a circle to the midpoint of a chord is perpendicular to the chord).

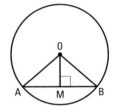

2. The angle at the centre of a circle is twice the angle at the circumference, both subtended by the same arc.

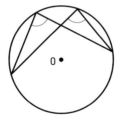

3. The angle in a semicircle is a right angle.

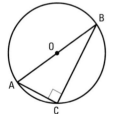

4. Angles in the same segment are equal.

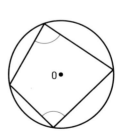

5. Opposite angles of a **cyclic quadrilateral** add up to 180°. (A cyclic quadrilateral is a quadrilateral that has all four vertices on the circumference of a circle.)

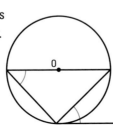

6. The angle between a tangent and a chord is equal to the angle in the alternate segment.

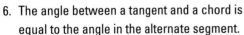

⚙ Review exercise

1 The diagram shows a circle centre O.

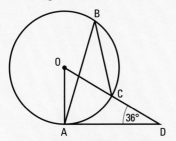

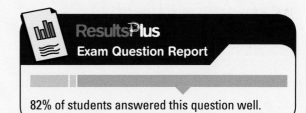

Results Plus
Exam Question Report

82% of students answered this question well.

A, B and C are points on the circumference. DCO is a straight line.
DA is a tangent to the circle. Angle ADO = 36°
Work out the size of angle ABC.

June 2009, adapted

2 In the diagram, A, B, C and D are points on the circumference of a circle, centre O.
Angle BAD = 70°. Angle BOD = x°. Angle BCD = y°.
   a **i** Work out the value of x.   **ii** Give a reason for your answer.
   b **i** Work out the value of y.   **ii** Give a reason for your answer.

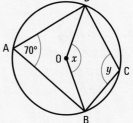

Results Plus
Exam Question Report

77% of students answered this question poorly as
they had not learnt circle theorems correctly.

June 2008

Find the size of each of the angles marked with a letter.
O is the centre of the circle, where marked.

3

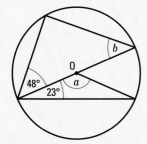

4

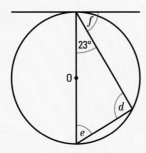

5

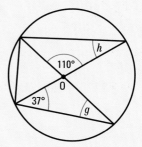

In questions 6–8 give reasons for your answers.

*6

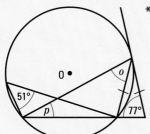

*7

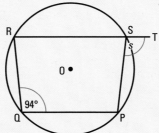

*8

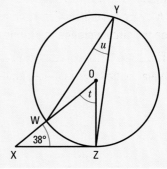

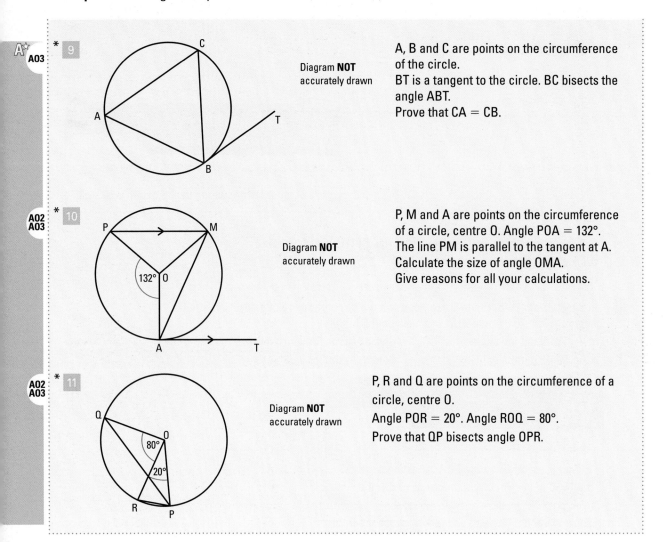

A03

* 9

Diagram **NOT** accurately drawn

A, B and C are points on the circumference of the circle.
BT is a tangent to the circle. BC bisects the angle ABT.
Prove that CA = CB.

A02
A03

* 10

Diagram **NOT** accurately drawn

P, M and A are points on the circumference of a circle, centre O. Angle POA = 132°.
The line PM is parallel to the tangent at A.
Calculate the size of angle OMA.
Give reasons for all your calculations.

A02
A03

* 11

Diagram **NOT** accurately drawn

P, R and Q are points on the circumference of a circle, centre O.
Angle POR = 20°. Angle ROQ = 80°.
Prove that QP bisects angle OPR.

32 ALGEBRAIC FRACTIONS AND ALGEBRAIC PROOF

The word algebra is derived from the Arabic word Al-Jabr. It first appeared in 820AD in the work of the Persian mathematician Al-Khwarizmi. Al-Khwarizmi became known as the father of algebra, creating and using algebraic proofs to solve the mathematical problems of the time.

⊙ Objectives

In this chapter you will:
- simplify algebraic fractions
- add and subtract algebraic fractions
- multiply and divide algebraic fractions
- prove a given result using algebra.

◈ Before you start

You need to be able to:
- add, subtract, multiply and divide fractions
- factorise algebraic expressions
- use the laws of indices.

32.1 Simplifying algebraic fractions

◎ Objective

○ You can simplify algebraic fractions.

❓ Why do this?

Doctors, engineers and scientists often have to use and simplify algebraic fractions in their jobs.

⬆ Get Ready

1. Factorise fully.

   a $x^2 + 5x + 4$ **b** $x^2 - x - 6$ **c** $2x^2 + 7x + 3$

🔑 Key Points

◉ Algebraic fractions, like numerical fractions, can often be simplified.

◉ To simplify an algebraic fraction, factorise the numerator and the denominator. Then divide the numerator and denominator by any common factors.

Example 1   Simplify fully $\dfrac{2x^2 + 4x}{x^2 + 3x + 2}$

$2x^2 + 4x = 2x(x + 2)$ ⟵   Factorise the numerator fully.

$x^2 + 3x + 2 = (x + 2)(x + 1)$ ⟵   Factorise the denominator.

$\dfrac{2x^2 + 4x}{x^2 + 3x + 2} = \dfrac{2x(x + 2)^1}{^1(x + 2)(x + 1)}$ ⟵   Write the fraction in fully factorised form.

$= \dfrac{2x}{(x + 1)}$ ⟵   Divide both the numerator and denominator by the common factor $(x + 2)$.

$= \dfrac{2x}{x + 1}$ ⟵   Write your answer without the bracket as there is only one factor in the denominator.

> **ResultsPlus**
> **Watch Out!**
>
> You cannot simplify an algebraic fraction until it has been expressed as a product of factors.

Example 2   Simplify fully $\dfrac{2x^2 - 5x - 3}{x^3 - 9x}$

$2x^2 - 5x - 3 = (2x + 1)(x - 3)$ ⟵   Factorise the numerator.

$x^3 - 9x = x(x^2 - 9)$ ⟵   Factorise the denominator.

$= x(x + 3)(x - 3)$ ⟵   Take out the common factor x. Factorise $(x^2 - 9)$ using the difference of two squares.

$\dfrac{2x^2 - 5x - 3}{x^3 - 9x} = \dfrac{(2x + 1)(x - 3)}{x(x + 3)(x - 3)}$ ⟵   Write the fraction in factorised form.

$= \dfrac{(2x + 1)(x - 3)^1}{x(x + 3)(x - 3)^1}$ ⟵   Divide both the numerator and denominator by any common factors.

$= \dfrac{2x + 1}{x(x + 3)}$ ⟵   Write the numerator without the bracket as there is only one factor left in the numerator.

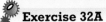

Exercise 32A

Questions in this chapter are targeted at the grades indicated.

A

A☆

1 Simplify fully.

a $\dfrac{2x^5}{x^2}$
b $\dfrac{x^2y}{3xy^2}$
c $\dfrac{x^2 - 5x}{2x}$
d $\dfrac{x^2 + 3x}{x + 3}$
e $\dfrac{2x - 4x^2}{2x - 1}$

2 Simplify fully.

a $\dfrac{x^2 + 4x + 3}{x^2 + 5x + 6}$
b $\dfrac{x^2 + 6x + 5}{x^2 + 5x}$
c $\dfrac{x^2 - 5x + 6}{x^2 + x - 12}$
d $\dfrac{x^2 - x - 12}{x^2 + 6x + 9}$

3 Simplify fully.

a $\dfrac{x^2 - 1}{x^2 - x}$
b $\dfrac{4x^2 + 24x}{x^2 - 36}$
c $\dfrac{2x^2 - 8}{x^2 + 4x + 4}$
d $\dfrac{3x^2 - 27}{3x^2 + 9x}$

4 Simplify fully.

a $\dfrac{2x^2 + 5x + 3}{3x^2 + 5x + 2}$
b $\dfrac{10x^2 - x - 3}{6x^2 - x - 2}$
c $\dfrac{9x^2 - 1}{9x^2 - 6x + 1}$
d $\dfrac{6x^2 + 5x - 1}{12x^2 + 16x - 3}$

5 Simplify fully.

a $\dfrac{x(x + 5)}{x^2 - 5x}$
b $\dfrac{x^2 - 10x + 25}{2x^2 - 50}$
c $\dfrac{8x^2 - 10x + 3}{8x^2 - 6x}$

d $\dfrac{6x^2 - 2x^3}{2x^3 + 6x^2}$
e $\dfrac{4 - x^2}{(x + 2)^2}$
f $\dfrac{16 - x^2}{x - 4}$

32.2 Adding and subtracting algebraic fractions

⊙ Objective

◉ You can add and subtract algebraic fractions.

? Why do this?

If you win two competitions in a tennis tournament, and you know what fraction of the prize money each one is worth, you can work out what the total prize money for the tournament was.

◈ Get Ready

1. Write down the lowest common multiple (LCM) of:

a 6 and 15
b $3x$ and $4x$
c $(x + 1)$ and $x(x + 1)$.

Key Points

◉ To add (or subtract) algebraic fractions we use a similar method to that used for adding and subtracting numerical fractions.

◉ If the denominators of the fractions are the same, add (or subtract) the numerators but do not change the denominator.

◉ To add (or subtract) algebraic fractions with different denominators, find a common denominator and write each fraction as an equivalent fraction with this denominator.

◉ To find the lowest common denominator of algebraic fractions, you may need to factorise the denominators first.

◉ To simplify your answers, you may have to factorise the numerator.

Example 3 — Add $\dfrac{2}{x} + \dfrac{3}{x}$

$\dfrac{2}{x} + \dfrac{3}{x} = \dfrac{5}{x}$ ← The denominators are the same so just add the numerators.

Example 4 — Subtract $\dfrac{5x}{7} - \dfrac{3x}{7}$

$\dfrac{5x}{7} - \dfrac{3x}{7} = \dfrac{2x}{7}$ ← Subtract the numerators.

Example 5 — Write $\dfrac{3}{2x} - \dfrac{1}{x}$ as a single fraction.

$\dfrac{3}{2x} - \dfrac{1}{x} = \dfrac{3}{2x} - \dfrac{2}{2x}$ ← Write each fraction with the same common denominator.

$= \dfrac{1}{2x}$ ← Subtract the numerators, but leave the denominator the same.

Example 6 — Simplify $\dfrac{x+2}{3} + \dfrac{x-1}{4}$

Common denominator = 12. ← Work out the lowest common denominator.

$\dfrac{x+2}{3} + \dfrac{x-1}{4} = \dfrac{4(x+2)}{12} + \dfrac{3(x-1)}{12}$ ← Write as equivalent fractions with the same denominator.

$= \dfrac{4(x+2) + 3(x-1)}{12}$ ← Add the two fractions.

$= \dfrac{4x + 8 + 3x - 3}{12}$ ← Expand the brackets.

$= \dfrac{7x + 5}{12}$ ← Simplify the numerator.

Example 7 — Write $\dfrac{3}{x-1} - \dfrac{2}{x+1}$ as a single fraction.

Common denominator = $(x-1)(x+1)$ ← Find a common denominator.

$\dfrac{3}{x-1} - \dfrac{2}{x+1} = \dfrac{3(x+1)}{(x-1)(x+1)} - \dfrac{2(x-1)}{(x-1)(x+1)}$ ← Convert each fraction to an equivalent fraction with the common denominator $(x-1)(x+1)$.

$= \dfrac{3(x+1) - 2(x-1)}{(x-1)(x+1)}$ ← Subtract the fractions.

$= \dfrac{3x + 3 - 2x + 2}{(x-1)(x+1)}$ ← Expand the brackets in the numerator.

$= \dfrac{x + 5}{(x-1)(x+1)}$ ← Simplify the numerator.

ResultsPlus
Watch Out!

There is no need to multiply out the brackets in the denominator.

Exercise 32B

1 Write as a single fraction in its simplest form.

a $\dfrac{2x}{3} + \dfrac{x}{3}$ **b** $\dfrac{x}{2} + \dfrac{3x}{2}$ **c** $\dfrac{3}{10x} + \dfrac{4}{10x}$

d $\dfrac{7x}{9} - \dfrac{3x}{9}$ **e** $\dfrac{4x}{5} - \dfrac{3x}{5}$ **f** $\dfrac{7}{3x} - \dfrac{2}{3x}$

2 Write as a single fraction in its simplest form.

a $\dfrac{x}{3} + \dfrac{x}{4}$ **b** $\dfrac{x}{5} + \dfrac{2x}{15}$ **c** $\dfrac{x}{2} - \dfrac{x}{8}$

d $\dfrac{3x}{2} - \dfrac{2x}{3}$ **e** $\dfrac{1}{3x} + \dfrac{1}{2x}$ **f** $\dfrac{4}{10x} - \dfrac{3}{20x}$

3 Simplify.

a $\dfrac{x}{2} + \dfrac{x+1}{3}$ **b** $\dfrac{x-3}{4} + \dfrac{x+2}{5}$ **c** $\dfrac{2x}{3} - \dfrac{5x}{9}$

d $\dfrac{1}{x+2} + \dfrac{1}{x+3}$ **e** $\dfrac{4}{x+2} - \dfrac{3}{x+1}$ **f** $\dfrac{1}{2x-1} - \dfrac{1}{2x+3}$

Example 8  Write $\dfrac{1}{x} - \dfrac{3}{x^2 + 3x}$ as a single fraction in its simplest form.

$\dfrac{1}{x} - \dfrac{3}{x^2 + 3x} = \dfrac{1}{x} - \dfrac{3}{x(x + 3)}$ ← Factorise the denominator $x^2 + 3x$.

Common denominator $= x(x + 3)$ ← Find the lowest common denominator.

$\dfrac{1}{x} - \dfrac{3}{x^2 + 3x} = \dfrac{x + 3}{x(x + 3)} - \dfrac{3}{x(x + 3)}$ ← Write each fraction with the same denominator.

$\qquad = \dfrac{x + 3 - 3}{x(x + 3)}$ ← Combine the fractions.

$\qquad = \dfrac{x}{x(x + 3)}$ ← Simplify the numerator.

$\qquad = \dfrac{1}{x + 3}$ ← Divide the numerator and denominator by x to simplify your answer.

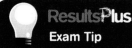

ResultsPlus
Exam Tip

$x \div x = 1$
$x(x + 3) \div x = x + 3$

Example 9 Simplify $\dfrac{1}{5x + 10} + \dfrac{1}{x^2 + 5x + 6}$

$5x + 10 = 5(x + 2)$ ← Factorise each denominator.
$x^2 + 5x + 6 = (x + 3)(x + 2)$

$\dfrac{1}{5x + 10} + \dfrac{1}{x^2 + 5x + 6} = \dfrac{1}{5(x + 2)} + \dfrac{1}{(x + 3)(x + 2)}$ ← Replace the denominators with the factorised expressions.

Common denominator $= 5(x + 3)(x + 2)$ ← Find the lowest common denominator of $5(x + 2)$ and $(x + 3)(x + 2)$.

$\dfrac{1}{5x + 10} + \dfrac{1}{x^2 + 5x + 6} = \dfrac{(x + 3)}{5(x + 3)(x + 2)} + \dfrac{5}{5(x + 3)(x + 2)}$ ← Write equivalent fractions with the common denominator.

$\qquad = \dfrac{x + 3 + 5}{5(x + 3)(x + 2)}$ ← Combine the fractions.

$\qquad = \dfrac{x + 8}{5(x + 3)(x + 2)}$

Exercise 32C

1 **a** Factorise **i** $2x + 2$ **ii** $6x + 6$.

b Write down the lowest common multiple of $2x + 2$ and $6x + 6$.

c Write $\dfrac{1}{2x + 2} + \dfrac{1}{6x + 6}$ as a single fraction in its simplest form.

2 Write down the lowest common multiple of each of the following pairs of expressions.

a $3x$ and $5x$ **b** $x + 2$ and $x + 3$

c x and $x(x - 1)$ **d** $x + 2$ and $(x + 1)(x + 2)$

e $2x - 6$ and $x - 3$ **f** $x + 1$ and $x^2 + x$

3 **a** Factorise $x^2 + 3x + 2$.

b Write $\dfrac{1}{x + 2} - \dfrac{1}{x^2 + 3x + 2}$ as a single fraction in its simplest form.

4 **a** Factorise $x^2 - 4$.

b Write $\dfrac{3}{x - 2} - \dfrac{2}{x^2 - 4}$ as a single fraction in its simplest form.

5 **a** Factorise $2x^2 - 3x + 1$.

b Write $\dfrac{1}{2x^2 - 3x + 1} + \dfrac{2}{2x - 1}$ as a single fraction in its simplest form.

6 Simplify $\dfrac{1}{2x + 6} - \dfrac{1}{x^2 + 4x + 3}$

7 Write $\dfrac{1}{4} + \dfrac{1}{2x} + \dfrac{1}{8(x + 1)}$ as a single fraction.

8 Express $\dfrac{3}{3 - x} - \dfrac{9}{9 - x^2}$ as a single fraction.

9 **a** Factorise **i** $x^2 + 9x + 20$ **ii** $x^2 + 11x + 30$.

b Write $\dfrac{4}{x^2 + 9x + 20} - \dfrac{1}{x^2 + 11x + 30}$ as a single fraction in its simplest form.

10 Show that $\dfrac{1}{4x^2 - 8x + 3} - \dfrac{1}{4x^2 - 1} = \dfrac{A}{(2x - 1)(2x + 1)(2x - 3)}$ and find the value of A.

32.3 Multiplying and dividing algebraic fractions

⊙ Objective

- You can multiply and divide algebraic fractions.

❓ Why do this?

Doctors and nurses need to multiply and divide algebraic fractions when calculating the drugs dosage to give their patients.

◆ Get Ready

1. Simplify.

a $\dfrac{ab}{b}$ **b** $\dfrac{x + 1}{2x + 2}$ **c** $\dfrac{(x + 1)(x + 3)}{(x + 1)^2}$

Key Points

- To multiply (or divide) algebraic fractions we use a similar method to that used for multiplying and dividing numerical fractions.
- To multiply fractions, multiply the numerators and multiply the denominators.
$$\frac{a}{b} \times \frac{c}{d} = \frac{ac}{bd}$$
- To divide fractions, multiply the first fraction by the reciprocal of the second.
$$\frac{a}{b} \div \frac{c}{d} = \frac{a}{b} \times \frac{d}{c} = \frac{ad}{bc}$$
- Simplify your answers if you can.
- You may need to factorise the numerator and/or the denominator before you multiply or divide algebraic fractions.

Example 10   Simplify $\frac{2x}{3} \times \frac{x}{4}$

$$\frac{2x}{3} \times \frac{x}{4} = \frac{2x \times x}{3 \times 4} = \frac{{}^{1}2x^2}{12_6}$$
    ← Multiply $2x$ by x. Work out 3×4.

$$= \frac{x^2}{6}$$
    ← Divide both the numerator and denominator by 2.

Example 11   Simplify $\frac{2x}{5y} \div \frac{x^2}{y}$

$$\frac{2x}{5y} \div \frac{x^2}{y} = \frac{2x}{5y} \times \frac{y}{x^2}$$
    ← Multiply the first fraction by the reciprocal of the second.

$$= \frac{2 \times \overset{1}{x} \times \overset{1}{y}}{5 \times \underset{1}{y} \times \underset{1}{x} \times x}$$
    ← Divide both the numerator and denominator by x and by y.

$$= \frac{2}{5x}$$

Example 12   Simplify $\frac{x + 1}{x + 2} \times \frac{(x + 2)^2}{3(x + 1)}$

$$\frac{x + 1}{x + 2} \times \frac{(x + 2)^2}{3(x + 1)} = \frac{\overset{1}{(x + 1)}\overset{1}{(x + 2)}(x + 2)}{3(x + 2)(x + 1)}$$
    ← Write down the *product of the two numerators* and the product of the two denominators.

$$= \frac{x + 2}{3}$$
    ← Simplify the fraction.

Example 13   Simplify $\frac{2x - 1}{4} \div \frac{4x - 2}{5}$

$$\frac{2x - 1}{4} \div \frac{4x - 2}{5} = \frac{2x - 1}{4} \div \frac{2(2x - 1)}{5}$$
    ← Factorise the numerator of the second fraction.

$$= \frac{2x - 1}{4} \times \frac{5}{2(2x - 1)}$$

$$= \frac{5(2x - 1)^1}{8(2x - 1)_1}$$
    ← Divide the numerator and denominator by $2x - 1$.

$$= \frac{5}{8}$$

Example 14 Simplify $\dfrac{2x + 1}{x^2 - 1} \times \dfrac{x + 1}{2x^2 - x - 1}$

$$\dfrac{2x + 1}{x^2 - 1} \times \dfrac{x + 1}{2x^2 - x - 1} = \dfrac{2x + 1}{(x - 1)(x + 1)} \times \dfrac{x + 1}{(2x + 1)(x - 1)}$$

← Factorise $x^2 - 1$ and $2x^2 - x - 1$.

$$= \dfrac{{}^1\!(2x + 1)(x + 1)^1}{(x - 1)(x + 1)_1(2x + 1)_1(x - 1)}$$

← Write down the product of the two fractions and simplify.

$$= \dfrac{1}{(x - 1)(x - 1)}$$

$$= \dfrac{1}{(x - 1)^2}$$

Write $(x - 1)(x - 1)$ as $(x - 1)^2$.

ResultsPlus

Exam Tip

Note that it is often preferable to leave the fraction in factorised form.

Exercise 32D

C

1 Write as a single fraction.

a $\dfrac{x}{3} \times \dfrac{x}{5}$ b $\dfrac{4}{y} \times \dfrac{3}{y}$ c $\dfrac{5x}{2} \times \dfrac{3y}{4}$ d $\dfrac{x}{3} \times \dfrac{x - 3}{4}$

B

2 Write as a single fraction in its simplest form.

a $\dfrac{3x}{5} \times \dfrac{10y}{12}$ b $\dfrac{4x}{9y} \times \dfrac{y}{2}$ c $\dfrac{2x^2}{y^2} \times \dfrac{3y}{x^2}$ d $\dfrac{x + 1}{x} \times \dfrac{2x}{x - 1}$

3 Write as a single fraction.

a $\dfrac{x}{9} \div \dfrac{x}{5}$ b $\dfrac{5x}{6} \div \dfrac{3}{y}$ c $x^2 y \div \dfrac{1}{y}$ d $\dfrac{2x}{x + 1} \div \dfrac{x + 1}{x + 2}$

4 Write as a single fraction in its simplest form.

a $\dfrac{3x}{2} \div \dfrac{2x}{9}$ b $\dfrac{5y^2}{8x} \div \dfrac{y^2}{x^2}$ c $\dfrac{7x}{12y} \div \dfrac{y}{6}$ d $\dfrac{x}{2} \div \dfrac{x - 5}{4}$

A

5 Write as a single fraction in its simplest form.

a $\dfrac{x + 1}{3} \times \dfrac{3x + 3}{2}$ b $\dfrac{x + 2}{x - 1} \times (x - 1)^2$ c $\dfrac{x}{x + 1} \times \dfrac{x + 1}{x + 2}$

d $\dfrac{x + 4}{9} \div \dfrac{2x + 8}{3}$ e $\dfrac{6}{3x - 1} \div \dfrac{2}{(3x - 1)^2}$ f $\dfrac{3x - 12}{4} \div \dfrac{x - 4}{x + 4}$

A☆

6 a Factorise $x^2 - 4$.

b Write $\dfrac{1}{x + 2} \times \dfrac{x^2 - 4}{x^2 + 4}$ as a single fraction in its simplest form.

7 a Factorise i $x^2 + 5x + 4$ ii $x^2 + 6x + 8$.

b Write $\dfrac{x + 3}{x^2 + 5x + 4} \div \dfrac{x + 1}{x^2 + 6x + 8}$ as a single fraction in its simplest form.

8 Write $\dfrac{x^2 - x}{x^2 + x} \times \dfrac{x^2 + 2x + 1}{x - 1}$ as a single fraction in its simplest form.

32.4 Algebraic proof

⊙ Objective	⟡ Why do this?
⊙ You can prove a given result using algebra.	Algebraic proofs are used to prove many key ideas (theorems) in life, from thermodynamics to quantum mechanics.

◈ Get Ready

1. n is an integer.

State whether each of the following must represent an even number, an odd number or either.

   a $2n$ **b** $n + 1$ **c** $2n - 1$ **d** n^2 **e** $3n$

◉ Key Points

◉ To demonstrate that a result is true, you can give details of a particular case. For example, to demonstrate that the sum of two odd numbers is even, you could choose any two odd numbers and show that their sum is even, such as $3 + 5 = 8$.

◉ To **prove** that a result is true, you must show that it will be true in all cases. For example, to prove that the sum of two odd numbers is even, you must choose two 'general' odd numbers and show that their sum will always be even. You could write

$(2m - 1) + (2n - 1) = 2m + 2n - 2 = 2(m + n - 1)$ which is an even number as it is a multiple of 2.

◉ In algebraic **proof** you will find the following points helpful.

Where n is an integer:

  ◉ Consecutive integers can be written in the form $n, n + 1, n + 2, n + 3, \ldots$ In some cases it is more useful to write them in a slightly different form, for example $n - 2, n - 1, n, n + 1, n + 2, \ldots$

  ◉ Any even number can be written in the form $2n$.

  ◉ Consecutive even numbers can be written in the form $2n, 2n + 2, 2n + 4, \ldots$

  ◉ Any odd number may be written in the form $2n - 1$ (alternatively any odd number may be written in various other forms, for example $2n + 1$).

  ◉ Consecutive odd numbers can be written in the form $2n - 1, 2n + 1, 2n + 3, \ldots$

Example 15   **a** Show that $(2n - 1)^2 + (2n + 1)^2 = 8n^2 + 2$.

       b Hence prove that the sum of the squares of any two consecutive odd numbers is even.

a $(2n - 1)^2 + (2n + 1)^2 = (2n - 1)(2n - 1) + (2n + 1)(2n + 1)$ ← Write out the expression.

$= 4n^2 - 4n + 1 + 4n^2 + 4n + 1$ ← Multiply out the brackets.

$= 8n^2 + 2$ ← Simplify the expression.

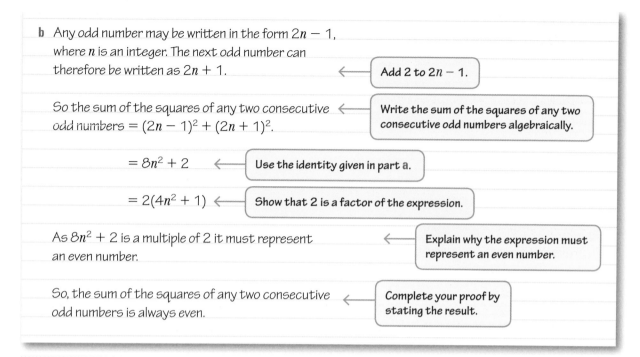

b Any odd number may be written in the form $2n - 1$, where n is an integer. The next odd number can therefore be written as $2n + 1$.

← Add 2 to $2n - 1$.

So the sum of the squares of any two consecutive odd numbers $= (2n - 1)^2 + (2n + 1)^2$.

← Write the sum of the squares of any two consecutive odd numbers algebraically.

$= 8n^2 + 2$

← Use the identity given in part **a**.

$= 2(4n^2 + 1)$

← Show that 2 is a factor of the expression.

As $8n^2 + 2$ is a multiple of 2 it must represent an even number.

← Explain why the expression must represent an even number.

So, the sum of the squares of any two consecutive odd numbers is always even.

← Complete your proof by stating the result.

Exercise 32E

A☆
A03

1 Prove that the sum of any odd number and any even number is odd.

A03

*2 Prove that half the sum of four consecutive numbers is odd.

A03

*3 Prove that the sum of any three consecutive numbers is a multiple of 3.

A03

4 **a** Prove that the product of any odd number and any even number is even.
  b Prove that the product of any two odd numbers is odd.
  c Prove that the product of any two even numbers is even.

A03

*5 Prove that for any two numbers the product of their difference and their sum is equal to the difference of their squares.

A03

*6 Prove that, if the difference of two numbers is 4, then the difference of their squares is a multiple of 8.

Chapter review

- Algebraic fractions, like numerical fractions, can often be simplified.
- To simplify an algebraic fraction, factorise the numerator and the denominator, then divide the numerator and denominator by any common factors.
- To add (or subtract) algebraic fractions we use a similar method to that used for adding and subtracting numerical fractions.
- If the denominator of the fractions are the same, add (or subtract) the numerators but do not change the denominator.
- To add (or subtract) algebraic fractions with different denominators, find a common denominator and write each fraction as an equivalent fraction with this denominator.

◉ To find the lowest common denominator of algebraic fractions, you may need to factorise the denominators first.

◉ To simplify your answers you may have to factorise the numerator.

◉ To multiply (or divide) algebraic fractions we use a similar method to that used for multiplying and dividing numerical fractions.

◉ To multiply fractions, multiply the numerators and multiply the denominators.
$$\frac{a}{b} \times \frac{c}{d} = \frac{ac}{bd}$$

◉ To divide fractions, multiply the first fraction by the reciprocal of the second.
$$\frac{a}{b} \div \frac{c}{d} = \frac{a}{b} \times \frac{d}{c} = \frac{ad}{bc}$$

◉ You may need to factorise the numerator and/or the denominator before you multiply or divide algebraic fractions.

◉ To demonstrate that a result is true, you can give details of a particular case.

◉ To **prove** that a result is true, you must show that it will be true in all cases.

◉ In algebraic **proof** you will find the following points helpful.

Where n is an integer:

◉ Consecutive integers can be written in the form $n, n + 1, n + 2, n + 3, \ldots$ In some cases it is more useful to write them in a slightly different form, for example $n - 2, n - 1, n, n + 1, n + 2, \ldots$

◉ Any even number can be written in the form $2n$.

◉ Consecutive even numbers can be written in the form $2n, 2n + 2, 2n + 4, \ldots$

◉ Any odd number may be written in the form $2n - 1$ (alternatively any odd number may be written in various other forms, for example $2n + 1$).

◉ Consecutive odd numbers can be written in the form $2n - 1, 2n + 1, 2n + 3, \ldots$

Review exercise

1 Here are the first 4 lines of a number pattern.

$1 + 2 + 3 + 4 \qquad = \qquad (4 \times 3) - (2 \times 1)$

$2 + 3 + 4 + 5 \qquad = \qquad (5 \times 4) - (3 \times 2)$

$3 + 4 + 5 + 6 \qquad = \qquad (6 \times 5) - (4 \times 3)$

$4 + 5 + 6 + 7 \qquad = \qquad (7 \times 6) - (5 \times 4)$

n is the first number in the nth line of the number pattern.

Show that the above number pattern is true for the four consecutive integers

$n, (n + 1), (n + 2)$ and $(n + 3)$

Nov 2007

A

2 Simplify fully.

a $\dfrac{4x^3}{8x}$

b $\dfrac{2(x + 1)^4}{6(x + 1)^2}$

c $\dfrac{x^2 + 5x}{x + 5}$

d $\dfrac{x^2 + 7x + 6}{x^2 + 8x + 12}$

e $\dfrac{x^2 - 2x}{x^2 + x - 6}$

f $\dfrac{x^2 - 25}{x^2 + 10x + 25}$

g $\dfrac{x^2 - 2x + 1}{x^3 - x}$

h $\dfrac{2x^2 + 7x - 4}{6x^2 + x - 2}$

A☆

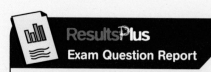

ResultsPlus

Exam Question Report

78% of students answered this sort of question poorly because they simplified at the wrong stage in the calculation.

A☆

3 Write as a single fraction in its simplest form.

a $\dfrac{3x}{10} + \dfrac{x}{5}$

b $\dfrac{3}{2x} + \dfrac{2}{3x}$

c $\dfrac{1}{5x - 3} + \dfrac{1}{5x + 3}$

d $\dfrac{3}{(x + 1)(x + 2)} + \dfrac{2}{x + 1}$

e $\dfrac{1}{(x + 1)^2} + \dfrac{1}{x + 1}$

f $\dfrac{1}{x^2 + 4x + 3} + \dfrac{1}{x^2 + 8x + 15}$

4 Simplify.

a $\dfrac{x}{2} \times \dfrac{4}{x}$

b $\left(\dfrac{x}{3}\right)^2 \times \dfrac{9}{x}$

c $\dfrac{x}{10} \div \dfrac{x}{4}$

d $\dfrac{2s}{3} \div \dfrac{1}{s}$

e $\dfrac{x - 1}{(x + 1)^2} \times \dfrac{x + 1}{x + 3}$

f $\dfrac{x^2 - x}{x^2 + x} \times \dfrac{x^2 - 1}{(x - 1)^2}$

g $\dfrac{x + 2}{5} \div \dfrac{3x + 6}{10}$

h $\dfrac{2x - 1}{x^2} \div \dfrac{2x - 1}{2x^2 + x}$

A03

5 a Show that $(n + 1)(n + 2) + n(n + 1) = 2(n + 1)^2$.

b Hence prove that for any three consecutive integers, the sum of the product of the last two and the product of the first two is always even.

A03

* **6** Prove that the difference of the squares of two consecutive odd numbers is a multiple of 8.

A03

7 Show that $25 - \dfrac{(x - 8)^2}{4} = \dfrac{(2 + x)(18 - x)}{4}$

June 2005

A03

8 The nth even number is $2n$.

a Explain why the next even number after $2n$ is $2n + 2$.

b Show algebraically that the sum of any 3 consecutive even numbers is always a multiple of 6.

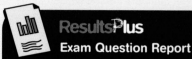

ResultsPlus
Exam Question Report

93% of students answered this question poorly because they did not follow the direction indicated in the early parts of the question.

June 2008, adapted

A03

* **9** Prove that $(3n + 1)^2 - (3n - 1)^2$ is a multiple of 4, for all positive integer values of n.

June 2009

A02
A03

10 Jim runs 16 km from home at a speed of x kph. He then runs the same distance back home at a speed 1 kph slower.

Work out an expression, in terms of x, for the total time Jim took to run from home and back.

Give your answer as a single fraction in its simplest form.

33 VECTORS

Netball players have to decide on the exact direction and power needed to pass a ball to a teammate without other players intercepting. The direction and strength of the throw can be described using vectors. Vectors are used to describe any quantity that requires both a direction and a size. Many physical problems can be explored and solved using vectors.

◎ Objectives

In this chapter you will:
- understand and use vector notation
- calculate, and represent graphically, the sum of two vectors, the difference of two vectors and a scalar multiple of a vector
- calculate the resultant of two vectors
- learn how to solve geometrical problems in two dimensions and apply vector methods for simple geometrical proofs.

⬦ Before you start

You should be able to:
- understand bearings
- plot points on a graph
- use Pythagoras' Theorem.

33.1 Vectors and vector notation

Objective

- You can understand and use vector notation.

Why do this?

You can describe journeys using vectors, for example, a trip from London to Brighton is a vector with magnitude 60 km and direction south.

Get Ready

1. The points A, B, C and D are the vertices of a parallelogram where A is (1, 0), B is (4, 0) and C is (5, 2). Draw points A, B and C on squared paper to find the coordinates of point D.

Key Points

- In mathematics, there are many quantities that need a **direction** as well as a size in order to describe them completely.

- For example, to describe a change in position or a **displacement**, it is necessary to give the direction of the movement as well as the distance moved.

- Similarly, when describing a force it is important to state the direction. This is because how an object moves when it is pushed or pulled will depend on the direction of the push or pull as well as the size or **magnitude** of the push or pull.

- Displacements and forces that need a magnitude and a direction to describe them are examples of vectors. In this chapter, only displacement vectors will be considered but the results apply to other vectors as well.

- As vectors need magnitude and direction to describe them, vectors are equal only when they have equal magnitudes and the same direction. For example:

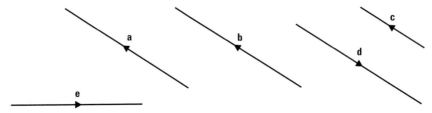

- The vectors **a** and **b** are **equal vectors**, that is **a** = **b**. They have the same magnitude and direction.

- The vectors **a** and **c** are not equal. Although they have the same direction, they do not have the same magnitude.

- The vectors **a** and **d** are not equal. Although they have the same magnitude and are parallel, they are in opposite directions and so do not have the same direction.

- The vectors **a** and **e** are not equal. They have the same magnitude but they do not have the same direction.

- The displacement from A to B is 4 cm on a bearing of 030°.

- This displacement is written $\overrightarrow{AB}$ to show that it is a vector and it has a direction from A to B.

- In the diagram the line from A to B is drawn 4 cm long in a direction of 030° and it is marked with an arrow to show that the direction is from A to B.

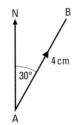

● Vectors can also be labelled with single bold letters such as **a**, **b** and **c**.
  For example, the vector **b** has been drawn on a grid.

● When hand writing the vector **a** you can use <u>a</u> to represent it.

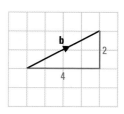

● The displacement represented by **b** can be described as 4 to the right and 2 up.

  As with translations this can be written as the column vector $\begin{pmatrix} 4 \\ 2 \end{pmatrix}$.

  So we can write **b** $= \begin{pmatrix} 4 \\ 2 \end{pmatrix}$.

Example 1

  a Point A has coordinates (1, 6) and point B has coordinates (4, 1).
     Write $\overrightarrow{AB}$ as a column vector.

  b The point C is such that $\overrightarrow{BC} = \begin{pmatrix} -2 \\ 3 \end{pmatrix}$. Find the coordinates of C.

a

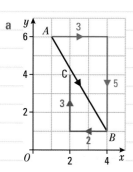

> Mark the points A and B on a grid.

> To move from A to B go 3 to the right and 5 down.

$\overrightarrow{AB} = \begin{pmatrix} 3 \\ -5 \end{pmatrix}$

b The coordinates of C are (2, 4).

> For $\overrightarrow{BC} = \begin{pmatrix} -2 \\ 3 \end{pmatrix}$, from B go 2 to the left and 3 up to find C.

Exercise 33A

Questions in this chapter are targeted at the grades indicated.

1 On squared paper draw and label the following vectors.

  a **a** $= \begin{pmatrix} 1 \\ 2 \end{pmatrix}$ **b** **b** $= \begin{pmatrix} 4 \\ -2 \end{pmatrix}$ **c** **c** $= \begin{pmatrix} -5 \\ -3 \end{pmatrix}$ **d** $\overrightarrow{AB} = \begin{pmatrix} -4 \\ 3 \end{pmatrix}$ **e** $\overrightarrow{CD} = \begin{pmatrix} 0 \\ 5 \end{pmatrix}$

2 The point A is (1, 3), the point B is (6, 9) and the point C is (5, −3).

  a Write as column vectors:

    i $\overrightarrow{AB}$ **ii** $\overrightarrow{BC}$ **iii** $\overrightarrow{AC}$

  b What do you notice about your answers in **a**?

3 The points A, B, C and D are the vertices of a quadrilateral where A has coordinates (2, 1),
$\overrightarrow{AB} = \begin{pmatrix} 2 \\ 4 \end{pmatrix}$, $\overrightarrow{BC} = \begin{pmatrix} 3 \\ 1 \end{pmatrix}$ and $\overrightarrow{CD} = \begin{pmatrix} 4 \\ -2 \end{pmatrix}$.

  a On squared paper draw quadrilateral ABCD.

  b Write as a column vector $\overrightarrow{AD}$.

  c What type of quadrilateral is ABCD?

  d What do you notice about $\overrightarrow{BC}$ and $\overrightarrow{AD}$?

A03

A

A
A03

4   The points A, B, C and D are the vertices of a parallelogram.
A has coordinates (0, 1), $\overrightarrow{AB} = \begin{pmatrix} 4 \\ 0 \end{pmatrix}$ and $\overrightarrow{AD} = \begin{pmatrix} 2 \\ 3 \end{pmatrix}$.

a   On squared paper draw the parallelogram ABCD.
b   Write as a column vector **i** $\overrightarrow{DC}$ **ii** $\overrightarrow{CB}$
c   What do you notice about **i** $\overrightarrow{AB}$ and $\overrightarrow{DC}$ **ii** $\overrightarrow{AD}$ and $\overrightarrow{CB}$?

5   Here are eight vectors.

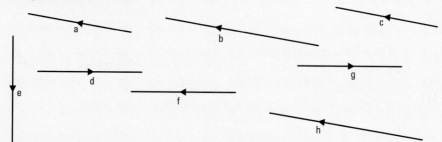

There are three pairs of equal vectors. Name the equal vectors.

33.2 The magnitude of a vector

◎ **Objective**

● You can calculate the magnitude of a vector.

◈ **Why do this?**

When designing high-speed trains, aerodynamic modellers use vectors of different magnitudes to represent air resistance and friction.

◈ **Get Ready**

1. Work out   a $\sqrt{24^2 + 7^2}$ b $\sqrt{(-8)^2 + (-5)^2}$.

◈ **Key Points**

◉ The magnitude of the vector **a** is written a or $|a|$.
◉ The magnitude of the vector $\overrightarrow{AB}$ is AB, that is, the length of the line segment AB.
◉ In general, the magnitude of the vector $\begin{pmatrix} x \\ y \end{pmatrix}$ is $\sqrt{x^2 + y^2}$.

Example 2   Find the magnitude of the vector $\mathbf{a} = \begin{pmatrix} 4 \\ -6 \end{pmatrix}$.

Give your answer: **i** as a surd and **ii** correct to 3 significant figures.

4

$\begin{pmatrix} 4 \\ -6 \end{pmatrix}$ means 4 to the right and 6 down.

a   6

Draw a right-angled triangle to show this.

$a^2 = 4^2 + 6^2 = 16 + 36$
$a^2 = 52$
i $a = \sqrt{52} = 2\sqrt{13}$
ii $a = 7.21$ (to 3 s.f.)

Use Pythagoras' Theorem to find the length, a, of the hypotenuse.

> **Example 3** Find the magnitude of the vector $\vec{AB} = \begin{pmatrix} -3 \\ -4 \end{pmatrix}$.

$AB = \sqrt{(-3)^2 + (-4)^2}$ ←   Substitute $x = -3$ and $y = -4$ into $\sqrt{x^2 + y^2}$.

$\quad = \sqrt{9 + 16}$

$AB = 5$

Exercise 33B

1 Work out the magnitude of each of these vectors. (Where necessary, answers may be left as surds.)

  a $\mathbf{a} = \begin{pmatrix} 5 \\ 12 \end{pmatrix}$ b $\mathbf{b} = \begin{pmatrix} 12 \\ -5 \end{pmatrix}$ c $\mathbf{c} = \begin{pmatrix} 1 \\ 3 \end{pmatrix}$

  d $\mathbf{d} = \begin{pmatrix} -5 \\ -7 \end{pmatrix}$ e $\vec{AB} = \begin{pmatrix} 8 \\ -15 \end{pmatrix}$ f $\vec{PQ} = \begin{pmatrix} -8 \\ 4 \end{pmatrix}$

2 In triangle ABC, $\vec{AB} = \begin{pmatrix} -20 \\ -15 \end{pmatrix}$ and $\vec{AC} = \begin{pmatrix} 24 \\ -7 \end{pmatrix}$.

  a Work out the length of the side AB of the triangle.

  b Show that the triangle is an isosceles triangle.

3 In quadrilateral ABCD, $\vec{AB} = \begin{pmatrix} 3 \\ 4 \end{pmatrix}$, $\vec{BC} = \begin{pmatrix} 5 \\ 0 \end{pmatrix}$, $\vec{CD} = \begin{pmatrix} -3 \\ -4 \end{pmatrix}$, $\vec{DA} = \begin{pmatrix} -5 \\ 0 \end{pmatrix}$.

What type of quadrilateral is ABCD?

A

A☆

A03

33.3 Addition of vectors

◎ Objectives

○ You can calculate, and represent graphically, the sum of two vectors.

○ You can calculate the resultant of two vectors.

⑦ Why do this?

When kayaking against a current you would need to allow for the strength and direction of the current in order to reach your destination.

⬙ Get Ready

1. Translate shape A by the vector $\begin{pmatrix} 2 \\ 0 \end{pmatrix}$. Label this new shape B.

Translate shape B by the vector $\begin{pmatrix} -1 \\ 2 \end{pmatrix}$. Label this new shape C.

What single translation will map shape A onto shape C?

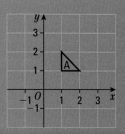

⬙ Key Points

◉ The two-stage journey from A to B and then from B to C has the same starting point and the same finishing point as the single journey from A to C.

That is, A to B followed by B to C is equivalent to A to C, or $\vec{AB}$ followed by $\vec{BC}$ is equivalent to $\vec{AC}$.

This is written as $\vec{AB} + \vec{BC} = \vec{AC}$.

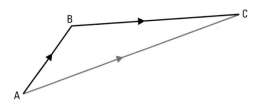

- Notice the pattern here AC + BC gives AC.
  This leads to the triangle law of vector addition.

This does not mean that AB + BC = AC.
The sum of the lengths of AB and BC is
not equal to the length of AC.

Triangle law of vector addition

- Let $\overrightarrow{AB}$ represent the vector **a** and $\overrightarrow{BC}$ represent the vector **b**.
  Then if $\overrightarrow{AC}$ represents the vector **c**,
  a + **b** = **c**.

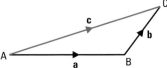

Parallelogram law of vector addition

- PQRS is a parallelogram.
  In a parallelogram, opposite sides are equal in length and are parallel.
- So since $\overrightarrow{PQ}$ and $\overrightarrow{SR}$ are also in the same direction $\overrightarrow{PQ} = \overrightarrow{SR}$ (= **a**).

  Similarly $\overrightarrow{PS} = \overrightarrow{QR}$ (= **b**).

  From the triangle law $\overrightarrow{PQ} + \overrightarrow{QR} = \overrightarrow{PR}$ so that $\overrightarrow{PR} = $ **a** + **b**.

  Hence $\overrightarrow{PR} = \overrightarrow{PQ} + \overrightarrow{PS}$ as $\overrightarrow{PQ} = $ **a** and $\overrightarrow{PS} = $ **b**.

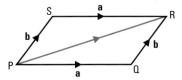

- So if in parallelogram PQRS, $\overrightarrow{PQ}$ represents the vector **a** and $\overrightarrow{PS}$ represents the vector **b**, the diagonal $\overrightarrow{PR}$ of the parallelogram represents the vector **a** + **b**.
- When **c** = **a** + **b** the vector **c** is said to be the **resultant vector** of the two vectors **a** and **b**.
- $\begin{pmatrix} a \\ b \end{pmatrix} + \begin{pmatrix} c \\ d \end{pmatrix} = \begin{pmatrix} a + c \\ b + d \end{pmatrix}$

Example 4 Find, by drawing, the sum of the vectors **a** and **b**.

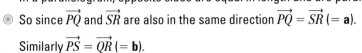

Use the triangle law of vector addition.

Move vector **b** to the end of vector **a** so that the arrows follow on.

Draw and label the vector **a** + **b** to complete the triangle.

a + **b** could also have been found by moving the vector **a** to the beginning of vector **b**. The answer is the same as the two triangles are congruent.

Example 5

In the quadrilateral ABCD, $\overrightarrow{AB} = \mathbf{a}$, $\overrightarrow{BC} = \mathbf{b}$ and $\overrightarrow{CD} = \mathbf{c}$.
Find the vectors **i** $\overrightarrow{AC}$ **ii** $\overrightarrow{AD}$.

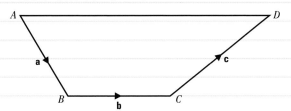

i $\overrightarrow{AC} = \overrightarrow{AB} + \overrightarrow{BC}$
so $\overrightarrow{AC} = \mathbf{a} + \mathbf{b}$

Use the triangle law of vector addition.
Make sure that the Bs follow each other.

ii $\overrightarrow{AD} = \overrightarrow{AC} + \overrightarrow{CD}$
so $\overrightarrow{AD} = (\mathbf{a} + \mathbf{b}) + \mathbf{c}$
$\overrightarrow{AD} = \mathbf{a} + \mathbf{b} + \mathbf{c}$

Use $\overrightarrow{AC} = \mathbf{a} + \mathbf{b}$.
Vector expressions like this can be treated as in ordinary algebra.
The brackets can be removed.

Example 6

$\overrightarrow{AB} = \begin{pmatrix} 3 \\ 5 \end{pmatrix}$ and $\overrightarrow{BC} = \begin{pmatrix} 8 \\ -4 \end{pmatrix}$

Find $\overrightarrow{AC}$.

$\overrightarrow{AC} = \overrightarrow{AB} + \overrightarrow{BC}$

Use the triangle law of vector addition.

$\overrightarrow{AC} = \begin{pmatrix} 3 \\ 5 \end{pmatrix} + \begin{pmatrix} 8 \\ -4 \end{pmatrix}$

Draw a sketch.

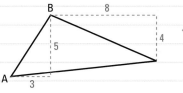

From A to B is 3 to the right.
From B to C is 8 to the right.
So from A to C is $3 + 8 = 11$ to the right.
From B to C is 4 down.
So from A to C is $5 + -4 = 1$ up.

$\overrightarrow{AC} = \begin{pmatrix} 11 \\ 1 \end{pmatrix}$

Example 7

$\mathbf{a} = \begin{pmatrix} 5 \\ -6 \end{pmatrix}$ and $\mathbf{b} = \begin{pmatrix} -3 \\ 4 \end{pmatrix}$ Find $\mathbf{a} + \mathbf{b}$.

$\mathbf{a} + \mathbf{b} = \begin{pmatrix} 5 \\ -6 \end{pmatrix} + \begin{pmatrix} -3 \\ 4 \end{pmatrix} = \begin{pmatrix} 5 + -3 \\ -6 + 4 \end{pmatrix}$

Add across.

$\mathbf{a} + \mathbf{b} = \begin{pmatrix} 2 \\ -2 \end{pmatrix}$

Exercise 33C

1 A vector **a** has magnitude 5 cm and direction 030°. A vector **b** has magnitude 7 cm and direction 140°.
Draw the vector **a** **a** **b** **b** **c** $\mathbf{a} + \mathbf{b}$.

2 Work out.

a $\begin{pmatrix} 2 \\ 6 \end{pmatrix} + \begin{pmatrix} 4 \\ 2 \end{pmatrix}$ **b** $\begin{pmatrix} 6 \\ 3 \end{pmatrix} + \begin{pmatrix} -2 \\ 5 \end{pmatrix}$ **c** $\begin{pmatrix} -5 \\ 8 \end{pmatrix} + \begin{pmatrix} 3 \\ -4 \end{pmatrix}$ **d** $\begin{pmatrix} 6 \\ 0 \end{pmatrix} + \begin{pmatrix} 3 \\ -5 \end{pmatrix}$ **e** $\begin{pmatrix} -5 \\ 3 \end{pmatrix} + \begin{pmatrix} -3 \\ -6 \end{pmatrix}$

A

A

3   $\overrightarrow{PQ} = \begin{pmatrix} 3 \\ 1 \end{pmatrix}$ $\overrightarrow{QR} = \begin{pmatrix} 7 \\ -6 \end{pmatrix}$

Work out $\overrightarrow{PR}$.

4   $\mathbf{p} = \begin{pmatrix} 3 \\ 6 \end{pmatrix}$ $\mathbf{q} = \begin{pmatrix} 1 \\ -3 \end{pmatrix}$ $\mathbf{r} = \begin{pmatrix} 4 \\ 7 \end{pmatrix}$

   a   Work out **i** $\mathbf{p} + \mathbf{q}$ **ii** $\mathbf{q} + \mathbf{p}$

   b   What do you notice?

   c   Work out **i** $(\mathbf{p} + \mathbf{q}) + \mathbf{r}$ **ii** $\mathbf{p} + (\mathbf{q} + \mathbf{r})$

   d   What do you notice?

A A02 A03

5   ABCDEF is a regular hexagon.

$\overrightarrow{AB} = \mathbf{n}$

   a   Explain why $\overrightarrow{ED} = \mathbf{n}$.

$\overrightarrow{BC} = \mathbf{m}$ $\overrightarrow{CD} = \mathbf{p}$

   b   Find **i** $\overrightarrow{AC}$ **ii** $\overrightarrow{AD}$.

   c   What is $\overrightarrow{FD}$?

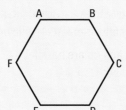

33.4 Parallel vectors

⊙ Objectives

- ○ You can calculate, and represent graphically, the difference of two vectors.
- ○ You can use the scalar multiple of a vector.

❓ Why do this?

Planes flying in formation would need to use parallel vectors.

⬙ Get Ready

1. Work out   **a** $\begin{pmatrix} 3 \\ 5 \end{pmatrix} + \begin{pmatrix} 3 \\ 5 \end{pmatrix}$ **b** $\begin{pmatrix} -2 \\ 0 \end{pmatrix} + \begin{pmatrix} -7 \\ -3 \end{pmatrix}$ **c** $\begin{pmatrix} -4 \\ 6 \end{pmatrix} + \begin{pmatrix} 4 \\ -6 \end{pmatrix}$.

⬙ Key Points

- ⊙ The ordinary rules of algebra state that $a + a = 2a$. This can also be applied to vectors. For example, here is the vector **a**.

  Here are $\mathbf{a} + \mathbf{a}$ and $2\mathbf{a}$.

- ⊙ $2\mathbf{a}$ is a vector in the same direction as **a** and with twice the magnitude.

  For $\mathbf{a} = \begin{pmatrix} 2 \\ 5 \end{pmatrix}$, $\mathbf{a} + \mathbf{a} = \begin{pmatrix} 2 \\ 5 \end{pmatrix} + \begin{pmatrix} 2 \\ 5 \end{pmatrix} = \begin{pmatrix} 2 + 5 \\ 5 + 5 \end{pmatrix} = \begin{pmatrix} 2 \times 2 \\ 2 \times 5 \end{pmatrix}$

  that is, $2\mathbf{a} = 2\begin{pmatrix} 2 \\ 5 \end{pmatrix} = \begin{pmatrix} 2 \times 2 \\ 2 \times 5 \end{pmatrix} = \begin{pmatrix} 4 \\ 10 \end{pmatrix}$.

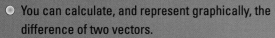

- ⊙ Similarly, $3\mathbf{a}$ is a vector in the same direction as **a** and with magnitude 3 times the magnitude of **a**.

  And $3\mathbf{a} = 3\begin{pmatrix} 2 \\ 5 \end{pmatrix} = \begin{pmatrix} 3 \times 2 \\ 3 \times 5 \end{pmatrix} = \begin{pmatrix} 6 \\ 15 \end{pmatrix}$.

- ⊙ The vector $\overrightarrow{AB}$ is the displacement from A to B, and $\overrightarrow{BA}$ is the displacement from B to A.

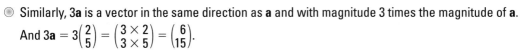

⊙ These displacements have the same magnitudes but are in opposite directions, so $\overrightarrow{AB}$ followed by $\overrightarrow{BA}$ is the zero displacement (**0**) as there is no overall change in position.
This is written $\overrightarrow{AB} + \overrightarrow{BA} = \mathbf{0}$.

⊙ Using the usual rules of algebra, it follows that $\overrightarrow{BA} = -\overrightarrow{AB}$.

⊙ A negative sign in front of a vector reverses the direction of the vector.
$$\overrightarrow{AB} = \begin{pmatrix} 3 \\ -5 \end{pmatrix} \text{ so } \overrightarrow{BA} = -\begin{pmatrix} 3 \\ -5 \end{pmatrix} = -1\begin{pmatrix} 3 \\ -5 \end{pmatrix} = \begin{pmatrix} -1 \times 3 \\ -1 \times -5 \end{pmatrix} = \begin{pmatrix} -3 \\ 5 \end{pmatrix}$$
showing that the reverse of 3 to the right and 5 down is 3 to the left and 5 up.

⊙ The vector $-\mathbf{a}$ has the same magnitude as $\mathbf{a}$ but is in the opposite direction.
The vector $-3\mathbf{a}$ has the same magnitude as $3\mathbf{a}$ but is in the opposite direction.
So the vector $-3\mathbf{a}$ has 3 times the magnitude as $\mathbf{a}$ but is in the opposite direction.
Vectors that are parallel either have the same direction or have opposite directions.

⊙ For any non-zero value of k, the vectors $\mathbf{a}$ and $k\mathbf{a}$ are parallel.
The number k is called a **scalar**; it has magnitude only.
If $\mathbf{a} = \begin{pmatrix} p \\ q \end{pmatrix}$ then $k\mathbf{a} = k\begin{pmatrix} p \\ q \end{pmatrix} = \begin{pmatrix} kp \\ kq \end{pmatrix}$

⊙ With the origin O, the vectors $\overrightarrow{OA}$ and $\overrightarrow{OB}$ are called the **position vectors** of the points A and B. In general, the point (p, q) has position vector $\begin{pmatrix} p \\ q \end{pmatrix}$.

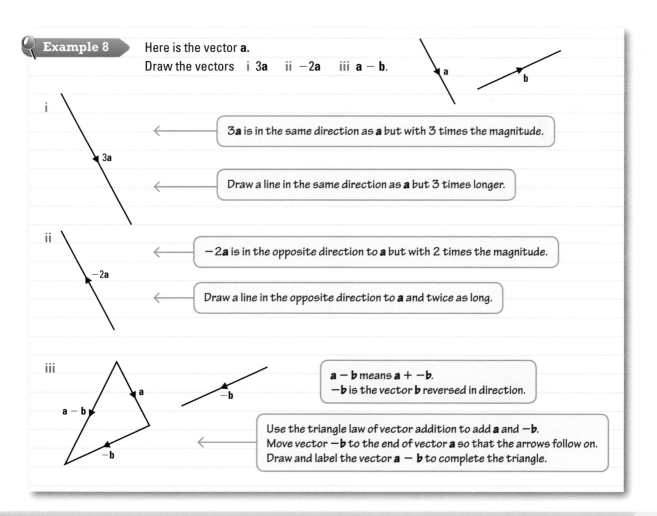

Example 8

Here is the vector **a**.
Draw the vectors   **i** 3**a**   **ii** −2**a**   **iii** **a** − **b**.

i

3**a** is in the same direction as **a** but with 3 times the magnitude.

Draw a line in the same direction as **a** but 3 times longer.

ii

−2**a** is in the opposite direction to **a** but with 2 times the magnitude.

Draw a line in the opposite direction to **a** and twice as long.

iii

a − **b** means **a** + −**b**.
−**b** is the vector **b** reversed in direction.

Use the triangle law of vector addition to add **a** and −**b**.
Move vector −**b** to the end of vector **a** so that the arrows follow on.
Draw and label the vector **a** − **b** to complete the triangle.

Example 9

With origin O, the points A, B, C and D have coordinates (1, 3), (2, 7), (-6, -10) and (-1, 10) respectively.

  a Write down as a column vector **i** $\overrightarrow{OA}$ **ii** $\overrightarrow{OB}$.
  b Work out **i** $\overrightarrow{AB}$ as a column vector   **ii** $\overrightarrow{CD}$ as a column vector.
  c What do these results show about AB and CD?

a i $\overrightarrow{OA} = \begin{pmatrix} 1 \\ 3 \end{pmatrix}$ ← From O to A is 1 across and 3 up.

ii $\overrightarrow{OB} = \begin{pmatrix} 2 \\ 7 \end{pmatrix}$ ← From O to B is 2 across and 7 up.

b i

Method 1

$\overrightarrow{AB} = \begin{pmatrix} 1 \\ 4 \end{pmatrix}$ ← A to B, that is, (1, 3) to (2, 7), is 1 across and 4 up.

Method 2

$\overrightarrow{AB} = \overrightarrow{AO} + \overrightarrow{OB}$

$\overrightarrow{AB} = -\overrightarrow{OA} + \overrightarrow{OB}$ $\overrightarrow{AO} = -\overrightarrow{OA}$ ← Another way to obtain $\overrightarrow{AB}$ is to use the triangle law of vector addition.

$\overrightarrow{AB} = -\begin{pmatrix} 1 \\ 3 \end{pmatrix} + \begin{pmatrix} 2 \\ 7 \end{pmatrix}$

$= \begin{pmatrix} -1 \\ -3 \end{pmatrix} + \begin{pmatrix} 2 \\ 7 \end{pmatrix}$

$= \begin{pmatrix} -1 + 2 \\ -3 + 7 \end{pmatrix}$

$\overrightarrow{AB} = \begin{pmatrix} 1 \\ 4 \end{pmatrix}$

ii $\overrightarrow{CD} = \begin{pmatrix} 5 \\ 20 \end{pmatrix}$ ← Using Method 1, C to D, that is, (-6, -10) to (-1, 10), is 5 to the right and 20 up.

c $\overrightarrow{CD} = \begin{pmatrix} 5 \\ 20 \end{pmatrix} = 5\begin{pmatrix} 1 \\ 4 \end{pmatrix}$

$\overrightarrow{CD} = 5\overrightarrow{AB}$

The lines CD and AB are parallel and the length of   ← **a** and k**a** are parallel vectors.
the line CD is 5 times the length of the line AB.

Example 10   Simplify **i** $3\mathbf{a} + 5\mathbf{b} + 2\mathbf{a} - 3\mathbf{b}$ **ii** $2\mathbf{a} + \frac{1}{2}(4\mathbf{a} - 2\mathbf{b})$.

i $3\mathbf{a} + 5\mathbf{b} + 2\mathbf{a} - 3\mathbf{b}$ ← $3\mathbf{a} + 2\mathbf{a} = 5\mathbf{a}$ | The ordinary rules of algebra can be applied to vector expressions like this.

$= 5\mathbf{a} + 2\mathbf{b}$ ← $5\mathbf{b} - 3\mathbf{b} = 2\mathbf{b}$

ii $2\mathbf{a} + \frac{1}{2}(4\mathbf{a} - 2\mathbf{b})$ ← $\frac{1}{2}(4\mathbf{a} - 2\mathbf{b}) = 2\mathbf{a} - \mathbf{b}$

$= 2\mathbf{a} + 2\mathbf{a} - \mathbf{b}$

$= 4\mathbf{a} - \mathbf{b}$

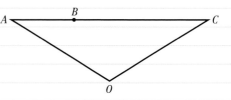

Example 11 ABC is a straight line where BC = 3AB.

$\overrightarrow{OA} = \mathbf{a}$ $\overrightarrow{AB} = \mathbf{b}$

Express $\overrightarrow{OC}$ in terms of $\mathbf{a}$ and $\mathbf{b}$.

$\overrightarrow{OC} = \overrightarrow{OA} + \overrightarrow{AC}$
$\overrightarrow{OC} = \overrightarrow{OA} + 4\overrightarrow{AB}$
$\overrightarrow{OC} = \mathbf{a} + 4\mathbf{b}$

> Use the triangle law of vector addition.

> As BC = 3AB, $\overrightarrow{AC} = 4\overrightarrow{AB}$.

Exercise 33D

1 The vector $\mathbf{a}$ has magnitude 4 cm and direction 130°.
The vector $\mathbf{b}$ has magnitude 5 cm and direction 220°.
Draw the vector   **a** $\mathbf{a}$ **b** $\mathbf{b}$ **c** $-\mathbf{b}$ **d** $\mathbf{a} - \mathbf{b}$.

2 Here is the vector $\mathbf{p}$.
Draw the vector   **a** $2\mathbf{p}$ **b** $-\frac{1}{2}\mathbf{p}$.

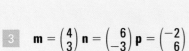

3 $\mathbf{m} = \begin{pmatrix} 4 \\ 3 \end{pmatrix}$ $\mathbf{n} = \begin{pmatrix} 6 \\ -3 \end{pmatrix}$ $\mathbf{p} = \begin{pmatrix} -2 \\ 6 \end{pmatrix}$

  a Find as a column vector. **i** $5\mathbf{m}$ **ii** $-2\mathbf{n}$ **iii** $4\mathbf{m} + 3\mathbf{p}$ **iv** $2\mathbf{m} - 4\mathbf{n} + 5\mathbf{p}$

  b Find   **i** the magnitude of the vector $\mathbf{m}$ **ii** the magnitude of the vector $2\mathbf{m} - \mathbf{p}$.

4 The points P, Q, R and S have coordinates $(-2, 5)$, $(3, 1)$, $(-6, -9)$ and $(14, -25)$ respectively.

  a Write down the position vector, $\overrightarrow{OP}$, of the point P.

  b Write down as a column vector. **i** $\overrightarrow{PQ}$ **ii** $\overrightarrow{RS}$.

  c What do these results show about the lines PQ and RS?

5 The point A has coordinates $(1, 3)$, the point B has coordinates $(4, 5)$, the point C has coordinates $(-2, -4)$. Find the coordinates of the point D where $\overrightarrow{CD} = 6\overrightarrow{AB}$.

6 $\overrightarrow{OA} = \mathbf{a}$ $\overrightarrow{OB} = \mathbf{b}$

  a Express $\overrightarrow{AB}$ in terms of $\mathbf{a}$ and $\mathbf{b}$.

  b Where is the point C such that $\overrightarrow{OC} = \frac{1}{2}\mathbf{b}$?

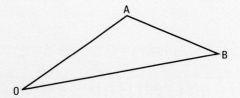

7 Here are five vectors.
$\overrightarrow{AB} = 2\mathbf{m} + 4\mathbf{n}$, $\overrightarrow{CD} = 6\mathbf{m} - 12\mathbf{n}$, $\overrightarrow{EF} = 4\mathbf{m} + 8\mathbf{n}$, $\overrightarrow{GH} = -\mathbf{m} - 2\mathbf{n}$, $\overrightarrow{IJ} = 6\mathbf{m} + 16\mathbf{n}$

  a Three of these vectors are parallel. Which are the parallel vectors?

  b Simplify. **i** $8\mathbf{p} + 5\mathbf{q} - 3\mathbf{p} - 8\mathbf{q}$ **ii** $2(2\mathbf{m} - 5\mathbf{n}) + \frac{2}{3}(3\mathbf{m} - 6\mathbf{n})$

A

A*

A02
A03

A02
A03

A02

8 Here is a regular hexagon ABCDEF. In the hexagon, FC is parallel to AB and twice as long.

$\overrightarrow{AB} = \mathbf{m}$

a Express $\overrightarrow{FC}$ in terms of $\mathbf{m}$.

$\overrightarrow{CD} = \mathbf{n}$

b Express $\overrightarrow{FD}$ in terms of $\mathbf{m}$ and $\mathbf{n}$.

$\overrightarrow{BC} = \mathbf{x}$

c Express $\overrightarrow{AC}$ in terms of $\mathbf{m}$ and $\mathbf{x}$.

The lines AC and FD are parallel and equal in length.

d Find an expression for $\mathbf{x}$ in terms of $\mathbf{m}$ and $\mathbf{n}$.

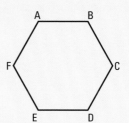

33.5 Solving geometric problems in two dimensions

Objectives

- You can solve geometric problems in two dimensions using vector methods.
- You can apply vector methods for simple geometric proofs.

Why do this?

Videogame programmers solve 2D and 3D geometric problems using vectors, in order to create virtual worlds.

Get Ready

1. The point A has coordinates (0, 4), the point B has coordinates (7, −3), and the point C has coordinates (−8, −1).
Write down as a column vector a $\overrightarrow{AB}$ b $\overrightarrow{AC}$ c $\overrightarrow{BC}$

Key Points

To solve geometric problems the following results are useful:

- Triangle law of vector addition so that $\overrightarrow{PQ} + \overrightarrow{QR} = \overrightarrow{PR}$.
- When $\overrightarrow{PQ} = \mathbf{a}$, $\overrightarrow{QP} = -\mathbf{a}$.
- When $\overrightarrow{PQ} = k\overrightarrow{RS}$, k is a scalar (number), the lines PQ and RS are parallel and the length of PQ is k times the length of RS.
- When $\overrightarrow{PQ} = k\overrightarrow{PR}$ then the lines PQ and PR are parallel. But these lines have the point P in common so that PQ and PR are part of the same straight line. That is, the points P, Q and R lie on the same straight line.

Example 14 In triangle OAB the point M is the midpoint of OA and the point N is the midpoint of OB.

$\overrightarrow{OA} = 2\mathbf{a}$ $\overrightarrow{OB} = 2\mathbf{b}$

i Express $\overrightarrow{AB}$ in terms of $\mathbf{a}$ and $\mathbf{b}$.

ii Express $\overrightarrow{MN}$ in terms of $\mathbf{a}$ and $\mathbf{b}$.

iii Explain what the answers in i and ii show about AB and MN.

Use the triangle law of vector addition.

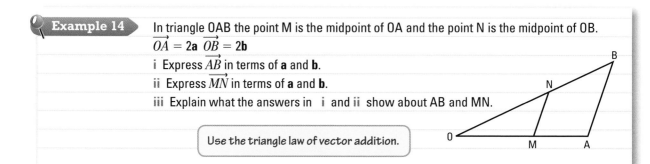

i $\overrightarrow{AB} = \overrightarrow{AO} + \overrightarrow{OB}$
$\overrightarrow{AB} = -2\mathbf{a} + 2\mathbf{b}$

$\overrightarrow{OA} = 2\mathbf{a}$ so $\overrightarrow{AO} = -2\mathbf{a}$.

ii $\overrightarrow{OM} = \frac{1}{2}2\mathbf{a} = \mathbf{a}$

M is the midpoint of OA so $\overrightarrow{OM} = \frac{1}{2}\overrightarrow{OA}$.

Similarly, $\overrightarrow{ON} = \mathbf{b}$
$\overrightarrow{MN} = \overrightarrow{MO} + \overrightarrow{ON}$
$\overrightarrow{MN} = -\mathbf{a} + \mathbf{b}$

$\overrightarrow{OM} = \mathbf{a}$ so $\overrightarrow{MO} = -\mathbf{a}$

Use the triangle law of vector addition.

iii $\overrightarrow{AB} = 2\overrightarrow{MN}$

$\overrightarrow{AB} = -2\mathbf{a} + 2\mathbf{b}$ and $\overrightarrow{MN} = -\mathbf{a} + \mathbf{b}$

This means that AB and MN are parallel and that the length of AB is twice the length of MN.

Example 15

OABC is a quadrilateral in which
$\overrightarrow{OA} = \mathbf{a}$, $\overrightarrow{OB} = \mathbf{a} + 2\mathbf{b}$ and $\overrightarrow{OC} = 4\mathbf{b}$.

i Find $\overrightarrow{AB}$ in terms of $\mathbf{a}$ and $\mathbf{b}$ and explain what this answer means.

ii Find $\overrightarrow{CB}$ in terms of $\mathbf{a}$ and $\mathbf{b}$.

D is the point such that $\overrightarrow{BD} = \overrightarrow{OC}$, and X is the midpoint of BC.

Find in terms of $\mathbf{a}$ and $\mathbf{b}$ iii $\overrightarrow{OD}$ iv $\overrightarrow{OX}$ and   v explain what these results mean.

i $\overrightarrow{AB}$ $= \overrightarrow{AO} + \overrightarrow{OB}$

Express $\overrightarrow{AB}$ in terms of known vectors using the triangle law of vector addition.

$\overrightarrow{AB}$ $= -\mathbf{a} + \mathbf{a} + 2\mathbf{b} = 2\mathbf{b}$

$\overrightarrow{OA} = \mathbf{a}$ so $\overrightarrow{OA} = -\mathbf{a}$.

$\overrightarrow{OC}$ $= 2\overrightarrow{AB}$

$\overrightarrow{OC} = 4\mathbf{b}$

OC and AB are parallel and the length of OC is twice the length of AB.

ii $\overrightarrow{CB}$ $= \overrightarrow{CO} + \overrightarrow{OB}$

Express $\overrightarrow{CB}$ in terms of known vectors using the triangle law of vector addition.

$\overrightarrow{CB}$ $= -4\mathbf{b} + \mathbf{a} + 2\mathbf{b}$
$= \mathbf{a} - 2\mathbf{b}$

$\overrightarrow{OC} = 4\mathbf{b}$ so $\overrightarrow{CO} = -4\mathbf{b}$.

$\overrightarrow{CB} = \overrightarrow{CO} + \overrightarrow{OA} + \overrightarrow{AB}$
could also have been used.

$\overrightarrow{BD} = \overrightarrow{OC}$ means that the point D is on AB extended so that BD and OC have the same length. Redraw the diagram with BD in and X the midpoint of BC.

iii $\overrightarrow{OD}$ $= \overrightarrow{OA} + \overrightarrow{AD}$

Use the triangle law of vector addition for $\overrightarrow{OD}$.

$\overrightarrow{OD}$ $= \mathbf{a} + 6\mathbf{b}$

$\overrightarrow{OA} = \mathbf{a}$, $\overrightarrow{BD} = \overrightarrow{OC} = 4\mathbf{b}$.

$\overrightarrow{AD} = \overrightarrow{AD} + \overrightarrow{BD} = 2\mathbf{b} + 4\mathbf{b} = 6\mathbf{b}$.

iv As X is the midpoint of BC

$$\overrightarrow{CX} \quad = \tfrac{1}{2}\overrightarrow{OD} \qquad \longleftarrow \boxed{\overrightarrow{CB} = \mathbf{a} - 2\mathbf{b}.}$$
$$\qquad = \tfrac{1}{2}(\mathbf{a} - 2\mathbf{b})$$

$$\overrightarrow{CX} \quad = \tfrac{1}{2}\mathbf{a} - \mathbf{b}$$
$$\overrightarrow{OX} \quad = \overrightarrow{OC} + \overrightarrow{CX} = 4\mathbf{b} + \tfrac{1}{2}\mathbf{a} - \mathbf{b} \qquad \longleftarrow \boxed{\text{Use the triangle law of vector addition for } \overrightarrow{OX}.}$$
$$\overrightarrow{OX} \quad = \tfrac{1}{2}\mathbf{a} + 3\mathbf{b}$$

$$\boxed{\overrightarrow{OD} = \mathbf{a} + 6\mathbf{b}}$$
$$\mathbf{v} \; \overrightarrow{OD} = 2\overrightarrow{OX} \qquad \longleftarrow$$
$$\boxed{\overrightarrow{OX} = \tfrac{1}{2}\mathbf{a} + 3\mathbf{b}}$$

So the lines OD and OX are parallel with the point O in common.

This means that OX and OD are part of the same straight line.

That is, OXD is a straight line such that the length of OD is 2 times the length of OX.

In other words, X is the midpoint of OD.

Exercise 33E

A

1 The points A, B and C have coordinates (2, 13), (5, 22) and (11, 40) respectively.

   a Find as column vectors   **i** $\overrightarrow{AB}$ **ii** $\overrightarrow{AC}$.

   b What do these results show about the points A, B and C?

A⭐ A02 A03

2 In triangle OAB, $\overrightarrow{OA} = \mathbf{a}$ and $\overrightarrow{OB} = \mathbf{b}$.

   a Find in terms of **a** and **b** the vector $\overrightarrow{AB}$.

   P is the midpoint of AB.

   b Find in terms of **a** and **b** the vector $\overrightarrow{AP}$.

   c Find in terms of **a** and **b** the vector $\overrightarrow{OP}$.

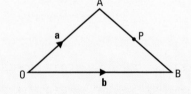

A02 A03

3 OACB is a parallelogram with $\overrightarrow{OA} = \mathbf{a}$ and $\overrightarrow{OB} = \mathbf{b}$.

   P is the midpoint of AB.

   a Use the result of question **3** to write down $\overrightarrow{OP}$ in terms of **a** and **b**.

   b Express $\overrightarrow{OC}$ in terms of **a** and **b**.

   Q is the midpoint of OC.

   c Express $\overrightarrow{QO}$ in terms of **a** and **b**.

   d What do your answers to **a** and **c** show about the points P and Q?

   e What property of a parallelogram has been proved in this question?

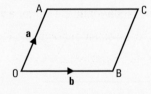

A02 A03

4 KLMN is a quadrilateral where $\overrightarrow{KL} = \mathbf{k}$, $\overrightarrow{LM} = \mathbf{m}$, $\overrightarrow{MN} = \mathbf{n}$ and $\overrightarrow{KN} = 3\mathbf{m}$.

   a What type of quadrilateral is KLMN?

   b Express **n** in terms of **k** and **m**.

5 OACB is a parallelogram with $\overrightarrow{OA} = \mathbf{a}$ and $\overrightarrow{OB} = \mathbf{b}$.

E is the point on AC such that $AE = \frac{1}{4}AC$.

F is the point on BC such that $BF = \frac{1}{4}BC$.

a Find in terms of $\mathbf{a}$ and $\mathbf{b}$.

   i $\overrightarrow{AB}$ ii $\overrightarrow{AE}$ iii $\overrightarrow{OE}$ iv $\overrightarrow{OF}$ v $\overrightarrow{EF}$

b Write down two geometric properties connecting EF and AB.

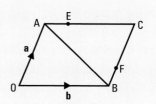

6 In triangle OMN, $\overrightarrow{OM} = \mathbf{m}$ and $\overrightarrow{ON} = \mathbf{n}$.

The point P is the midpoint of MN and Q is the point such that $\overrightarrow{OQ} = \frac{3}{2}\overrightarrow{OP}$.

a Find in terms of $\mathbf{m}$ and $\mathbf{n}$.   i $\overrightarrow{OP}$ ii $\overrightarrow{OQ}$ iii $\overrightarrow{MQ}$

The point R is such that $\overrightarrow{OR} = 3\overrightarrow{ON}$.

b Find in terms of $\mathbf{m}$ and $\mathbf{n}$ the vector $\overrightarrow{MR}$.

c Explain why MQR is a straight line and give the value of $\dfrac{MR}{MQ}$.

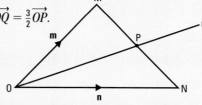

7 In the diagram $\overrightarrow{OR} = 6\mathbf{a}$, $\overrightarrow{OP} = 2\mathbf{b}$ and $\overrightarrow{PQ} = 3\mathbf{a}$.

The point M is on PQ such that $\overrightarrow{PM} = 2\mathbf{a}$.

The point N is on OR such that $\overrightarrow{ON} = \frac{1}{3}\overrightarrow{OR}$.

The midpoint of MN is the point S.

a Find in terms of $\mathbf{a}$ and/or $\mathbf{b}$ the vector $\overrightarrow{NM}$.

b Find in terms of $\mathbf{a}$ and/or $\mathbf{b}$ the vector $\overrightarrow{OS}$.

T is the point such that $\overrightarrow{QT} = \mathbf{a}$.

c Find in terms of $\mathbf{a}$ and $\mathbf{b}$ the vector $\overrightarrow{OT}$.

d Give a geometric fact about the point S and the line OT.

e When $\mathbf{a} = \begin{pmatrix} 8 \\ 2 \end{pmatrix}$ and $\mathbf{b} = \begin{pmatrix} 3 \\ 15 \end{pmatrix}$ find the length of QR.

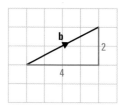

Chapter review

◉ A **vector** needs both a **magnitude** and a **direction** to describe it completely.

◉ Vectors are equal only when they have equal magnitudes and the same direction.

◉ Vectors can be labelled with single bold letters such as $\mathbf{a}$, $\mathbf{b}$ and $\mathbf{c}$.

◉ When hand-writing the vector $\mathbf{a}$ you can use $\underline{a}$ to represent it.

◉ The displacement represented by $\mathbf{b}$ can be described as 4 to the right and 2 up.

   This can be written as the column vector $\begin{pmatrix} 4 \\ 2 \end{pmatrix}$.

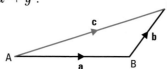

◉ The magnitude of the vector $\mathbf{a}$ is written a or $|a|$.

◉ The magnitude of the vector $\overrightarrow{AB}$ is AB, that is, the length of the line segment AB.

◉ In general, the magnitude of the vector $\begin{pmatrix} x \\ y \end{pmatrix}$ is $\sqrt{x^2 + y^2}$.

◉ $\overrightarrow{AB} + \overrightarrow{BC} = \overrightarrow{AC}$, or $\mathbf{a} + \mathbf{b} = \mathbf{c}$.

   This is the triangle law of vector addition.

⊚ PQRS is a parallelogram.

$\vec{PQ} = \vec{SR} = \mathbf{a}$, $\vec{PS} = \vec{QR} = \mathbf{b}$.

From the triangle law, $\vec{PQ} + \vec{QR} = \vec{PR}$ so that $\vec{PR} = \mathbf{a} + \mathbf{b}$.

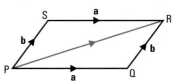

⊚ So the diagonal $\vec{PR}$ of the parallelogram represents the vector $\mathbf{a} + \mathbf{b}$.

This is the parallelogram law of vector addition.

⊚ When $\mathbf{c} = \mathbf{a} + \mathbf{b}$ the vector $\mathbf{c}$ is said to be the **resultant** of the two vectors $\mathbf{a}$ and $\mathbf{b}$.

⊚ $\begin{pmatrix} a \\ b \end{pmatrix} + \begin{pmatrix} c \\ d \end{pmatrix} = \begin{pmatrix} a + c \\ b + d \end{pmatrix}$

⊚ $\vec{AB} + \vec{BA} = \mathbf{0}$ (the zero displacement).

⊚ $\vec{BA} = -\vec{AB}$.

⊚ A negative sign in front of a vector reverses the direction of the vector.

⊚ For any non-zero value of k, the vectors $\mathbf{a}$ and $k\mathbf{a}$ are parallel.

The number k is called a **scalar**; it has magnitude only.

If $\mathbf{a} = \begin{pmatrix} p \\ q \end{pmatrix}$ then $k\mathbf{a} = k\begin{pmatrix} p \\ q \end{pmatrix} = \begin{pmatrix} kp \\ kq \end{pmatrix}$

⊚ With the origin O, the vectors $\vec{OA}$ and $\vec{OB}$ are called the **position vectors** of the points A and B.

In general, the point (p, q) has position vector $\begin{pmatrix} p \\ q \end{pmatrix}$.

⊚ To solve geometric problems the following results are useful:

⊚ Triangle law of vector addition, so that $\vec{PQ} + \vec{QR} = \vec{PR}$.

⊚ When $\vec{PQ} = \mathbf{a}$, $\vec{QP} = -\mathbf{a}$.

⊚ When $\vec{PQ} = k\vec{RS}$, k is a scalar (number), the lines PQ and RS are parallel and the length of PQ is k times the length of RS.

⊚ When $\vec{PQ} = k\vec{PR}$ then the lines PQ and PR are parallel. But these lines have the point P in common so that PQ and PR are part of the same straight line. That is, the points P, Q and R lie on the same straight line.

Review exercise

A

1 The diagram shows two vectors $\mathbf{a}$ and $\mathbf{b}$.

$\vec{PQ} = \mathbf{a} + 2\mathbf{b}$

Use the resource sheet to draw the vector $\vec{PQ}$ on the grid.

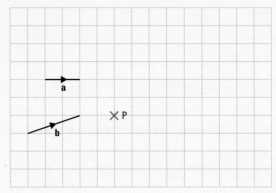

March 2005

A☆
A02
A03

2 **a** A is the point $(1, 3)$. $\vec{AB} = \begin{pmatrix} 3 \\ 2 \end{pmatrix}$

Find the coordinates of B.

b C is the point $(4, 3)$. BD is a diagonal of the parallelogram ABCD.

Express $\vec{BD}$ as a column vector.

c $\vec{CE} = \begin{pmatrix} 1 \\ -3 \end{pmatrix}$

Calculate the length of AE.

3 OAB is a triangle.

$\overrightarrow{OA} = \mathbf{a}$ $\overrightarrow{OB} = \mathbf{b}$

a Find the vector $\overrightarrow{AB}$ in terms of **a** and **b**.

P is the point on AB such that AP : PB = 3 : 2.

b Show that $\overrightarrow{OP} = \frac{1}{5}(2\mathbf{a} + 3\mathbf{b})$.

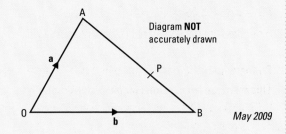

Diagram **NOT** accurately drawn

May 2009

4 $\overrightarrow{OX} = 2\mathbf{a} + \mathbf{b}$ $\overrightarrow{OY} = 4\mathbf{a} + 3\mathbf{b}$

a Express the vector $\overrightarrow{XY}$ in terms of **a** and **b**.

Give your answer in its simplest form.

XYZ is a straight line.

XY : YZ = 2 : 3

b Express the vector $\overrightarrow{OZ}$ in terms of **a** and **b**.

Give your answer in its simplest form.

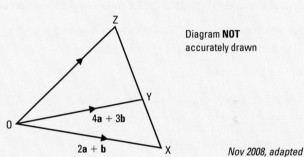

Diagram **NOT** accurately drawn

Diagram **NOT** accurately drawn

Nov 2008, adapted

5 a A is the point (1, 4) and B is the point (−3, 1).

   i Write $\overrightarrow{AB}$ as a column vector.

   ii Find the length of the vector $\overrightarrow{AB}$.

   b D is the point such that $\overrightarrow{BD}$ is parallel to $\begin{pmatrix} 0 \\ 1 \end{pmatrix}$ and the length of $\overrightarrow{AD}$ = the length of $\overrightarrow{AB}$.

   O is the point (0, 0).

   Find $\overrightarrow{OD}$ as a column vector.

   c C is the point such that ABCD is a rhombus.

   AC is a diagonal of the rhombus.

   Find the coordinates of C.

6 OABC is a parallelogram.

P is the point on AC such that AP = $\frac{2}{3}$AC.

$\overrightarrow{OA} = 6\mathbf{a}$ $\overrightarrow{OC} = 6\mathbf{c}$

a Find the vector $\overrightarrow{OP}$.

   Give your answer in terms of **a** and **c**.

The midpoint of CB is M.

b Prove that OPM is a straight line.

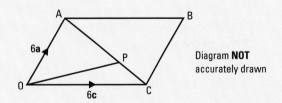

Diagram **NOT** accurately drawn

June 2004

7 CDEF is a quadrilateral with

$\overrightarrow{CD} = \mathbf{a}, \overrightarrow{DE} = \mathbf{b}$ and $\overrightarrow{FC} = \mathbf{a} - \mathbf{b}$.

a Express $\overrightarrow{CE}$ in terms of **a** and **b**.

b Prove that FE is parallel to CD.

M is the midpoint of DE.

c Express $\overrightarrow{FM}$ in terms of **a** and **b**.

X is the point on FM such that FX : XM = 4 : 1.

d Prove that C, X and E lie on the same straight line.

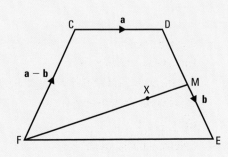

A02
A03
A★

A02
A03

A02
A03

A02
A03

A02
A03

8 PQRS is a kite.

The diagonals PR and QS intersect at M.

$\overrightarrow{PM} = 4\mathbf{p}$ $\overrightarrow{QM} = \mathbf{q}$

$\overrightarrow{MR} = \mathbf{p}$ $\overrightarrow{QM} = \overrightarrow{MS}$

a Find expressions in terms of **p** and/or **q** for

i $\overrightarrow{PR}$

ii $\overrightarrow{QS}$

iii $\overrightarrow{PQ}$.

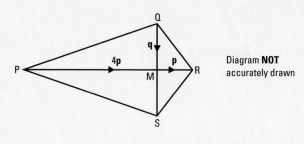

Diagram **NOT** accurately drawn

SR and PQ are extended to meet at point T.

Q is the midpoint of PT.

b Find $\overrightarrow{RT}$ in terms of **p** and **q**.

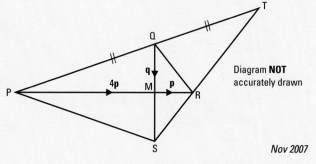

Diagram **NOT** accurately drawn

Nov 2007

9 OPQ is a triangle.

R is the midpoint of OP.

S is the midpoint of PQ.

$\overrightarrow{OP} = \mathbf{p}$ $\overrightarrow{OQ} = \mathbf{q}$

a Find $\overrightarrow{OS}$ in terms of **p** and **q**.

b Show that RS is parallel to OQ.

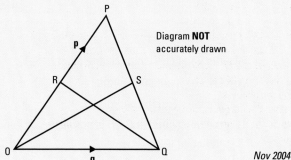

Diagram **NOT** accurately drawn

Nov 2004

10 OPQR is a trapezium with PQ parallel to OR.

$\overrightarrow{OP} = 2\mathbf{b}$ $\overrightarrow{PQ} = 2\mathbf{a}$ $\overrightarrow{OR} = 6\mathbf{a}$

M is the midpoint of PQ and N is the midpoint of OR.

a Find the vector $\overrightarrow{MN}$ in terms of **a** and **b**.

b X is the midpoint of MN and Y is the midpoint of QR.

Prove that XY is parallel to OR.

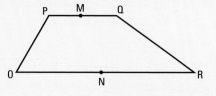

June 2005

11 ABCDEF is a regular hexagon.

$\overrightarrow{AB} = \mathbf{a}$ $\overrightarrow{BC} = \mathbf{b}$ $\overrightarrow{AD} = 2\mathbf{b}$

a Find the vector $\overrightarrow{AC}$ in terms of **a** and **b**.

$\overrightarrow{AC} = \overrightarrow{CX}$

b Prove that AB is parallel to DX.

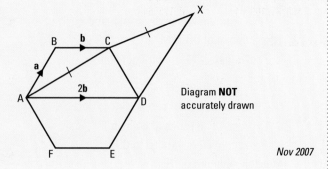

Diagram **NOT** accurately drawn

Nov 2007

Answers

Chapter 1 Answers

1.1 Get Ready

1 a Neither   b Factor   c Factor   d Multiple
   e Neither   f Factor
2 a No   b Yes   c No   d Yes   e No

Exercise 1A

1 Yes, for example $2 + 3 = 5$ is prime.
2 4
3 $n = 2, m = 3, p = 7$
4 a 2, 24   b 5, 10   c 2, 20   d 6, 18
5 a $24 = 2^3 \times 3, 60 = 2^2 \times 3 \times 5$
   b 12   c 120
6 a $72 = 2^3 \times 3^2, 120 = 2^3 \times 3 \times 5$
   b 24   c 360
7 a 18, 180   b 18, 216   c 12, 480   d 36, 720
8 a 6   b 2520
9 a 40   b 126 000
10 a 11, 13, 17 and 19 are prime numbers between 10 and 20.
   b 23, 29, 31 and 37 are prime numbers between 20 and 40.
   c 37, 41, 43, 47, 53, 59, 61 and 67 are prime numbers between 34 and 68.
11 Every 2 minutes
12 3 boxes of burgers and 4 packets of buns
13 No. If one of the prime numbers is 2 you will get an even number.

1.2 Get Ready

1 a 36   b 8   c 9

Exercise 1B

1 a 1, 4, 9, 16, 25, 36, 49, 64, 81, 100, 121, 144, 169, 196, 225
   b 1, 8, 27, 64, 125
2 a i 64, 1, 49, 9   ii 64, 1, 8
   b i 4, 16   ii 125, 27
   c i 64, 81, 144   ii 125, 64
   d i 100, 81, 169, 64   ii 125, 64

Exercise 1C

1 a 9   b 49   c 64
   d 1000   e 121
2 a 6   b 4   c 9
   d 1   e 8
3 a 36   b −8   c 81
   d −1   e 144
4 a 2   b −3   c −1
   d 4   e 10
5 a 17   b 50   c 250
   d 14   e 1   f 8
   g 34   h −11   i 9
   j 50   k 10   l 6

1.3 Get Ready

1 a 18   b 16   c 10

Exercise 1D

1 a 25   b 13   c 6   d 4
   e −5   f 8   g 6   h 2
   i 1   j 5   k 14   l 52
   m 32   n 15   o 8   p 70
2 a 49   b 25   c 243   d 123
   e 72   f 69   g 7   h −7
3 a 25   b 4   c 8   d 31

1.4 Get Ready

a 20 000   b 100   c 49

Exercise 1E

1 a 31   b 6.4   c 17   d 1.5
   e 32
2 a 39.36   b 32.65   c 5.76   d 155.125
3 a 219.5   b 305.7   c 22.6   d 410.9
4 a 5.17   b 5.34   c 3.16   d 1.67
5 a 2.77   b 7.68   c 205   d 455 000
6 a 0.917   b 1.08   c 8.67   d 15.8

Exercise 1F

1 a 0.25   b 1.6   c 0.156 25   d $2^3 = 8$

1.5 Get Ready

1 32
2 125
3 729

Exercise 1G

1 a 6^{12} b 4^5 c 7^6
   d 5^6 e 3^{10}
2 a 100 000   b 125   c 64
   d 9   e 64
3 a 5   b 3   c 5
   d 4   e 9
4 a 3^4 b 5^9 c 2^6
   d 6^5 e 4^2
5 a 9   b 16   c 16
   d 10 000   e 49
6 a 3   b 5   c 2
   d 4   e 3

Review exercise

1 a 96   b 33   c 23
2 a −4   b 4   c −2   d 4
   e −2
3 a −2   b 2   c −16

4 a 100 cans   **b** 14 (with 4 cans spare)
  c 200 cans   **d** 14 (with 8 cans spare)
5 a 5 arrangements: 1 by 36, 2 by 18, 3 by 12, 4 by 9, 6 by 6
  b 3 arrangements: 1 by 18, 2 by 9, 3 by 6
  c 3 arrangements: 1 by 12, 2 by 6, 3 by 4
6 64 is the next number which is both a square number and a cube number.
7 a 5 760 000   **b** 5 760 000   **c** 160
  d 57 600   **e** 2304
8 a 3   **b** 30   **c** 0
9 a 3^2 **b** 4   **c** 2^{12} **d** 5^2
10 a $2^2 \times 3^2 \times 7$
   b $2^3 \times 3^3 \times 7$
11 9, 15
12 $84 = 2^2 \times 3 \times 7$
   $168^2 = 2^6 \times 3^2 \times 7^2$
13 120 000 miles
14 1.258 048 316
15 3
16 a False, e.g. $3 + 5 = 8$
   b False, e.g. $4 + 9 = 13$
   c False, e.g. $5 - 3 = 2$
   d False, e.g. $2 \times 3 = 6$
   e True
17 a $2^7 = 128$ **b** 2020

Chapter 2 Answers

2.1 Get Ready

1 $4a$ **2** $8c$ **3** $2p^2$

Exercise 2A

1 a $7x + 4y$ **b** $10w + 2z$ **c** $4p + 5q$
  d $3a + b$ **e** $6c - 2d$ **f** $2m - 3n$
  g $4e - 7f$ **h** $2x + 10y + 2$ **i** $-2p + 3q - 5$
  j $13 - 5b - 4a$
2 $5x - 9$
3 $5x + 13y$

2.2 Get Ready

1 $4x + 2y + 12$ **2** $6y + 2x - 4$ **3** $8x + 2y + 6$

Exercise 2B

1 a 1   **b** 5   **c** 16   **d** 15
2 a -4 **b** -19 **c** 5   **d** 11
  e 17   **f** 11

2.3 Get Ready

1 4^{11} **2** 7^7 **3** 6^6

Exercise 2C

1 a m^5 **b** $6p^2$ **c** $20q^3$
2 a a^{11} **b** n^4 **c** x^6 **d** y^9
3 a $12p^6$ **b** $12a^5$ **c** $5b^9$ **d** $18n^3$
4 a $20t^8u^5$ **b** $6x^6y^7$ **c** $7a^5b^6$ **d** $8c^2d^9$
  e $24m^6n^5$

Exercise 2D

1 a a^3 **b** b^4 **c** c^3 **d** d
2 a $2q^2$ **b** $3p^5$ **c** $4x$ **d** $10y^7$
3 a $5a^2b^4$ **b** $5pq^3$ **c** $4c^2d^4$ **d** $3x^5$
  e $10m^2n$

Exercise 2E

1 a a^{14} **b** b^{15} **c** c^9 **d** d^{16}
2 a $4p^6$ **b** $81q^8$ **c** $25x^8$ **d** $\dfrac{m^{12}}{8}$
3 a $16x^{12}y^8$ **b** $49e^{10}f^6$ **c** $125p^{15}q^3$ **d** $\dfrac{8x^9}{27y^6}$

2.4 Get Ready

1 a^{18} **2** $27y^{15}$ **3** $\dfrac{4a^2}{b^6}$

Exercise 2F

1 a $\dfrac{1}{a}$ **b** $\dfrac{1}{b^2}$ **c** $\dfrac{1}{c^2}$ **d** $\dfrac{1}{d^3}$
2 a $\dfrac{1}{e^6}$ **b** $\dfrac{1}{f^8}$ **c** x^2 **d** y
3 a 1   **b** 1   **c** $\dfrac{1}{5p^2q^4}$
  d $\dfrac{1}{27c^9d^3}$ **e** $\dfrac{9r^4}{4p^6q^2}$

Exercise 2G

1 a $3a^2$ **b** $2c^{\frac{1}{2}}$ **c** $\dfrac{3e}{f^3}$ **d** $10x^{\frac{3}{2}}y^{\frac{5}{2}}$
2 a $\dfrac{1}{a^2}$ **b** $\dfrac{1}{2c}$ **c** $\dfrac{1}{2x^{\frac{9}{5}}y}$ **d** $\dfrac{1}{x^{\frac{1}{2}}y^{\frac{3}{2}}}$

2.5 Get Ready

1 12, 14, 16   **2** 34, 39, 44,   **3** 11, 13, 15

Exercise 2H

1 a add 3   **b** 14, 17   **c** 29
2 a add 6   **b** 20, 26   **c** 50
3 a subtract 7   **b** $-9, -16$ **c** -44
4 a add 1 to the difference of consecutive terms
  b 15, 21   **c** 55
5 a add 2 to the difference of consecutive terms
  b 20, 30   **c** 90

2.6 Get Ready

1 a add 3   **b** 13, 16   **c** 28
2 a add 3   **b** 11, 14   **c** 23
3 a subtract 6   **b** 94, 88   **c** 70

Exercise 2I

1 a i 2   **ii** -2
  b i 4   **ii** -11
  c i -5 **ii** 19
2 a $6n - 5$ **b i** 67   **ii** 295
3 a $4n + 3$ **b i** 63   **ii** 403
4 a $37 - 5n$ **b i** -63 **ii** -963
5 $7n + 11 = 103$ has no integer solution
6 a $4n + 3$ **b** $4n + 3 = 453$ has no integer solution

Review exercise

1 **a** $5x - 5y$ **b** $6m - 10n$
2 **a** $240B + 114A$, $B =$ British stamp, $A =$ Australian stamp
   b $375B + 212A$
3 **a** 3   **b** 8   **c** 7   **d** -30 **e** 62
4 **a** y^3 **b** $3x^2$ **c** z^8 **d** p^7 **e** $16a^7$
5 **a** a^3 **b** b^5 **c** $7p^3$ **d** $8x^3$ **e** $8a$
6 **a** subtract 3   **b** 87, 84   **c** 69
7 **a** 5   **b** -8
8 **a** $216 - 12n$ **b** **i** 60   **ii** -972
9 The nth term in the sequence is $3n + 2$. If $3n + 2 = 140$, $n = 46$. So 140 is the 46th term in the sequence.
10 **a** 55 cans
   b Students' proofs
   c 19 high (with 10 cans spare)
11 **a** $5\frac{1}{2}$ hours   **b** 3.03 hours = 3 hours 2 mins
   c Naismith's formula is for fit experienced walkers.
12 **a** a^{20} **b** $9b^8$ **c** $27e^{15}f^3$
13 Three consecutive even numbers are $2n, 2n + 2, 2n + 4$. Their sum $= 2n + 2n + 2 + 2n + 4 = 6n + 6$, which is always a multiple of 6.
14 $\frac{6x^2y}{4y^3} = \frac{3x^2}{2y^2}$. Squared numbers cannot be negative.
15 **a** $\frac{3p^2}{2y}$ **b** $\frac{1}{4q^{\frac{3}{2}}}$ **c** $\frac{2y}{x^2}$
16 64 cubes = 8 have 0 sides painted
   24 have 1 side painted
   24 have 2 sides painted
   8 have 3 sides painted

Sides of cube	0	1	2	3
n by n by n	$(n-2)^3$	$6(n-2)^2$	$12(n-2)$	8

Chapter 3 Answers

3.1 Get Ready

1 $\frac{8}{9}$ 2 $\frac{19}{8}$ 3 $9\frac{2}{5}$

Exercise 3A

1 **a** $\frac{8}{11}$ **b** $\frac{5}{9}$ **c** $\frac{11}{15}$ **d** $\frac{2}{5}$
2 **a** $\frac{7}{10}$ **b** $\frac{10}{21}$ **c** $\frac{29}{35}$ **d** $\frac{8}{9}$
   e $\frac{13}{20}$ **f** $\frac{13}{18}$ **g** $\frac{7}{18}$ **h** $\frac{3}{4}$
3 **a** $\frac{1}{4}$ **b** $\frac{1}{12}$ **c** $\frac{19}{40}$ **d** $\frac{2}{9}$
   e $\frac{1}{4}$ **f** $\frac{5}{12}$ **g** $\frac{1}{10}$ **h** $\frac{1}{18}$
4 **a** $1\frac{11}{40}$ **b** $1\frac{11}{20}$ **c** $\frac{1}{6}$ **d** $\frac{19}{20}$
   e $1\frac{11}{15}$ **f** $1\frac{5}{8}$ **g** $1\frac{1}{5}$ **h** $1\frac{5}{18}$

Exercise 3B

1 **a** $8\frac{1}{4}$ **b** $5\frac{3}{10}$ **c** $11\frac{5}{42}$ **d** $18\frac{3}{20}$
2 $7\frac{1}{12}$ miles   3 $1\frac{22}{35}$ lb

Exercise 3C

1 **a** $1\frac{1}{4}$ **b** $2\frac{3}{8}$ **c** $\frac{3}{4}$ **d** $3\frac{1}{3}$
2 **a** $\frac{3}{4}$ **b** $1\frac{7}{12}$ **c** $2\frac{24}{35}$ **d** $3\frac{4}{9}$

3 $3\frac{3}{8}$ kg
4 $2\frac{7}{8}$ pints

3.2 Get Ready

1 32   2 45   3 $\frac{19}{5}$ 4 $4\frac{2}{3}$

Exercise 3D

1 **a** $\frac{3}{10}$ **b** $\frac{3}{20}$ **c** $\frac{6}{11}$ **d** $\frac{2}{9}$
   e $\frac{4}{21}$ **f** $\frac{9}{20}$ **g** $\frac{3}{10}$ **h** $\frac{15}{32}$
2 **a** $\frac{2}{3}$ **b** $\frac{3}{4}$ **c** $3\frac{3}{5}$ **d** 15
3 **a** 21 kg   **b** $6\frac{2}{3}$ m   **c** $7\frac{1}{2}$ litres   **d** $7\frac{1}{2}$ pints
4 21
5 £32.65
6 **a** $\frac{5}{12}$ **b** $\frac{4}{5}$ **c** $4\frac{1}{8}$ **d** 7
   e 3   **f** $4\frac{3}{8}$ **g** 10   **h** $22\frac{1}{2}$
7 $14\frac{5}{8}$ minutes
8 $20\frac{5}{8}$ lb

3.3 Get Ready

1 $\frac{6}{35}$ 2 $\frac{1}{6}$ 3 $\frac{24}{7}$

Exercise 3E

1 **a** $\frac{5}{12}$ **b** $\frac{3}{16}$ **c** $2\frac{2}{3}$ **d** $1\frac{1}{2}$
   e $\frac{3}{4}$ **f** $1\frac{1}{5}$ **g** $1\frac{11}{14}$ **h** $\frac{5}{6}$
2 **a** $\frac{1}{2}$ **b** 28   **c** $2\frac{1}{12}$ **d** $2\frac{4}{13}$
   e 2   **f** $3\frac{2}{3}$ **g** 6   **h** $1\frac{7}{8}$
3 16
4 36
5 $11\frac{2}{11}$ days

3.4 Get Ready

1 $\frac{7}{8}$ 2 $\frac{4}{45}$ 3 $3\frac{5}{8}$

Exercise 3F

1 **a** $\frac{5}{6}$ **b** $\frac{1}{6}$
2 880 m^3
3 £5.60
4 £78
5 525
6 $\frac{3}{8} \times 36 = 13\frac{1}{2}$ which is not a whole number
7 $4\frac{1}{12}$ hours
8 $\frac{1}{2}$
9 $\frac{7}{8}$ km
10 192
11 £8.40

Review exercise

1 **a** $\frac{2}{3}$ **b** $\frac{4}{5}$ **c** $\frac{2}{3}$ **d** $\frac{2}{3}$ **e** $\frac{4}{11}$
2 **a** $\frac{17}{5}$ **b** $5\frac{1}{6}$

3

Name	Hourly rate	Hourly rate at time and a half	Hourly rate at double time
Aaron	£8.50	£12.75	£17.00
Chi	£12.00	£18.00	£24.00
Mahmood	£14.40	£21.60	£28.80

4   **a** $\frac{13}{20}$ **b** $1\frac{7}{12}$ **c** $5\frac{7}{12}$ **d** $84\frac{13}{24}$

5   **a** £374   **b**   6 hours

6   **a** $\frac{1}{12}$ **b** $3\frac{13}{30}$ **c** $1\frac{11}{12}$

7   **a** $4\frac{13}{18}\,\text{m}^2$ **b** 9 m   **c** $1\frac{1}{6}$ m

8   Yes, the part is $6\frac{2}{16}$ cm long.

9   720 000 m²

10 $3\frac{1}{2}$ miles

11   **a** $8\frac{7}{16}$ inches   **b** $11\frac{1}{4}$ inches

12 $a = \frac{8}{15}, b = \frac{1}{3}, c = \frac{1}{5}, d = \frac{4}{15}$

13   64

14   350

15   **a** $\frac{1}{4}$ **b** $\frac{1}{8}$ **c** $\frac{1}{4}$ **d** $\frac{3}{8}$

Chapter 4 Answers

4.1 Get Ready

1   8.02, 8.09, 8.092, 8.2, 8.29, 8.9, 8.92

2   **a** $\frac{1}{3}$ **b** $\frac{7}{12}$ **c** $\frac{7}{8}$

Exercise 4A

1   0.8, 0.85, $\frac{86}{100}, \frac{9}{10}$, 0.98

2   **a** terminating   **b** terminating   **c** recurring
     d recurring   **e** terminating   **f** recurring

3   Mitch is correct, as there is a factor of 3 in the denominator.

4.2 Get Ready

1   **a** 18.79   **b** 5.18   **c** 32.74

Exercise 4B

1   **a** 0.12   **b** 0.0012   **c** 0.04   **d** 0.0063

2   **a** 2.536   **b** 1.263   **c** 0.043 38   **d** 2.52
     e 13.02   **f** 0.504   **g** 0.046 72   **h** 0.323

3   15p

4   £6.86

5   **a** 60   **b** 14   **c** 640   **d** 65
     e 25   **f** 2040   **g** 0.05   **h** 0.092

6   **a** 2.31   **b** 642   **c** 41.3   **d** 42.2

7   £26.13

8   5

4.3 Get Ready

1   **a** 0.5772   **b** 160.3̇   **c** £1.41

Exercise 4C

1   **a** 6.4   **b** 5.7   **c** 16.9
     d 0.1   **e** 1.0

2   **a** 5.67   **b** 8.06   **c** 0.13
     d 3.04   **e** 0.08

3   **a** 6.446   **b** 0.079   **c** 5.079
     d 6.008   **e** 0.020

4.4 Get Ready

1   **a** 3   **b** 0   **c** 9

2   e.g. 438, 48, 6798

Exercise 4D

1   **a** 3900   **b** 230   **c** 46
     d 6.5   **e** 5.1   **f** −0.43

2   **a** 2500   **b** 39.0   **c** 4.90
     d 4.09   **e** 0.0110

3   **a** 3000   **b** 40   **c** 3
     d 8   **e** 20   **f** 1

4.5 Get Ready

1   **a** 5000   **b** 20   **c** −7

2   **a** 600   **b** 15 000   **c** 540 000

3   **a** 60   **b** 300   **c** 0.025

Exercise 4E

1   **a** 4200   **b** 7000   **c** 6000
     d 200 000   **e** 80 000

2   **a** 35   **b** 2   **c** 10   **d** 5   **e** 6

3   **a** 8000, overestimate   **b** 4, overestimate
     c 10, underestimate   **d** 300, overestimate

4   **a** 50, overestimate   **b** 4, underestimate
     c 40, overestimate   **d** 8, overestimate

5   2400

4.6 Get Ready

1   **a** 0.3   **b** 0.05   **c** 0.001

2   **a** 6   **b** 150   **c** 0.054

3   **a** 6000   **b** 30 000   **c** 25

Exercise 4F

1   **a** 2.4   **b** 0.15   **c** 1   **d** 0.0018

2   **a** 20   **b** 50   **c** 100   **d** 4
     e 0.2

3   **a** 1.5, underestimate   **b** 0.2, overestimate
     c 5, underestimate   **d** 200, underestimate

4   **a** 400, overestimate   **b** 210, overestimate
     c 30, underestimate   **d** 4000, underestimate

5   0.25

4.7 Get Ready

1   **a** 60   **b** 600   **c** 6000

2   **a** 30   **b** 3   **c** 0.3

Exercise 4G

1   **a** 1792   **b** 1792   **c** 17.92   **d** 0.017 92

2   **a** 146.4   **b** 1.464   **c** 0.1464   **d** 0.014 64

3   **a** 348   **b** 3480   **c** 34.8   **d** 348

4 a 128.8 **b** 230 **c** 56 **d** 2
5 a 0.026 **b** 340 **c** 0.034 **d** 100
6 a 13 **b** 13 **c** 13 000 **d** 0.065

4.8 Get Ready

1 a terminating **b** recurring **c** recurring

Exercise 4H

1 $\frac{7}{9}$ **2** $\frac{34}{99}$ **3** $\frac{305}{333}$
4 $\frac{2}{11}$ **5** $\frac{317}{999}$ **6** $\frac{1}{18}$
7 $\frac{323}{990}$ **8** $\frac{347}{495}$ **9** $\frac{7}{30}$
10 $6\frac{83}{99}$ **11** $2\frac{7}{66}$ **12** $7\frac{317}{900}$

4.9 Get Ready

1 a 6.1 **b** 7.0 **c** 6.5 **d** 6.5
2 a 0.3 **b** 0.3 **c** 0.3 **d** 0.3

Exercise 4I

1 a 84.5, 83.5 **b** 84.05, 83.95 **c** 84.005, 83.995
2 a 0.95, 0.85 **b** 0.905, 0.895 **c** 0.095, 0.085
3 a 118.5 cm **b** 117.5 cm
4 a 6450 g **b** 6350 g
5 a 48.05 l **b** 47.95 l
6 a 1.005 m **b** 0.995 m

4.10 Get Ready

1 10.1
2 2.1
3 a e and f **b** c and f **c** e and f **d** c and f

Exercise 4J

1 a 535 **b** 26 612.25 **c** 229 920.25
2 a 6.88 **b** 10.823 575 **c** 24.35
3 a 11 275 568.625 **b** 151 474.75
4 a 11 **b** 1.3606…
5 a 0.8 **b** 1.3137…
6 a −31.3375 **b** −29.8275 **c** −30 (1 s.f.)
7 2.05×10^{17} (3 s.f.), 2.28×10^{17} (3 s.f.)
8 a 50 **b** 15.5%

Review exercise

1 £153.90
2 $\frac{2}{3}$
3 $0.47, \frac{12}{25}, \frac{3}{5}, \frac{31}{50}$
4 £42.96
5 a 4780 **b** 107 **c** 3.23×10^{15}
d 7000 **e** 57.0
6 a 46 **b** 31 **c** 0.046
d 20 **e** 4.1
7 a 400 **b** 40 **c** 1×10^{18}
d 0.005 **e** −3
8 a 3.1 **b** 0.6 **c** 2.1 **d** 4.0
9 a $\frac{77}{400}$ **b** $\frac{77}{4} = 19\frac{1}{4}$ **c** $\frac{770\,000}{4} = 192\,500$
10 75 923 1p pieces

11 a Students' checks
b

	A	B	C	Best deal
John	£56.75	£62.27	£64.33	Avery Energy
Vijay	£192.37	£186.17	£280.42	Brawn Power

12 a 3000 **b** 4000 **c** 24
d 350 000 **e** 360
13 a 5 **b** 5 **c** 80
d 250 **e** 1
14 36 minutes
15 a $\frac{3}{1} = £3$ **b** $\frac{30}{1.5} = £20$ **c** $\frac{1500}{150} = £10$
d $\frac{10}{2.5} = £4$ **e** $\frac{300}{50} = £6$ **f** $\frac{2000}{10} = £200$
16 a £21 **b** 32
17 a 75, underestimate **b** 150, underestimate
c 3.6, overestimate **d** 6000, overestimate
e 45 000, overestimate
18 Volume $\approx 3 \times 8 \times 9 = 216\text{ m}^3$.
Number of people $= \frac{216}{4} = 54$
19 £370.22
20 $\frac{1}{3}$
21 a 17.1 **b** 31.95 **c** 60.1425 **d** 24.5025
22 a 200 **b** 645 **c** 1.8163… **d** 0.0638…
23 lower bound for length is 199.5 cm, so rod may fit into slot of length 199.8 cm
24 8.75 km/l
25 LB of cylinder's capacity = 325 ml, so the cylinder always contains more than stated on the label.

Chapter 5 Answers

5.1 Get Ready

1 $a = 135°, b = 45°, c = 135°$ **2** 40° **3** 48°

Exercise 5A

1 $a = 63°$ (corresponding angles)
$b = 49°$ (corresponding angles)
$c = 68°$ (angles in a triangle add up to 180°)
2 $p = 113°$ (corresponding angles and angles on a straight line)
$q = 67°$ (corresponding angles)
$r = 113°$ (alternate angles or angles on a straight line)
3 $l = 81°$ (vertically opposite angles)
$m = 54°$ (alternate angles)
$n = 45°$ (angles of a triangle add up to 180°)
4 $y = 58°$ (alternate angles)
$z = 58°$ (alternate angles and angles on a straight line)
5 $g = 57°$ (isosceles triangle and alternate angles)
$h = 180 - 2 \times 57 = 66°$ (angles of a triangle add up to 180° and alternate angles)
$k = 114°$ (angles of a triangle add up to 180° and angles on a straight line)
6 $a = 50°$ (2 sets of alternate angles)
7 a a and p, b and q, c and s, d and r
b a and r, b and s, c and q, d and p
c a and b, b and d, d and c, c and a, p and q, q and r, r and s, s and p
The angles are on a straight line.
8 Angle BAC = Angle DCE = 56°, so they are corresponding angles

Answers

5.2 Get Ready

1 $a = 50°, b = 80°$ **2** $c = 28°, d = 28°$
3 $e = 60°$

Exercise 5B

1 angles in a triangle, angles on a straight line
2 same angle, same angle, angles in a triangle, angles in a triangle
3 a $b = d$ (alternate angles)
    $a = c$ (corresponding angles)
    so $a + b = c + d$
  b The exterior angle is equal to the sum of the opposite two interior angles.

5.3 Get Ready

$j = 143°$

Exercise 5C

1 141° (angle sum of quadrilateral / equilateral triangle)
2 126° (angles on a straight line / exterior angle of a triangle)
3 a $a = 132°$ (symmetry), $b = 37°$ (angle sum of quadrilateral)
  b 66°, 114°, 114° (symmetry / angle sum of quadrilateral)
4 113° (vertically opposite angles / angles on a straight line / angles at a point / angle sum of a quadrilateral)

5.4 Get Ready

  a alternate angles   **b** 90°

Exercise 5D

1 alternate angles
2

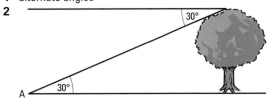

5.5 Get Ready

1 a East   **b** South   **c** North-west

Exercise 5E

1 a 073°   **b** 225°   **c** 243°
2 a 070°   **b** 218°   **c** 102°
3 249°
4 312°
5 a 111°   **b** 239°

5.6 Get Ready

$a = 43°$ (alternate angles)
$b = 72°$ (opposite angles)
$c = 65°$ (angles sum of a triangle)
$d = 64°$ (angles on a straight line)
$e = 58°$ (angle sum of an isosceles triangle)
$f = 54°$ (opposite angles of a parallelogram)
$g = 126°$ (angles at the end of a parallelogram)

Exercise 5F

1 a 124° (isosceles triangle / angle sum of a triangle / vertically opposite angles)
  b 56° (angles on a straight line / corresponding angles)
2 L = 70°, M = 55°, N = 55° (alternate angles / angles on a straight line)
3 a $p = 57°$ (exterior angle of a triangle / angle sum of a quadrilateral)
  b $q = 117°$ (exterior angle of a triangle / corresponding angles)
4 a alternate angles
  b $b = d$ (alternate angles)
    $a = c$, part **a**
    so $a + b = c + d$
  c Opposite angles in a parallelogram are equal.
5 $a + b + c + d = 360°$ (angles in a quadrilateral)
  $a + c = 180°$ (given in question)
  so $b + d = 180°$

5.7 Get Ready

1 Equilateral triangle
2 A square is a quadrilateral with **equal** sides and **equal** angles.

Exercise 5G

1 a

Polygon	Number of sides (n)	Number of diagonals from one vertex	Number of triangles formed	Sum of interior angles
Triangle	3	0	1	180°
Quadrilateral	4	1	2	360°
Pentagon	5	2	3	540°
Hexagon	6	3	4	720°
Heptagon	7	4	5	900°
Octagon	8	5	6	1080°
Nonagon	9	6	7	1260°
Decagon	10	7	8	1440°

  b i $n - 3$ **ii** $n - 2$ **iii** $(n - 2) \times 180°$
2 Angles not equal

Exercise 5H

1 a 15   **b** 18   **c** 160°
2 a 120°   **b** 144°   **c** 168°
3 a 142°   **b** 103°
4 360 is not divisible by 25
5 a $a = 32°, b = 30°, c = 42°, d = 63°, e = 44°, f = 27°, g = 59°, h = 63°$
  b 360°
6 a 135° (angle sum of isosceles triangle / angles on a straight line)
  b i The interior angles are all the same.
    ii Not all the sides are the same length.

Exercise 5I

1 a i 110°   **ii** 143°
  b i 50°   **ii** 36°
2 a 72°   **b** 45°   **c** 30°   **d** 14.4°
3 a 24   **b** 3960°
4 144°, 98°, 129°, 128°, 107° and 114°
5 Exterior angle = 12°, interior angle = 168°, angle BCA = 6°
(angle sum of isosceles triangle)
6 Angle BCD = 180 − e (angles on a straight line)
So angle BCO = (180 − e) ÷ 2 (by symmetry)
And angle CBO = (180 − e) ÷ 2 (isosceles triangle)
So c = 180 − {(180 − e) ÷ 2} − {(180 − e) ÷ 2} (angles in a triangle) giving $c = e$

Review exercise

1 88 + 96 ≠ 180, so the lines are not parallel. Ben is right.
2 Angles on a straight line add up to 180°,
but 120 + 50 ≠ 180.
3 a $x = 30°$
  b vertically opposite angles
  c Angles around a point add up to 360°,
    but 125 + 135 + 125 ≠ 360.
4 a i $x = \dfrac{180 - 54}{2} = 63°$
    ii angle sum of an isosceles triangle
  b $y = 54 + 63 = 117°$ (exterior angle of a triangle)
    or
    $y = 180 − 63 = 117°$ (angles on a straight line)

5

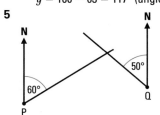

6 a $y = 58°$
  b alternate angles
7 $x = 130°$ (angles on a straight line)
  $y = 50°$ (alternate angles)
8 $x = 180 − (360 − 50 − 119 − 105) = 94°$ (angle sum of a
quadrilateral / angles on a straight line)
9 a $x = 180 − 2 × 52 = 76°$
  b angle sum of an isosceles triangle
10 angle ABQ = 90° (angle in a square)
angle ABC = $180 − \dfrac{360}{6} = 120°$ (exterior angle of a regular
polygon / angles on a straight line)
$x = 360 − 90 − 120 = 150°$ (angles around a point)
11 a angle BDA = 180 − 127 = 53° (angles on a straight
    line)
    angle BAD = 180 − 2 × 53 = 74° (angle sum of an
    isosceles triangle)
    angle DAC = 90 − 74 = 16°
  b angle DCA = 180 − 127 − 16 = 37° (angle sum of a
    triangle)
12 angle in equilateral triangle = 60°
base angle in isosceles triangle = $\dfrac{180 - 57}{2} = 61.5°$
(angle sum of a triangle)
$p = 360 − 60 − 61.5 = 238.5°$ (angles around a point)

13 a i $w = 25°$ (base angles of an isosceles triangle)
    ii $x = 180 − 2 × 25 = 130°$ (angle sum of a triangle)
  b angle SQR = 180 − 130 = 50° (angles on a straight
    line)
    $y = \dfrac{180 - 50}{2} = 65°$ (angle sum of an isosceles
    triangle)
14 a 030°
  b

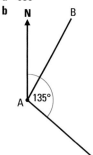

15 exterior angle = $\dfrac{360}{10} = 36°$
$x = 180 − 36 = 144°$ (angles on a straight line)
16 a $x = \dfrac{180 - 120}{2} = 30°$ (angle sum of an isosceles
    triangle)
  b angle ABD = 180 − 30 = 150° (angles on a straight
    line)
    $y = 360 − 150 − 54 − 108 = 48°$ (angle sum of a
    quadrilateral)
17 angle ACB = angle ABC = $x + 20$ (base angles of an
isosceles triangle)
angle BAC + $2(x + 20) = 180°$ (angle sum of a triangle)
angle BAC = $140 − 2x$
18 angle EFD = $360 − 2x − 65 − x − 90 = 205 − 3x$
(angles around a point)
angle FED = angle EFD = $205 − 3x$
(base angles of an isosceles triangle)
19 a interior angle = 180 − exterior angle
    (angles on a straight line)
    interior angle = $180 − \dfrac{2}{3} ×$ interior angle
    interior angle = 108°
  b exterior angle = $72° = \dfrac{360}{n}$
    $n = 5$
20 Angle $CAB = 360 − x − 90 − 80 − x$
(Angles at a point add up to 360°)
$= 190 − 2x = $ angle ABC
(Base angles in an isosceles triangle are equal)
Angle $BCD = 190 − 2x + 190 − 2x$
(exterior angle of a triangle = sum of two interior opposite
angles)
$= 380 − 4x = 4(95° − x)$
21 Angle $PQR = $ angle $PRQ = (180° − 20°) ÷ 2 = 80°$
(base angles of an isosceles triangle are equal and Angles
in a triangle add up to 180°)
Angle $PQY = 80° − 60° = 20°$
Since angle $PQY = 20°$ and angle $QPY = 20°$, triangle PQY
is isosceles.
Thus $PY = QY$
(sides opposite the equal angles in an isosceles triangle
are equal)
22 5 minutes and 27 seconds past 1

Chapter 6 Answers

6.1 Get Ready

a Ask each classmate how much lunch money they get and calculate the mean of this data.
b Use an internet air ticketing site or the Manchester airport site.
c Search the internet for a government site that gives information on voting figures.

Exercise 6A

1 **a** secondary   **b** secondary   **c** primary
2 **a** quantitative   **b** qualitative   **c** quantitative
   d quantitative
3 **a** discrete   **b** continuous   **c** continuous
   d discrete
4 **a** Drug A is effective at curing malaria OR Drug A is not effective at curing malaria.
   b Collect data on patients that have been treated with Drug A and those that haven't.

6.2 Get Ready

If the answers are written down: 15 seconds
If the answers are given orally (i.e. one student at a time): number of students in class × 15 seconds.

Exercise 6B

1 For example, rolling a dice or using the random function on a calculator or from a random number table
2 A fraction of the population is chosen at random.
3 For example, assign the numbers 1 to 60 to the workers, then take the first eight different numbers under 61 that are generated by the calculator
(21, 32, 54, 34, 26, 45, 35, 22)

6.3 Get Ready

1 **a** $\frac{15}{25} = \frac{3}{5}$ **b** $\frac{10}{25} = \frac{2}{5}$
2 **a** 80   **b** $\frac{3}{8}$

Exercise 6C

1 23 boys, 27 girls
2 Randomly select 15 employees with less than six months' experience and 40 employees with more than six months' experience.
3

	Office workers	Factory floor workers	Managers
Females	5	25	1
Males	8	49	2

They should be picked by simple random sampling.

6.4 Get Ready

1 **a** 4   **b** 6
2 ⅢⅡ Ｉ

Exercise 6D

1 **a**

Vehicle	Frequency
Car	ⅢⅡ ⅢⅡ ⅢⅡ ⅢⅡ ⅡⅠ
Bus	ⅢⅡ
HGV	ⅢⅠ
Bike	ⅢⅠ
Motorbike	Ｉ

b Motorbike   **c** Car

2

Number of DVDs bought	Frequency
0–3	9
4–7	11
8–11	6
12–15	4

3 **a**

Weight	Frequency
$57 \leqslant w < 60$	7
$60 \leqslant w < 63$	3
$63 \leqslant w < 66$	5
$66 \leqslant w < 69$	5
$w \geqslant 69$	4

b $57 \leqslant w < 60$ **c** $60 \leqslant w < 63$

6.5 Get Ready

Tallying

Exercise 6E

1 It is a biased question.
2 A: open, B: closed, C: open, D: closed
3 **a** No option for dissatisfied customers.
   New suitable question:
   What do you think of the new amusements?
   Very good ☐   Good ☐   Satisfactory ☐   Poor ☐
   b Options overlap.
   New suitable question:
   How much money would you normally expect to pay for each amusement?
   £5–£7 ☐   £7.01–£8 ☐   more than £8 ☐
   c Not clear what the options mean.
   New suitable question:
   How often do you visit the park each year?
   0–2 times ☐   3–5 times ☐   6–8 times ☐
   more than 8 times ☐
4 Do you like the new layout? Yes/No

6.6 Get Ready

1 **a** 15 girls   **b** 6 students

Exercise 6F

1 a

	Plain	Salt and Vinegar	Cheese and Onion	Total
Males	7	7	14	28
Females	5	6	12	23
Total	12	13	26	51

b 13 **c** 51

2 a

	Orange Juice	Grapefruit Juice	Total
Men	22	8	30
Women	18	12	30
Total	40	20	60

b 40

3 a

	Supervisors	Office staff	Shop floor workers	Total
Males	10	3	82	95
Females	2	11	38	51
Total	12	14	120	146

b 51 **c** 146

Exercise 6G

1 Yes, only one area sampled
2 A: Biased. Not everyone in the hospital's area has a chance of being asked.
B: Biased. Only people with phones have a chance of being asked and only in 10 towns, the sample is too small.
C: Not biased.
D: Biased. Only people already using the recycling facility are being asked.
3 Students' own discussions

6.8 Get Ready

The internet, supermarkets, high street shops

Exercise 6H

1 a 41 000 tonnes **b** Cars **c** 2004
d Cars
2 a 6.3 days **b** May **c** May **d** May
3 a 799 **b** Stays the same
c More dairy **d** Numbers have decreased

Review exercise

1 a It does not allow for sending no text messages. It does not include a time frame, e.g. per week.
b It only includes people of one age.
2 a The categories overlap. It does not include a time frame, e.g. per day. It does not allow for people who use their computer for more than 6 hours.
b It only includes people of one age.
3 a It does not allow for never visiting the cinema. It is hard to decide what the categories mean. It does not include a time frame, e.g. per month.

b On average, how many times do you go to the cinema each month?
0–1 times ☐ 2–3 times ☐ 4–5 times ☐
more than 5 times ☐
4 On average, how many emails do you send each week?
0–5 ☐ 6–10 ☐ 11–15 ☐ 16–20 ☐
more than 20 ☐
5 It only includes women. It does not include people who never go to the cinema.

6

Animal	Tally	Frequency
Lions		
Tigers		
Elephants		
Monkeys		
Giraffes		

7

Country	Tally	Frequency
France	ЦШ	5
Spain	ЦШ II	7
England	IIII	4
Italy	IIII	4

8 a The first question does not allow for people who never visit the park and it is hard to decide what each category means.
The second question has overlapping categories.
b On average, how often do you go to the County Park each month?
Never ☐ 1–3 times ☐ 4–6 times ☐
more than 6 times ☐
How old are you?
0–10 years ☐ 11–20 years ☐ Over 20 years ☐
9 a Biased, because it only includes those working on the night shift.
b Not biased, because it uses a simple random sample.
c Biased because the question starts with 'Do you agree…'
10 6 girls
11 12 boys
12 a 5 students **b** 26 students
13 19 students
14

	London	York	Total
Boys	23	14	37
Girls	19	24	43
Total	42	38	80

Chapter 7 Answers

7.1 Get Ready

1 **a** 150 g **b** 3 cm **c** 5 m **d** 300 ml

Exercise 7A

1 **a** 600 cm **b** 21 cm **c** 510 cm
  d 84 cm **e** 5.9 cm **f** 48.3 cm
  g 300 000 cm **h** 6700 cm
2 **a** 3000 kg **b** 8200 kg **c** 6 kg
  d 0.9 kg **e** 0.43 kg **f** 4.7 kg
3 **a** 2 litres **b** 7 litres **c** 5.9 litres
  d 45 litres
4 7 litres

Exercise 7B

1 8.8 pounds
2 50 kg
3 4 litres
4 8 gallons
5 17.5 pints
6 360 cm
7 60 miles
8 96 km
9 585p per gallon

Exercise 7C

1 **a** 80 ounces **b** 6 pounds
2 180 inches
3 **a** 132 pounds **b** 60 kg

7.2 Get Ready

1 60 minutes
2 $\frac{1}{4}$
3 0.6
4 18 minutes
5 5 hours 42 minutes
6 7 hours 42 minutes

Exercise 7D

1 **a** 13 km/litre **b** 6 litres
2 **a** 240° **b** 30 seconds
3 **a** 60 litres/min **b** 18 minutes 30 seconds
4 0.0625 litre

7.3 Get Ready

1 **a** 18 km/l **b** 6.5 km/l **c** 13.2 km/l

Exercise 7E

1 4.24 km/h 2 10 km/h
3 1 hour 12 minutes 4 306 km/h
5 The speed for the 100 m was 10.32 m/s. The speed for the 200 m was 10.36 m/s. The 200 m race was won with the faster average speed.

7.4 Get Ready

1 **a** 8.3 **b** 1.3 **c** 26.7 **d** 12.9

Exercise 7F

1 250 g/cm^3 2 0.777 cm^3 3 1.96 g/cm^3
4 The aluminium block has the greater mass by 155 kg.

Review exercise

1 **a** 110 m, 70 g, 40 litres **b** 400 cm **c** 1.5 kg
2 No, 1.5 km is 1500 m.
3 80 km/hour
4 432 miles/hour
5 John's speed: 20 km/hour. Kamala's speed: 21 km/hour. Kamala had the greater average speed.
6 320 seconds
7 1.14 g/cm^3
8 193 g

Chapter 8 Answers

8.1 Get Ready

1 Not necessarily, one could be an enlargement of the other.
2 No
3 No

Exercise 8A

1 QS is the hypotenuse of both triangles
  angle QPS = angle SRQ = 90° (given)
  PS = QR (given)
  So the triangles are congruent (RHS)
2 angle YZX = angle WVX (alternate angles)
  angle ZYX = angle VWX (alternate angles)
  YZ = WV (given)
  So the triangles are congruent (AAS)
  And YX = XW, so X is the mid-point of WY
3 PQ = PR (given)
  QS = RT (given)
  angle PQS = angle PRT (isosceles triangle)
  So triangle PQS is congruent to triangle PRT (SAS)
  Therefore PS = PT and so triangle PST is isosceles
4 Let the point where the line from L cuts the base at right angles be X.
  Now LM = LN (given)
  LX is common to both triangles
  angle LXM = angle LXN = 90° (given)
  So triangles LMX and LNX are congruent (RHS)
  As the triangles are congruent then MX = NX, so the line from L bisects the base.
5 AD = DB (as D is the midpoint of AB)
  angle ADE = angle DBF (corresponding angles)
  angle DAE = angle BDF (corresponding angles)
  So the triangles ADE and DBF are congruent (AAS)

8.2 Get Ready

When you fold it over, the star fits exactly on top of itself. When you rotate the star, it fits exactly on top of itself in four different positions.

Exercise 8B

1 a Yes **b** Yes

 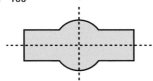

c No **d** No
e Yes **f** No

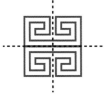

2 a No rotational symmetry
b Rotational symmetry of order 2
c Rotational symmetry of order 3
d Rotational symmetry of order 8
e Rotational symmetry of order 2
f No rotational symmetry

3 a

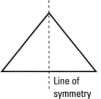

Line of symmetry

b Isosceles triangle

4 a

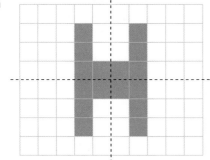

b

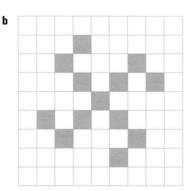

5 a e.g.

b e.g.

c e.g.

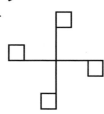

8.3 Get Ready

1 a i An isosceles triangles has two sides the same length and two angles the same.
ii An equilateral triangle has all three sides the same length and all three angles the same.
b Yes
2 A polygon with four sides.

Exercise 8C

1 a

b It is also an isosceles triangle.
2 a No. It could be a rectangle, square, rhombus, parallelogram, trapezium or isosceles trapezium.
b No. It could be a rectangle, rhombus or parallelogram.
c No. It could be a rectangle or rhombus.
d Yes. It is a rectangle.
3 e.g.

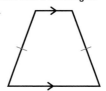

4 e.g.

Rotational symmetry of order 2

8.4 Get Ready

A and C

Exercise 8D

1 a Similar **b** Not similar
2 All corresponding angles are equal so the pentagons are similar
(angle CDE = angle HIJ = 150°)

Answers

Exercise 8E

1 15 cm
2 a 5.14 cm   **b** 0.448 m
3 15 cm

8.5 Get Ready

By inspection

Exercise 8F

1 a i AC and FE, BC and DF, AB and DE
    ii angle ABC = angle EDF, angle BCA = angle DFE,
        angle BAC = angle DEF
  b i JK and GH, KL and HI, JL and GI
    ii angle KJL = angle HGI, angle JKL = angle GHI,
        angle KLJ = angle HIG
  c i PN and MN, QN and ON, PQ and MO
    ii angle NPQ = angle NMO, angle PQN = angle NOM,
        angle PNQ = angle MNO
  d i SR and WU, ST and WT, RT and UT
    ii angle STR = angle UTW, angle RST = angle TWU,
        angle SRT = angle TUW
2 a 5.525 cm   **b** 7.225 cm
3 a angle DEB = angle ACB (given)
    angle DBE = angle ABC (same angle)
    Therefore angle BDE = angle BAC (angles in a triangle)
    So triangle ABC is similar to triangle DBE
  b 6.46 cm   **c** 2.26 cm   **d** 1.99 cm
4 a angle BDE = angle BAC (given)
    angle DBE = angle ABC (same angle)
    Therefore angle DEB = angle ACB (angles in a triangle)
    So triangle ABC is similar to triangle DBE
  b 6.36 cm   **c** 10 cm
5 a angle ABM = angle CDM (alternate angles)
    angle BAM = angle DCM (alternate angles)
    angle AMB = angle CMD (vertically opposite angles)
    So triangle ABM is similar to triangle CDM
  b i 8 cm   **ii** 12 cm

Review exercise

1 a 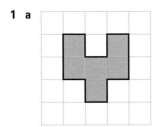   **b**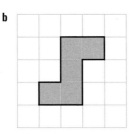

2 A and C
3 Y and Z
4 a angle ADB = angle ADC = 90° (given)
    AB = AC (sides of equilateral triangle)
    AD is common
    So triangle ADC is congruent to triangle ADB (RHS)
  b BD = CD (corresponding sides)
    BD + CD = BC = AB
    So BD = $\frac{1}{2}$ BC

5 AB = BC (given)
    AD = CD (given)
    BD is common
    So triangle ADB is congruent to triangle CDB (SSS)
6 angle ACB = angle CED (vertically opposite angles)
    angle CAB = angle CED (alternate angles)
    angle ABC = angle EDC (alternate angles)
    So triangle ABC is similar to triangle EDC
  a 12 cm   **b** 9 cm
7 12.5 cm
8 a angle HJI = angle GJF (same angle)
    angle IHJ = angle FGJ (corresponding angles)
    angle HIJ = angle GFJ (corresponding angles)
    So triangle HIJ is similar to triangle GFJ
  b 3.43 cm
9 angle CAD = angle BAE (same angle)
    angle ACD = angle ABE (corresponding angles)
    angle ADC = angle AEB (corresponding angles)
    So triangle ACD is similar to triangle ABE
  a 2 cm   **b** 5.25 cm

Chapter 9 Answers

9.1 Get Ready

1 a $10x$ **b** $-12x^2$ **c** $2x^2$
  d $x - 4$ **e** $x^2 + 5x + 6$
  f $x^2 - 3x + 2$

Exercise 9A

1 a $2x + 6$ **b** $3p - 6$ **c** $4m + 4n$
  d $15 - 3q$ **e** $4x + 2y - 6$ **f** $10c + 5$
  g $4x^2 - 8$ **h** $3n^2 - 6n + 3$
2 a $y^2 + 2y$ **b** $g^2 - 3g$ **c** $2x^2 + 10x$
  d $4n - n^2$ **e** $ab + ac$ **f** $3s^2 - 4s$
  g $6t^2 + 3t$ **h** $4x^3 - 12x^2$
3 a $-2m - 6$ **b** $-6x - 6$ **c** $-m^2 - 5m$
  d $-8y^2 - 12y$ **e** $-5p + 10$ **f** $-3q + 3q^2$
  g $-2s^2 + 6s$ **h** $-12mn - 3n^2 + 15n$

Exercise 9B

1 a $8t - 3$ **b** $9p + 6$ **c** $11w + 6$
  d $7d - 2$ **e** $5a + 3b$ **f** $5x + 3y + 5$
2 a $y + 20$ **b** $9a - 6$ **c** $-4x - 15$
  d $q^2 - 3$ **e** $-5n$ **f** $11m^2 + 2m$
3 a $t - 16$ **b** $x + 19$ **c** $g^2 + g$
  d $13c^2 - 22c$ **e** $4s^2 + 14s - 2$ **f** $p^2 + q^2$
4 a $3s - 4$ **b** $3m + 18$ **c** $5f^2 - 3f$
  d $n^2 + 4n$ **e** $2x - x^2 + xy$ **f** $2p^2 + 5p$

9.2 Get Ready

1 a 2   **b** 5   **c** 4   **d** $3y$

Exercise 9C

1 a $3(x + 2)$ **b** $2(y - 1)$ **c** $5(p + 2q)$
  d $7(2t - 1)$ **e** $2(4s + t)$ **f** $9(a + 2b)$
  g $5(3u + v + 2w)$ **h** $t(x - y)$
  i $c(a - 1)$ **j** $3(2x^2 + 3x + 1)$

k $2p(p-1)$ **l** $q(q-1)$ **m** $x(4x+3)$
n $h(2-5h)$ **o** $p(p^2+2)$ **p** $s^2(1+s)$
2 a $5x(y+t)$ **b** $3a(d-2c)$ **c** $2p(3q+2h)$
d $4y(2x-1)$ **e** $2p(2q+s+4t)$
f $mn(1-k)$ **g** $2x(x+2)$ **h** $12s(s-2)$
i $2f^2(3+f)$ **j** $y^2(y^2+1)$ **k** $cd(3d-5c)$
l $ab(a^2+b^2)$ **m** $2pr(4q+5s)$ **n** $7ab(2a-b+3)$
o $5x^2y(3-7y)$ **p** $3y(3y+1)$

Exercise 9D

1 a $(x+3)(x+5)$ **b** $(x-y)(x+y)$
c $p(p+1)$ **d** $(2t-s)(2t+s+1)$
e $(a-5)(a-7)$ **f** $2(d+1)(d+1)$
2 a $2(y+2)(y+4)$ **b** $5(x-1)(3x-5)$
c $2(p+5)(4p+25)$ **d** $3(q+1)(2q+5)$
e $7(a+b)(a-b-2)$ **f** $2x(x+1)(2x-3)$

9.3 Get Ready

1 $24\,\text{cm}^2$ **2** $x \times (x+2)$

Exercise 9E

1 a $x^2+7x+12$ **b** x^2+3x+2
c $x^2-3x-10$ **d** y^2+y-6
e y^2-y-2 **f** x^2-5x+6
g $a^2-9a+20$ **h** x^2+4x+4
i $p^2+8p+16$ **j** $k^2-14k+49$
k $a^2+2ab+b^2$ **l** $a^2-2ab+b^2$
2 a $2x^2+3x+1$ **b** $3x^2-2x-1$
c $2x^2+11x+12$ **d** $3y^2-8y-3$
e $2p^2+7p+3$ **f** $6t^2+7t+2$
g $6s^2+19s+10$ **h** $4x^2+4x-15$
i $12y^2+5y-2$ **j** $6a^2-7a+2$
k $9x^2+12x+4$ **l** $4k^2-4k+1$
3 a $x^2+3xy+2y^2$ **b** $x^2+xy-2y^2$
c $x^2-xy-2y^2$ **d** $x^2-3xy+2y^2$
e $6p^2+7pq-3q^2$ **f** $6s^2-7st+2t^2$
g $4a^2+12ab+9b^2$ **h** $4a^2-12ab+9b^2$

9.4 Get Ready

1 a 1 and -6, 6 and -1, 2 and -3, 3 and -2
b 1 and 15, -1 and -15, 3 and 5, -3 and -5
2 5 and 2
3 -3 and -5

Exercise 9F

1 a $3, 5$ **b** $-6, -4$ **c** $-6, -3$
d $4, -2$ **e** $-4, 2$ **f** $-3, 3$
2 a $(x+3)(x+5)$ **b** $(x+1)(x+7)$ **c** $(x+4)(x+5)$
d $(x-5)(x-1)$ **e** $(x-8)(x-1)$ **f** $(x-1)^2$
g $(x-3)(x+6)$ **h** $(x-6)(x+3)$ **i** $(x-4)(x+7)$
j $(x-4)(x+3)$ **k** $(x-4)(x+6)$ **l** $(x-2)(x+2)$
m $(x-9)(x+9)$

Exercise 9G

1 a $(x-6)(x+6)$ **b** $(x-7)(x+7)$
c $(y-12)(y+12)$ **d** $(5-y)(5+y)$
e $(w-50)(w+50)$ **f** $(100-a)(100+a)$

g $(x-1)(x+3)$ **h** $y(18-y)$
i $4ab$
2 a 2800 **b** 50 **c** 0.75 **d** $20\,000$
3 a $(2x-7)(2x+7)$ **b** $(3y-1)(3y+1)$
c $(11t-20)(11t+20)$ **d** $-(q+1)(q+3)$
e $8t$ **f** $4(p+q)$
g $4(5p-q+2)(5p+q+3)$
h $100st$
4 a $3(x-2)(x+2)$ **b** $5(y-5)(y+5)$
c $10(w-10)(w+10)$ **d** $4(p-4q)(p+4q)$
e $3(2a-3b)(2a+3b)$ **f** $8x$

Exercise 9H

1 a $(5x+1)(x+3)$ **b** $(2x+1)(x+5)$
c $(3x+1)(x+1)$ **d** $(4x+1)(2x+1)$
e $(3x+2)(2x+3)$ **f** $(6x-1)(x-1)$
g $(5x-2)(x-1)$ **h** $(4x-1)(3x-2)$
i $(4x+3)(2x-1)$ **j** $(2x+3)(x-5)$
k $(7x+2)(x-3)$ **l** $(3x+2)(x-4)$
m $(2y+1)(2y+5)$ **n** $(6y-1)(y-2)$
o $(3y-5)(2y-5)$
2 a $2(3x+4)(x+1)$ **b** $3(2y-1)(y-2)$
c $5(x+2)(x-1)$
3 a $(x-y)(x+2y)$ **b** $(x+y)(2x+5y)$
c $(3x-2y)(2x+3y)$

Review exercise

1 $5x-2$
2 a $5(m+2)$ **b** $y(y-3)$
3 a $(a+b)(x+y)$ **b** $(a-b)(c+d)$
4 x^2-x-12
5 a a^2+4a+4 **b** c^2-6c+9
c d^2+2d+1 **d** $x^2+2xy+y$
6 a $x^2+15x+50$ **b** $y^2+18y+81$
c x^2-2x-8 **d** x^2-x-6
e t^2-7t+6 **f** $2x^2+11x+12$
g $6p^2+p-1$ **h** $4c^2-d^2$
i $16y^2-8y+1$
7 a $(t+5)(t+6)$ **b** $(x+7)^2$ **c** $(p+5)(p-3)$
d $(y-6)^2$ **e** $(x-4)(x-1)$ **f** $(s-8)(s+8)$
8 a $(x+6)(x+7)$ **b** 11×17
9 $2(Y+3)$, 26 red flowers
10 a $(x-20)(x+20)$ **b** $(3t-2)(3t+2$
c $(10-y)(10+y)$ **d** $(5-2p)(5+2p)$
11 a 41 **b** 1.99 **c** 16
12 8000
13 Three consecutive numbers are $n, n+1, n+2$.
$(n+1)(n+2) - n(n+1) = (n^2+3n+2) - (n^2+1)$
$= 2n+2 = 2(n+1)$
14 $6(x+2)$
15 a Each team plays 3 games at home against the other teams.
So total number of games $= 4 \times 3 = 12$
b 380 **c** $a^2 - a = a(a-1)$
16 a $(2x+1)(x+2)$ **b** $(2w-1)(w+3)$
c $(3a+2)(a+4)$ **d** $(3z-2)(10z-1)$
e $(8y-1)(y+3)$ **f** $(3p+q)(2p-q)$

17 Let the top left number in the 2 by 2 square be n.

n	$n + 1$
$n + 6$	$n + 7$

Difference of products from opposite corners
$= (n + 1)(n + 6) - n(n + 7) = (n^2 + 7n + 6) - (n^2 + 7n)$
$= 6$

Chapter 10 Answers

10.1 Get Ready

1 5 cm, 45 cm²

Exercise 10A

1 a 40 cm² **b** 18 m² **c** 35 cm²
d 54 mm² **e** 30 cm² **f** 54 cm²
2 Area 15 cm², area 25 cm², base 6 cm, area 32 cm²,
height 8 cm
3 a 5 cm **b** 12 cm

Exercise 10B

1 a 56 cm² **b** 96 m² **c** 75 cm² **d** 98 cm²

10.2 Get Ready

a $A = lw$
b $A = l^2$
c $A = \frac{1}{2}bh$
d $A = bh$
e $A = \frac{1}{2}(a + b)h$

Exercise 10C

1 a 28 m **b** 37 m²
2 60
3 30
4 6 cm
5 a i 32 m² **ii** 14 m²
   b 24 m **c** 13 **d** 9
6 a 66 cm **b** 234 cm²
7 30 cm²
8 100 cm²

10.3 Get Ready

1

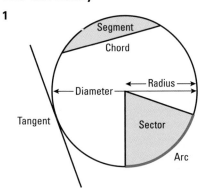

Exercise 10D

1 a 22.0 cm **b** 40.5 mm **c** 17.6 cm
   d 126 cm **e** 68.8 m
2 a 145 cm **b** 25.1 cm
3 6.00 cm
4 a 210 cm **b** 2.1 km **c** 2859
5 a 7.54 cm
   b i 45.2 cm **ii** 2.51 cm
6 a 408 cm **b** 14.6 cm
7 71.7 cm

Exercise 10E

1 a 201 cm² **b** 507 cm² **c** 2550 mm²
   d 297 cm² **e** 499 m²
2 a 452 cm² **b** 54.1 cm² **c** 0.709 m²
   d 2680 mm² **e** 262 cm²
3 a 19.6 m² **b** 18.8 m² **c** £36.80
4 a 10 800 m² **b** £2480
5 70.7 cm²
6 13 cm
7 193 cm²

10.4 Get Ready

1 a **b** **c** **d**

Exercise 10F

1

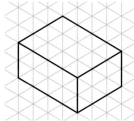

2 Any six out of:

3 a Cylinder   **b** Cone
c Triangular-based pyramid   **d** Square-based pyramid

4 a e.g.

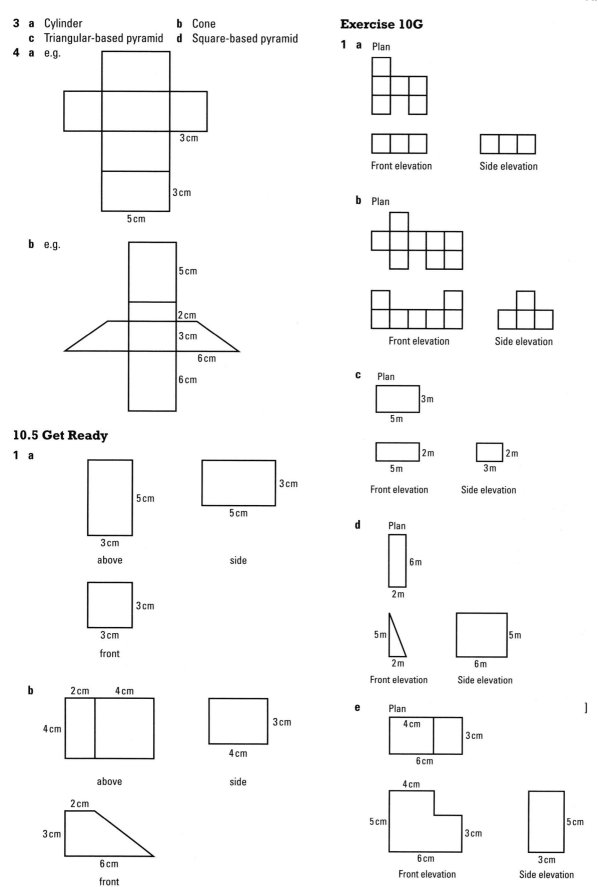

b e.g.

10.5 Get Ready

1 a

above   side

front

b

above   side

front

Exercise 10G

1 a Plan

Front elevation   Side elevation

b Plan

Front elevation   Side elevation

c Plan

Front elevation   Side elevation

d Plan

Front elevation   Side elevation

e Plan

Front elevation   Side elevation

f Plan

5 cm

4 cm

2 cm

Front elevation

4 cm

5 cm

Side elevation

g Plan

4 cm

5 cm

8 cm

Front elevation

5 cm

8 cm

Side elevation

2 a

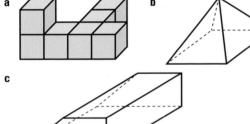

b

c

10.6 Get Ready

1 a 192 m³   **b** 576 cm³

Exercise 10H

1 a 396 cm³   **b** 378 cm³   **c** 204 cm³
2 400 cm³

10.7 Get Ready

1 a a^3 **b** $2a^3$ **c** a^3

Exercise 10I

1 a 78 cm³   **b** 2250 mm³   **c** 0.498 75 m³
   d 216 cm³
2 a 225 cm³   **b** 10 500 cm³   **c** 80.43 cm³
   d 84 000 cm³
3 9 cm
4 Area $= \frac{1}{2}(x + 3x) \times 2x \times 2x = 8x^3$ cm³
5 $h = 4.5y$

10.8 Get Ready

1 a 28.3 cm² (3s.f.)   **b** 19.6 cm² (3s.f.)
   c 78.5 cm² (3s.f.)

Exercise 10J

1 a 251 cm³   **b** 13 600 000 mm³
   c 236 cm³   **d** 8930 cm³
2 a 360 π cm³   **b** 650 π cm³
   c 0.101 25 π m³
3 1600 π m³
4 114.4 cm³
5 31.92 mm
6 3.56 m³ = 3560 litres

Review exercise

1

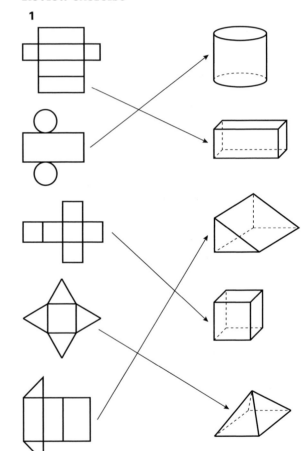

2 20 cm³
3 70 cm²
4 a

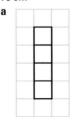

b

5 5 cm
6 150 boxes

7 **a** 5 cm   **b** **i** 19 boxes   **ii** 8 chocolates
8 **a** 118.12 cm²   **b** 83.52 cm³
9 **a**

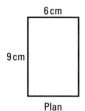

6 cm

9 cm

Plan

4.5 cm

6 cm

Front elevation

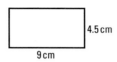

4.5 cm

9 cm

Side elevation

     b 189 cm²
10 Skirting board: $3 \times 4 + 2 \times 2 + 1$, therefore cheapest
     $= 3 \times £30.50 + 1 \times £18.75 + 1 \times £14.00 = £124.25$
     Coving: 20 m, therefore cheapest
     $= 6 \times £27.50 + 1 \times £22.00 = £187.00$
     Total $= £311.25$
11 Area $= 28$ m², therefore $\frac{550}{28} = £19.64$ to spend per m²
     Amy can afford Natural Twist with either underlay or
     Medium Blend with Cushion.
12 Paving $= 20 \times £40 = £800$, grass $= 110 \times £15 = £1650$,
     total $= £2450$
13 $2 \times \pi \times 2.55 = 16$ m $= 32$ roses, so cost $= £134.40$
14 Area $= 4 \times 4 - (4 \times 1 + 0.5 \times 1 \times 1.8 + \pi \times 1^2)$
     $= 7.958$ m²
     Cost $= 7.958 \times 4.60 = £36.61$
15 **a** Volume of cup $= \dfrac{\pi \times 10 \times 4.5^2}{4} = 159$ ml, so the cup
     can hold 150 ml.
   b Volume of squash required $= 30 \times 150 \times 3 = 13\,500$ ml
     $= 13.5$ litres
     1 bottle makes $0.8 \times 7 = 5.6$ litres
     $\frac{13.5}{5.6} = 3.2$, so 3 bottles are needed.
   c £3.75
16 **a** 700 cm³   **b** 13.51 kg
17 Volume $= \frac{25}{2}(1 + 3) \times 10 = 500$ m³,
     so time $= 250$ minutes $= 4$ hours 10 minutes
18 **a** 308.5 m
   b Add 45.8 m to the straights making each one 105.8 m or
     add 29.1 m to the diameter of the bends, making each
     diameter 89.1 m
19 $200 - 50 \times \pi \times 1^2 = 42.9$ cm²
20 Volume of oil $= \pi \times 60^2 \times 180 = 2\,036\,752$ cm³,
     so mass of oil $= 8754$ kg
     Surface area of tank $= 2\pi r^2 + 2\pi rh = 90\,478$ cm²,
     so mass of tank $= 253$ kg
     Total mass $= 9007$ kg $= 9$ tonnes
21 **a** 315 cm³   **b** 0.6 g/cm³

Chapter 11 Answers

11.1 Get Ready

1 **a** 1 2 3 4 5 5 7 8 8 9 10 12   **b** 3.5, 3.5, 4.5, 4.6, 6.2, 8.7, 12.5

Exercise 11A

1 2
2 16 litres
3 **a** 16   **b** 16
4 227

11.2 Get Ready

1 **a** 44   **b** 20.4

Exercise 11B

1 3
2 **a** 136   **b** 134
3 £1292
4 **a** 10   **b** 20   **c** 24

11.3 Get Ready

a Median $= 20$
   Mode $= 20$
b Mean $= 21$ to 2 s.f.

Exercise 11C

1 **a** Mean £47, mode £17, median £22
   b The median is best. The mean has been affected by one
     high value and the mode is the lowest value.
2 One advantage from: Is the most popular measure. Can
   be used for further calculations. Uses all the data. One
   disadvantage from: Affected by extreme values. Actual
   value may not exist.
3 **a** 28   **b** 38
   c 64 (to the nearest whole person)
   d The mean because it is the highest average.

11.4 Get Ready

Number	1	2	3	4	5	6
Frequency	1	4	4	3	3	1

Exercise 11D

1 **a** Frequency $\times$ number of siblings: 0, 8, 18, 12, 12, 0, 12, 7
     Total $f = 30$
     Total $f \times x = 69$
   b 2   **c** 2   **d** 2.3
2 **a** 104   **b** 104   **c** 104
3 **a** 17   **b** 18   **c** 18.5

11.5 Get Ready

Class interval	Frequency
1–4	11
5–9	5
10–14	7
15–19	7

Exercise 11E

1 a $70.0 \leqslant x < 70.1$ **b** $69.9 \leqslant x < 70.0$
2 a £281–£320 **b** £321–£360
3 a $0.45 \leqslant x < 0.50$ **b** $0.40 \leqslant x < 0.45$

11.6 Get Ready

a 60 **b** 0.8 **c** 0.000 55

Exercise 11F

1 42.5 **2** 181.0 seconds **3** 54.23 seconds

11.7 Get Ready

1 14 16 16 18 21 23 27 32 38 43 45 49
2 36.0 kg 43.4 kg 43.5 kg 49.9 kg 56.2 kg 56.2 kg

Exercise 11G

1 Median
2 a $Q_1 = 50, Q_2 = 62, Q_3 = 70$ **b** 20 **c** 42
3 a $Q_1 = 8, Q_2 = 13, Q_3 = 19$ **b** 11 **c** 19
4 a $Q_1 = 32, Q_2 = 45, Q_3 = 52$ **b** 20 **c** 47

Review exercise

1 a 4.5 **b** 3.6 **c** 4
2 a Mode = £4, median = £5, mean = £9
 b The median, because the mode is close to the lowest value and the mean is affected by the single large amount of £38.
3 a One advantage from: unaffected by extreme values; can be used with qualitative data.
 One disadvantage from: may be more than one mode; may not be a mode.
 b Advantage: not influenced by extreme values.
 Disadvantage: actual value may not exist.
 c One advantage from: can be used for further calculations; uses all the data.
 Disadvantage: affected by extreme values.
4 a 86 **b** 3.6
5 a The mode is 12, the number of rooms that occurs most frequently. Ali has given the maximum number of rooms.
 b 6.3 to 1 d.p.
6 2.4 to 1 d.p.
7 45
8 7.7
9 a 10.5–10.7 **b** 10.2–10.4
10 a

Class Interval	Frequency (f)	Class mid-point	$f \times x$
$26 \leqslant w < 29$	4	27.5	110
$29 \leqslant w < 32$	7	30.5	213.5
$32 \leqslant w < 35$	15	33.5	502.5
$35 \leqslant w < 38$	12	36.5	438
$38 \leqslant w < 41$	2	39.5	79
Totals	40		1343

 b 33.6 kg to 3 s.f.

11 84.8 mm to 3 s.f.
12 19 minutes
13 a $30 < t \leqslant 40$ **b** 27.3 minutes
14 13.0 minutes to 3 s.f.
15 Year 11 – 26 pets (Year 9 – 76 pets, Year 10 – 72 pets)
16 a $Q_1 = 42$ kg, $Q_2 = 47$ kg, $Q_3 = 49$ kg
 b 7 kg **c** 11 kg
17 a 18 minutes **b** 15 minutes
 c Before: $Q_1 = 14$ minutes, $Q_3 = 22$ minutes
 After: $Q_1 = 11$ minutes, $Q_3 = 19$ minutes
 d Before: interquartile range = 8 minutes
 After: interquartile range = 8 minutes
 e Before the introduction of the traffic management scheme the mean time taken to travel to work was higher than after the scheme was introduced. The interquartile range stayed the same, so the spread of times was similar before and after the scheme started.
18 Mean $= \dfrac{300\,000}{10} = £30\,000$. The owner could say that the average salary is £30 000 and it is high enough.
 Mode/median = £10 000. The workers could say that the average salary is £10 000 and too low.
19 If 1 dog in 100 had three legs, then the mean number of legs $= \dfrac{399}{100} = 3.99$
 The majority of dogs have four legs.

Chapter 12 Answers

12.1 Get Ready

Students' accurate drawings

Exercise 12A

1–5 Students' drawings
6 The sum of the two shorter sides is less than the longest side.

12.2 Get Ready

1

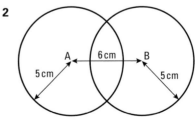

4 cm

2

A   6 cm   B
5 cm   5 cm

3

Exercise 12B

1

2 a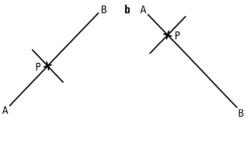

3
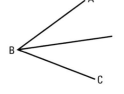

12.3 Get Ready

1

2,3

Exercise 12C

1 a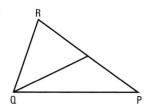
b

2 a
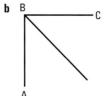

b
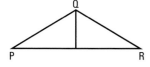

3 a 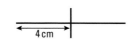 **b** **c**

d **e**

4
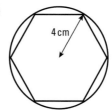

5
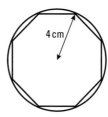

12.4 Get Ready

1
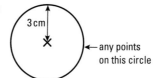

2
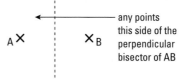

3
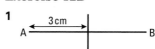

Exercise 12D

1
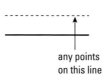

2
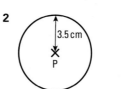

3 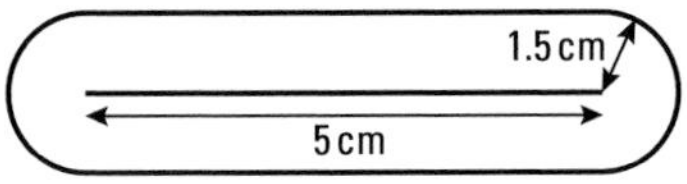
1.5 cm
5 cm

4 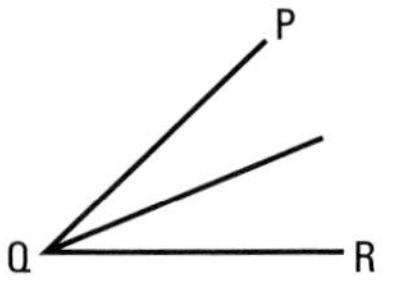
P
Q
R

12.5 Get Ready

1 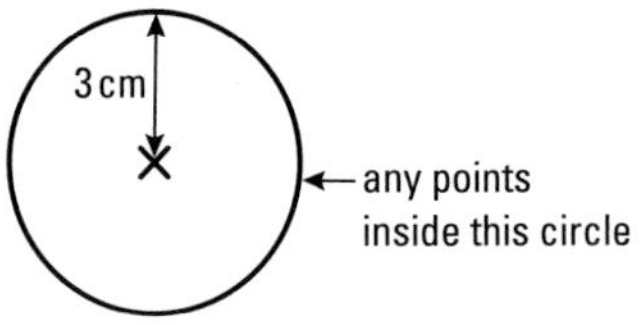
3 cm
any points
inside this circle

2 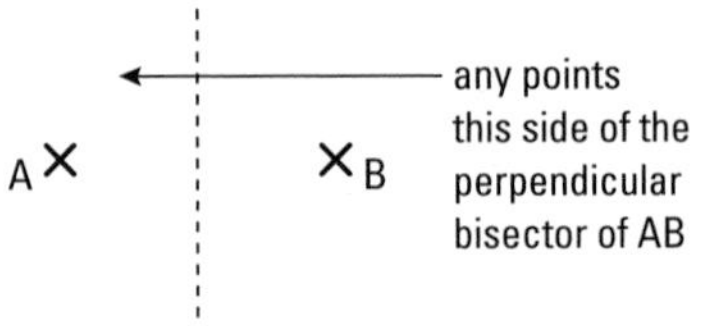
A
B
any points
this side of the
perpendicular
bisector of AB

3 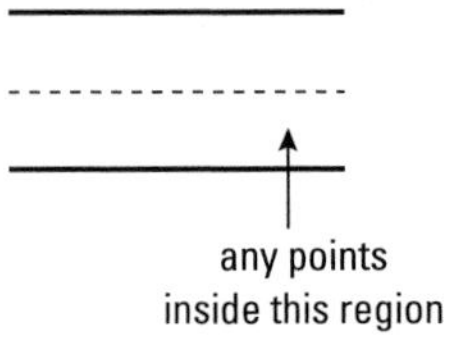
any points
inside this region

Exercise 12E

1 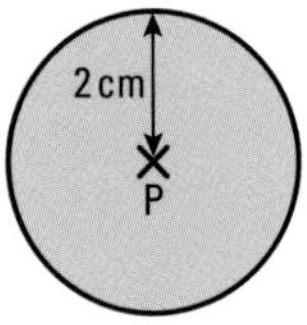
2 cm
P

2 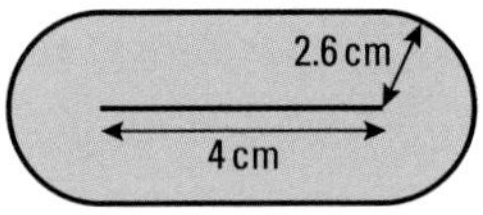
2.6 cm
4 cm

3 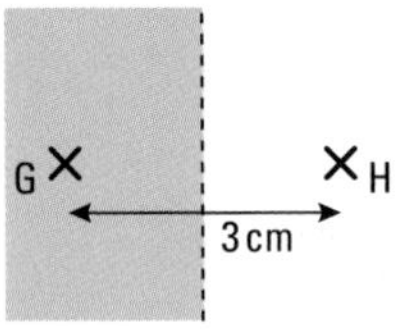
G
H
3 cm

4 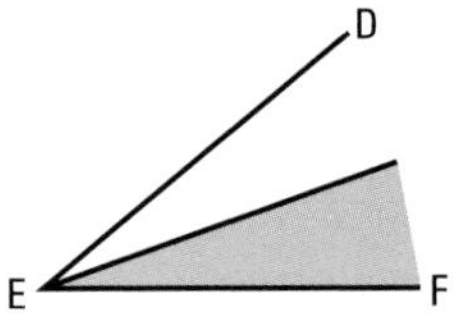
D
E
F

5 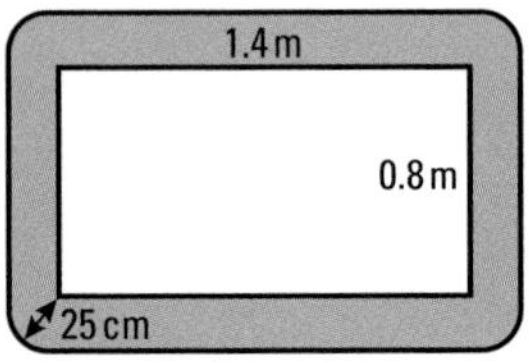
1.4 m
0.8 m
25 cm

12.6 Get Ready

1 **a** 50 km **b** 2.5 km
2 **a** 400 000 cm **b** 30 000 cm

Exercise 12F

1 **a** 1 cm represents 2 km
b **i** 6 km **ii** 9.5 km **iii** 4.2 km
2 **a** 13.5 cm **b** 312 km
3 **b** 14.4 km
4

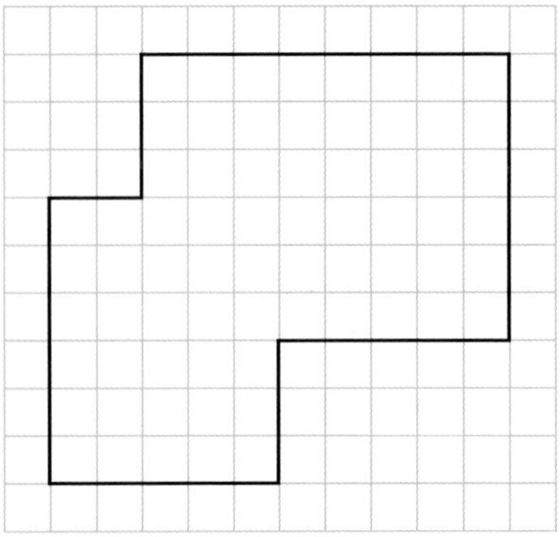

5 **a** 1 : 50 **b** 10 cm
6 **a** 1 : 5 000 000 **b** 3.66 cm

Review exercise

1 By inspection
2 By inspection
3 By inspection
4 By inspection
5 **a** 250 mm **b** 324 m
6

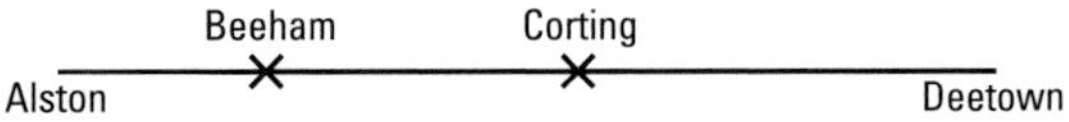
Beeham
Corting
Alston
Deetown

7

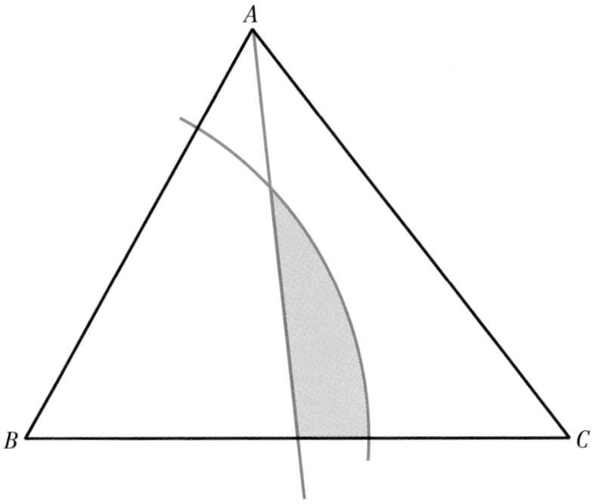
A
B
C

8 Construction of angle of 30° at P
9 **a**

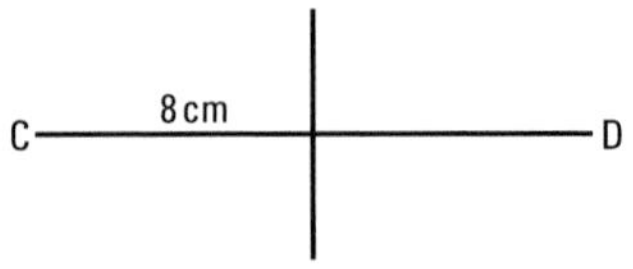
8 cm
C
D

b

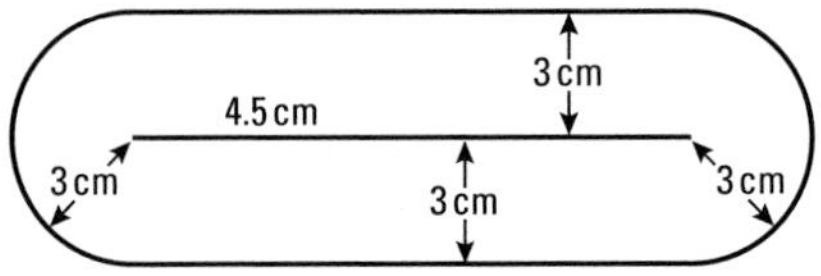
3 cm
4.5 cm
3 cm
3 cm
3 cm

10 a Students' drawings
b 034° **c** 258°
11 Construction of bisector of angle ABC
12

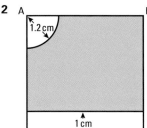

13 Construction of perpendicular bisector of a line 7 cm long
14 Construction of perpendicular to the line ST from a point above the line M
15 a

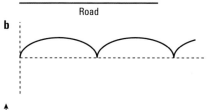

b

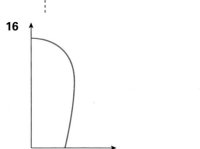

16

Chapter 13 Answers

13.1 Get Ready

1 a $4 + 2p + 5q$ **b** $-3 + 6z$ **c** $12m^2 + 36m$

Exercise 13A

1 $a = 4$ **2** $b = 7$
3 $c = 2.5$ **4** $d = 1.8$
5 $e = -1.5$ **6** $f = -4$
7 $g = 6$ **8** $h = 0.5$
9 $k = -2.6$ **10** $m = -3.5$

13.2 Get Ready

1 a $x = 6$ **b** $b = 4$ **c** $q = 8$

Exercise 13B

1 $x = 3$ **2** $y = 3$
3 $x = 2.25$ **4** $y = 2$
5 $x = 0.25$ **6** $w = -4$
7 $z = -0.25$ **8** $x = -\frac{1}{9}$
9 $x = 1$ **10** $y = \frac{13}{3}$

13.3 Get Ready

1 a $x = 8$ **b** $a = 2$ **c** $b = 2$

Exercise 13C

1 $a = 2.5$ **2** $b = -2.5$
3 $c = 3$ **4** $d = -2$
5 $e = 1$ **6** $f = -\frac{9}{10} = 0.9$
7 $x = 15$ **8** $x = -2$
9 $x = -3$ **10** $x = \frac{3}{5} = 0.6$

13.4 Get Ready

1 a $x = 2$ **b** $x = -\frac{6}{7}$ **c** $x = 0.9$

Exercise 13D

1 $p = 20$ **2** $q = 10$
3 $m = 30$ **4** $x = 24$
5 $y = \frac{53}{8}$ **6** $x = 33$
7 $n = \frac{5}{42}$ **8** $t = -44$
9 $x = 6.5$ **10** $y = \frac{76}{43}$

13.5 Get Ready

1 a $a = -2$ **b** $b = -\frac{1}{2}$ **c** $x = 11\frac{2}{3}$

Exercise 13E

1 $x = 5$
2 31
3 Jessica 80%, Mason 60%, Zach 70%
4 $x = 8$
5 $23\frac{1}{4}$ hours
6 a $(x - 4) = \frac{3(x + 6)}{5}$ **b** 19 cm
7 149 units

13.6 Get Ready

1 a 6 **b** 2 **c** 4

Exercise 13F

1 formula **2** identity
3 expression **4** equation
5 formula **6** formula
7 expression **8** equation
9 formula **10** equation
11 equation **12** identity

Review exercise

1 $t = 4.5$
2 $x = 4.5$
3 a $x = 2.5$ **b** $y = -2.5$
4 58°
5 57 cm
6 Uzma £18, Hajra £38, Mabintou £76
7 A £8, B £12, C £4
8 $x = \frac{44}{3} = 14\frac{2}{3}$
9 $x = 5.5$
10 $y = 10$

Chapter 14 Answers

14.1 Get Ready

1 a i $\frac{1}{5}$ **ii** 0.2 **b i** $\frac{1}{8}$ **ii** 0.125
c i $\frac{3}{5}$ **ii** 0.6 **d i** $\frac{7}{40}$ **ii** 0.175

Exercise 14A

1 a £180 **b** 6 kg **c** £1.62 **d** 4.96 kg
e £6 **f** 45 **g** 2.52 km **h** £52.50
2 12
3 £10
4 1058
5 $\frac{7}{22}$

14.2 Get Ready

1 £75 **2** £16 **3** $\frac{1}{4}$

Exercise 14B

1 a 1.64 **b** 1.03 **c** 1.14 **d** 1.4
e 1.134 **f** 1.125 **g** 1.15 **h** 1.0236
2 a 1.4 **b** £21.56
3 Helen £12 504, Tom £25 008, Sandeep £33 344
4 £621
5 a £144 **b** 70 kg **c** 2.784 m
d £1370.20 **e** 128.52 cm

Exercise 14C

1 a 0.93 **b** 0.8 **c** 0.84 **d** 0.73
e 0.944 **f** 0.975 **g** 0.9275 **h** 0.992
2 a £255 **b** £34 **c** £1020
3 77.9 kg
4 £748
5 a £5840 **b** £4672

14.3 Get Ready

1 £12.75 **2** 1300 **3** $\frac{9}{40}$

Exercise 14D

1 a 50% **b** 25% **c** 40% **d** 20%
e 25% **f** 25% **g** 60% **h** 12.5%
2 90%
3 45%
4 a 20% **b** 60% **c** 11.25% **d** 8.75%

Exercise 14E

1 a +50% **b** +60% **c** −12% **d** −8%
2 15%
3 40%
4 Shop C – as percentage increases are
A 5.19%, B 4.77%, C 5.50%
5 20% profit
6 2.7%

14.4 Get Ready

1 2^5 **2** 0.3 **3** 1.25

Exercise 14F

1 a 1.728 **b** 0.6561 **c** 1.0608 **d** 0.52
2 £1102.50
3 a 1.1136 **b** £66 816
4 No. It is the same as an increase of 68%.
5 8 years

14.5 Get Ready

1 1.15 **2** 0.85 **3** 1.04 **4** 0.96

Exercise 14G

1 £24 000 **2** £280 **3** £620
4 £180 **5** 421 000 **6** £270

Review exercise

1 a £240 **b** 5 kg **c** 10.5 kg **d** £10.50
2 16.7%
3 18 years
4 A (A £510, B £512, C £517.50)
5 CompuSystems (Able £23 000, Beta £23 400,
CompuSystems £24,240, Digital £24 000)
6 72%
7 £7800
8 a 62.5% **b** $\frac{1}{4}$
9 19.9%
10 a It will be worth 32.8% of its original value.
b 0.64
11 £275
12 £8400
13 £1600
14 £665
15 B (A 9.2% over two years, B 9.2025% over two years)

Chapter 15 Answers

15.1 Get Ready

1 a $x = 7$ **b** $x = 1$
2 a $y = 1$ **b** $y = 1.5$

Exercise 15A

1 a

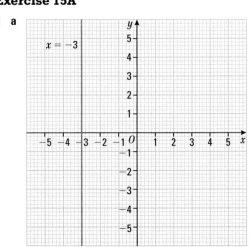

b

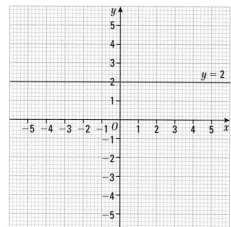

e

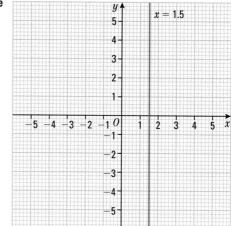

c

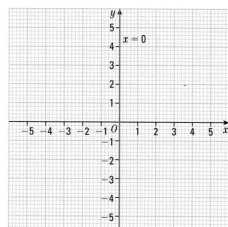

f

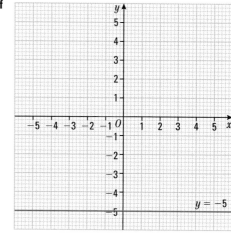

2 a $x = -4$　**b** $y = 2$　**c** $y = 5$　**d** $x = -\frac{1}{2}$
3 a $(1, 3)$　　**b** $(-4, 2)$　**c** $\left(-\frac{1}{2}, 3\right)$
4 22 units, 28 units squared

Exercise 15B

1 a

x	-2	-1	0	1	2	3	4
y	8	6	4	2	0	-2	-4

d

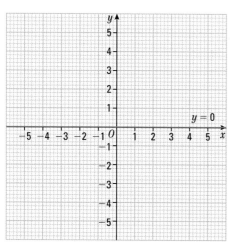

b

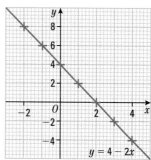

Answers

2 a

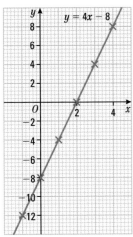

b i (2, 0)   **ii** (0, −8)
c i 2   **ii** 3.5

3 a

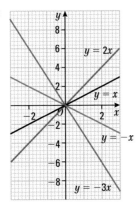

b all pass through (0,0)

4 a

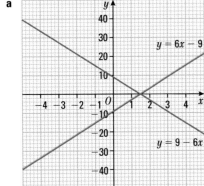

b (1.5, 0)

5

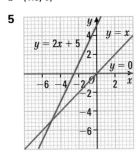

6.25 square units

Exercise 15C

1

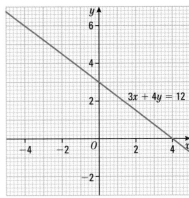

2 a

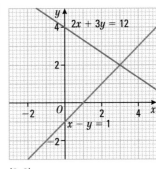

b They are parallel to each other.

3

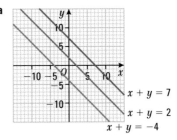

(3, 2)

4 a

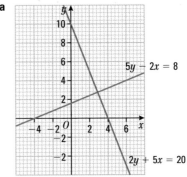

b They intersect at right angles.
c e.g. $2x + 3y = 6$ and $3x − 2y = 12$

15.2 Get Ready

a 5   **b** 5.5   **c** 17.25   **d** 0.225
e 1   **f** −8.5

Exercise 15D

1 **a** (1, 2) **b** (3, −0.5) **c** (−1.5, 3)
 d (−2.5, −0.5) **e** (−4, 0.5) **f** (−0.5, −2)
 g (2.5, −0.5) **h** (0.5, 2.5) **i** (−2, 3)
 j (4.5, −0.5)
2 **a** (1.5, 0.5) **b** (4, 1) **c** (1.5, 1)
 d (3.5, 2.5)
3 **a** (4, 4) **b** (−2, 2.5) **c** (1.5, −3.5)
 d (−3, 4.5) **e** (2, −2.5) **f** (2.5, −1.5)

15.3 Get Ready

1

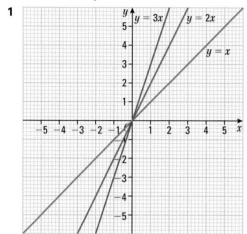

2

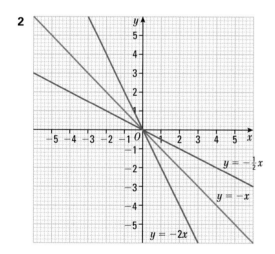

3

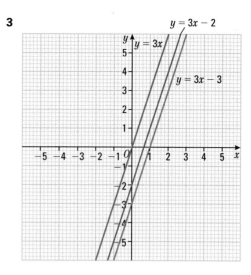

Exercise 15E

1 **a** 30 **b** 10 **c** $\frac{10}{3}$
 d $-\frac{2}{5}$ **e** $-\frac{1}{5}$ **f** $-\frac{1}{40}$
2 **a** 2 **b** −2
 c 4 **d** $\frac{3}{4}$
3 **a**

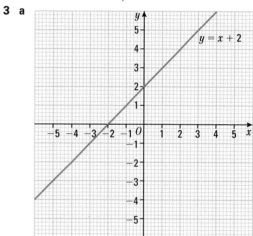

 b

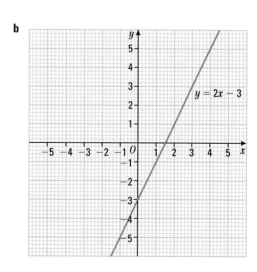

c

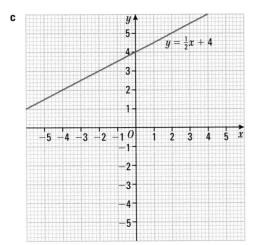

d

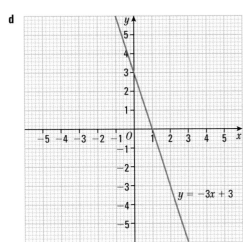

e

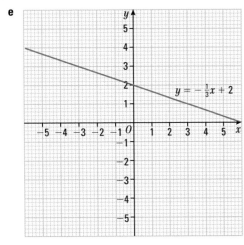

4 e.g. (3, 2)

5 **a** $-\frac{1}{2}$ **b** (0, 4)

Exercise 15F

1 **a** Gradient = 40. This represents the extra time needed (in minutes) for each extra kilogram of chicken.

   b Cooking time = 40 minutes per kilogram plus 20 minutes

   c You can't have a negative weight of chicken.

2 **a** 50°F

b Gradient = 1.8. This represents the number of degrees Fahrenheit for each degree Celsius.

3 **a** A 15, B 10, C 4

   b The gradients represent speed. A is the car (the fastest vehicle), B is the lorry, C is the cycle (the slowest vehicle).

4 **a**

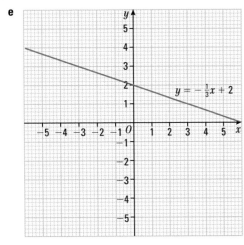

   b $-\frac{5}{6}$

   c The depth of water is going down by $\frac{5}{6}$ cm per second. The swimming pool is being emptied.

15.4 Get Ready

1 $2, \frac{1}{2}, -1$

   The gradient is the same as the coefficient of x.

2 $-1, 2, 2$

   The y-intercept is the same as the value of the number in the equation.

Exercise 15G

1 **A** $y = 2x + 8$ **B** $y = \frac{1}{3}x + 2$ **C** $y = 5 - \frac{1}{2}x$

   D $y = 4 - x$ **E** $y = -2x - 6$

2 $y = 2x + 5$

3 **a i** 4   **ii** 1

   b i 3   **ii** -4

   c i $\frac{2}{3}$ **ii** 4

   d i -0.4 **ii** (0, 4)

   e i $1\frac{1}{3}$ **ii** (0, -4)

   f i $\frac{1}{2}$ **ii** (0, 0)

4 $y = 5x - 2$

5 $y = 3x - 10$

15.5 Get Ready

1 **a** $-\frac{1}{2}, 2$ **b** $-\frac{1}{3}, 3$ **c** $\frac{2}{3}, \frac{2}{3}$

   In **a** and **b** the product of the gradients is -1.

   In **c** the gradients of the parallel lines are the same.

Exercise 15H

1 **a** $-\frac{1}{3}$ **b** $\frac{1}{4}$ **c** -5 **d** $-\frac{1}{3}$ **e** 6

2 **a** $y = 2x + c$ for any value of c except 5

   b $y = \frac{1}{3}x + c$ for any value of c except -1

   c $y = c - x$ for any value of c except 4

3 **a** $y = c - x$ for any value of c

   b $y = c - \frac{1}{3}x$ for any value of c

   c $y = 2x + c$ for any value of c

4 $y = 4x + 3$

5 $2x + y = 0$

6 $y = -4x$

7 $y = x - 3$

15.6 Get Ready

1 C 14:30, D 15:15, E 15:54

Exercise 15I

1 a 4 km **b** 75 minutes **c** $5\frac{1}{3}$ km/h

d 6 minutes each time **e** 10 km/h

2 a £2.75 **b** 2 kg **c** £12.20

3 a 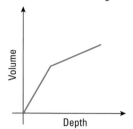 **b**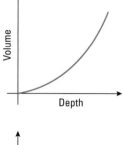

c 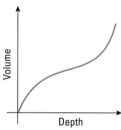 **d**

4 A **b**, B **a**, C **c**

5

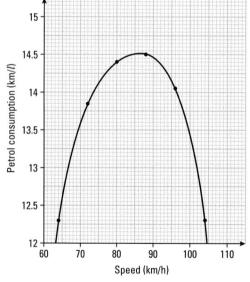

a 13.5 km/l **b** 74 km/h, 97 km/h

Review exercise

1 a

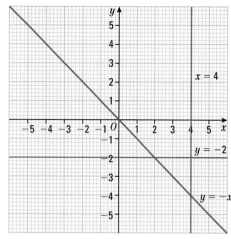

b 2 square units

2

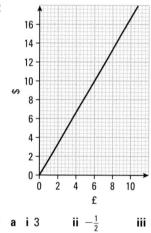

3 a i 3 **ii** $-\frac{1}{2}$ **iii** 0

b i $\frac{1}{2}$ **ii** -3

4

C	0	20	40	60	80	100
F	30	70	110	150	190	230

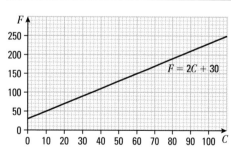

Answers

5 a

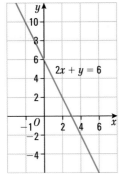

$2x + y = 6$

b i -2 **ii** 6

6 a $120\,km$ **b** $0.5\,hour$ **c** $80\,km/h$
d $18{:}52$ **e** $48\,km/h$

7 B (A gradient $= 2$, B gradient $= 4$, C gradient $= 2.5$)

8 a

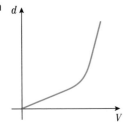

b

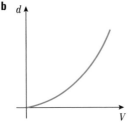

c

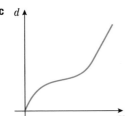

d

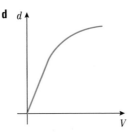

9 a No **b** $y = \frac{1}{2}x + 5$ **c** $y = -2x + 9$

10

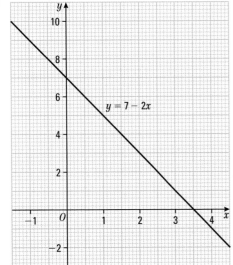

$y = 7 - 2x$

11 a 4 **b** $\left(-\frac{3}{4}, 0\right)$ **c** 4.5

12 a From the top on the right-hand side: B, A, C
b £40
c C would be cheapest. (A £32, B £30, C £25)

13

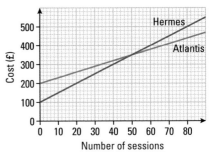

If Abbie plans to go to the health club more than 50 times a year, she should choose Atlantis.

14 $(3, 2)$

15

Equation of line	Gradient	y-intercept
$y = 2x + 5$	2	5
$y = 7x - 3$	7	-3
$y = 6 - x$	-1	6
$y = \frac{2}{3}x - 1$	$\frac{2}{3}$	-1
$y = 3 - x$	-4	3

16 A: The temperature stays constant.
B: The temperature rises at a constant rate.
C: The temperature rises at a constant rate and then falls at a faster constant rate.
D: The temperature stays the same and then falls at a constant rate.
E: The temperature rises at a constant rate, stays the same for a period of time and then continues to rise at the same constant rate.
F: The temperature rises at a constant rate, stays the same for a period of time and then falls at the same rate at which it rose.

17 a Perimeter $= 2 \times (2x + y) = 24$, so $2x + y = 12$
b

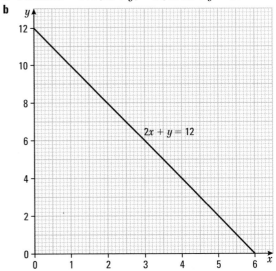

$2x + y = 12$

c $x = 3$

18 $k = 7$. $x = 3$ in $y = 3x - 2$ gives $y = 7$, so $(3, 7)$ also lies on $y = 3x - 2$

19 $y = 5 - 2x$

20 a $-\frac{1}{2}$ **b** $y = 2x - 5$

Chapter 16 Answers

16.1 Get Ready

1 18 km

Exercise 16A

1 a 1:3 **b** 1:4 **c** 1:3.5 **d** 1:0.5
e 1:0.3 **f** 1:0.6 **g** 1:8 **h** $1:\frac{8}{15}$
2 1:5
3 a $\frac{4}{9}$ **b** 4:5 **c** 1:1.25
4 1:375
5 a $1:\frac{1}{6}$ **b** 1:0.2 **c** 1:0.02 **d** 1:40

16.2 Get Ready

1 a 2:3 **b** 1:5 **c** 8:7 **d** 9:200

Exercise 16B

1 a 10 g **b** 30 g **c** 250 g
2 a 10 kg **b** 15 kg
3 108.5 km
4 £1080
5 1.53 m
6 250

16.3 Get Ready

1 20.65 **2** 6 **3** 241.5

Exercise 16C

1 a £4.26:£10.65 **b** 360 g:240 g
c £14.21:£56.84:£99.47 **d** 6.3 m:12.6 m:15.75 m
2 60°, 50°, 70°
3 £36.50
4 $\frac{7}{16}$
5 18
6 £52

16.4 Get Ready

1 15 **2** 50p **3** 40 mins **4** 2.75

Exercise 16D

1 a 3 hours **b** 7 hours
2 £64.05 **3** £33.75 **4** £4.50
5 £3.24 **6** 70 cm

Exercise 16E

1 a 160 g **b** 150 g
2 2 hours 20 minutes
3 a $320 **b** £340
4 £120
5 America by £10.25
6 $58.50

16.5 Get Ready

1 a 52 **b** 16 **c** 36

Exercise 16F

1 a 4 days **b** 5 days
2 2 hours **3** 9 hours **4** 9 days
5 192 cm **6** 8

Review exercise

1 3:2
2 21:4
3 5:2
4 a 0.5 m **b** 16 m
5 45 litres
6 10
7 110
8 Local professional (local $652.50, USA $684)
9 Small bottle (large 0.25 p/g, small 0.22 p/g)
10 2.33 pm
11 1.5 km^2
12 a B **b** D
c A and B, as they lie on the same line through the origin
d D

Multiplication

1 £249.50
2 Yes, her Nan will have to give her £35.80.
3 £48.46

Area

1 e.g.

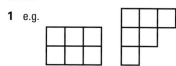

2 £210
3 £248
4 e.g. 1 cm by 36 cm, perimeter 74 cm; 2 cm by 18 cm, perimeter 40 cm; 3 cm by 12 cm, perimeter 30 cm; 4 cm by 9 cm, perimeter 26 cm; 6 cm by 6 cm, perimeter 24 cm
5 £8

Averages and range

1 a Add 2 **b** the same **c** double
d double **e** 1 2 3 4 5 8 19
2 72
3 The new mean is two larger.
4

Answers

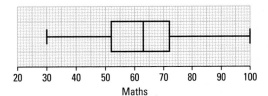

Maths

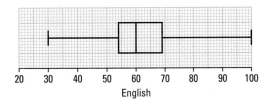

English

The median for Maths (62) is larger than the Median for English (60).
The Maths marks are more varied as The IQR is 20 compared with the English IQR of 15.

5 a May's average hours of sunshine are higher than June so May had more hours of sunshine than June
The number of hours of sunshine were more variable in June as the range was bigger.

b i The mean and range take into account all of the data.
ii The median and IQR are not affected by extreme values.

Price comparisons

1 Ahmed should use Cogas if he uses fewer than 3800 units per annum, otherwise he should use Ourgas.

2 a Cable: $C = 30 + 5m$; Broadband: $C = 6.5m$
b Broadband is cheaper for up to 20 months.

3 a ACars: $C = 60 + 0.32x$; BMotors: $C = 50 + 0.4x$
b ACars is cheaper for more than 125 miles a day.

4 Quick Delivery for parcels up to 1.5 kg and Parcels Fly for parcels over 1.5 kg

5 Pete's Mix if more than 3.33 m³, otherwise Concrete Sue

Intepreting and displaying data

1

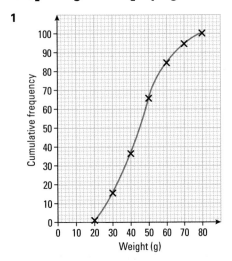

11 pebbles

2 10 workers

3 Take a random stratified sample.
Stratified by school 9 from Avon, 10 from Moorside, 14 Heaton, 12 Moortop, 15 Brambell
Students must be chosen randomly (names in a hat)
Or
Take a systematic sample $\frac{5720}{60}$ students $= 95$
Put numbers 1–95 in a hat and draw out a number.
Go systematically through each school numbering all students and selecting the students with numbers divisible by the number drawn.
Or
Put all names in a hat and draw out 60 students – this sample may not contain students from every school.

4 75

Probability

1 $\frac{1}{3}$, 33% or $0.\dot{3}$

2 When Mona has eaten one sweet there are 19 sweets left in the bag. There are still x chocolates. So the probability of Sam eating a chocolate is x divided by 19 (to be formatted correctly in the book).

3 The probability of the first sock picked being x, depends on the total number of socks in the drawer (i.e. the number of grey socks + the number of black socks) and the number of black socks at the start. When one black sock has been removed, the number of black socks and the total number of socks will both be lower by one. Therefore the probability of picking a black sock the second time will not be x.

4 $\frac{3}{8}$, 0.375 or 37.5%

5 20%, $\frac{1}{5}$ or 0.2

6 0.1, $\frac{1}{10}$ or 10%

7 0.625, $\frac{5}{8}$ or 62.5%

Trigonometry

1 14 cm

2 12.0 cm

3 From corner of shed to top of roof at opposite end $= \sqrt{6^2 + 2^2 + 7^2} = 9.43$ ft, so the bean poles will fit in the shed.

Climbing Snowdon

1 The 9 am train is cheaper by £6. (£46 compared to £52)

2 Rashid: time $= 2.75$ hours, distance $= 7$ km, speed $= 2.55$ km/h
Chelsea: time $= 3$ hours, distance $= 9$ km, speed $= 3$ km/h
Rashid is not the faster walker.

3 a

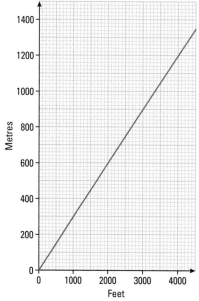

b Height of Snowdon = 3560 × 0.3048 = 1085 m
Ben Nevis is higher.

Music sales

1 Sales in £ millions:

	2006	2007	2008	2009
Downloads	69	115	169	166
CDs	134	135	98	95
Total	203	250	267	261

The total sales between 2006 and 2008 increase each year and then decrease slightly in 2009.
CD sales stay approximately the same from 2006 to 2007 and then decrease, whereas download sales increase significantly each year before steadying out in 2009.

2 Various answers are possible, for example, two frequency polygons.

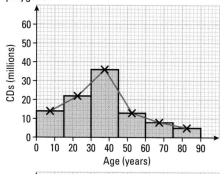

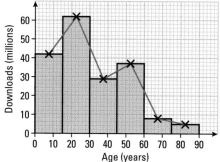

3 Suitable questions for a data collection sheet, for example:
Are you male or female?   Male ☐   Female ☐
How old are you?
0–14 ☐   15–29 ☐   30–44 ☐   45–59 ☐   60–74 ☐
75–89 ☐   90 or over ☐
How much do you spend on music each month?
£0–4.99 ☐   £5–9.99 ☐   £10–14.99 ☐   £15–19.99 ☐
£20 or more ☐
What kind of music do you buy most often?
Pop ☐   Classical ☐   Jazz ☐   Folk ☐   Other ☐

Communication

1 Students' comparisons of costs for different numbers of months. For more than 18 months, 'pay as you go' is cheaper, otherwise monthly contract is cheaper.
2 215.04 seconds
3 13 cm by 19 cm (285 ppi by 285 ppi)
15 cm by 23 cm (247 ppi by 235 ppi)
20 cm by 30 cm (185 ppi by 180 ppi)
Ranji should print the photo at 13 cm by 19 cm.

Energy efficiency

1 100 mm: Space Combi (Economy roll £200, Easy Roll £175, Space Combi £40)
150 mm: Space Blanket (medium) (Economy roll £300, Space Blanket (medium) £168)
200 mm: Space Combi (Economy roll £400, Space Blanket (thick) £210, Space Combi £80)
2 £15.18
3 100 weeks

Going on holiday

1 155 points
2 England (England €478.80, Malta €483)
3 16:56

All at sea

1 12 664 km
2

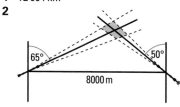

3 12:29

Chapter 17 Answers

17.1 Get Ready

1 arc, mean
   a Start at square K, go 1 to the right, then 1 down and stop. Then go 2 up, and 1 to the left and stop. Then go 4 to the right, and 3 down and stop.

Answers

b Start at square C, go 1 up, then 1 left and stop. Then go 2 up and 1 to the right and stop. Then go 3 down and stop. Then go 1 to the left and 2 up and stop. Then go 3 to the right and 2 down and stop.

c Start at square A, go 3 to the right and stop. Then go 1 up and then 1 down and stop.

d Students' examples

Exercise 17A

1 a $\begin{pmatrix} 5 \\ 5 \end{pmatrix}$ **b** $\begin{pmatrix} 2 \\ -4 \end{pmatrix}$ **c** $\begin{pmatrix} 4 \\ 0 \end{pmatrix}$

   d $\begin{pmatrix} 0 \\ 4 \end{pmatrix}$ **e** $\begin{pmatrix} -2 \\ 3 \end{pmatrix}$ **f** $\begin{pmatrix} -2 \\ -5 \end{pmatrix}$

2
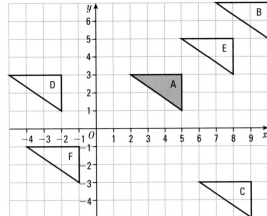

3 a
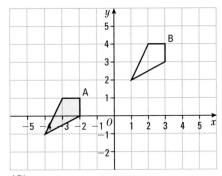

  b $\begin{pmatrix} 5 \\ 3 \end{pmatrix}$

4 a, b
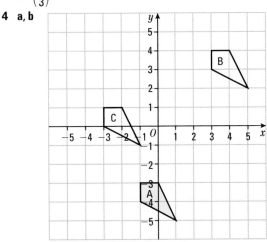

  c $\begin{pmatrix} -2 \\ 4 \end{pmatrix}$ **d** $\begin{pmatrix} 2 \\ -4 \end{pmatrix}$

17.2 Get Ready

A: $x = 4$, B: $y = -3$, C: $y = x$, D: $y = -x$

Exercise 17B

1 a, b
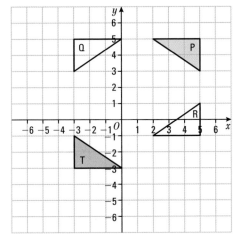

  c Reflection in the line $y = 1$

2 a, b
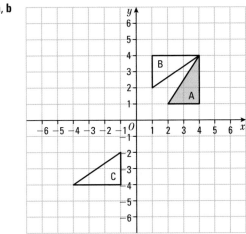

  c Reflection in the line $y = x$

3 a
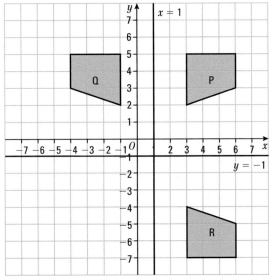

  b Reflection in the line $x = 1$

17.3 Get Ready

1 a 9   **b i** 90° clockwise   **ii** 180°   **iii** 150° clockwise
2 a 90° clockwise
  b 90° anticlockwise
  c 180°
  d 150° anticlockwise

Exercise 17C

1

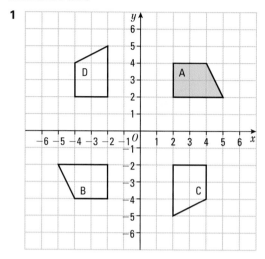

2

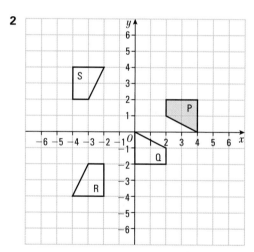

3 a i 90° anticlockwise about the origin
    ii 180° about the origin
    iii 90° clockwise about the origin
  b 90° clockwise about the origin
  c 180° about the origin
4 a i 180° about (3, 5)
    ii 90° clockwise about (2, 1)
    iii 180° about (0, 0)
    iv 90° anticlockwise about (0, 4)
    v 90° anticlockwise about (−2, 2)
  b 90° clockwise rotation about (2, 8)
  c i Translation $\begin{pmatrix} 6 \\ 10 \end{pmatrix}$ **ii** Translation $\begin{pmatrix} 4 \\ 0 \end{pmatrix}$

17.4 Get Ready

1,2

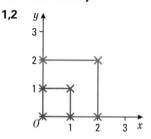

3 They are both squares.
They have a common vertex at (0, 0).
The second square is twice as large as the first square.

Exercise 17D

1 a 48 cm, 52 cm, 20 cm
  b The perimeter is also 4 times as long.
2 a, b

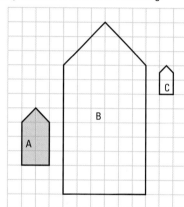

  c 6
3 a P is 4 cm by 2 cm, Q is 8 cm by 4 cm, R is 12 cm by 6 cm.
  b i 12 cm   **ii** 24 cm   **iii** 36 cm
  c i 8 cm²   **ii** 32 cm²   **iii** 72 cm²
  d i 2, same as scale factor
    ii 3, same as scale factor
  e i 4, same as scale factor squared
    ii 9, same as scale factor squared
  f 96 cm

Exercise 17E

1 a

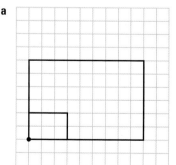

Answers

b

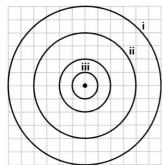

2 a, b

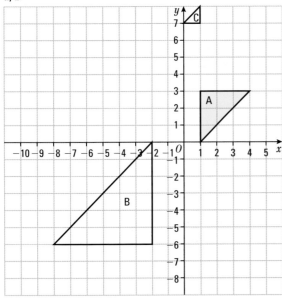

c 6

3 a

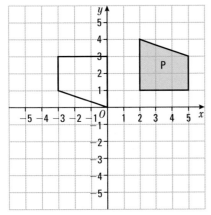

b Rotation 180° about (1, 2)

Exercise 17F

1 a, b

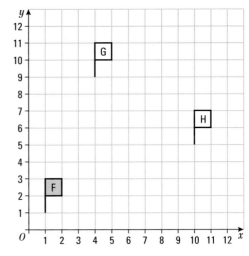

c Translation $\begin{pmatrix} 9 \\ 4 \end{pmatrix}$

d Translation $\begin{pmatrix} -9 \\ -4 \end{pmatrix}$

2 a, b

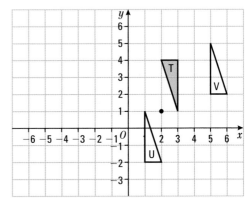

c Rotation 180° about (4, 3)

3 a, b

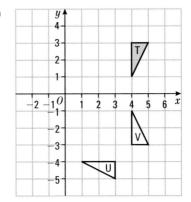

c Reflection in the x-axis

4 Reflection in the line $x = 5$

5 Reflection in the x-axis

Review exercise

1

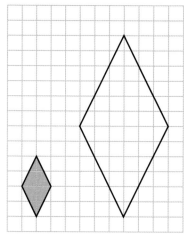

2 a

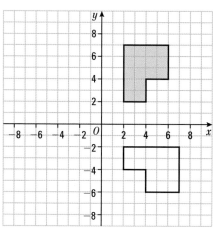

b Translation $\begin{pmatrix} 3 \\ -1 \end{pmatrix}$

3 a

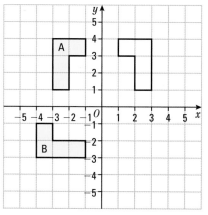

b 90° anticlockwise rotation about the origin

4

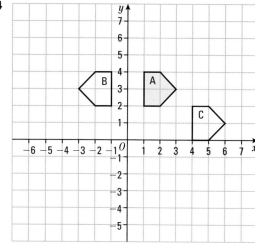

5 Students' tile designs

6 90° clockwise rotation about $(-2, 3)$

7

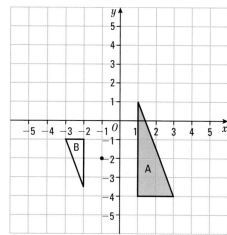

8 a Reflection in the line $y = x$

b

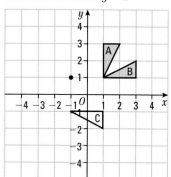

9 a

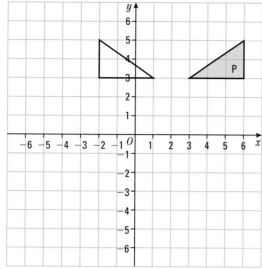

b Translation $\begin{pmatrix} 5 \\ -4 \end{pmatrix}$

10

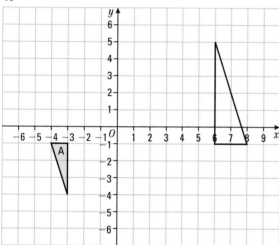

11 180° rotation about (1, 0)
12 180° rotation about the origin

Chapter 18 Answers

18.1 Get Ready

1 360
2 90
3 a 120   **b** 45   **c** 60

Exercise 18A

1

2

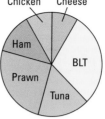

3

Exercise 18B

1 a Golf   **b** Athletics   **c** 45   **d** 30
2

Shop A   **Shop B**

18.3 Get Ready

a 50, 54, 65, 72
b 4.0, 4.3, 4.4, 4.6
c 0.01, 0.1, 0.11, 0.12

Exercise 18C

1 a 36   **b** 28   **c** 47   **d** 18, 36   **e** 18
2 a

0	4	5	6	7	7	8	9	9	9	9
1	0	2	4	5	7					
2	1	4	8							
3	0									

Key 2 | 1 stands for 21
  b 9 minutes   **c** 9 minutes   **d** 26 minutes
  e 7 minutes, 17 minutes   **f** 10 minutes
3 a

5	2	6	9					
6	3	3	4	5	5	8	8	8	9
7	2	4	4	4	4			
8	2	3	5	8				
9	2	4						

Key 7 | 2 stands for 72
  b 74 km   **c** 69 km   **d** 42 km
  e 64 km, 82 km   **f** 18 km

18.4 Get Ready

a B occurs more frequently than A, which occurs more frequently than C.

b The red category accounts for a higher proportion of the total than the green or blue categories. Red: 50%, green and blue: 25% each.

Exercise 18D

1 a 30°C   **b** 33°C   **c** G   **d** C and F   **e** A
2 a Food   **b** Pension   **c** 25%
3 a 70 g   **b** 10 g   **c** Fruitbix
4 a Saturday   **b** Thursday   **c** 40

18.5 Get Ready

a 3   **b** 16   **c** 25

Exercise 18E

1

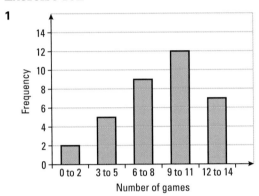

2 a $135 \leqslant w < 140$ **b** $140 \leqslant w < 145$
c

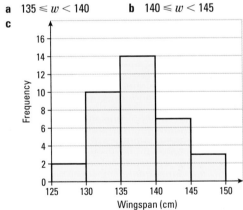

3 a $139 \leqslant w < 141$ **b** $137 \leqslant w < 139$
c

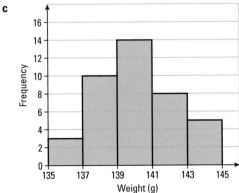

18.6 Get Ready

a 5   **b** 17.5   **c** 115.5

Exercise 18F

1

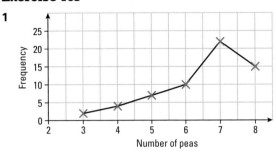

2 a $70 \leqslant d < 80$
b, c

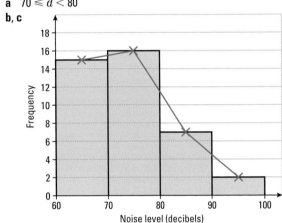

3

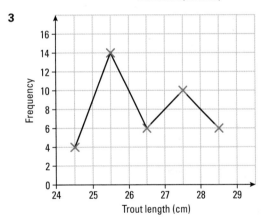

4 Girls, because their mode is 5 minutes and the boys' mode is 9 minutes.

Exercise 18G

1 a

Lifetime (l hours)	Frequency	Class width	Frequency density
$10 \leqslant l < 15$	4	5	0.8
$15 \leqslant l < 20$	10	5	2
$20 \leqslant l < 25$	20	5	4
$25 \leqslant l < 30$	15	5	3
$30 \leqslant l < 40$	6	10	0.6

Answers

b

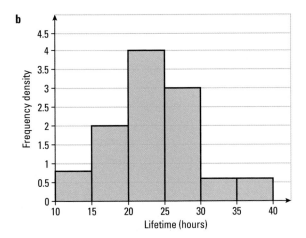

2

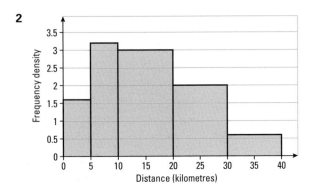

25

3 a

Age (y years)	Frequency
$0 < y \leqslant 5$	10
$5 < y \leqslant 10$	28
$10 < y \leqslant 20$	49
$20 < y \leqslant 40$	48
$40 < y \leqslant 70$	60

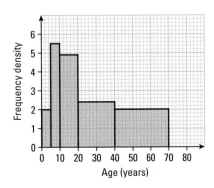

b 101

Exercise 18H

1 a Cumulative frequency: 3, 10, 20, 35, 43, 48, 50
b

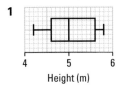

2 a 11 **b** 19 **c** 33
3 a i 22 **ii** 74
   b 110
   c 18%

18.9 Get Ready

a 9 and 12
b 18 and 20

Exercise 18I

1 a 42 **b** 33, 50 **c** 17 **d** 70
2 a £6000 **b** £4000, £7500 **c** £3500
3 a Median = £242 000, Q_1 = £222 000, Q_3 = £256 000
   b Range = £120 000, interquartile range = £34 000

18.10 Get Ready

Median = 15
Lower quartile = 8
Upper quartile = 25
Interquartile range = 17

Exercise 18J

1

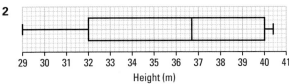

2

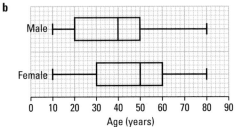

3 a Male: min = 10, Q_1 = 20, Q_2 = 40, Q_3 = 50, max = 80
   Female: min = 10, Q_1 = 30, Q_2 = 50, Q_3 = 60, max = 80
b

The female members are older on average (higher median). The interquartile range is the same for males and females, but the range is slightly greater for females so their ages vary slightly more.

Review exercise

1

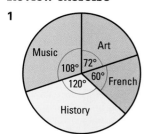

2

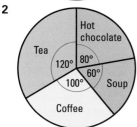

3 a 14 **b** 5

c

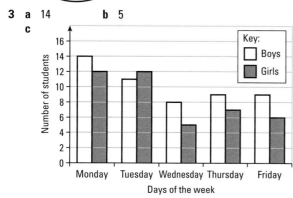

d Tuesday

4 a

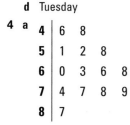

Key 5 | 1 means 51 kg

b 5 **c** 87 − 46 = 41 kg

5 30 mm

6

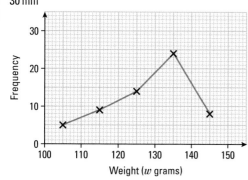

7

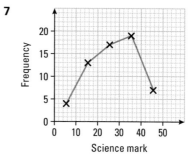

8 James. The angle for the combined proportion must be between the Year 9 and Year 10 angles.

9 On average, the prices are lower at Peter's garage than at John's garage, as the median is lower. The range and interquartile range are both smaller for Peter's garage so the spread is less than for John's garage. Both the cheapest and the most expensive cars come from John's garage.

10 a 88 people **b** 38 years **c** 54 − 22 = 32 years

11 a

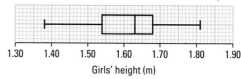

b On average, the girls are taller than the boys, as the median is higher, but their heights are more variable as the range is larger. The interquartile ranges are the same so the variation in the middle of each group is similar.

12 From a cumulative frequency chart, the median for the men is £240. On average, the men spent more than the women.

13 a 23 kg is the value of Q_3. The heaviest bag weighs 29 kg.
   b 17 kg **c** 13 kg **d** 60 bags

14 a 21 **b** 10
   c Most of the high and low scores are scored by the boys. More of the girls scored an average mark.

15

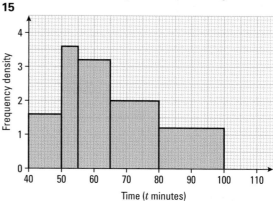

16 a

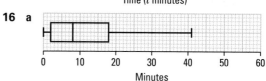

b The delays were greater on Saturday, as the median and the quartiles are all higher. There was also more variation in the delays on Saturday, since both the range and the interquartile range are larger.

17 a

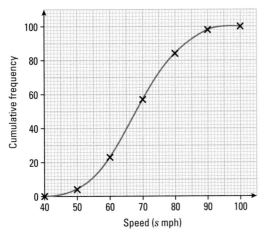

b 88 mph   **c** $76 - 61 = 15$ mph

b

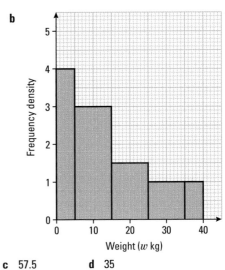

c 57.5   **d** 35

18 a

Distance (d km)	Frequency
$0 < d \leqslant 5$	15
$5 < d \leqslant 10$	20
$10 < d \leqslant 20$	25
$20 < d \leqslant 40$	20
$40 < d \leqslant 60$	10

b

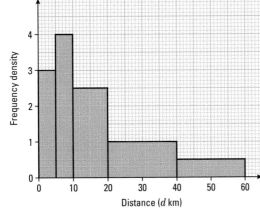

19 a

Weight (w kg)	Frequency	Frequency density
$0 < w \leqslant 5$	20	4
$5 < w \leqslant 15$	30	3
$15 < w \leqslant 25$	15	1.5
$25 < w \leqslant 35$	10	1
$35 < w \leqslant 40$	5	1

Chapter 19 Answers

19.1 Get Ready

1 a 2.8   **b** 0.4   **c** -0.8 **d** -2.6

Exercise 19A

1 a

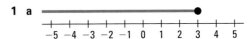

b

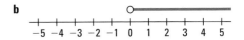

c

d

e

2 a $x \leqslant 4$ **b** $x > -1$ **c** $x \leqslant 5$ and $x > -2$
d $x < 0$ and $x > -3$ **e** $x \leqslant 3$ and $x \geqslant -5$
f $x < 5$ and $x \geqslant 1$

19.2 Get Ready

1 $x = 10$
2 $x = \frac{1}{2}$
3

4

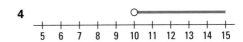

Exercise 19B

1 a $x > 4$

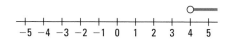

b $x \leq 1$

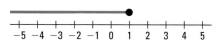

c $x \leq -2$

d $x > 1.6$

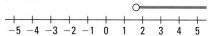

2 a $x < 4.5$ **b** $x > 4$ **c** $x \leq 2.5$ **d** $x > 6.5$

3 a $x \leq 3.25$ **b** $x > 23$ **c** $x \geq -\frac{6}{11}$ **d** $x \geq -\frac{1}{19}$

19.3 Get Ready

1 $x \geq 4$ **2** $x > 4$ **3** $x \leq 2$ **4** $x > 3$

Exercise 19C

1 a $-1, 0, 1, 2, 3, 4, 5$
  b $-4, -3, -2, -1, 0, 1$
  c $1, 2, 3$
  d $-5, -4, -3, -2, -1, 0, 1, 2, 3, 4$

2 a $-3, -2, -1, 0, 1, 2, 3$
  b $-4, -3, -2, -1, 0, 1, 2, 3, 4, 5, 6, 7$
  c $0, 1, 2, 3, 4$
  d $-3, -2, -1, 0, 1, 2, 3, 4, 5, 6, 7, 8, 9$

3 a $-2, -1, 0, 1, 2, 3$
  b $-1, 0, 1, 2, 3, 4$
  c $-1, 0, 1, 2, 3, 4$
  d $-7, -6, -5, -4, -3, -2, -1, 0, 1, 2, 3, 4$

4 a $-2, -1, 0, 1, 2, 3, 4, 5, 6$
  b $-4, -3, -2, -1, 0, 1, 2, 3, 4, 5, 6, 7$
  c $0, 1, 2, 3$
  d $-4, -3, -2, -1, 0, 1, 2, 3, 4, 5, 6, 7, 8, 9, 10$

19.4 Get Ready

1

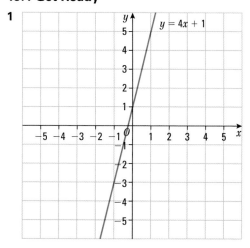

2

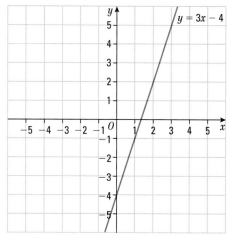

3

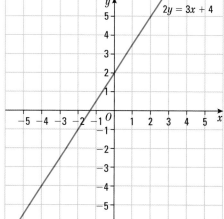

Exercise 19D

1 $(1, 1), (1, 0), (1, -1), (0, 0), (0, -1), (-1, -1)$

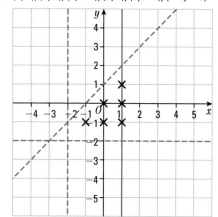

Answers

2 a

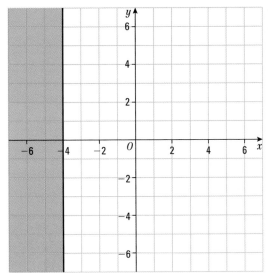

b

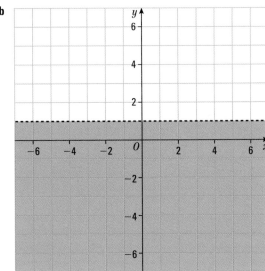

c

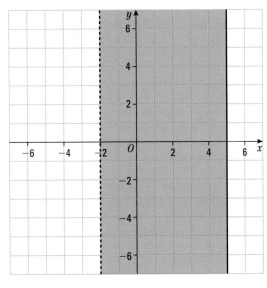

d

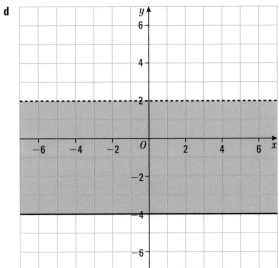

3 a

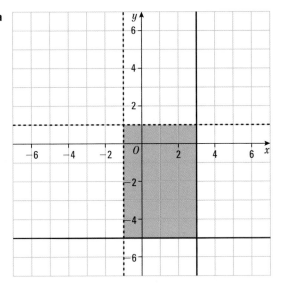

b

c

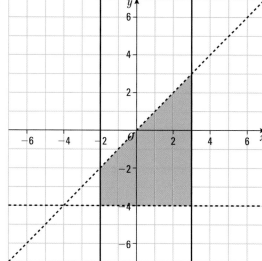

d

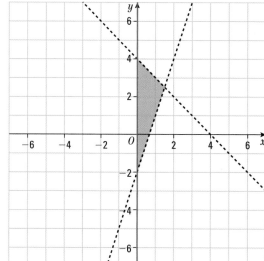

4 a $y = 3, y = 2x + 3, x + y = -1$
  b $y < 3, y \geqslant 2x + 3, x + y > -1$
  c $(-2, 2), (-1, 2)$ and $(-1, 1)$

5 a $y = 2, x = 3, y = \frac{1}{2}x - 3, y = -x$
  b $y \leqslant 2, x < 3, y > \frac{1}{2}x - 3, y \geqslant -x$
  c -1

19.5 Get Ready

1 a 5347.76 **b** $\frac{8}{3}$, or $2\frac{2}{3}$

Exercise 19E

1 a 120 **b** -25
2 a 65 **b** 5
3 a 1570 **b** 120.576
4 a 120 miles **b** 8 m/s
5 a 5 hours 5 minutes **b** 18 lb
6 a 12 **b** 9

19.6 Get Ready

a 3200 **b** 9

Exercise 19F

1 a $C = 50d + 90$ **b** €790
2 $P = 5x + 3y + z$
3 $T = \dfrac{45m + 75f}{60} = \dfrac{3m + 5f}{4}$
4 $P = 4x + 4y$
5 a $T = 2a + 5c$ **b** $P = 2a + 3c$

19.7 Get Ready

1 60 miles **2** 1.8×10^{17}

Exercise 19G

1 $x = \dfrac{y - 3}{5}$ **2** $d = \dfrac{c + 2}{5}$

3 $a = \dfrac{v - u}{t}$ **4** $x = \dfrac{P - y}{5y}$

5 $m = \dfrac{4 - E}{3}$ **6** $g = \dfrac{f + 30}{3}$

7 $x = 7T - 2$ **8** $n = \dfrac{2Y + 3m}{3}$

9 $p = \dfrac{3W - 2y}{2y}$ **10** $w = 4 - 3A$

19.8 Get Ready

1 $u = v - at$ **2** $R = \dfrac{V}{I}$ **3** $m = \dfrac{E}{c^2}$

Exercise 19H

1 $R = \sqrt{\dfrac{A}{\pi}}$ **2** $x = P^2 + y$

3 $u = \sqrt{v^2 - 2as}$ **4** $g = \dfrac{2s}{T^2}$

5 $x = \sqrt{\dfrac{75 - y^2}{50}}$ **6** $g = \dfrac{f + 4}{4}$

7 $m = \dfrac{1}{T - 3}$ **8** $p = \dfrac{q - 12}{q + 4}$

9 $c = \dfrac{a - 3}{b + 7}$ **10** $T = \dfrac{7}{2W^2 - 3}$

Review exercise

1 9
2 a $b = 3a$ **b** $c = a + 5$
3 a £130 **b** $C = 90 + 0.5m$
4 0.6
5 a 8 **b** $p = \dfrac{S - 3q}{4}$
6 a $1 < x < 4$ **b** $-1 < x \leqslant 5$ **c** $x < 2$
7 a

$$\begin{array}{c} \overset{\circ}{\longleftrightarrow}\!\overset{\circ}{} \\ \underset{-5\ -4\ -3\ -2\ -1\ \ 0\ \ 1\ \ 2\ \ 3\ \ 4\ \ 5}{} \end{array}$$

b

$$\begin{array}{c} \bullet\!\!\!\longrightarrow \\ \underset{-5\ -4\ -3\ -2\ -1\ \ 0\ \ 1\ \ 2\ \ 3\ \ 4\ \ 5}{} \end{array}$$

c

$$\begin{array}{c} \longleftarrow\!\!\!\circ \\ \underset{-5\ -4\ -3\ -2\ -1\ \ 0\ \ 1\ \ 2\ \ 3\ \ 4\ \ 5}{} \end{array}$$

d

$$\begin{array}{c} \bullet\!\!\!-\!\!\!\bullet \\ \underset{-5\ -4\ -3\ -2\ -1\ \ 0\ \ 1\ \ 2\ \ 3\ \ 4\ \ 5}{} \end{array}$$

8 $-3, -2, -1, 0, 1$

9 **a** $t < 5.5$ **b** 5
10 $-3, -2, -1, 0, 1, 2$
11 $p < 5$
12 $0 < x < 4$
13 **a** $L = x + (x + 4) + 2(x + 4) = x + x + 4 + 2x + 8$
     $= 4x + 12$
   b $4x + 12 < 50$ **c** $0 < x < 9.5$
14

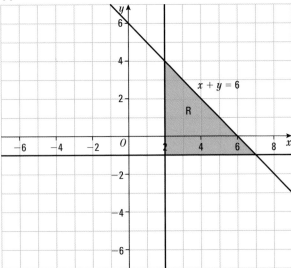

15 **a** $\frac{10}{7}$ **b** $u = \frac{vf}{v - f}$
16 $x = \frac{4y + 15}{5 + 3y}$
17 **a** 49.6   **b** $r = \frac{P - 2a}{\pi + 2}$
18 **a** Any two of: (1, 1), (1, 2), (2, 1)
   b

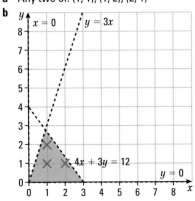

19 **a** $x \geqslant 2, y \geqslant 3, 2x + 3y < 20$
   b

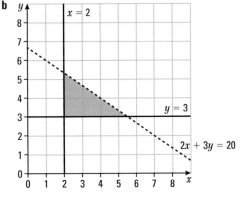

Chapter 20 Answers

20.1 Get Ready

1 **a** 144   **b** 54.76
2 **a** 1681   **b** 276.93
3 **a** 225   **b** 102.55
4 **a** 25   **b** 12
5 **a** 6.44   **b** 8.91

Exercise 20A

1 **a** 13 cm   **b** 10 cm   **c** 10.4 cm   **d** 11.6 cm
2 **a** 12.6 cm   **b** 11.7 cm   **c** 14.4 cm   **d** 20.8 cm

Exercise 20B

1 **a** 7 cm   **b** 35 cm   **c** 12.3 cm   **d** 8.03 cm
2 **a** 15.2 cm   **b** 5.77 cm
   c **i**

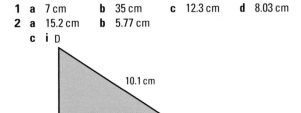

    ii 6.42 cm

20.2 Get Ready

The rounded lengths are 21 cm and 30 cm. The diagonal is approximately 36 cm.

Exercise 20C

1 **a** 15.0 m   **b** 6.92 cm   **c** 18.7 cm
2 3.47 m
3 52.4 cm^2
4 **a** 3.20 m
   b **i** 4.33 m   **ii** 2.33 m
   c 8.66 m
5 Triangles 1, 2, 4 and 6 are right-angled triangles.
6 2.85 cm
7 **a** 17 inches
   b Height = 18 inches, width = 32 inches

20.3 Get Ready

a 26   **b** 19.3   **c** 10.3

Exercise 20D

1 **a** 10   **b** 26   **c** $\sqrt{212} = 14.6$
   d $\sqrt{424} = 20.6$ **e** 29   **f** $\sqrt{117} = 10.8$
2 **a** **i** 5   **ii** $\sqrt{50} = 7.07$ **iii** 5
   b $5^2 + 5^2 = 50$, so triangle is right-angled at A.
3 **a** 13
   b B, C, E lie on the circle.

20.4 Get Ready

1 **a** p **b** r **c** q

2 a e **b** f **c** d
3 a AC   **b** BC   **c** AB

Exercise 20E

1 a 0.3420   **b** 0.9542   **c** 0.5   **d** 0.9461
   e 1   **f** 15.8945   **g** -0.7408 **h** 0.0699
   i 0.7965   **j** 0.2538   **k** -0.3739 **l** 0.0471
2 a 53.1   **b** 25.5   **c** 60   **d** 43.8
   e 58.4   **f** 63.8   **g** 2.7   **h** 60
   i 45

Exercise 20F

1 a sine   **b** cosine   **c** tangent   **d** cosine

Exercise 20G

1 a 37.7°   **b** 46.2°   **c** 19.7°   **d** 40.1°
   e 47.1°   **f** 43.6°
2 a 36.3°   **b** 38.9°   **c** 32.3°
3 a 41.0°   **b** 57.2°   **c** 98°

20.5 Get Ready

1 a

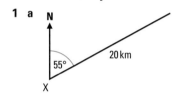

b

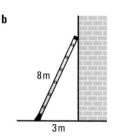

2 cosine

Exercise 20H

1 a 5.47 cm   **b** 17.1 cm   **c** 11.6 cm
   d 26.1 cm   **e** 10.4 cm   **f** 11.3 cm
2 a 18.1 cm   **b** 13.4 m   **c** 10.8 cm
3 a i 9.96 cm   **ii** 8.36 cm
   b 5.30 cm   **c** 13.7 cm   **d** 68 cm²

Exercise 20I

1 a i 1.45 m   **ii** 5.44 m   **iii** 2.54 m
   b 65.0°
2 a 16.8 m   **b** 64.5°   **c** 24.5°
3 a 46.9 m   **b** 88.3 m
4 a 2.87 cm   **b** 16.5°
5 a i 13.0 km   **ii** 10.9 km
   b i 7.73 km   **ii** 23.8 km
   c i 20.7 km   **ii** 34.7 km
   d 40.4 km, 059.2°
6 a 3.84 cm   **b** 16.8 cm   **c** 43 cm²

Review exercise

1 6.36 cm
2 6.71 cm
3 230 km
4 19.4 cm
5 20.6°
6 a 51.3°   **b** 10.5 cm
7 21.7 m
8 033°
9 116 cm
10 a Alan
   b $5\frac{1}{3}$ seconds. Alan's time $= \dfrac{60 + 80}{5} = 28$ seconds,
      Bhavana's time $= \dfrac{\sqrt{60^2 + 80^2}}{3} = 33\frac{1}{3}$ seconds
11 a

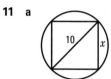

    $x^2 + x^2 = 10^2$, so $x = 7.07$ cm
   b

    $x = 10$ cm
12

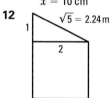

   a Area $= 4 \times 2.24 = 8.96$ m², so he needs to buy two
      rolls of felt.
   b Two rolls of felt cost £24.00, two tins of adhesive cost
      £13.98, so total cost $=$ £37.98
13 No. It depends on the diagonal of the desk:
    $d^2 = 2.3^2 + 1.6^2$, $d = 2.80$ m, but the room is only 2.75 m
    wide.

Chapter 21 Answers

21.1 Get Ready

1

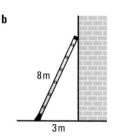

2 a -4 **b** 4
3 a 2   **b** 18

Answers

Exercise 21A

1 a

x	-3	-2	-1	0	1	2	3
y	6	1	-2	-3	-2	1	6

b

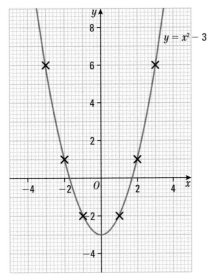

c $x = 0$ **d** $(0, -3)$

2 a

x	-3	-2	-1	0	1	2	3
y	-5	0	3	4	3	0	-5

b

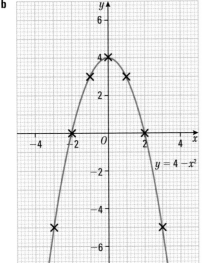

c $(0, 4)$ **d** $x = 2$ and $x = -2$

3 a

x	-3	-2	-1	0	1	2	3
y	20	10	4	2	4	10	20

b

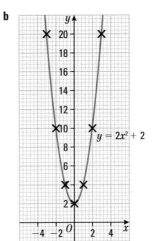

c **i** $y = 6.5$ **ii** $x = 2.1$ and $x = -2.1$

4 a

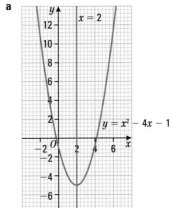

i $x = 4.2$ and $x = -0.2$ **ii** $x = 2$

b

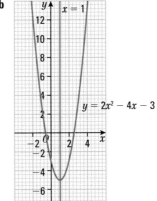

i $x = 2.6$ and $x = -0.6$ **ii** $x = 1$

c

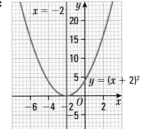

i touches the x-axis at -2 **ii** $x = -2$

d

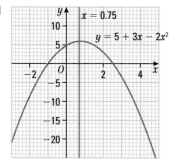

$x = 0.75$

$y = 5 + 3x - 2x^2$

i $x = 2.5$ and $x = -1$ **ii** $x = 0.75$

Exercise 21B

1 a $x = 0$ and $x = 3$ **b** $x = -1$ and $x = 4$
c $x = 2.8$ and $x = -1.3$ **d** $x = 3.6$ and $x = -1.6$
2 a $x = 0.4$ and $x = 2.6$ **b** $x = 0.4$ and $x = 2.6$
c $x = 3.1$ and $x = -1.6$ **d** $x = 3.2$ and $x = -1.2$

3 a

x	-2	-1	0	1	2	3	4
y	-7	-2	1	2	1	-2	-7

b

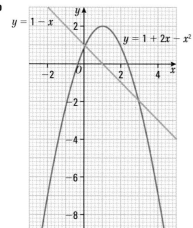

$y = 1 - x$

$y = 1 + 2x - x^2$

c $x = 0$ and $x = 3$

4 a

x	-3	-2	-1	0	1	2	3
y	32	16	6	2	4	12	26

b

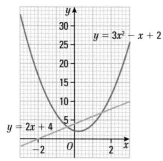

$y = 3x^2 - x + 2$

$y = 2x + 4$

c $x = 1.5$ and $x = -0.5$

21.2 Get Ready

1 1, 8, 27, 64, 125
2 a 1 000 000 **b** -1000

Exercise 21C

1 a

x	-3	-2	-1	0	1	2	3
y	-25	-6	1	2	3	10	29

b

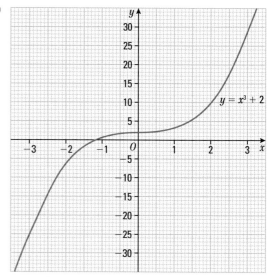

$y = x^3 + 2$

c $y = 17.6$

2 a

x	-4	-3	-2	-1	0	1	2	3	4
y	-28	0	10	8	0	-8	-10	0	28

b

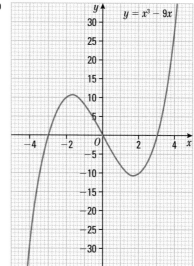

$y = x^3 - 9x$

c $x = -3$, $x = 0$ and $x = 3$

3 a

x	-3	-2	-1	0	1	2	3	4
y	$+45$	$+4$	-7	0	$+13$	$+20$	$+9$	-32

b

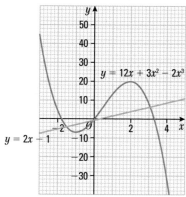

c $x = -1.5$, $x = -0.1$ and $x = 3.1$

4 i C   **ii** A   **iii** D   **iv** B

21.3 Get Ready

1 a $\frac{1}{4}$ **b** 4   **c** 0.4   **d** 2.5

2 As the value of x gets bigger, the value of $\frac{1}{x}$ gets smaller.

Exercise 21D

1 a

x	0.2	0.4	0.5	1	2	4	5	10	20
y	25	12.5	10	5	2.5	1.25	1	0.5	0.25

b

x	-20	-10	-5	-4	-2	-1	-0.5	-0.4	-0.2
y	-0.25	-0.5	-1	-1.25	-2.5	-5	-10	-12.5	-25

c

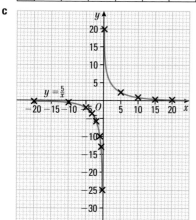

2

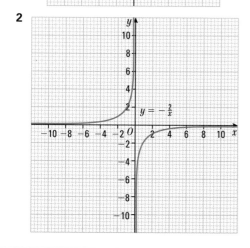

3 a

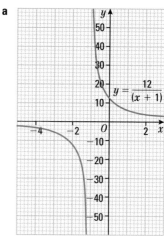

b $x = -1$

21.4 Get Ready

1 a 81   **b** 1   **c** $\frac{1}{9}$

2 a $x = 4$ **b** $x = 2$ **c** $x = 3$

Exercise 21E

1 a

x	-3	-2	-1	0	1	2	3
y	0.04	0.111	0.33	1	3	9	27

b

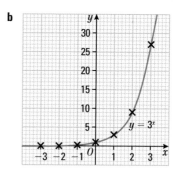

c i 5.2   **ii** 2.5

2 A is $y = \left(\frac{1}{2}\right)^x$, B is $y = 3^{-x}$, C is $y = 5^x$, D is $y = 2^x$

3 a 10   **b** 320   **c** 17 minutes

4 $p = 0.625$, $q = 16$

21.5 Get Ready

1 a 4.6   **b** 2.2   **c** 1.5

2 When $x = 1$, $x + \frac{1}{x} = 1 + \frac{1}{1} = 2$

3 a -2 **b** 2

Exercise 21F

1 a 1 and 2   **b** 0 and 1   **c** 0 and 1, -1 and 0, 4 and 5

    d 2 and 3

2 a 1.7   **b** -1.3 **c** 4.8 and 0.2

3 a 3.04   **b** 2.17   **c** 4.72

4 a $x(x + 2)(x - 2) = x^3 - 4x$ **b** $x = 8.6$

    c 10.6 cm, 6.6 cm, 8.6 cm

Review exercise

1 a

t	0	2	4	6	8	10
s	0	49	116	201	304	425

b

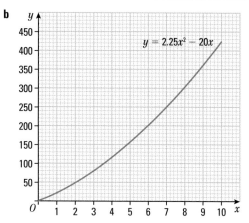

$y = 2.25x^2 - 20x$

c i 80.25 m   **ii** 9.6 seconds

2 a

x	−4	−3	−2	−1	0	1	2
y	8	3	0	−1	0	3	8

b

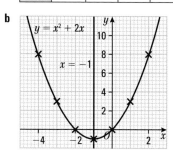

$y = x^2 + 2x$
$x = -1$

c $x = -1$
d i $y = 1.25$ **ii** $y = 1.6$ and $y = -3.6$

3 a

x	−3	−2	−1	0	1	2	3
y	−10	−4	0	2	2	0	−4

b

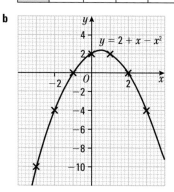

$y = 2 + x - x^2$

c i $x = -1$ and $x = 2$ **ii** 2.8 and −1.8
d (0.5, 2.25)

4 a $x^2 - 3x - 2 = x - 2$
$x^2 - 3x - 2 - x + 2 = 0$
$x^2 - 4x = 0$
b $x = 0$ and $x = 4$ **c** $y = 2 - x$

5 a When $x = -0.5$, $x^2 + \dfrac{2}{x} + 3 = -0.75$

When $x = -1$, $x^2 + \dfrac{2}{x} + 3 = 2$
So there must be a solution to $x^2 + \dfrac{2}{x} + 3 = 0$
between $x = -0.5$ and $x = -1$.
b 0.60

6 a

x	−2	−1	0	1	2	3	4
y	−8	1	0	−5	−8	−3	16

b

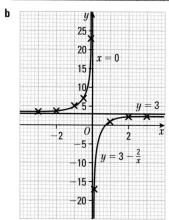

$y = x^3 - 2x^2 - 4x$

c i $x = 0$, $x = 3.2$ and $x = -1.2$
ii $x = -1.8$, $x = 1$ and $x = 2.8$

7 a

x	−3	−2	−1	−0.5	−0.1	0.1	1	2	3
y	3.7	4	5	7	23	−17	1	2	2.3

b

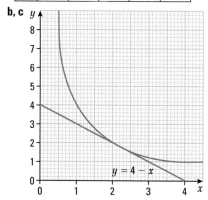

$x = 0$
$y = 3$
$y = 3 - \dfrac{2}{x}$

c $x = 0$, $y = 3$

8 a

x	0.5	1	2	3	4
y	8	4	2	1.3	1

b, c

$y = 4 - x$

d $x = 2$

9 a

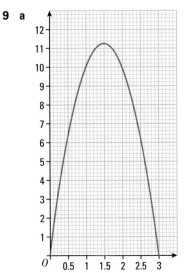

b 1.5 seconds
c 2.85 seconds

10 a

b i 22 metres
  ii 9 metres and 71 metres
  iii 82 metres

11 a

$y = x^2 + 3x$

b 3.2 cm
12 7.63 cm
13 a C   **b** D   **c** A   **d** E   **e** B
14 $k = 0.25$

Chapter 22 Answers

22.1 Get Ready

1 a $x = -1.5$ **b** $y = 4$
2 a -4 **b** 21

Exercise 22A

1 $x = 4, y = 1$ **2** $x = 5, y = 3$
3 $x = 1, y = -2$ **4** $x = 2, y = 1$
5 $x = 1, y = 4$ **6** $x = 1, y = 2$
7 $x = -1, y = -1$ **8** $x = -2, y = -1$
9 $x = 0.5, y = 2$ **10** $x = 0.5, y = -4$
11 $x = 3, y = 1$ **12** $x = 2, y = -0.5$
13 $x = -3, y = 5$ **14** $x = 1, y = -3$
15 $x = 0.5, y = -1$

22.2 Get Ready

1 $x = 2, y = 3$
2 $x = 1, y = 2$
3 $x = 3, y = 1$

Exercise 22B

1 7, 12
2 £18.50
3 £8.80
4 Adult = £4.50, child = £3
5 14 g
6 a $4a = 3b + 1$ or $4a - 3b = 1$
      $2a - 3 = 5 - b$ or $2a + b = 8$
  b $a = 2.5$ cm, $b = 3$ cm
7 Ann = £9/day, Mary = £12/day
8 $x = 3, y = 2$

22.3 Get Ready

1 $4a = 3b + 4$
    $3a = 24 - 3b$
2 $a = 4, b = 4$
3 56

Exercise 22C

1 a B $y = 2x$, A $x + y = 3$, C $x - 2y = 3$
  b i $x = 1, y = 2$ **ii** $x = 3, y = 0$ **iii** $x = -1, y = -2$
2 a $x = 2, y = 0$ **b** $x = 3, y = 3$ **c** $x = 2, y = 8$

22.4 Get Ready

1 a $x(x - 4)$ **b** $y(2y + 5)$ **c** $(2x - 1)(x + 3)$
  d $(3y + 4)(2y - 5)$
2 a 8   **b** 1   **c** 149

Exercise 22D

1 a 0, 4   **b** 3, -5 **c** 0.5, 2.25
  d 0, -2 **e** 0, 1   **f** 0, 1.75
2 a 2, 4   **b** $-1, -6$ **c** 3, -4
  d 3   **e** 9, -4 **f** 4, -4
  g -5 **h** 10, -10
3 a $-5, -\frac{1}{5}$ **b** $3, \frac{2}{3}$ **c** $\frac{1}{2}, -4$
  d $-3, \frac{1}{5}$
4 a 3, -2 **b** 5, -2 **c** 7, -3
  d 7, -5 **e** 3, -2.5 **f** $4\frac{1}{2}, -\frac{2}{3}$
  g 8, -1 **h** $1, -\frac{9}{16}$

22.5 Get Ready

1 $x^2 + 6x + 9$
2 $x^2 - 10x + 25$
3 $x^2 + 2ax + a^2$

Exercise 22E

1 a $(x + 2)^2 - 4$ **b** $(x + 5)^2 - 25$
  c $(x + 6)^2 - 36$ **d** $(x - 1)^2 - 1$
  e $(x - 7)^2 - 49$ **f** $(x - 12)^2 - 144$
  g $(x + 0.5)^2 - 0.25$ **h** $(x - 1.5)^2 - 2.25$
  i $(x + 2)^2 + 3$ **j** $(x + 4)^2 + 1$
  k $(x + 5)^2 - 45$ **l** $(x - 3)^2 + 2$
  m $(x - 10)^2 - 20$ **n** $(x - 13)^2 - 170$
  o $(x - 0.5)^2 + 1.25$ **p** $(x + 2.5)^2 - 11.25$
2 a $2(x + 3)^2 - 18$ **b** $2(x - 1)^2 + 3$
  c $3(x - 2)^2 - 2$ **d** $5(x + 5)^2 - 25$
3 a $p = 4, q = 8$ **b** 8
4 a $(0, 10)$ **b** $(-3, 1)$
5 a $r = 5, s = 2$ **b** 5 **c** $x = 2$

22.6 Get Ready

1 $x = 2, x = -2$
2 $x = 3, x = -2$
3 $x = 7, x = -4$

Exercise 22F

1 a $x = 3 \pm \sqrt{11}$ **b** $x = -2 \pm \sqrt{3}$
  c $x = -5 \pm \sqrt{37}$ **d** $x = 1 \pm 2\sqrt{2}$
  e $x = \dfrac{3 \pm \sqrt{15}}{2}$ **f** $x = \dfrac{-6 \pm \sqrt{21}}{5}$
2 a $x = -0.68, x = -7.32$ **b** $x = 8.27, x = 0.73$
  c $x = 2.37, x = -3.37$ **d** $x = 0.87, x = -2.87$
  e $x = 0.88, x = -0.38$ **f** $x = 0.93, x = -0.43$

22.7 Get Ready

1 $x = \pm\sqrt{14} - 2$
  $x = 1.74, x = -5.74$
2 $x = \pm\sqrt{14} - 3$
  $x = 0.74, x = -6.74$
3 $x = \pm\sqrt{8} + 1$
  $x = 3.83, x = -1.83$

Exercise 22G

1 $-0.586, -3.41$ **2** $-0.807, -6.19$
3 $0.606, -6.61$ **4** $2.70, -3.70$
5 $5.32, -1.32$ **6** $4.30, 0.697$
7 $-0.293, -1.71$ **8** $1.54, 0.260$
9 $1.64, -0.811$ **10** $0.322, -0.622$
11 $1.39, 0.360$ **12** $3.45, -1.45$
13 $0.372, -5.37$ **14** $5.11, -4.11$
15 $1.72, -1.52$

22.8 Get Ready

1 $2x^2 + 9x - 18$
2 $6x^2 - 18x + 12$
3 $2x + 12$

Exercise 22H

1 $x = 0, x = 5$ **2** $x = 2, x = -1.8$
3 $x = 1.5, x = -2$ **4** $x = 1, x = -\dfrac{5}{9}$
5 $x = 0.5, x = -0.6$ **6** $x = 2, x = -0.75$
7 $x = 2.30, x = -1.30$ **8** $x = 4.48, x = -3.73$
9 $x = 5.32, x = -1.32$ **10** $x = 3.27, x = -4.27$
11 $x = -0.576, x = -10.4$ **12** $x = 3.41, x = 0.586$

22.9 Get Ready

1 $4x - 2x^2$
2 $x^2 + 2x + 1$
3 $8 + 10x - 3x^2$

Exercise 22I

1 $4, -6$
2 4
3 12 and 48
4 7.22 m
5 a $x^2 + (x + 1)^2 = 41$
    $2x^2 + 2x - 40 = 0$
    $x^2 + x - 20 = 0$
  b $4, 5$ or $-4, -5$
6 12.35 cm, 15.72 cm and 20 cm
7 10 cm by 10 cm
8 14 m or 16 m
9 $x = 5$
10 54 km/h

22.10 Get Ready

1 a, c, d and **e**
2, 3

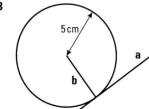

Exercise 22J

1

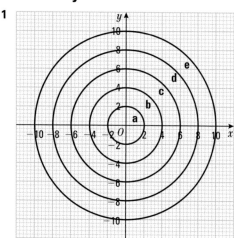

Answers

2

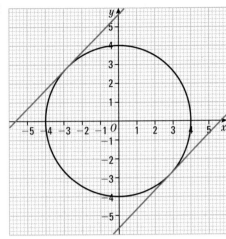

(2.83, −2.83) and (−2.83, 2.83)

3 a $y = 1, y = 11$ **b** $y = -9, y = 1$
  c $x = -2, x = 8$ **d** $x = -10, x = 0$

4 $y = x + 12.58, y = x - 4.48$

22.11 Get Ready

1

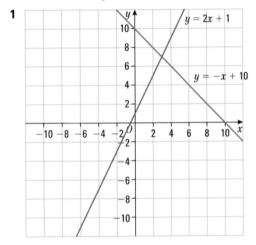

$x = 3, y = 7$

2

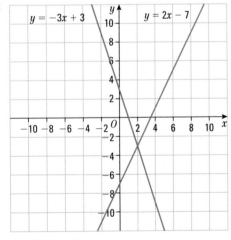

$x = 2, y = -3$

Exercise 22K

  a   graphs of the equations
  b   :

1 $x = 2, y = 2; x = -3, y = -3$
2 $x = 4, y = 7; x = -2, y = 1$
3 $x = 3, y = -0.5; x = -1, y = 1.5$
4 $x = 1, y = 1; x = -1.67, y = 6.33$
5 $x = 1, y = 2; x = 2, y = -1$
6 $x = 3.5, y = 21; x = -0.5, y = -3$

22.12 Get Ready

1

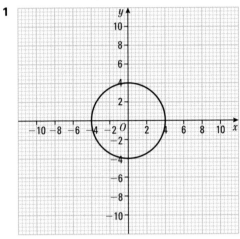

2

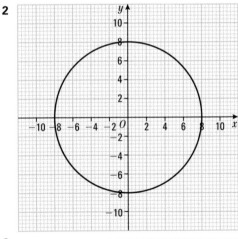

3

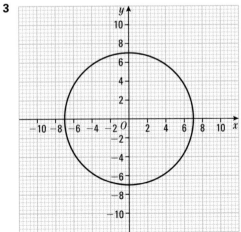

Exercise 22L

1 $x = 2.7, y = 5.4; x = -2.7, y = -5.4$
2 $x = 1.2, y = 4.2; x = -4.2, y = -1.2$
3 $x = 6.8, y = -1.8; x = -1.8, y = 6.8$
4 **a** $x = 2, y = 3; x = -3, y = -2$
   b $x = 4, y = -2; x = -2, y = 4$
   c $x = 2.2, y = 5.4; x = -3, y = -5$
5 **a** $x = 0.449, y = 4.45; x = -4.45, y = -0.449$
   b $x = 1.49, y = 5.46; x = -2.09, y = -5.26$
   c $x = 5.63, y = 8.26; x = -3.23, y = -9.46$

Review exercise

1 **a** $x = \pm 3$ **b** $x = \pm 6$ **c** $x = \pm 7.35$
2 **a** $y = \pm 2$ **b** $t = \pm 2$ **c** $p = \pm 3$
3 3 cm
4 **a** $y = 1$ and $y = 6$ **b** $t = 2$ and $t = -2$
   c $p = 1$ and $p = -4$
5 **a** $x = \frac{4}{3}, y = \frac{25}{3}$ **b** $x = -1, y = 1$
   c $x = -2, y = -5$
6 **a** $x = 2, y = 2$ **b** $x = 0.8, y = 2.4$
7 29.12 m
8 **a** $h = 0$ when $t = 0$ or 6, so the ball was in the air for 6 seconds.
   b By symmetry of graph, maximum height is when $t = 3$, giving height of 45 m
   c $h = 25$ when $t = 1$ or $t = 5$, so the ball was 25 m above the ground after 1 second and 5 seconds.
9 **a** $k = 0.5$ and $k = 5$ **b** $m = -0.5$ and $m = 1.5$
10 **a** $p = 4, q = -11$ **b** $x = -4 \pm \sqrt{11}$
11 $x = 6.27$ and $x = -1.27$
12 **a, b** $x = 2.41$ and $x = -0.414$
13 **a** $(x - 1)^2 + 2$ **b** 2   **c** $x = 1$
   d

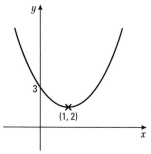

14 $4(x + 3)^2 - 36.$ $a = 4, p = 3, q = -36$
15 **a** $x = 2, y = 2$ **b** $x = 0, y = 2$
   c $x = -4, y = 3$
16 $F = 12.5, g = 5$
17 $y = 2x - 3$ and $2y + x = -6$
18 $x = -5, y = 8$
19 **a** $p = 4, q = 7$ **b** $(4, 7)$
20 50p
21 $x = 1, y = 0$ and $x = -1.33, y = 9.33$
22 **a** $5(x - 1) = (4 - 3x)(x + 2)$
       $5x - 5 = 8 - 2x - 3x^2$
       $3x^2 + 7x - 13 = 0$
   b $x = 1.22$ and $x = -3.55$
23 **a** $5x + x(2x + 1) = 95$
       $2x^2 + 6x - 95 = 0$
   b $x = 5.55$ and $x = -8.55$

24 **a** Time for first 10 km $= \dfrac{10}{x}$
       Time for second 10 km $= \dfrac{10}{x - 1}$
       Total time = 4, so $\dfrac{10}{x} + \dfrac{10}{x - 1} = 4$
   b $10(x - 1) + 10x = 4x(x - 1)$
       $4x^2 - 24x + 10 = 0$
   c $x = 5.55$ and $x = 0.450$
   d 4.55 km/h
25 **a** $x = -4.30, y = 0.697$ and $x = -0.697, y = 4.30$
   b $x = 0.350, y = 6.70$ and $x = -5.15, y = -4.30$
26 $x = 3.41$ and $x = 0.586$
27 $x^2 + y^2 = 16$ is a circle with radius 4, centre the origin, so it intersects the x- and y-axes at ± 4. The point $(1, 2)$ must lie inside this circle. Therefore any straight line that passes through $(1, 2)$ must intersect the circle twice.

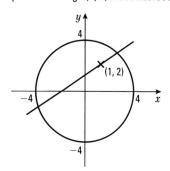

28 $x = 1.83, y = 4.65$ and $x = -2.63, y = -4.25$

Chapter 23 Answers

23.1 Get Ready

1 **a** 6   **b** $\frac{1}{6}$ **c** $\frac{1}{9}$

Exercise 23A

1 **a** 1.75 cm   **b** 17.6 cm   **c** 64.9 m
2 **a** 55.4 cm^2 **b** 9.25 cm^2 **c** 230 cm^2
3 **a** 19.3 cm   **b** 41.6 m
4 70°
5 5 cm
6 **a** 1710 m^2 **b** 48.9 m
7 **a** 57.9 cm   **b** 235 cm^2

23.2 Get Ready

1 **a** $10a$ **b** $3b$
2 $x = 4.5$

Exercise 23B

1 **a** 9π cm   **b** π m   **c** 16π m
2 **a** 4π cm^2 **b** 100π m^2 **c** 30.25π cm^2
3 **b** 16π cm
4 **a** 8 cm   **b** 64π cm^2
5 **a** 36π cm^2 **b** 4π cm   **c** $36 + 4\pi$ cm
6 Perimeter $= \frac{1}{2}\pi Y + \frac{1}{2}\pi X + \frac{1}{2}\pi(X + Y) = \pi(X + Y)$

Answers

23.3 Get Ready

1 a 100   **b** 10 000   **c** 1 000 000
2 a 500   **b** 130 000   **c** 7 600 00
3 a 70   **b** 0.35   **c** 0.049

Exercise 23C

1 a 45 239 cm^2 **b** 4.5239 m^2
2 a 14 cm^2 **b** 1400 mm^2
3 a 40 000 cm^2 **b** 69 000 cm^2 **c** 6 cm^2
   d 0.47 cm^2
4 a 5 000 000 m^2 **b** 300 000 m^2 **c** 4 m^2
   d 0.056 m^2
5 a 1000   **b** 1 000 000   **c** 8 300 000 mm^2
6 a 44 800 cm^2 **b** 3.6 cm^2

23.4 Get Ready

1 a 2.211 cm^2 **b** 14.56 mm^2 **c** 65.0 mm^2 (to 3 s.f.)

Exercise 23D

1 a 60 m^3 **b** 480 cm^3 **c** 320 cm^3
2 16 400 000 m^3
3 a 100 cm^3 **b** 125 cm^3 **c** 0.0804 cm^3
4 $53\frac{1}{3}\pi$ cm^3
5 1590 cm^3

23.5 Get Ready

1 a 10.24π **b** 32.2 cm^2 (to 3 s.f.)
2 a 64   **b** 216   **c** 1331

Exercise 23E

1 268 cm^3 **2** 4090 m^3 **3** 8.31 cm
4 17.9 cm   **5** 1940 cm^3 **6** 5.04 m

23.6 Get Ready

1 a 6.28 cm^3 (to 3 s.f.)   **b** 576 cm^3

Exercise 23F

1 a 223 cm^3 **b** 177 m^3 **c** 52.5 m^3 **d** 812 cm^3
2 120π cm^3 **3** $54 000\pi$ cm^3 **4** 1610 cm^3
5 731.25π cm^3 **6** $\frac{8}{3}\pi r^3$

23.7 Get Ready

1 a 1 000 000 000   **b** 4620   **c** 91.875 m^3

Exercise 23G

1 a 2 000 000 cm^3 **b** 6 750 000 cm^3 **c** 0.45 cm^3
   d 0.0068 cm^3
2 a 7000 mm^3 **b** 3750 mm^3 **c** 25 mm^3
3 a 0.075 m^3 **b** 0.0008 m^3 **c** 0.000 125 m^3
4 a 0.83 l **b** 5.6 l **c** 1000 l **d** 0.003 54 l
5 720 000 litres
6 1277 cubes
7 Volume A = 12.5 m^3 and volume C = 0.375 m^3 so shape A
   has the largest volume.

23.8 Get Ready

1 15 cm^2 **2** 36.3 mm^2 (to 3 s.f.)

Exercise 23H

1 a 10 950 cm^2 **b** 11 700 cm^2 **c** 3524 cm^2
   d 3 m^2 **e** 684 cm^2 **f** 1392 cm^2
2 £94

23.9 Get Ready

1 3.2 cm   **2** 3.6 cm   **3** 64.2 mm

Exercise 23I

1 a 240.3 cm^2 **b** 3.3 m^2
2 a 503 cm^2 **b** 5890 mm^2 **c** 302 cm^2
3 a 100π cm^2 **b** 80π cm^2 **c** 356π cm^2
4 a 192 cm^2 **b** 924 cm^2

23.10 Get Ready

(1.5, 3)

Exercise 23J

1 O (0, 0, 0), A (4, 0, 0), B (4, 0, 3), C (0, 0, 3), D (0, 2, 3), E (0, 2, 0),
   F (4, 2, 0), G (4, 2, 3)
2

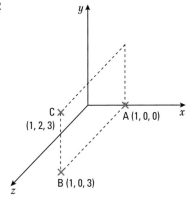

3 a O (0, 0, 0), A (4, 0, 0), B (4, 0, −3), C (0, 0, −3), D (0, −2, 0),
   E (4, −2, 0), F (4, −2, −3), G (0, −2, −3)
   b i (0, −2, −1.5)   **ii** (4, −1, −1.5)
4 (−2, −3, 0), (−2, −3, −1), (−2, 0, −1)

Review exercise

1 96 cm^2
2 110 cm^2
3 8 000 000 cm^3
4 a 4.5 cm^2 **b** 60 000 cm^2
5 6.54 km^2
6 a 905 m^3 **b** 4.92 m
7 9.3 cm
8 $h = \frac{10}{9}x^3 - 3x^2$
9 10.9 cm^2
10 $12 + 4\pi$
11 a $\frac{7\pi}{2}$ **b** $4 + \frac{7\pi}{2}$
12 vol = 2.7π litres
13 a (5, 2, 0)   **b** (2.5, 1, 3)

14 $h = 9x$

15 1700 cm³

16 Yes. Volume of a pyramid $= \frac{1}{3} \times$ base area $\times$ height and Volume of a cone $= \frac{1}{3} \times$ base area $\times$ height, so if the base areas and volumes are the same, then the heights must also be equal.

17 18.3 cm² (3 s.f.)

Chapter 24 Answers

24.1 Get Ready

a 0.05   **b** 0.4

Exercise 24A

1 **a** 700 m
  b They stopped twice; once for 1 minute then again for 2 minutes.
  c 9.5 minutes   **d** 420 m
2 **a** Month 1; 95 m   **b** 80 m
  c Months 6 and 8. These correspond to June and August when it doesn't rain much.
3 **a** 18:00 hours, because people are making their evening meals
  b 102 000 kilowatts   **c** 12:00, 16:36, 20:00
  d 22:00

24.2 Get Ready

a A 63.5, B 64.25   **b** C 10.75   **c** D 54.9   **d** E 16.6

Exercise 24B

1 **a**

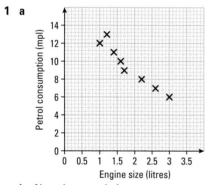

  b Negative correlation
  c The greater the engine size, the lower the petrol consumption.
2 **a**

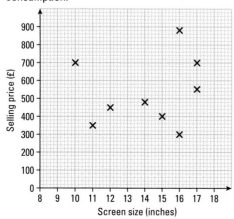

b No correlation
c Screen size and selling price are not related.

3 **a**

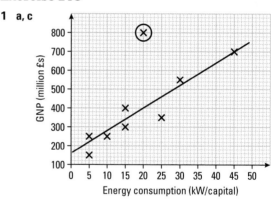

  b Negative correlation
  c The more years have passed since restocking, the smaller the number of fish.
4 **a** Related – the lighter the car, the faster it will go.
  b Related – the longer the motorway, the greater the number of petrol stations.
  c Unrelated
  d Related – the greater the number of bicycles sold, the greater the number of cycle helmets sold.

Exercise 24C

1 **a, c**

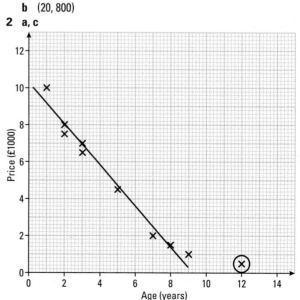

  b (20, 800)
2 **a, c**

Answers

b Car prices cannot go negative, so the relationship only holds for cars up to 9 years old. Cars older than this will still be worth a small amount.

3 a, b, d

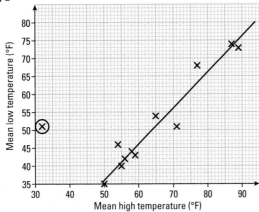

c Ottawa. The high and low temperatures are the wrong way round.

Exercise 24D

1 a 44   **b** 65
2 a i 65°F   **ii** 78°F
  b i 45 degrees   **ii** 17 degrees
3 a

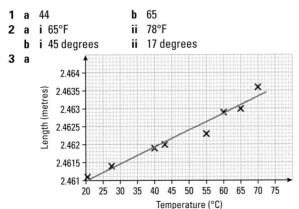

  b i 41.5°C   **ii** 63°C
  c 2.461 45 m

Review exercise

1

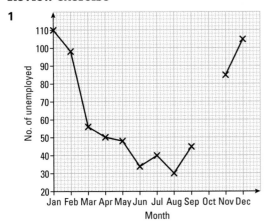

  b 65
  c August, because more people go to the seaside in the summer holidays.

2 a, b

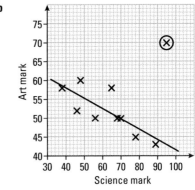

c Negative correlation
d The higher the science mark achieved by a student, the lower the art mark.

3 a, d

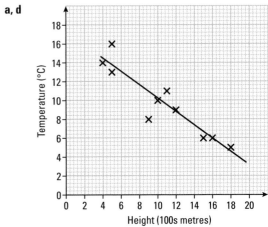

b Negative correlation
c The greater the height about sea level, the lower the temperature.

4 a, c

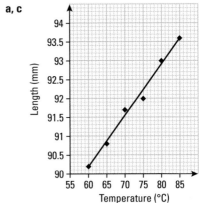

b Positive. As temperature rises so does the length.
d 0.14
e For every rise of 10 the length increase by approximately 0.14 mm
5 a The higher the price, the fewer the number of cameras sold.

b

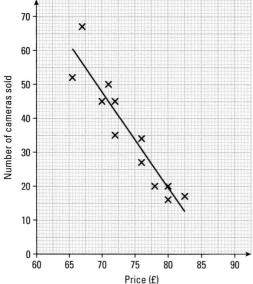

6 a As a car gets older, its value decreases.

b

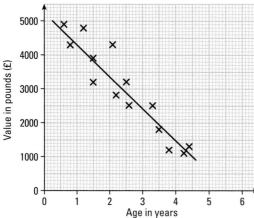

c £2450

d 1.9 years

7 a The larger the weight of a child, the greater its height.

b

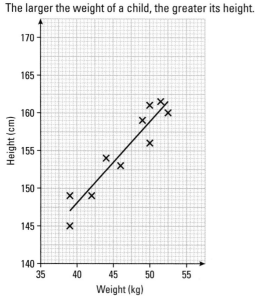

8 a, c

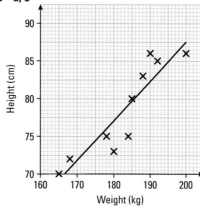

b Positive correlation

c The taller the athlete, the greater his or her weight. Using the line of best fit above, 74.2 kg

9 a, b, c

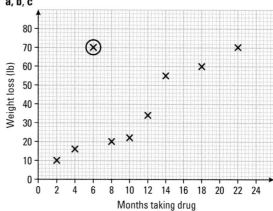

c This is unlikely to be a genuine piece of data as the weight loss is so high in only 6 months.

d Possitve correlation

e The drug is effective as the longer that the drug was taken, the greater the weight loss.

10 a Possitive correlation

b The greater the amount of fertiliser used, the higher the crop yield.

c 20 200 kg

d 2.5 kg per 80 m^2

e No, because there will be a limit on how high the crop yield can be, regardless of how much fertiliser is used.

11 a, b

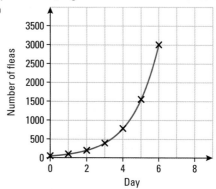

c $y = k^x$

12 **a** For every 100 m rise in height above sea level there is 1° drop in temperature.
OR For every 1 m rise in height above sea level there is 0.01° drop in temperature.
b At sea level the air temperature is 19°.

Chapter 25 Answers

25.1 Get Ready

1 $\frac{1}{16}$ **2** $\frac{9}{25}$ **3** -8

Exercise 25A

1 **a** 1   **b** $\frac{1}{8}$ **c** $\frac{1}{5}$ **d** 1
 e $-\frac{1}{8}$ **f** $\frac{1}{81}$ **g** $\frac{1}{10\,000}$ **h** 1
 i $\frac{1}{9}$ **j** 1   **k** 1   **l** $\frac{1}{1\,000\,000}$
2 **a** 3   **b** $\frac{7}{2}$ **c** 49   **d** 64
 e 16   **f** $15\frac{5}{8}$ **g** 1   **h** $\frac{5}{9}$
 i $\frac{25}{49}$ **j** $\frac{27}{64}$ **k** 10 000   **l** 125

25.2 Get Ready

1 **a** 1000   **b** $\frac{1}{100}$
2 10^4
3 23 500

Exercise 25B

1 **a** 7×10^5 **b** 6×10^2 **c** 2×10^3
 d 9×10^8 **e** 8×10^4
2 **a** 600 000   **b** 10 000   **c** 800 000
 d 300 000 000   **e** 70
3 **a** 4.3×10^4 **b** 5.61×10^5 **c** 5.6×10^1
 d 3.47×10^1 **e** 6×10^1
4 **a** 39 600   **b** 68 000 000   **c** 8020
 d 57   **e** 9.23
5 7×10^9
6 4×10^4

Exercise 25C

1 **a** 5×10^{-3} **b** 4×10^{-2} **c** 7×10^{-6}
 d 9×10^{-1} **e** 8×10^{-4}
2 **a** 0.000 06   **b** 0.08   **c** 0.000 000 5
 d 0.3   **e** 0.000 000 01
3 **a** 4.7×10^{-3} **b** 9.87×10^{-1} **c** 8.034×10^{-4}
 d 1.5×10^{-4} **e** 6.01×10^{-1}
4 **a** 0.000 084 3   **b** 0.0201   **c** 0.000 000 42
 d 0.078 54   **e** 0.000 94
5 **a** 4.57×10^5 **b** 2.3×10^{-3} **c** 3×10^{-4}
 d 2.356×10^6 **e** 7.82×10^{-1} **f** 8.9×10^4
 g 2×10^2 **h** 5.26×10^{-3} **i** 6.034×10^3
 j 8.73×10^6
6 **a** 0.000 412   **b** 3000   **c** 20 650 000
 d 0.000 004   **e** 327 000 000   **f** 0.75
 g 156.23   **h** 0.000 000 512   **i** 270 000
 j 0.612
7 1×10^{-6}
8 6.25×10^{-2} mm

Exercise 25D

1 **a** 4.5×10^4 **b** 9.8×10^{-1}
 c 3.4×10^1 **d** 1.86×10^{12}
2 **a** 9×10^2 **b** 4.5×10^4
 c 3.708×10^{-13} **d** 6×10^{-10}
3 **a** In standard form
 b 8.9×10^8 **c** 1.32×10^{-4} **d** 5.6×10^8
 e 6×10^{-4} **f** In standard form
 g 4.005×10^{-12} **h** 9.08×10^{18}
 i In standard form   **j** 4.6×10^5
 k 6.7×10^1 **l** 4×10^0
4 6 290 000, 6.3×10^6, 63.4×10^5, 0.637×10^7
5 0.000 033, 3.35×10^{-5}, 0.034×10^{-2}, 37×10^{-4}

Exercise 25E

1 **a** 8×10^{11} **b** 9×10^8 **c** 1.2×10^{-1}
 d 4.3×10^{-5} **e** 2.5×10^8 **f** 1×10^{-4}
2 **a** 4×10^{10} **b** 2.5×10^{-9} **c** 1.6×10^{13}
 d 4.9×10^{-15}
3 **a** 3.6×10^8 **b** 3.2×10^8 **c** 3.5×10^3
 d 1×10^{14}
4 3.2×10^0 mm^2
5 Time $= \dfrac{\text{Distance}}{\text{Speed}} = \dfrac{1.5 \times 10^8}{3 \times 10^5} = 5 \times 10^2$ seconds
 $= 500$ seconds $= 8$ mins 20 secs
6 160 m

Exercise 25F

1 **a** 2.1×10^8 **b** 2.4×10^{-5} **c** 5.2×10^0
 d 4.416×10^{-3} **e** 2.684×10^4 **f** 9.84×10^3
 g 9.6×10^{13} **h** 9.84×10^{-17} **i** 9.6×10^{-15}
2 **a** 5.38×10^4 **b** 6.20×10^{14} **c** 5.38×10^{16}
 d 6.20×10^{-6}
3 **a** 7.45×10^3 **b** 1.36×10^{19} **c** 4.24×10^5
 d 3.28×10^{12}
4 **a** 6×10^{-5} **b** 1.99×10^{-5} **c** 1.6×10^{-5}
5 $1 : 30.1$
6 6.28×10^{24}

25.3 Get Ready

1 10   **2** 2   **3** -3

Exercise 25G

1 **a** 3   **b** 7   **c** 10
 d 2   **e** $\frac{1}{2}$
2 **a** 3   **b** 10   **c** -4
 d 5   **e** $\frac{1}{2}$
3 **a** $\frac{1}{2}$ **b** $\frac{1}{2}$ **c** $\frac{1}{5}$
 d 2   **e** $\frac{3}{2}$
4 **a** 9   **b** 100   **c** 16
 d 8   **e** 125
5 **a** $\frac{1}{25}$ **b** $\frac{1}{1000}$ **c** $\frac{1}{3}$
 d $\frac{1}{4}$ **e** $\frac{1}{512}$ **f** $\frac{1}{625}$
 g $\frac{2}{25}$
6 **a** $n = -1$ **b** $n = 6$ **c** $n = -\frac{1}{2}$
 d $n = \frac{5}{2}$ **e** $n = \frac{11}{3}$

25.4 Get Ready

1 1, 4, 9, 16, 25, 36, 49, 64, 81, 100
2 a 6   **b** 10
3 $\sqrt{9}, \sqrt{64}$

Exercise 25H

1 a 2   **b** 3   **c** 5   **d** 4
2 a $10\sqrt{2}$ **b** $4\sqrt{2}$ **c** $2\sqrt{5}$ **d** $2\sqrt{7}$
3 $x = \pm\sqrt{30}$
4 a $3 + 2\sqrt{3}$ **b** $5 + 3\sqrt{3}$ **c** $3 + \sqrt{5}$
   d $-5 + \sqrt{7}$ **e** $7 - 4\sqrt{3}$ **f** $27 + 10\sqrt{2}$
5 $2\sqrt{10}$ cm
6 a 12 cm   **b** 4 cm^2
7 $3 + 2\sqrt{2}$ cm^2

Exercise 25I

1 a $\dfrac{\sqrt{2}}{2}$ **b** $\dfrac{\sqrt{5}}{5}$ **c** $\dfrac{\sqrt{10}}{2}$
   d $\sqrt{2}$ **e** $\dfrac{2\sqrt{3}}{3}$
2 a $1 + \sqrt{2}$ **b** $-1 + 3\sqrt{2}$ **c** $1 + 2\sqrt{5}$
   d $-1 + 4\sqrt{3}$ **e** $1 + 2\sqrt{7}$
3 $\dfrac{3\sqrt{6}}{2}$ cm^2
4 a $x = 3 \pm \sqrt{7}$ **b** $x = -5 \pm \sqrt{11}$
5 a 1.5 cm^2 **b** $\sqrt{22}$ cm

Review exercise

1 a 1   **b** $\frac{1}{4}$ **c** 1   **d** $\frac{1}{8}$
2 a 1   **b** 1   **c** $\frac{1}{9}$ **d** 1
3 a 3   **b** 3   **c** $\frac{1}{2}$ **d** 8
4 a 3   **b** 10   **c** 2   **d** 4
5 a $\frac{1}{3}$ **b** $\frac{1}{7}$ **c** $\frac{1}{5}$ **d** $\frac{1}{2}$
6 a 2   **b** $\frac{1}{2}$
7 a Mars   **b** 1.4×10^9 km
   c Saturn   **d** Neptune
   e Minimum distance from Earth to Mars =
      $2.3 \times 10^8 - 1.5 \times 10^8 = 8 \times 10^7$ km
      Speed $= \dfrac{8 \times 10^7}{365 \times 24} = 9132$ km/h
8 a **i** 7.9×10^3 **ii** 3.5×10^{-4}
   b 5×10^7
9 9.3×10^8
10 $n = \frac{5}{2}$
11 a $k = \frac{3}{2}$ **b** $m = 16$ **c** $\dfrac{\sqrt{2}}{32}$
12 9.43×10^{12}
13 $x = 5.8 \times 10^{-4}$
14 a 1×10^{-9} **b** 2×10^8
15 a $x = -2$ **b** $x = -4$ **c** $x = 3$ **d** $x = -\frac{1}{2}$
16 Using $a^2 - b^2 = (a + b)(a - b)$
   $\dfrac{1}{\sqrt{2} + 1} + \dfrac{1}{\sqrt{3} + \sqrt{2}} + \dfrac{1}{\sqrt{4} + \sqrt{3}} + \ldots + \dfrac{1}{10 + \sqrt{99}}$
   $= \dfrac{\sqrt{2} - 1}{2 - 1} + \dfrac{\sqrt{3} - \sqrt{2}}{3 - 2} + \dfrac{\sqrt{4} - \sqrt{3}}{4 - 3} + \ldots + \dfrac{10 - \sqrt{99}}{100 - 99}$
   $= -1 + 10 = 9$

Chapter 26 Answers

26.1 Get Ready

1 a 9   **b** 16   **c** 90
2 a 196   **b** 625   **c** 384.16

Exercise 26A

1 720 cm^2 **2** 270 cm^2 **3** 12 cm^2
4 1125 cm^2

Exercise 26B

1 a 18   **b** 15
2 8 cm

26.2 Get Ready

1 a 7   **b** 10
2 a 15.625   **b** 1331

Exercise 26C

1 270 cm^3 **2** 135 cm^3 **3** 1327 cm^3 **4** 2.25 cm^3

Exercise 26D

1 3 cm
2 a 18   **b** 7
3 a 28   **b** 1.2
4 1.95 cm
5 10 cm

26.3 Get Ready

1 2 : 9   **2** 4 : 5   **3** 1 : 6   **4** 1 : 1

Exercise 26E

1 1000 cm^2 **2** 3.91 cm^3 **3** 2950 cm^2
4 444 cm^2 **5 a** 292 cm^2 **b** 208 cm^2
6 0.101 m^2

Review exercise

1 2.25 cm
2 1.456 cm
3 12.96 cm^2
4 a 20 cm   **b** 6 cm
5 a 320 cm^2 **b** 7.5 cm^3
6 a 8 cm   **b** 96 cm^3
7 3800 cm^2
8 a 1 : 4   **b** 1 : 8
9 a All the edges of a cube are the same length, so the
      ratio between the edges of two cubes is a constant.
   b 1 : 6.25
   c The edges of a cuboid can be in any ratio, so two
      cuboids may not have the same ratio between every
      pair of edges.
10 a 11.9 cm   **b** 25.6 cm^2

Chapter 27 Answers

27.1 Get Ready

1 a 50p **b** 75p **c** £1 **d** £1.25
2 a 20 seconds **b** 40 seconds
c 80 seconds **d** 160 seconds

Exercise 27A

1 a

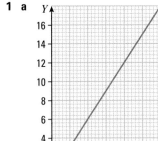

b Yes, the graph is a straight line through the origin
2 $k = 0.4$
3 a $\frac{9}{4} = \frac{22.5}{10} = \frac{36}{16} = \frac{45}{20}$ **b** $k = 2.25$
c 72
4 a

b $Y = \frac{1}{2}t$
c $Y = 50$
5 a i £3.60 **ii** £5.40 **iii** £7.20
b $C = 0.45v$ **c** 13 cm³

27.2 Get ready

1 a $x = 18$ **b** $x = 72$ **c** $x = 75$

Exercise 27B

1 $k = 1.5$
2 $k = 0.7$
3 a $P \propto h$, so $P = kh$
  When $P = 40.5$, $h = 18$, giving $k = \frac{9}{4}$
  So $P = \frac{9}{4}h$
b 72 **c** 12
4 a $V \propto c$, so $V = kc$
  When $V = 10$, $c = 0.04$, giving $k = 250$
  So $V = 250c$
b 0.036 amps
5 a $E = \frac{7}{150}m$ **b** 21 mm **c** 1125 g
6 $66\frac{2}{3}$ cm³

27.3 Get Ready

1 a $V = k$ **b** $V = kT$

Exercise 27C

1 a i $M \propto n$ **ii** $M = kn$
  b i $L \propto h^2$ **ii** $L = kh^2$
  c i $P \propto t^3$ **ii** $P = kt^3$
  d i $Q \propto \sqrt{y}$ **ii** $Q = k\sqrt{y}$
  e i $W \propto \sqrt[3]{x}$ **ii** $W = k\sqrt[3]{x}$
  f i $A \propto \frac{1}{b}$ **ii** $A = \frac{k}{b}$
  g i $H \propto \frac{1}{g^2}$ **ii** $H = \frac{k}{g^2}$
  h $U \propto \frac{1}{f^2}$ **ii** $U = \frac{k}{f^2}$
  i $E \propto \frac{1}{w^3}$ **ii** $E = \frac{k}{w^3}$
  j $V \propto \frac{1}{\sqrt{r}}$ **ii** $V = \frac{k}{\sqrt{r}}$
2 a $N \propto d^2$ **b** $N = kd^2$
3 $H = kc^2$
4 $Q = kt^3$
5 F is inversely proportional to the square of r
6 T is proportional to the square root of l

27.4 Get Ready

1 a 225 **b** 512 **c** 121 **d** 9 **e** 8 **f** 64

Exercise 27D

1 0.8
2 16
3 $Z = 1.28b^2$
4 $L = 1.875v^3$
5 4.5
6 a 11.25 m **b** 14.1 m/s
7 2.51 mm
8

Mass (grams)	Length (cm)
20	5
34.56	6
43.94	6.5
50	6.79

27.5 Get ready

1 a $P = \frac{1}{T}$

Exercise 27E

1 45
2 a 2.5 **b** 6
3 a 3.6 **b** 2.67
4 2.67 m³
5 a 384 hertz **b** 0.96 m
6 a 3.27 **b** 6.32
7 a $S = \frac{8000}{f^2}$ **b** 500

Review exercise

1 $0.75 \div 0.5 = 2.25 \div 1.5 = 5.25 \div 3.5 = 6.75$
  $\div 4.5 = 9.0 \div 6.0 = 1.5$
2 a 1600 **b** 560

3 **a** $T \propto b$ **b** R is directly proportional to a
   c $P \propto m^2$ **d** $Z \propto g^3$
   e H is inversely proportional to y

4 **a** $S = 1.5p$
   b

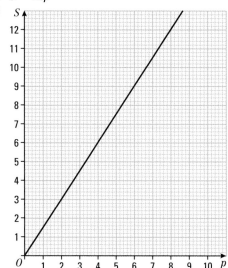

5 £70 125
6 Yes, the pressure will be 6 bars.
7 10.2 units
8 **a** $\dfrac{D}{t^2} = \dfrac{5}{1^2} = \dfrac{20}{2^2} = 5$

t	0	1	2	3	4	5
D	0	5	20	45	80	125

   b 1.73 seconds

9

Radius (r mm)	Resistance (R ohms)
10	500
15	222.22
17.5	163.27
14.14	250

10 **a** $X \propto \sqrt{h}$, so $X = k\sqrt{h}$
     When $X = 3$, $h = 16$, giving $k = 0.75$
     So $X = 0.75\sqrt{h}$
   b 3.75   **c** 64

11 **a** $T \propto \dfrac{1}{\sqrt{m}}$, so $T = \dfrac{k}{\sqrt{m}}$
     When $T = 20$, $m = 50$, giving $k = 100\sqrt{2}$
     So $T = 100\sqrt{\dfrac{2}{m}}$
   b 88.9 g

12 **a** A   **b** C
13 **a** 540   **b** 1.82
14 **a** $q = \dfrac{k}{t^2}$ **b** 5.44
15 51.6

16 **a** $y = kz$
   b $u = kv^2$ $v = \dfrac{l}{\sqrt{w}}$
                    $v^2 = \dfrac{l^2}{w}$
     $u = k \times \dfrac{l^2}{w}$
     $uw = kl^2$
     kl^2 is a constant so uw is a constant.

Chapter 28 Answers

28.1 Get Ready

1   **2**    **3**

Exercise 28A

1 **a** $\frac{2}{7}$ **b** $\frac{4}{7}$ **c** $\frac{1}{7}$
2 **a** $\frac{1}{6}$ **b** $\frac{1}{2}$ **c** $\frac{1}{13}$
   d $\frac{1}{4}$ **e** $\frac{1}{26}$
3 **a** $\frac{5}{9}$ **b** $\frac{4}{9}$ **c** 0
4 **a** $\frac{1}{2}$ **b** $\frac{1}{2}$ **c** $\frac{1}{4}$
   d $\frac{1}{2}$ **e** $\frac{3}{8}$
5 $\frac{1}{500}$
6 **a** $\frac{1}{3}$ **b** $\frac{4}{9}$ **c** $\frac{2}{9}$ **d** $\frac{2}{3}$
7 **a** $\frac{2}{11}$ **b** $\frac{1}{11}$ **c** $\frac{3}{11}$
   d $\frac{4}{11}$ **e** 0
8 **a**

	Left-handed	Right-handed	Total
Boys	47	135	182
Girls	61	119	180
Total	108	254	362

   b **i** $\frac{91}{181}$ **ii** $\frac{54}{181}$ **iii** $\frac{119}{181}$
9 **a** $\frac{1}{6}$ **b** $\frac{7}{12}$
10 0.3

Exercise 28B

1 **a** **i** $\frac{1}{36}$ **ii** $\frac{1}{9}$ **iii** $\frac{1}{6}$ **b** $\frac{1}{6}$ **c** $\frac{1}{9}$
2 **a**

				Dice		
Coin	H	(1,H)	(2,H)	(3,H)	(4,H)	(5,H)	(6,H)
	T	(1,T)	(2,T)	(3,T)	(4,T)	(5,T)	(6,T)
		1	2	3	4	5	6

   b **i** $\frac{1}{12}$ **ii** $\frac{1}{2}$ **iii** $\frac{1}{6}$

3 **a**

		Spinner A			
		1	**2**	**3**	**4**
Spinner B	**1**	0	1	2	3
	2	1	0	1	2
	3	2	1	0	1
	4	3	2	1	0

   b **i** $\frac{1}{4}$ **ii** $\frac{1}{8}$ **iii** 0

Answers

4 a

Spinner	1	(1,1)	(2,1)	(3,1)	(4,1)	(5,1)	(6,1)
	2	(1,2)	(2,2)	(3,2)	(4,2)	(5,2)	(6,2)
	3	(1,3)	(2,3)	(3,3)	(4,3)	(5,3)	(6,3)
		1	2	3	4	5	6
				Dice			

b i $\frac{1}{6}$ ii $\frac{1}{9}$ iii $\frac{1}{3}$

5 a

Box B	R	(R,R)	(Bu,R)	(Y,R)	(Ba,R)
	Y	(R,Y)	(Bu,Y)	(Y,Y)	(Ba,Y)
	Bu	(R,Bu)	(Bu,Bu)	(Y,Bu)	(Ba,Bu)
		R	Bu	Y	Ba
			Box A		

b i $\frac{1}{12}$ ii $\frac{1}{4}$ iii $\frac{3}{4}$

6 $\frac{3}{5}$

7 a $\frac{1}{6}$ **b** $\frac{1}{10}$ **c** $\frac{1}{60}$
d $\frac{1}{20}$ **e** $\frac{1}{12}$

28.2 Get Ready

1 a 0.4   **b** 0.65   **c** $\frac{3}{5}$ **d** $\frac{1}{6}$

Exercise 28C

1 a $\frac{3}{10}$ **b** $\frac{2}{5}$ **c** $\frac{7}{10}$
2 0.9
3 a 0.5   **b** 0.5   **c** 0.65   **d** 0.7
4 a $\frac{3}{13}$ **b** $\frac{3}{13}$ **c** $\frac{1}{13}$
d $\frac{6}{13}$ **e** $\frac{4}{13}$ **f** $\frac{4}{13}$
5 a $\frac{1}{2}$ **b** $\frac{7}{18}$ **c** $\frac{2}{9}$ **d** $\frac{2}{3}$
6 a $\frac{1}{26}$ **b** $\frac{1}{2}$ **c** $\frac{5}{52}$
d $\frac{7}{26}$ **e** $\frac{7}{26}$
7 0.35
8 The events are not mutually exclusive. The probability is $\frac{1}{36}$.

Exercise 28D

1 $\frac{11}{72}$
2 $\frac{4}{5}$
3 0.35
4 0.25
5 a $\frac{1}{13}$ **b** $\frac{12}{13}$
6 a i 0.65   ii 0.85
   b $0.35 + 0.15 + 0.5 = 1$
7 a $\frac{2}{9}$ **b** $\frac{7}{9}$ **c** $\frac{5}{9}$
8 0.6

28.3 Get Ready

1 a $\frac{1}{2}$ **b** $\frac{3}{4}$ **c** $\frac{8}{9}$ **d** $\frac{3}{13}$

Exercise 28E

1 $\frac{9}{20}$
2 $\frac{13}{15}$

---

3 a $\frac{77}{150}$
   b No, red is almost three times as likely as orange.
4 a Students' own work
   b Drop the drawing pin more times, because the greater the number of trials, the more accurate the estimated probability.
5 $\frac{33}{200}$, because the greater the number of trials, the more accurate the estimated probability.
6 Students' own work

28.4 Get Ready

1 a 50   **b** 18   **c** 30   **d** 40.5

Exercise 28F

1 50
2 10
3 30
4

Colour of peg	Red	White	Yellow
Expected number	25	125	100

5 30
6 No, the draw is random so he could be unlucky.
7 a 20   **b** 65   **c** 80
8 The doctor's estimate is a bit high, as the results from the 240 patients suggests a probability of 0.083.

28.5 Get Ready

1 a $\frac{1}{6}$ **b** $\frac{2}{7}$ **c** $\frac{2}{5}$ **d** $\frac{3}{16}$

Exercise 28G

1 0.32
2 $\frac{1}{2}$
3 0.6
4 a 0.65   **b** 0.85   **c** 0.5525
5 a $\frac{1}{4}$ **b** $\frac{1}{16}$ **c** $\frac{3}{169}$
d $\frac{1}{676}$ **e** $\frac{1}{2704}$
6 a $\frac{3}{44}$ **b** $\frac{2}{11}$ **c** $\frac{15}{22}$

28.6 Get Ready

1 a $\frac{3}{4}$ **b** $\frac{8}{15}$ **c** $\frac{13}{20}$ **d** $\frac{55}{63}$

Exercise 28H

1 a

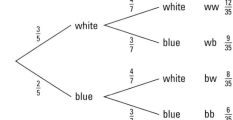

   b i $\frac{12}{35}$ ii $\frac{6}{35}$ iii $\frac{18}{35}$

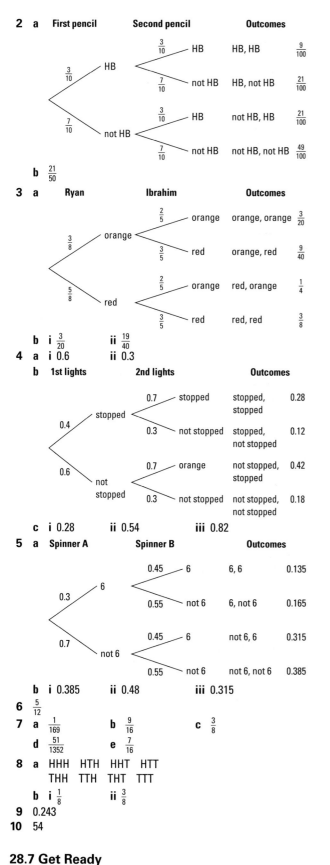

2 a

First pencil	Second pencil	Outcomes	
HB	$\frac{3}{10}$ HB	HB, HB	$\frac{9}{100}$
	$\frac{7}{10}$ not HB	HB, not HB	$\frac{21}{100}$
not HB	$\frac{3}{10}$ HB	not HB, HB	$\frac{21}{100}$
	$\frac{7}{10}$ not HB	not HB, not HB	$\frac{49}{100}$

First pencil: $\frac{3}{10}$ HB, $\frac{7}{10}$ not HB

b $\frac{21}{50}$

3 a

Ryan	Ibrahim	Outcomes	
orange ($\frac{3}{8}$)	$\frac{2}{5}$ orange	orange, orange	$\frac{3}{20}$
	$\frac{3}{5}$ red	orange, red	$\frac{9}{40}$
red ($\frac{5}{8}$)	$\frac{2}{5}$ orange	red, orange	$\frac{1}{4}$
	$\frac{3}{5}$ red	red, red	$\frac{3}{8}$

b i $\frac{3}{20}$ **ii** $\frac{19}{40}$

4 a i 0.6 **ii** 0.3

b

1st lights	2nd lights	Outcomes	
stopped (0.4)	0.7 stopped	stopped, stopped	0.28
	0.3 not stopped	stopped, not stopped	0.12
not stopped (0.6)	0.7 orange	not stopped, stopped	0.42
	0.3 not stopped	not stopped, not stopped	0.18

c i 0.28 **ii** 0.54 **iii** 0.82

5 a

Spinner A	Spinner B	Outcomes	
6 (0.3)	0.45 6	6, 6	0.135
	0.55 not 6	6, not 6	0.165
not 6 (0.7)	0.45 6	not 6, 6	0.315
	0.55 not 6	not 6, not 6	0.385

b i 0.385 **ii** 0.48 **iii** 0.315

6 $\frac{5}{12}$

7 a $\frac{1}{169}$ **b** $\frac{9}{16}$ **c** $\frac{3}{8}$

d $\frac{51}{1352}$ **e** $\frac{7}{16}$

8 a HHH HTH HHT HTT
THH TTH THT TTT

b i $\frac{1}{8}$ **ii** $\frac{3}{8}$

9 0.243

10 54

28.7 Get Ready

1 $\frac{3}{9} = \frac{1}{3}$

Exercise 28I

1 b i $\frac{7}{15}$ **ii** $\frac{8}{15}$ **iii** $\frac{7}{15}$

2 b i $\frac{5}{33}$ **ii** $\frac{35}{66}$ **iii** $\frac{15}{22}$

3 a $\frac{3}{14}$ **b** $\frac{4}{7}$

4 $\frac{121}{315}$

5 0.74

6 0.1325

7 a $\frac{2}{7}$ **b** $\frac{2}{7}$

8 a $\frac{1}{221}$ **b** $\frac{1}{17}$ **c** $\frac{13}{17}$

9 $\frac{37}{49}$

10 0.7816

Review exercise

1 a $\frac{4}{11}$ **b** $\frac{4}{11}$ **c** $\frac{7}{11}$

d $\frac{7}{11}$ **e** $\frac{8}{11}$ **f** $\frac{3}{11}$

2 a $\frac{23}{45}$ **b** $\frac{3}{10}$ **c** $\frac{8}{45}$

d $\frac{37}{90}$ **e** $\frac{8}{45}$

3 $\frac{7}{18}$

4 a $\frac{1}{6}$ **b** $\frac{1}{4}$ **c** $\frac{5}{12}$

5 a 97 or 98 **b** 7 or 8 **c** 30

6 Roll the dice 100 times, recording the results. If the dice is fair, each number should occur approximately 17 times. If Amy wanted to be more confident in her results, she could roll the dice more times.

7 90

8 a If Megan bought all the tickets, she is definitely right. Otherwise, she could be right, but there is no way of knowing because the winning tickets are drawn at random.

b The winning tickets are chosen at random, so she may just have been lucky.

9 a i $\frac{b}{12}$ **ii** $\frac{(12-b)}{12}$ **b** 3

10 $\frac{3}{4}$

11 0.5525

12 a 96

b The original set of marked rabbits came from throughout the warren. The marked rabbits spread throughout the warren after being returned. The dye did not come off any rabbits.

13 $\frac{5}{144}$

14 a

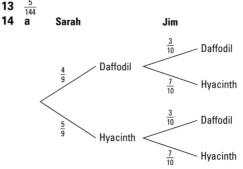

Sarah	Jim
Daffodil ($\frac{4}{9}$)	$\frac{3}{10}$ Daffodil
	$\frac{7}{10}$ Hyacinth
Hyacinth ($\frac{5}{9}$)	$\frac{3}{10}$ Daffodil
	$\frac{7}{10}$ Hyacinth

b $\frac{2}{15}$

15 a No. Horse C is more likely to win, because of the four possible outcomes, two are a head and tail, but two heads and two tails are each only one outcome.

b No. Horse 7 is more likely to win, because of the 36 possible outcomes from rolling two dice, a score of 7 occurs more often than any other score.

16 Number of possible selections of six numbers =
$\frac{40 \times 39 \times 38 \times 37 \times 36 \times 35}{6 \times 5 \times 4 \times 3 \times 2 \times 1} \approx 4$ million
So P(choosing correct six numbers) $= \frac{1}{4}$ million

17 $\left(\frac{1}{6}\right)^6 = \frac{1}{46\,656} = 2.1 \times 10^{-5}$

18 $\frac{\pi}{4}$

19 $\frac{1}{5}$

20 a $\frac{1}{12}$ **b** $\frac{5}{18}$ **c** $\frac{13}{18}$

21 Students' investigations

Chapter 29 Answers

29.1 Get Ready

1 ABF, ABE, AEF, BEF, BCF, BCG, BFG, CFG, CDG, CDH, CHG, DHG, ADE, ADH, AEH, DEH, ABC, ABD, ACD, BCD, EFG, EFH, EGH, FGH

Exercise 29A

1 a i 8.94 cm **ii** 13.6 cm **iii** 15.3 cm **iv** 15.8 cm
  b i 58.4° **ii** 72.9°
2 $AG^2 = AC^2 + CG^2 = AB^2 + BC^2 + CG^2$
3 No. The longest needle that could fit in the box is 14 cm.
4 5 cm
5 a i 14.1 cm **ii** 7.07 cm
  b 13.2 cm **c** 61.9° **d** 56.2° **e** 14.1 cm
  f 70.5°

29.2 Get Ready

The diagrams could show the same pole viewed from different sides.

Exercise 29B

1 69.3°
2 a i 51.3° **ii** 49.1° **iii** 38.7° **b** 90°
3 a 469 m **b** 171 m **c** 985 m **d** 848 m
4 a 15.6 m
  b i 26.6° **ii** 36.9° **iii** 22.6°
5 a 86.5 cm **b** 63.9°
6 a

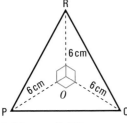

  b i 45° **ii** 45°
  c 1692 cm³

29.3 Get Ready

1 a $(-u, v)$ **b** $(-u, -v)$ **c** $(u, -v)$
2 a 0.5 and 0.5 **b** 0.64 and −0.64

c sin 30° and sin 150° give the same answer.
cos 50° and cos 130° give the same value, but opposite signs.

Exercise 29C

1 a

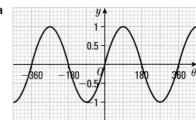

b

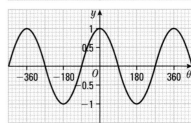

2 a 30, 150 **b** 84.3, 275.7
3 a $\sin\theta° = \frac{1}{3}$ so $\theta = 19.5$
  b $\theta = 19.5, 160.5, 379.5, 520.5$
4 a $\cos\theta° = -0.3$ so $\theta = 107.5$
  b $\theta = -252.5, -107.5, 107.5, 252.5$

29.4 Get Ready

1 a 30 cm² **b** 54 cm² **c** 18.4 cm²

Exercise 29D

1 a 21.9 cm² **b** 29.2 cm² **c** 15.4 cm²
  d 30.0 m²
2 16.8 m
3 33.3°
4 a 11.4 cm² **b** 11.4 cm²
  c The answers are the same because sin 25° = sin 155°.
5 a 45° **b** 12.7 cm² **c** 102 cm²
6 3.41 cm²

29.5 Get Ready

1 a 0.866 **b** 13.6

Exercise 29E

1 a 8.06 cm **b** 7.19 cm **c** 6.35 cm
  d 9.01 cm **e** 15.0 cm
  f $f = 6.06$ cm, $g = 11.4$ cm

29.6 Get Ready

1 52.0°

Exercise 29F

1 a 45.0° **b** 63.6° **c** 23.6°
  d $d = 43.4°, e = 63.7°$
2 a 13.3 cm **b** 39.3° **c** 154.7°

3 a 9.72 cm   **b** 9.89 cm   **c** 37.9 cm^2
4 a 66.0°   **b** 7.63 cm
5 a i 42°   **ii** 55°
  b 9.79 km   **c** 11.9 km   **d** 6.55 km

29.7 Get Ready

1 a 11   **b** 3   **c** 22.4
2 3.82

Exercise 29G

1 a 8.80 cm   **b** 12.6 cm   **c** 5.01 cm
  d 8.42 cm   **e** 15.3 cm   **f** 16.1 cm
2 12.6 cm

29.8 Get Ready

1 a 50.8°   **b** 72.3°   **c** 54.8°

Exercise 29H

1 a 54.7° **b** 81.2° **c** 46.0° **d** 131.2°
2 103.6°
3 a 68°   **b** 91.0 m
4 a 22.4 cm **b** 127.7° **c** 161 cm^2
5 20.9 km
6 7.71 cm, 11.9 cm

29.9 Get Ready

1 a cosine rule **b** sine rule
  c sine rule or cosine rule

Exercise 29I

1 a 70.5°, 59.0°, 50.5°   **b** 42.4 cm^2
2 a 9.68 cm **b** 43.1° **c** 17.7 cm
3 a 7.32 cm **b** 6.65 cm **c** 78.7°
4 a 6.76 cm **b** 75° **c** 13.1 cm **d** 19.9 cm
5 a 039° (or 321°)   **b** 148° (or 212°)
6 50°
7 a 70°   **b** 4.43 m
8 a 73.2° **b** 26.2 m **c** 67.2° **d** 22.6°
9 20.5 km

Review exercise

1 13 cm
2 85.5°
3 5.89 cm
4 a 9.11 cm **b** 19.2°
5 18.3 cm
6 76.3°
7 a $\dfrac{\sqrt{3}}{2}$ **b** $-\dfrac{\sqrt{3}}{2}$
8 42° and 318°
9 10.9 cm^2
10 a 105.7° **b** 10.7 cm^2
11 22.2°
12 a 21.8 cm **b** 66.4 cm^2
13 11.1 cm
14 a 41.4° **b** 17.8 cm
15 a 4.51 cm **b** 37°

Chapter 30 Answers

30.1 Get Ready

1 18   **2** $x = \frac{11}{4}$ **3** $y = (t + 1)^2$

Exercise 30A

1 a 12   **b** 0   **c** 48
  d 1   **e** −4   **f** 8
2 a 96   **b** 6.4   **c** 4
  d 28
3 a 8   **b** 2
4 a i −4 **ii** −3 **iii** 0 **b** 4, −4
5 a i 18 **ii** −12 **iii** −30 **b** 3, −4
6 a i −3 **ii** −4 **iii** 10 **b** 0, 4 **c** −1, 5
7 a 16   **b** $x^2 - 4$ **c** $(x - 4)^2$
8 a −8   **b** $4(2x + 1)$ **c** $12(x + 1)$

30.2 Get Ready

1 (2, 5)   **2** (4, 3)   **3** (−1, 3)

Exercise 30B

1 a

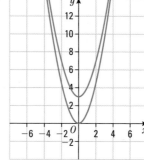

b

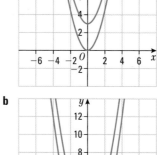

c

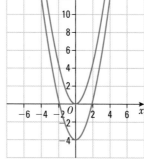

Answers

d
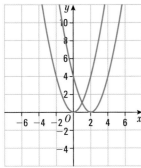

2 a i
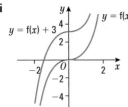

ii
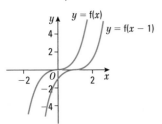

b i (0, 3) **ii** (1, 0)

3 a
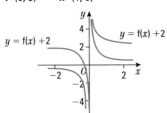

b (2, 2.5)
c i $y = f(x) + 2$ **ii** $y = \frac{1}{x} + 2$
d
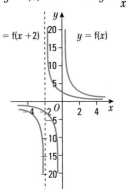

e $p = -1, q = 4$ **f i** $y = f(x+2)$ **ii** $y = \frac{4}{x+2}$
4 a (3, 2) **b** (1, 5)
5 a Translation by +3 units parallel to the y-axis.
  b $y = f(x) + 3$ **c** $y = 4x - x^2 + 3$
6 a Translation by +2 units parallel to the x-axis.
  b $y = f(x - 2)$
  c $y = (x - 2)^2$

7 a $a = 2, b = 5$ **b** Translation of $\begin{pmatrix} -2 \\ 5 \end{pmatrix}$
  c, d

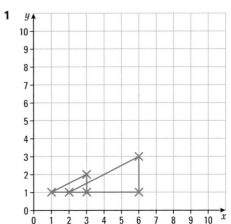

8 Translation by $\begin{pmatrix} 0 \\ -6 \end{pmatrix}$

30.3 Get Ready

1
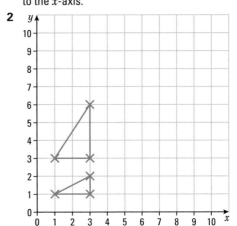

The first triangle has been stretched by a factor of 2 parallel to the x-axis.

2

The first triangle has been stretched by a factor of 3 parallel to the y-axis.

Exercise 30C

1

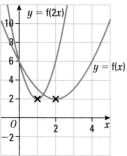

(2, 2) is mapped to the point (1, 2)

2 a

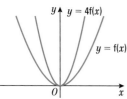

b $y = 4x^2$

c Stretch of magnitude 4 parallel to the y-axis

Stretch of magnitude $\frac{1}{2}$ parallel to the x-axis

3 a

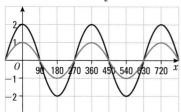

$y = 2\cos x°$

b

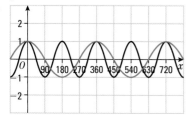

$y = \cos(2x°)$

4 a

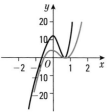

b (0, 4) is mapped to (0, 12). $\left(-\frac{2}{3}, 0\right)$ is mapped to $\left(-\frac{2}{3}, 0\right)$

5 a

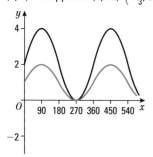

b

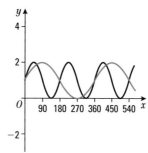

c

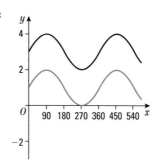

6 a

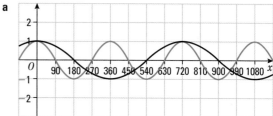

b 1 solution

7 a Stretch of magnitude 2 parallel to the x-axis.

b

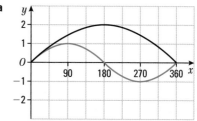

c **i** $y = f\left(\frac{x}{2}\right)$ **ii** $y = \frac{x}{2}\left(\frac{x}{2} - 4\right)$

8 a $p = 4$, $q = -11$

b (0, 5)

c Stretch parallel to the y-axis of magnitude 0.6.

$0.6\,(x^2 - 8x + 5)$

Exercise 30D

1 a

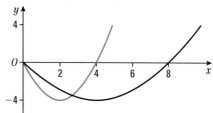

b $y = 2\sin\left(\frac{x}{2}\right)$

Answers

2 a

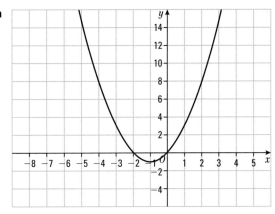

b $y = 2f\left(\dfrac{x}{2}\right)$

c

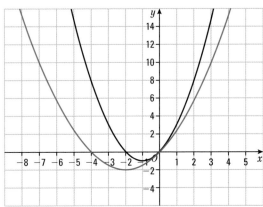

3 a, b

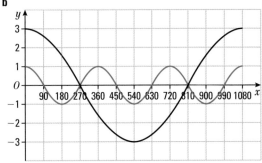

4 a $4\left(\dfrac{x^2}{16} + 3\right)$ **b** $4\left(\dfrac{4}{x} + 1\right)$

c $4 \times 2^{\frac{x}{4}} = 2^{\frac{x}{4} + 2}$ **d** $8\sin\left(\dfrac{x}{2}\right)$

30.4 Get Ready

1

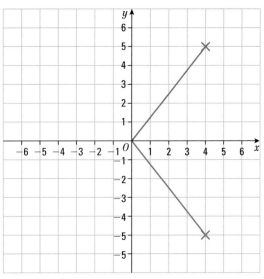

2

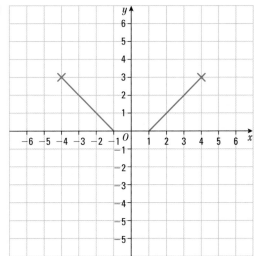

3

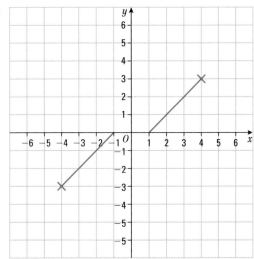

Exercise 30E

1 a

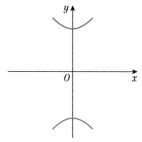

b $y = -(x^2 + 3)$

2 a

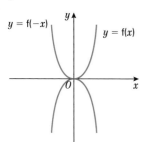

$y = f(-x)$ $y = f(x)$

b $p = -2, q = 8$ **c** $y = -x^3$

3 a

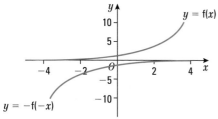

$y = f(x)$

$y = -f(-x)$

b $(r = -1, t = -2)$ **c** $y = -2^{-x}$

Review exercise

1 a 6   **b** 11   **c** $a = 0$

2 $f(x - 1) = (x - 1)^2 + 3(x - 1) = x^2 + x - 2$
   $= (x + 2)(x - 1)$

3

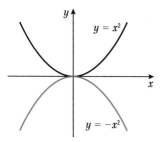

$y = x^2$

$y = -x^2$

4 a

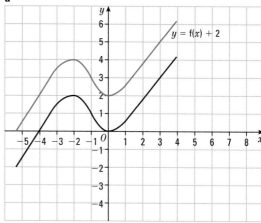

$y = f(x) + 2$

b

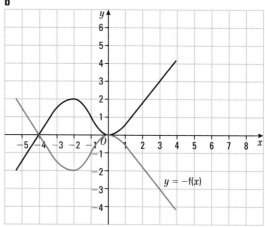

$y = -f(x)$

5 $f(x - 4)$

6 a, b

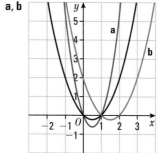

a

b

c $x = 1$ and $x = 2$

7 a $(180, 0)$

b

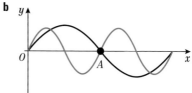

A

8

9 Translation by $\begin{pmatrix} 4 \\ -5 \end{pmatrix}$

10 **a, b** Stretch parallel to the y-axis of magnitude 4
Stretch parallel to the x-axis of magnitude 0.5

11 **a** $p = 2, q = 12$
b $(2, 12)$
c Stretch parallel to the y-axis of magnitude 2
d $y = 2f(x)$

Chapter 31 Answers

31.1 Get Ready

a $20°$ **b** $65°$ **c** $40°$

Exercise 31A

1 $a = 25°$
2 $b = 33°$
3 $c = 40°$
4 $d = 19°$ (angles on a line, isosceles triangle in a circle)
5 $e = 60°, f = 40°, g = 100°$ (equilateral triangle, isosceles triangle in a circle)
6 $h = 54°, i = 126°, j = 27°$ (right-angled triangle, angles on a line, isosceles triangle in a circle)

31.2 Get Ready

1 $a = 33°, b = 57°, c = 90°$

Exercise 31B

1 $a = 50°$ **2** $b = 59°$
3 $c = 136°$ **4** $d = 66°, e = 24°$
5 $f = 60°$ **6** $g = 107°$

31.3 Get Ready

1 **a** $28°$ **b** $49°$ **c** $57°$

Exercise 31C

1 $a = 83°$
2 $b = 90°, c = 52°$
3 $d = 64°, e = 38°$
4 $f = 110°, g = 55°$
5 $h = 30°$
6 $i = 72°$
7 $j = 49°$
The line drawn from the centre of a circle to the midpoint of a chord is perpendicular to that chord.

8 $k = 68°$
The angle at the centre of a circle is twice the angle at the circumference, both subtended by the same arc.
9 $l = 90°$
The angle in a semicircle is a right angle.
10 **a** angle DAB $= 65°$
The angle at the centre of a circle is twice the angle at the circumference, both subtended by the same arc.
b reflex angle DOB $= 230°$
Angles about a point add up to $360°$
c angle BCD $= 115°$
The angle at the centre of a circle is twice the angle at the circumference, both subtended by the same arc so reflex angle BCD is twice angle BCD.

31.4 Get Ready

1 **a** $93°$ **b** $94°$ **c** $46°$

Exercise 31D

1 $a = 40°$
2 $b = 82°$
3 $c = 72°$
4 $d = 50°, e = 100°, f = 40°$
5 $g = 58°$
6 $h = 74°$
7 $i = 71°$
Angles in the same segment are equal so angle SQR $= 31°$, and angles in a triangle add up to $180°$.
8 $j = 35°$
Opposite angles of a cyclic quadrilateral add up to $180°$ and the base angles of an isosceles triangle are equal.
9 $k = 57°$
The angle between a tangent and a chord is equal to the angle in the alternate segment and the angles on a straight line (or in a triangle) add up to $180°$.

Review exercise

1 angle OAD $= 90°$ (angle between tangent and radius)
angle AOD $= 180 - 90 - 36 = 54°$ (angle sum in a triangle)
angle ABC $= \frac{1}{2} \times 54 = 27°$ (angle at centre is twice angle at circumference)
2 **a** **i** $140°$
ii The angle at the centre of a circle is twice the angle at the circumference, both subtended by the same arc.
b **i** $110°$
ii The opposite angles of a cyclic quadrilateral add up to $180°$.
3 $a = 134°, b = 42°$
4 $d = 90°, e = 67°, f = 67°$
5 $g = 33°, h = 33°$
6 $o = 31°$ (angles in the same segment are equal)
$p = 26°$ (angles in an isosceles triangle add up to $180°$; angle between a tangent and a chord is equal to the angle in the alternate segment).
7 $s = 94°$ (angles on a straight line add up to $180°$; opposite angles of a cyclic quadrilateral add up to $180°$)

8 $t = 52°$ (angle between a tangent and a radius is a right angle; angles in a triangle add up to 180°)

$u = 26°$ (angle at the centre of a circle is twice the angle at the circumference, both subtended by the same arc)

9 angle CBT = angle CBA (BC bisects angle ABT)

angle CBT = angle CAB (angle between tangent and chord is equal to angle in alternate segment)

angle CBA = CAB

So triangle CAB is isosceles and CA = CB

10 angle PMA = $\frac{1}{2} \times 132 = 66°$ (angle at the centre is twice angle at circumference)

angle MAT = 66° (alternate angles, PM parallel to AT)

angle OAT = 90° (angle between tangent and radius)

angle OAM = 90 − 66 = 24°

OA = OM (radii), so triangle AOM is isosceles

angle OMA = 24° (base angles of isosceles triangle)

11 OR = OP (radii), so triangle ORP is isosceles

angle OPR = $\frac{180 - 20}{2} = 80°$ (angle sum of isosceles triangle)

OQ = OP (radii), so triangle OQP is isosceles

angle OPR = $\frac{180 - 100}{2} = 40°$ (angle sum of isosceles triangle)

So angle OPQ = $\frac{1}{2} \times$ angle OPR and QP bisects angle OPR

Chapter 32 Answers

32.1 Get Ready

1 a $(x + 4)(x + 1)$
  b $(x - 3)(x + 2)$
  c $(2x + 1)(x + 3)$

Exercise 32A

1 a $2x^3$ **b** $\frac{x}{3y}$ **c** $\frac{x - 5}{2}$ **d** x
  e $-2x$

2 a $\frac{x + 1}{x + 2}$ **b** $\frac{x + 1}{x}$ **c** $\frac{x - 2}{x + 4}$ **d** $\frac{x - 4}{x + 3}$

3 a $\frac{x + 1}{x}$ **b** $\frac{4x}{x - 6}$ **c** $\frac{2(x - 2)}{x + 2}$ **d** $\frac{x - 3}{x}$

4 a $\frac{2x + 3}{3x + 2}$ **b** $\frac{5x - 3}{3x - 2}$ **c** $\frac{3x + 1}{3x - 1}$ **d** $\frac{x + 1}{2x + 3}$

5 a $\frac{x + 5}{x - 5}$ **b** $\frac{x - 5}{2(x + 5)}$ **c** $\frac{2x - 1}{2x}$
  d $\frac{3 - x}{x + 3}$ **e** $\frac{2 - x}{x + 2}$ **f** $-(x + 4)$

32.2 Get Ready

1 a 30   **b** $12x$ **c** $x(x + 1)$

Exercise 32B

1 a x **b** $2x$ **c** $\frac{7}{10x}$ **d** $\frac{4x}{9}$
  e $\frac{x}{5}$ **f** $\frac{5}{3x}$

2 a $\frac{7x}{12}$ **b** $\frac{x}{3}$ **c** $\frac{3x}{8}$ **d** $\frac{5x}{6}$
  e $\frac{5}{6x}$ **f** $\frac{1}{4x}$

3 a $\frac{5x + 2}{6}$ **b** $\frac{9x - 7}{20}$ **c** $\frac{x}{9}$

d $\frac{2x + 5}{(x + 2)(x + 3)}$ **e** $\frac{x - 2}{(x + 2)(x + 1)}$
f $\frac{4}{(2x - 1)(2x + 3)}$

Exercise 32C

1 a i $2(x + 1)$ **ii** $6(x + 1)$
  b $6(x + 1)$ **c** $\frac{2}{3(x + 1)}$

2 a $15x$ **b** $(x + 2)(x + 3)$
  c $x(x - 1)$ **d** $(x + 1)(x + 2)$
  e $2(x - 3)$ **f** $x(x + 1)$

3 a $(x + 1)(x + 2)$ **b** $\frac{x}{(x + 1)(x + 2)}$

4 a $(x - 2)(x + 2)$ **b** $\frac{3x + 4}{(x - 2)(x + 2)}$

5 a $(2x - 1)(x - 1)$ **b** $\frac{1}{(x - 1)}$

6 $\frac{x - 1}{2(x + 1)(x + 3)}$

7 $\frac{2x^2 + 7x + 4}{8x(x + 1)}$

8 $\frac{3x}{(3 - x)(3 + x)}$

9 a i $(x + 4)(x + 5)$ **ii** $(x + 5)(x + 6)$
  b $\frac{3x + 20}{(x + 4)(x + 5)(x + 6)}$

10 $\frac{4}{(2x - 1)(2x + 1)(2x - 3)}$

32.3 Get Ready

1 a a **b** $\frac{1}{2}$ **c** $\frac{x + 3}{x + 1}$

Exercise 32D

1 a $\frac{x^2}{15}$ **b** $\frac{12}{y^2}$ **c** $\frac{15xy}{8}$ **d** $\frac{x(x - 3)}{12}$

2 a $\frac{xy}{2}$ **b** $\frac{2x}{9}$ **c** $\frac{6}{y}$ **d** $\frac{2(x + 1)}{x - 1}$

3 a $\frac{5}{9}$ **b** $\frac{5xy}{18}$ **c** $x^2 y^2$ **d** $\frac{2x(x + 2)}{(x + 1)^2}$

4 a $\frac{27}{4}$ **b** $\frac{5x}{8}$ **c** $\frac{7x}{2y^2}$ **d** $\frac{2x}{x - 5}$

5 a $\frac{(x + 1)^2}{2}$ **b** $(x + 2)(x - 1)$ **c** $\frac{x}{x + 2}$
  d $\frac{1}{6}$ **e** $3(3x - 1)$ **f** $\frac{3(x + 4)}{4}$

6 a $(x - 2)(x + 2)$ **b** $\frac{x - 2}{x^2 + 4}$

7 a i $(x + 1)(x + 4)$ **ii** $(x + 2)(x + 4)$
  b $\frac{(x + 2)(x + 3)}{(x + 1)^2}$

8 $x + 1$

32.4 Get Ready

1 a even   **b** either   **c** odd   **d** either
  e either

Exercise 32E

1 $(2n - 1) + 2m = 2(m + n) - 1$

2 $\frac{1}{2}[n + (n + 1) + (n + 2) + (n + 3)] = \frac{1}{2}[4n + 6] = 2n + 3$

3 $n + (n + 1) + (n + 2) = 3n + 3 = 3(n + 1)$

4 **a** $(2n - 1)2m = 2[m(2n - 1)]$

 b $(2n - 1)(2m - 1) = 4mn - 2m - 2n + 1$
$$= 2(2mn - m - n) + 1$$

 c $(2n)(2m) = 4mn = 2(2mn)$

5 $(m - n)(m + n) = m^2 - nm + mn - n^2 = m^2 - n^2$

6 $(n + 4)^2 - n^2 = n^2 + 8n + 16 - n^2 = 8n + 16 = 8(n + 2)$

Review exercise

1 $n + (n + 1) + (n + 2) + (n + 3) = 4n + 6$
$(n + 3)(n + 2) - n(n + 1) = (n^2 + 5n + 6) - (n^2 + n)$
$= 4n + 6$

2 **a** $\frac{x^2}{2}$ **b** $\frac{(x + 1)^2}{3}$ **c** x **d** $\frac{x + 1}{x + 2}$

 e $\frac{x}{x + 3}$ **f** $\frac{x - 5}{x + 5}$ **g** $\frac{x - 1}{x(x + 1)}$ **h** $\frac{x + 4}{3x + 2}$

3 **a** $\frac{x}{2}$ **b** $\frac{13}{6x}$ **c** $\frac{10x}{(5x - 3)(5x + 3)}$

 d $\frac{2x + 7}{(x + 1)(x + 2)}$ **e** $\frac{x + 2}{(x + 1)^2}$

 f $\frac{2}{(x + 1)(x + 5)}$

4 **a** 2 **b** x **c** $\frac{2}{5}$ **d** $\frac{2s^2}{3}$

 e $\frac{x - 1}{(x + 1)(x + 3)}$ **f** 1 **g** $\frac{2}{3}$ **h** $\frac{2x + 1}{x}$

5 **a** $(n + 1)(n + 2) + n(n + 1) = n^2 + 3n + 2 + n^2 + n$
$= 2n^2 + 4n + 2$
$= 2(n^2 + 2n + 1)$
$= 2(n + 1)^2$

 b $2(n + 1)^2$ is always even.

6 $(2n + 1)^2 - (2n - 1)^2 = (4n^2 + 4n + 1) - (4n^2 - 4n + 1)$
$= 8n$

7 $\frac{100 - (x^2 - 16x + 64)}{4} = \frac{36 + 16x - x^2}{4} = \frac{(2 + x)(18 - x)}{4}$

8 **a** $4 - 2 = 2, 6 - 4 = 2, 8 - 6 = 2, \ldots$
The difference between each pair of consecutive even numbers is 2. Therefore if the nth even number is $2n$, the next even number must be $2n + 2$.

 b $2n + (2n + 2) + (2n + 4) = 6n + 6 = 6(n + 1)$, which must be a multiple of 6

9 $(3n + 1)^2 - (3n - 1)^2 = (9n^2 + 6n + 1) - (9n^2 - 6n + 1)$
$= 12n = 4 \times 3n$, which must be a multiple of 4

10 $\frac{16(2x - 1)}{x(x - 1)}$

Chapter 33 Answers

33.1 Get Ready

1 $(2, 2)$

Exercise 33A

1

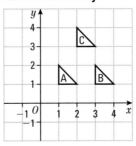

2 **a** **i** $\begin{pmatrix} 5 \\ 6 \end{pmatrix}$ **ii** $\begin{pmatrix} -1 \\ -12 \end{pmatrix}$ **iii** $\begin{pmatrix} 4 \\ 6 \end{pmatrix}$

 b They add up to $\begin{pmatrix} 0 \\ 0 \end{pmatrix}$.

3 **a**

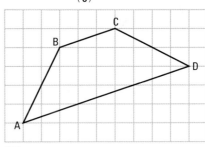

 b $\begin{pmatrix} 9 \\ 3 \end{pmatrix}$

 c trapezium

 d They are parallel and the length of AD is 3 times the length of BC.

4 **a**

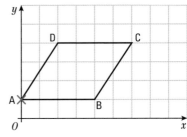

 b **i** $\begin{pmatrix} 4 \\ 0 \end{pmatrix}$ **ii** $\begin{pmatrix} -2 \\ -3 \end{pmatrix}$

 c **i** They are the same.
 ii They are parallel with the same length, but in opposite directions.

5 **a** and **c**, **b** and **h**, **d** and **g**

33.2 Get Ready

1 **a** 25 **b** 9.43

Exercise 33B

1 **a** 13 **b** 13 **c** $\sqrt{10}$ **d** $\sqrt{74}$
 e 17 **f** $4\sqrt{5}$

2 **a** 25
 b length $AC = \sqrt{(7^2 + 24^2)} = 25$, so $AB = AC$ and the triangle is isosceles.

3 rhombus

33.3 Get Ready

Translation by $\begin{pmatrix} 1 \\ 2 \end{pmatrix}$

Exercise 33C

1

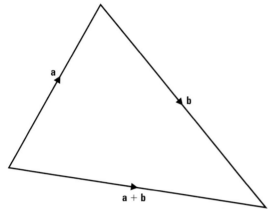

d

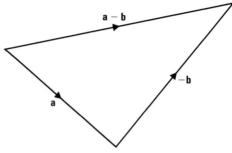

2 a $\begin{pmatrix} 6 \\ 8 \end{pmatrix}$ **b** $\begin{pmatrix} 4 \\ 8 \end{pmatrix}$ **c** $\begin{pmatrix} -2 \\ 4 \end{pmatrix}$ **d** $\begin{pmatrix} 9 \\ -5 \end{pmatrix}$ **e** $\begin{pmatrix} -8 \\ -3 \end{pmatrix}$

3 $\begin{pmatrix} 10 \\ -5 \end{pmatrix}$

4 a i $\begin{pmatrix} 4 \\ 3 \end{pmatrix}$ **ii** $\begin{pmatrix} 4 \\ 3 \end{pmatrix}$ **b** They are the same.

　c i $\begin{pmatrix} 8 \\ 10 \end{pmatrix}$ **ii** $\begin{pmatrix} 8 \\ 10 \end{pmatrix}$ **d** They are the same.

5 a $\overrightarrow{ED}$ is parallel to $\overrightarrow{AB}$, in the same direction and has the same length.

　b i n + m **ii** n + m + p **c** n + m

33.4 Get Ready

1 a $\begin{pmatrix} 6 \\ 10 \end{pmatrix}$ **b** $\begin{pmatrix} -9 \\ -3 \end{pmatrix}$ **c** $\begin{pmatrix} 0 \\ 0 \end{pmatrix}$

Exercise 33D

1 a **b**

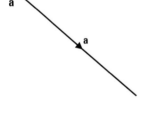

c

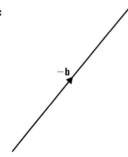

2 a **b**

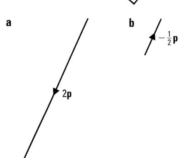

3 a i $\begin{pmatrix} 20 \\ 15 \end{pmatrix}$ **ii** $\begin{pmatrix} -12 \\ 6 \end{pmatrix}$ **iii** $\begin{pmatrix} 10 \\ 30 \end{pmatrix}$ **iv** $\begin{pmatrix} -26 \\ 48 \end{pmatrix}$

　b i 5 **ii** 10

4 a $\begin{pmatrix} -2 \\ 5 \end{pmatrix}$ **b i** $\begin{pmatrix} 5 \\ -4 \end{pmatrix}$ **ii** $\begin{pmatrix} 20 \\ -16 \end{pmatrix}$

　c They are parallel and the length of RS is 4 times the length of PQ.

5 (16, 8)

6 a b − a **b** midpoint of OB

7 a $\overrightarrow{AB}, \overrightarrow{EF}$ and $\overrightarrow{GH}$

　b i 5p − 3q **ii** 6m − 14n

8 a 2m **b** 2m + n **c** m + x **d** x = m + n

33.5 Get Ready

1 a $\begin{pmatrix} 7 \\ -7 \end{pmatrix}$ **b** $\begin{pmatrix} -8 \\ -5 \end{pmatrix}$ **c** $\begin{pmatrix} -15 \\ 2 \end{pmatrix}$

Exercise 33E

1 a i $\begin{pmatrix} 3 \\ 9 \end{pmatrix}$ **ii** $\begin{pmatrix} 9 \\ 27 \end{pmatrix}$

　b ABC is a straight line such that the length of AC is 3 times the length of AB.

2 a b − a **b** $\frac{1}{2}$(b − a) **c** $\frac{1}{2}$(a + b)

3 a $\frac{1}{2}$(a + b) **b** a + b **c** $\frac{1}{2}$(a + b)

　d They are the same point.

　e The diagonals of a parallelogram bisect each other.

4 a trapezium **b** n = 2m − k

5 a i b − a **ii** $\frac{1}{4}$b **iii** a + $\frac{1}{4}$b **iv** $\frac{1}{4}$a + b

　v $\frac{3}{4}$(b − a)

　b EF and AB are parallel. The length of EF is $\frac{3}{4}$ times the length of AB.

6 a i $\frac{1}{2}$(m + n) **ii** $\frac{3}{4}$(m + n) **iii** $\frac{1}{4}$(3n − m)

　b 3n − m

　c MQ and MR are parallel with the point M in common, so MQ and MR are part of the same straight line.

　$\dfrac{MR}{MQ} = 4$

Answers

7 a 2**b**   **b** 2**a** + **b**   **c** 4**a** + 2**b** = 2(2**a** + **b**)
 d S is the midpoint of OT.
 e 30

Review exercise

1

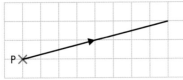

2 a (4, 5)   **b** $\begin{pmatrix} -3 \\ -4 \end{pmatrix}$ **c** 5

3 a **b** − **a**
 b $\overrightarrow{AP} = \frac{3}{5}(\mathbf{b} - \mathbf{a})$
  $\overrightarrow{OP} = \overrightarrow{OA} + \overrightarrow{AP} = \mathbf{a} + \frac{3}{5}(\mathbf{b} - \mathbf{a}) = \frac{1}{5}(2\mathbf{a} + 3\mathbf{b})$

4 a 2(*a* + *b*)
 b 7*a* + 6*b*

5 a i $\begin{pmatrix} -4 \\ -3 \end{pmatrix}$ **ii** 5
 b $\begin{pmatrix} -3 \\ 7 \end{pmatrix}$ **c** (−7, 4)

6 a 2(**a** + 2**c**)
 b $\overrightarrow{OM} = 6\mathbf{c} + 3\mathbf{a} = 3(\mathbf{a} + 2\mathbf{c}) = \frac{3}{2}\overrightarrow{OP}$
  OP and OM are parallel with the point O in common, so
  OP and OM are part of the same straight line.

7 a **a** + **b**
 b $\overrightarrow{FE} = \overrightarrow{FC} + \overrightarrow{CD} + \overrightarrow{DE} = \mathbf{a} - \mathbf{b} + \mathbf{a} + \mathbf{b} = 2\mathbf{a}$
  So FE is parallel to CD.
 c $2\mathbf{a} - \frac{1}{2}\mathbf{b}$
 d $\overrightarrow{FX} = \frac{4}{5}(2\mathbf{a} - \frac{1}{2}\mathbf{b}) = \frac{8}{5}\mathbf{a} - \frac{2}{5}\mathbf{b}$
  $\overrightarrow{CX} = (\frac{8}{5}\mathbf{a} - \frac{2}{5}\mathbf{b}) - (\mathbf{a} - \mathbf{b}) = \frac{3}{5}(\mathbf{a} + \mathbf{b})$
  $\overrightarrow{CE} = \mathbf{a} + \mathbf{b}$
  CX and CE are parallel with the point C in common, so
  CX and CE are part of the same straight line.

8 a i 5**p**   **ii** 2**q**   **iii** 4**p** − **q**
 b 3**p** − 2**q**

9 a $\frac{1}{2}(\mathbf{p} + \mathbf{q})$
 b $\overrightarrow{RS} = -\overrightarrow{OR} + \overrightarrow{OS} = \mathbf{p} + \frac{1}{2}(\mathbf{p} + \mathbf{q}) = \frac{1}{2}\mathbf{q}$
  So RS is parallel to OQ.

10 a 2**a** − 2**b**
 b $\overrightarrow{QR} = -2\mathbf{a} - 2\mathbf{b} + 6\mathbf{a} = 4\mathbf{a} - 2\mathbf{b}$
  $\overrightarrow{XY} = -\frac{1}{2}\overrightarrow{MN} + \overrightarrow{MQ} + \frac{1}{2}\overrightarrow{QR} = -\frac{1}{2}(2\mathbf{a} - 2\mathbf{b}) + \mathbf{a}$
  $+ \frac{1}{2}(4\mathbf{a} - 2\mathbf{b}) = 2\mathbf{a}$
  So XY is parallel to OR.

11 a **a** + **b**
 b $\overrightarrow{AX} = 2(\mathbf{a} + \mathbf{b})$
  $\overrightarrow{DX} = -\overrightarrow{AD} + \overrightarrow{AX} = -2\mathbf{b} + 2(\mathbf{a} + \mathbf{b}) = 2\mathbf{a}$
  So AB is parallel to DX.

Index

Index

Index